Principles of Nuclear Magnetic Resonance Microscopy

Principles of Nuclear Magnetic Resonance Microscopy

Paul T. Callaghan

Department of Physics and Biophysics
Massey University
New Zealand

CLARENDON PRESS · OXFORD

Oxford University Press, Walton Street, Oxford OX2 6DP
Oxford New York Toronto
Delhi Bombay Calcutta Madras Karachi
Kuala Lumpur Singapore Hong Kong Tokyo
Nairobi Dar es Salaam Cape Town
Melbourne Auckland Madrid
and associated companies in
Berlin Ibadan

Oxford is a trade mark of Oxford University Press

Published in the United States
by Oxford University Press Inc., New York

First published as paperback with corrections 1993

British Library Cataloguing in Publication Data
A catalogue record for this book
is available from the British Library

Library of Congress Cataloging in Publication Data
Callaghan, Paul T.
Principles of nuclear magnetic resonance microscopy
Paul T. Callaghan.
Includes bibliographical references and index.
1. Nuclear magnetic resonance spectroscopy. 2. Magnetic resonance imaging. I. Title.
QP519.9.N83C35 1991 538′.362—dc20 91-8439
ISBN 0-19-853997-5 (pbk)

Printed in Great Britain by
Bookcraft (Bath) Ltd
Midsomer Norton, Avon

PREFACE

Since its inception in 1973, nuclear magnetic resonance imaging (MRI) has been largely confined to medical applications and, in particular, whole-body scanning. Consequently MRI technology has developed a reputation for being invariably exotic and expensive. However, in 1986 several researchers reported applications of MRI at small scale and at high spatial resolution. This work was mostly performed using standard NMR spectrometers designed for chemistry, biochemistry, and materials science, in which some simple but ingenious modifications had been made. The key to this development was the installation of sophisticated computing facilities in NMR laboratories during the 1980s.

NMR microscopy involves the acquisition of the NMR signal in the presence of a magnetic field gradient, a process known as **k**-space acquisition. A dynamic analogue of NMR imaging is the pulsed gradient spin echo (PGSE) experiment, sometimes termed **q**-space imaging, in which the self-correlation function of the molecule containing the nuclear spins can be obtained. This type of experiment can be used to study the spectrum of molecular motion as well as the morphology in porous systems. While it is not customary to group together these two apparently very different applications of magnetic field gradients, it is clear that there are common physical principles governing the imaging of static displacements via **k**-space and dynamic displacements via **q**-space and a theme of this book concerns the linkages between these methods.

The resolution in **k**-space microscopy is not particularly good by comparison with that achievable in the optical microscope. Indeed, given the expense of modern NMR equipment, one might properly ask what virtue this new method possesses. Such virtue partly lies in the non-invasive nature of the magnetic resonance imaging method but, whereas this is an overriding concern in medicine, it is less imperative, for example, in dealing with plants and somewhat superfluous when dealing with inanimate materials. It is from the range of contrast available that NMR microscopy gains its value. Every measurement that is possible in conventional NMR spectroscopy is *in principle* possible as a spatially resolved contrast in the study of heterogeneous systems, the *caveat* being that the price paid in achieving spatial resolution is always a loss in sensitivity. This means that in practice only some of the simplest features of the nuclear spin behaviour are available as a signature in the image. These features, however, include the chemical shift, the nuclear spin relaxation times, the dipolar couplings, and the spectrum of translational motions.

Despite there being no medical images displayed in this book it is clear that many of the methods developed specifically for large-scale imaging are equally applicable in microscopy. Indeed the potential for pulse sequence variation is somewhat richer in non-medical imaging because of the removal of constraints relating to patient safety, although many of these constraints must still apply in studies of small animals. A number of imaging applications given as examples here do not exhibit resolution which is strictly 'microscopic'. Their inclusion is justified partly by the possibilities which they suggest for studies at higher spatial resolution, and partly because many readers of this book will be interested in unusual applications of NMR imaging without specifically requiring submillimetre voxels.

By contrast with **k**-space imaging the measurement of nuclear spin translation via the pulsed gradient spin echo method can achieve a spatial resolution some two orders of magnitude better. Furthermore, the method is one of the very few means by which we can observe the motion of molecules without an invasive tag, thus placing the method alongside neutron scattering as a window on 'self-motion' in condensed matter. Despite having been proposed over two decades ago and having been used with great success by a handful of groups around the world, the pulsed gradient spin echo NMR method has not become available as a routine technology and is little understood outside a small group of practitioners. Through the **q**-space formalism the role of PGSE NMR as an imaging method in its own right is explored in this book.

It is clear that **k**-space and **q**-space NMR imaging have many potential applications in biological and materials science and, because of the enormous number of NMR spectrometers in research use around the world and the large community of NMR spectroscopists, the NMR microscopy method has both the equipment and expertise base to be widely used. NMR microscopy offers a cheap method of trialling imaging and localized spectroscopy methods using very small animals such as mice and rats. This feature makes it an attractive option in medical research. In plant physiology the method offers unrivalled opportunities for studying transport processes, and, in chemical physics, the possibility of studying molecular dynamics, especially in liquid crystals and high polymers, is especially interesting. Recent advances in dynamic NMR microscopy make possible the study of heterogeneous phases and fluid systems in shear. NMR microscopy has applications in a wide variety of industries and in particular those concerned with petrochemicals, polymers, biotechnology, food processing, and natural product processing.

Chapters 1 and 2 of this book are attempts to provide an introduction to the principles both of imaging and of NMR for those who are inexperienced in these areas of science. Chapter 3 provides a detailed description of the influence of magnetic field gradients on the evolution of the nuclear spin magnetization and the spatial information which can be gleaned from their use. In Chapter 4 the specific limits to spatial resolution in **k**-space microscopy are discussed while Chapter 5 reviews the various applications of **k**-space microscopy which have been reported both in the scientific literature and at recent international meetings since the inception of this area of research in the mid-1980s. Chapters 6 and 7 both deal with the underlying principles behind the PGSE method, showing how different descriptions of molecular motion are possible according to the gradient modulation employed. Special prominence is given to the narrow gradient pulse spin echo method, the so-called **q**-space mode of imaging, because of the usefulness of the average propagator which is 'reconstructed' by this method. Chapter 7 examines how molecular boundaries can be probed via the spectrum of restricted motion, examining also the role of local geometry on relaxation times. In Chapter 8 various motional contrast schemes in imaging are compared and the special uses of combined **k**-space and **q**-space imaging explored.

This book is concerned with the underlying principles governing NMR microscopy and in particular the information which can be derived by using magnetic field gradients. In practical terms NMR microscopy represents a new area of application of the laboratory NMR spectrometer, requiring some additional hardware and software, but none the less embodying much that is familiar to the experienced spectroscopist. While initial work was the result of home-made attachments, since 1987 microscopy options have become available as standard spectrometer 'add-ons' and, in view of the quality and sophistication of these, the prospective microscopist would probably more wisely spend his or her time developing new measurement strategies and understanding their results, than in building gradient coil sets or r.f. modulators. For that reason the final chapter dealing with instrumentation is necessarily brief, in recognition of the fact that this is a rapidly changing field of research and that today's technology will soon be yesterday's.

While the text is aimed at researchers, the level is that appropriate to a graduate student in chemistry, physics, or engineering. The biologist may wish to skip the mathematical aspects but should be able to follow the principles and appreciate the applications.

I wish to thank the many authors who have given permission to reproduce examples of their work and to acknowledge a number of colleagues who have assisted by way of valuable discussions, including Myer Bloom, Ella Campbell, David Cory, David Doddrell, Michel Decorps, Dieter Gross, Axel Haase, Laurie Hall, Colin Jenner, Ken Jolley, Rainer Kimmich, Winfried Kuhn, Paul Lauterbur, John Lelievre, Alex Mackay, David Macgowan, Peter Mansfield, Ken Packer, Jim Pope, Neil Rennie, Vasil Sarafis, and Fernando Zelaya. I am especially grateful to my 'NMR Imaging' research students Craig Eccles, Yang Xia, and Andrew Coy for their beautiful experimental work and for teaching me a great deal. Thanks for helpful advice are also due to other students who have worked in my group including Craig Trotter, Peter Daivis, Te Whitinga Huirua, Tina Baker, and Craig Rofe and to group visitors Ted Garver, Wilmer Miller, Phil Back, Ken Jeffrey, and Sharon Umbach. Much of this book was written during periods of leave in Melbourne and Oxford and I am grateful to Tony Klein, Geoff Opat, Nick Stone, and Milo and Irina Shott for their hospitality, and to Geoff Malcolm, David Parry, Di Reay, and other Massey University colleagues for their support. Finally I wish to thank Sue, Catherine, and Chris for their patience and constant encouragement while this book was being written.

Palmerston North P.T.C.
May 1991

ACKNOWLEDGEMENTS

I am indebted to the following colleagues, authors, and publishers for permission to reproduce figures.

R. L. Armstrong, Fig. 5.24
P. J. Back, Fig. 9.6
S. J. Blackband, Fig. 5.11
R. Blinc, Fig. 6.6
R. A. Brooks, Fig. 1.8
K. R. Brownstein, Fig. 7.18
D. Burstein, Fig. 8.1
D. G. Cory, Fig. 5.44, 5.45, 5.46, 7.5
A. Coy, Fig. 7.14
G. Di Chiro, Fig. 1.8
D. M. Doddrell, Fig. 3.12(c)
R. Dykstra, Fig. 9.2(b)
R. R. Ernst, Fig. 1.7, 5.38
J. Frahm, Fig. 8.2
J. P. Freyer, Fig. 5.12
A. N. Garroway, Fig. 5.43, 5.46, 7.5
G. Gassner, Fig. 5.9, 5.10, 5.31
D. Gross and V. Lehman, Fig. 4.19, 4.21, 9.4
L. D. Hall, Fig. 5.5, 5.6
B. E. Hammer and G. Belfort, Fig. 8.13
L. G. Harrison, Fig. 5.5, 5.6
R. M. Henkelman, Fig. 5.1
G. A. Johnson, Fig. 5.7, 5.8, 5.21
R. I. Joseph and D. I. Hoult, Fig. 3.9(b)
J. L. Koenig, Fig. 5.13
R. A. Komorowski, Fig. 5.17, 5.18
W. Kuhn, Fig. 5.2, 5.16, 5.19
M. J. Kushmerick, Fig. 5.36
R. K. Lambert, Fig. 5.34
M. Lipsicus, Fig. 7.19
A. L. Mackay, Fig. 5.23
P. Mansfield, Fig. 5.14
G. Mateescu, Fig. 5.4, 5.37
M. M. Mattingley, Fig. 5.11, 5.16, 5.19
J. B. Miller, Fig. 5.43, 5.46
L. J. Neuringer, Fig. 5.36

K. J. Packer, Fig. 6.11, 7.3
J. M. Pope, Fig. 5.25, 5.28
R. G. Ratcliffe, Fig. 5.3
A. Rigamonti, Fig. 5.15
W. P. Rothwell, Fig. 5.22
H. Rumpel, Fig. 5.28
V. Sarafis, Fig. 5.28
E. O. Stejskal, Fig. 7.2
C. E. Tarr, Fig. 7.18
W. S. Veeman, Fig. 5.44, 5.45
Y. Xia, Fig. 8.9, Plate 4
F. O. Zelaya, Fig. 6.11

Academic Press
from *Journal of Magnetic Resonance*, Fig. 3.9(b), 3.12(c), 3.29, 4.4, 4.6, 4.8, 4.9, 4.12, 4.15, 4.16, 5.1, 5.23, 5.43, 5.44, 8.2
from *Magnetic Resonance in Medicine*, Fig. 5.7, 5.12, 7.5, Plate 1
from *Journal of Colloid and Interface Science*, Fig. 7.3, 7.6

American Chemical Society
from *Macromolecules*, Fig. 5.17, 5.18
from *Journal of Physical Chemistry*, Fig. 6.9, 7.15

American Institute of Chemical Engineers
From *American Institute of Chemical Engineers Journal*, Fig. 8.15

American Institute of Physics
from *Applied Optics*, Fig. 5.22
from *Applied Physics Letters*, Fig. 7.19

American Physical Society,
from *Journal of Chemical Physics*, Fig. 6.5, 6.7, 6.8, 7.2
from *Physical Review Letters*, Fig. 6.6
from *Physical Review*, Fig. 7.18

American Physiological Society
from *Journal of Applied Physiology*, Fig. 5.34

Bruker Analytische Messtechnik
from *Bruker Report*, Fig. 5.16

Cambridge University Press
from *Quarterly Reviews of Biophysics*, Fig. 1.7

The Company of Biologists Ltd
from *Journal of Cell Science*, Fig. 5.5, 5.6

Elsevier
from *Journal of Molecular Liquids*, Fig. 5.15
from *Chemical Physics Letters*, Fig. 5.45
from *Colloids and Surfaces*, Fig. 6.11

The Institute of Physics
from *Physics in Medicine and Biology*, Fig. 1.8
from *Journal of Physics C*, Fig. 5.14

Hüthig and Wepf Verlag
From *Die Makromolekular Chemie*, Fig. 8.12

J. B. Lippincott
from *Investigative Radiology*, Fig. 5.8

Macmillan
from *Nature*, Fig. 5.11

Oxford University Press
from *Journal of Experimental Biology*, Fig. 5.3
from *Principles of Nuclear Magnetic Resonance in 1 and 2 Dimensions* by R. R. Ernst, G. Bodenhausen, and A. Wokaun, Fig. 5.38

Pergamon Press
from *Magnetic Resonance Imaging*, Fig. 5.36

Rockefeller University Press
from *Biophysical Journal*, Fig. 6.10

Taylor and Francis
from *Molecular Physics*, Fig. 5.45

Society for Applied Spectroscopy
from *Applied Spectroscopy*, Fig. 5.13

VCH Verlagsgesellschaft
From *Angewandte Chemie*, Fig. 5.2

The plates section is situated between pages 428 and 429

CONTENTS

1
PRINCIPLES OF IMAGING

1.1 Introduction

Until the discovery of X-rays by Roentgen in 1895 our ability to view the spatial organization of matter depended on the use of visible light with our eyes being used as primary detectors. Unaided, the human eye is a remarkable instrument, capable of resolving separations of 0.1 mm on an object placed at the near point of vision and, with bifocal vision, obtaining a depth resolution of around 0.3 mm. However, because of the strong absorption and reflection of light by most solid materials, our vision is restricted to inspecting the appearance of surfaces. 'X-ray vision' gave us the capacity, for the first time, to see inside intact biological, mineral, and synthetic materials and observe structural features.

The early X-ray photographs gave a planar representation of absorption arising from elements right across the object. In 1972 the first X-ray CT scanner was developed with reconstructive tomography being used to produce a two-dimensional absorption image from a thin axial layer.[1] The mathematical methods used in such image reconstruction were originally employed in radio astronomy by Bracewell[2] in 1956 and later developed for optical and X-ray applications by Cormack[3] in 1963. A key element in the growth of tomographic techniques has been the availability of high speed digital computers. These machines have permitted not only the rapid computation of the image from primary data but have also made possible a wide variety of subsequent display and processing operations. The principles of reconstructive tomography have been applied widely in the use of other radiations. In 1973, Lauterbur[4] reported the first reconstruction of a proton spin density map using nuclear magnetic resonance. (NMR), and in the same year Mansfield and Grannell[5] independently demonstrated the Fourier relationship between the spin density and the NMR signal acquired in the presence of a magnetic field gradient. Since that time the field has advanced rapidly to the point where magnetic resonance imaging (MRI) is now a routine, if expensive, complement to X-ray tomography in many major hospitals. Like X-ray tomography, conventional MRI has a spatial resolution coarser than that of the unaided human eye with volume elements of order $(1\ \text{mm})^3$ or larger. Unlike X-ray CT however, where resolution is limited by the beam collimation, MRI can in principle achieve a resolution considerably finer than 0.1 mm and, where the resolved volume elements are smaller than $(0.1\ \text{mm})^3$, this method of imaging may be termed microscopic.

Although MRI medical scanners had been available since the early 1980s, it was not until 1986 that the first NMR micrographs were produced. This book is concerned with the application of NMR imaging to mineral, synthetic, and biological materials at high spatial resolution. There is much in common between the methods used in magnetic resonance microscopy and its medical counterpart, MRI. However, the range of microscopic applications is potentially enormous and, as yet, largely unexplored. While its resolution is considerably coarser than optical and electron microscopy, magnetic resonance microscopy offers several major advantages. Its potential for studying dynamical processes in living systems gives it special importance and its sensitivity to the rotational and translational mobility of molecules is the basis of some unique applications in materials science and especially polymer science. In the following pages the technique of magnetic resonance microscopy will be described in the context of other imaging methods and its principles will be explained and illustrated using recent applications.

1.2 Reciprocal space and Fourier transformation

1.2.1 *Conjugate variables*

Oscillatory behaviour is characterized by both amplitude and phase. A convenient representation for such behaviour is given by complex numbers written in the polar form. In the time domain we may describe this oscillation by $A\exp(i\omega t)$ where A is the amplitude and ωt the phase. The real and imaginary parts of $A\exp(i\omega t)$ oscillate in quadrature as cosine and sine variations, respectively. The angular frequency ω and time t in this example have inverse dimensions and are known as conjugate variables. In the case of spatial variation, the equivalent expression is $A\exp(i\boldsymbol{\kappa}\cdot\mathbf{r})$ where the wave-vector $\boldsymbol{\kappa}$ and spatial coordinate $\mathbf{r}$ are conjugate. In wave motion both oscillatory components are present and the product $A\exp(i\boldsymbol{\kappa}\cdot\mathbf{r} - i\omega t)$ describes a plane wave propagating in the direction of $\boldsymbol{\kappa}$. Note that $\omega = 2\pi/T$ and $|\boldsymbol{\kappa}| = 2\pi/\lambda$ where T and λ are the period and wavelength, respectively.

While only the real parts of such functions can represent physical variables, the use of complex numbers allows a description in which the numbers represent actual phasors. In linear systems where the superposition principle holds, the effect of superposing phasors is simply described by the mathematics of complex number addition. It is interesting to consider the relationships which prevail when such a superposition occurs. A simple example, familiar to all physics undergraduates, is the interference of coherent light originating from two slits, the so-called Young's interference experiment. This example is illustrated in Fig. 1.1 where the slits receive light of identical

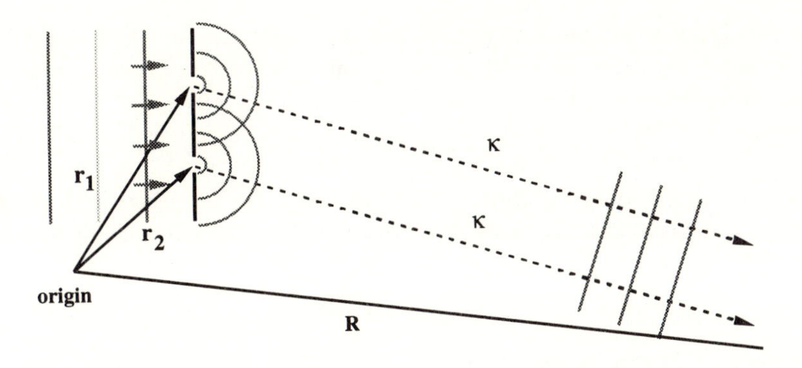

Fig. 1.1 The superposition of waves from two point sources.

phase (incident plane waves) and act as sources of waves which co-add at some point sufficiently far removed that the two outgoing wavevectors are effectively parallel (far field approximation).

Neglecting the time variation, the superposed amplitude at a distant point at **R** is given by:

$$E(\boldsymbol{\kappa}) = E_0\,\mathrm{e}^{\mathrm{i}\boldsymbol{\kappa}\cdot(\mathbf{R}-\mathbf{r}_1)} + E_0\,\mathrm{e}^{\mathrm{i}\boldsymbol{\kappa}\cdot(\mathbf{R}-\mathbf{r}_2)}$$
$$= E_0\,\mathrm{e}^{\mathrm{i}\boldsymbol{\kappa}\cdot\mathbf{R}}(\mathrm{e}^{-\mathrm{i}\boldsymbol{\kappa}\cdot\mathbf{r}_1} + \mathrm{e}^{-\mathrm{i}\boldsymbol{\kappa}\cdot\mathbf{r}_2}). \quad (1.1)$$

Note that identical polarization is presumed so that E_0 is represented by a scalar. The phasor superposition in the bracketed term leads to interference effects. To appreciate these in the practical case of an *energy*-sensitive detector, we must obtain the light intensity from the amplitude by calculating the square of the absolute value namely,

$$|E(\boldsymbol{\kappa})|^2 = |E_0\,\mathrm{e}^{\mathrm{i}\boldsymbol{\kappa}\cdot\mathbf{R}}|^2|\,\mathrm{e}^{-\mathrm{i}\boldsymbol{\kappa}\cdot\mathbf{r}_1} + \mathrm{e}^{-\mathrm{i}\boldsymbol{\kappa}\cdot\mathbf{r}_2}|^2$$
$$= 2|E_0|^2[1 + \cos\boldsymbol{\kappa}\cdot(\mathbf{r}_1 - \mathbf{r}_2)]. \quad (1.2)$$

This yields the familiar result that the intensity varies between maxima and minima corresponding to angles θ given by $d \sin\theta = n\lambda$ and $(n+\frac{1}{2})\lambda$, respectively, where d is the slit spacing and n is an integer.

This simple result is the basis of X-ray diffraction in which the slits are replaced by scattering centres forming part of a periodic crystal lattice. The interference pattern bears a distinct mathematical relation to the original structure although interpretation is complicated by the fact that intensities are detected and absolute phase information is lost. The comparison between intensity and phase-dependent information is the comparison between eqns. (1.1) and (1.2). Solving the ‘phase problem’ is central to such methods.

We will ignore this particular problem and imagine an 'ideal' technique where eqn (1.1) represents the detection scheme. In the case of a lattice of scattering centres we have

$$E(\boldsymbol{\kappa}) = E_0 \, e^{i\boldsymbol{\kappa}\cdot\mathbf{R}}\left[\Sigma_i \, e^{-i\boldsymbol{\kappa}\cdot\mathbf{r}_i}\right] \tag{1.3}$$

where the sum is over the lattice positions. Suppose we describe the lattice as a sum of discrete point densities represented by the Dirac delta function sum

$$\rho(\mathbf{r}) = \Sigma_i \, \delta(\mathbf{r} - \mathbf{r}_i) \tag{1.4}$$

Then eqn (1.3) can be rewritten

$$E(\boldsymbol{\kappa}) = E_0 \, e^{i\boldsymbol{\kappa}\cdot\mathbf{R}} \iiint \rho(\mathbf{r}) \, e^{-i\boldsymbol{\kappa}\cdot\mathbf{r}} \, d\mathbf{r} \tag{1.5}$$

where the triple integral sign represents the sum over all three dimensions of the spatial coordinates. The factor in front of the integral merely represents the amplitude of the wave and the phase shift in travelling directly from the source to the detector. The essential feature of the method is contained within the integral which we will call the 'signal', $S(\boldsymbol{\kappa})$. $S(\boldsymbol{\kappa})$ and $\rho(\mathbf{r})$ are bound by a special relationship, the Fourier transformation (FT). The practical significance of this is seen by considering the conjugate nature of the transformation:

$$S(\boldsymbol{\kappa}) = \iiint \rho(\mathbf{r}) \, e^{-i\boldsymbol{\kappa}\cdot\mathbf{r}} \, d\mathbf{r} \tag{1.6}$$

$$\rho(\mathbf{r}) = \frac{1}{2\pi} \iiint S(\boldsymbol{\kappa}) \, e^{i\boldsymbol{\kappa}\cdot\mathbf{r}} \, d\boldsymbol{\kappa}. \tag{1.7}$$

Eqns (1.6) and (1.7) are of fundamental significance in image science. What we seek is an accurate representation of the object structural density, $\rho(\mathbf{r})$. This function may represent the density of atomic scatterers in X-ray diffraction, the photon absorber density in X-ray CT or the nuclear spin density in magnetic resonance imaging. However the primary signal may be detected in conjugate (or 'reciprocal') space as either the amplitude $S(\boldsymbol{\kappa})$ or as the intensity $|S(\boldsymbol{\kappa})|$.[2] Even in optical microscopy where we apparently obtain our image directly in real space, that image is generated via an intermediate conjugate form and an understanding of the various image artefacts requires an appreciation of system influences in conjugate space.

The basic imaging scheme is shown in Fig. 1.2. In diffraction and computerized tomography applications the reconstruction process, and sometimes part of the detection process as defined in Fig. 1.2, is undertaken by

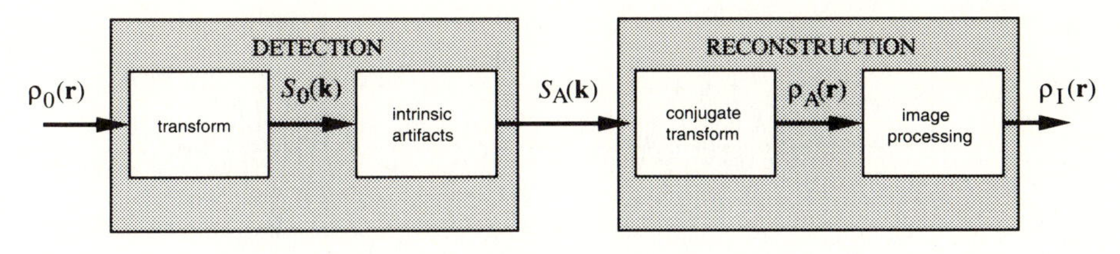

Fig. 1.2 Schematic representation of the imaging process.

mathematical calculation. This enables aspects of image processing to be incorporated in order to faithfully reproduce $\rho(\mathbf{r})$. Fig. 1.2 applies equally to analogue systems of which the optical microscope is a simple example, although in this case the opportunities for image processing are somewhat limited. In all systems, however, the final image density, ρ_I, will contain artefacts intrinsic to the particular detection method employed. In particular, where digital analysis is involved, the use of a finite space of points will automatically limit the spatial frequencies present in the final image. To this extent the maximum $\boldsymbol{\kappa}$-space magnitude used in the reconstruction process will determine the ultimate resolution.

An appreciation of image science depends on an understanding of the various relationships involved in Fourier transformation. These are thoroughly described in several excellent texts[6-8] but a brief summary of salient points is useful.

1.2.2 *Cyclic frequency*

The conjugate transformations written in eqns (1.6) and (1.7) contain an asymmetry because of the 2π factor. It is customary in engineering applications to rewrite the transformation in symmetric form using the spatial (or temporal) cyclic frequency k (or f) where $\kappa = 2\pi k$ ($\omega = 2\pi f$). Note that we use a bold-face symbol to represent a vector and the plain text symbol to represent a scalar magnitude. Thus

$$F(\mathbf{k}) = \iiint f(\mathbf{r})\, \mathrm{e}^{-\mathrm{i}2\pi\mathbf{k}\cdot\mathbf{r}}\, \mathrm{d}\mathbf{r} \tag{1.8}$$

$$f(\mathbf{r}) = \iiint F(\mathbf{k})\, \mathrm{e}^{\mathrm{i}2\pi\mathbf{k}\cdot\mathbf{r}}\, \mathrm{d}\mathbf{k}, \tag{1.9}$$

or alternatively,

$$\begin{aligned} \mathfrak{F}\{f(\mathbf{r})\} &= F(\mathbf{k}) \\ \mathfrak{F}^{-1}\{F(\mathbf{k})\} &= f(\mathbf{r}) \\ F(\mathbf{k}) &\Leftrightarrow f(\mathbf{r}). \end{aligned} \tag{1.10}$$

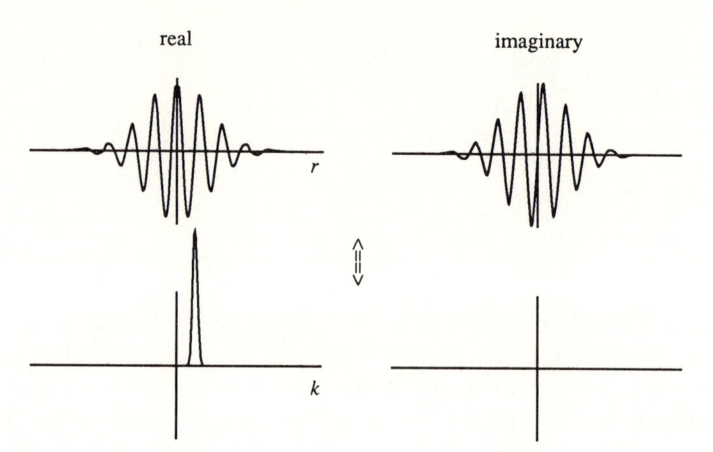
(a)
real
imaginary
r
k

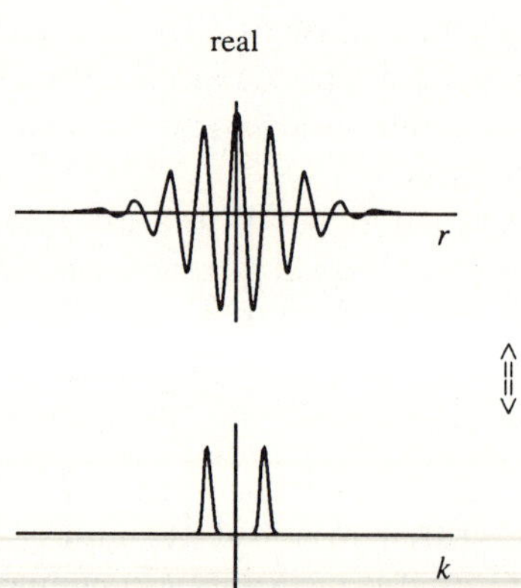
(b)
real
r
k

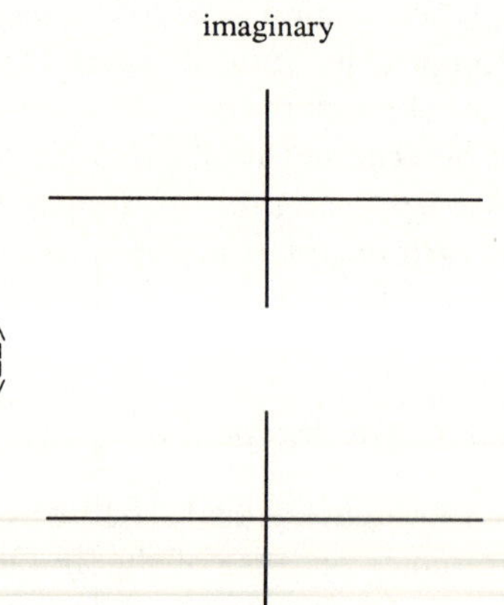
imaginary

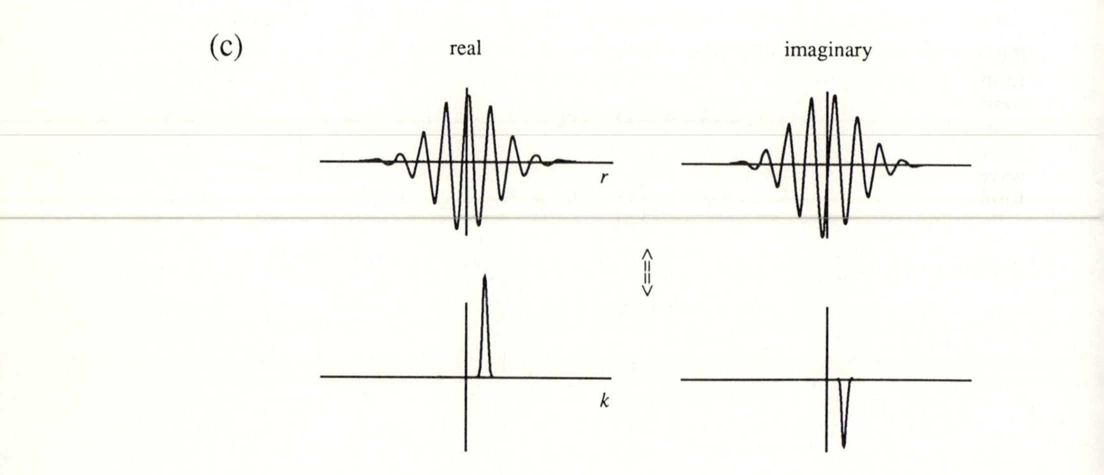
(c)
real
imaginary
r
k

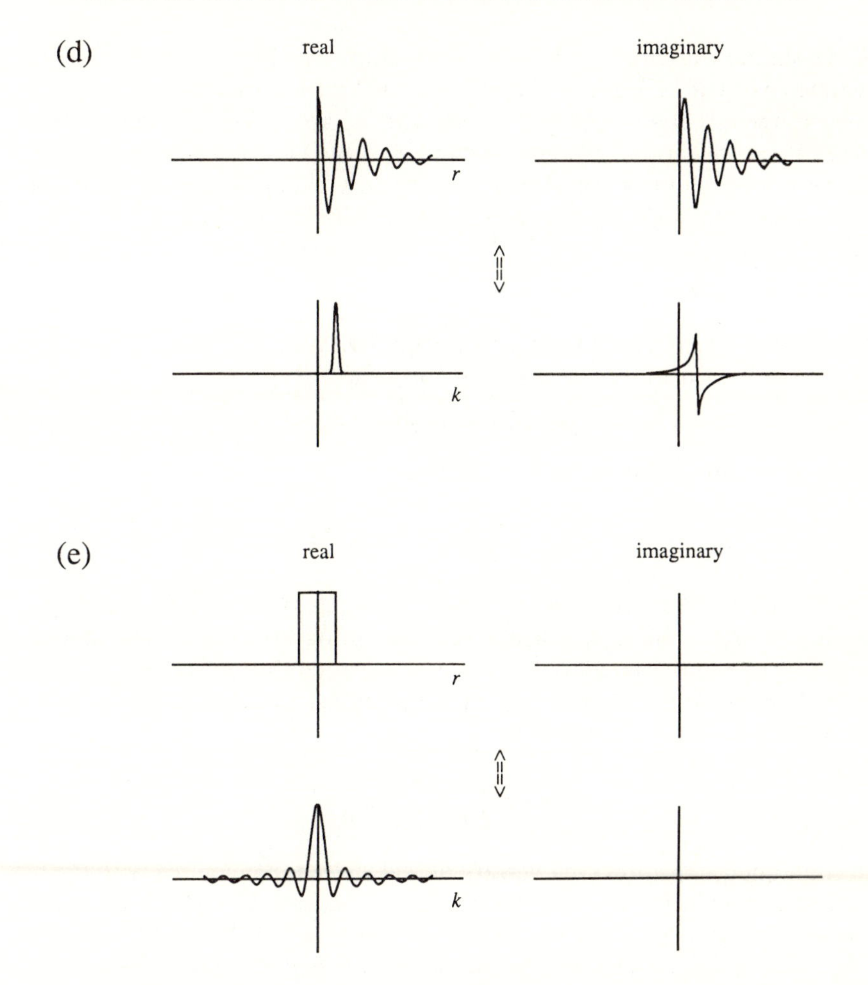

Fig. 1.3 Fourier relationships between simple examples of functions. (a) Oscillatory wave-form with symmetric real and antisymmetric imaginary parts. Note that the imaginary component of the transform is zero. (b) As in (a) for the real part of the wave-form but with zero imaginary signal. Positive and negative frequencies cannot be distinguished and both are present in the transform. (c) As in (a) but with the original wave-form phase shifted. (d) Initial wave-form sampled only in the positive domain leading to a dispersion component in the transform. (e) The Heaviside hat function and its Fourier pair, the sinc function.

Some simple examples of FT pairs are shown in Fig. 1.3. Note that the symmetry of eqns 1.8 and 1.9 imply that the Fourier transformation and its inverse are essentially equivalent operations, apart from a sign reversal. While it is conventional to refer to eqn 1.8 as the Fourier transform and 1.9 as the inverse Fourier transform, we can interchange these labels at will. Each operation is clearly the inverse of the other.

1.2.3 *Convolution theorem*

The Fourier transformation of a product of two functions is the convolution of the two transformed functions in conjugate space:

$$f(\mathbf{r}) \otimes g(\mathbf{r}) \Leftrightarrow F(\mathbf{k})\, G(\mathbf{k}) \tag{1.11}$$

where the convolution integral is given by

$$f(\mathbf{r}) \otimes g(\mathbf{r}) = \iiint f(\mathbf{r}')\, g(\mathbf{r} - \mathbf{r}')\, \mathrm{d}\mathbf{r}'. \tag{1.12}$$

Convolution effects are always apparent when the signal is sampled in conjugate space, since in practice such sampling involves a finite domain whereas the Fourier integral is taken over all of conjugate space. The effect of such truncation is that the signal is multiplied by a hat function, which in one dimension is described by

$$H(\mathbf{k}) = \begin{cases} 1 & |\mathbf{k}| \le k_{\max} \\ 0 & |\mathbf{k}| > k_{\max}. \end{cases} \tag{1.13}$$

In consequence the observed density in real space is convoluted with the hat function transform, the sin x/x or sinc function. The effect of such a convolution is to limit the resolution as shown in Fig. 1.4(a). It is obvious that such a limitation will result, since the higher spatial frequencies, truncated by observation, contained the information necessary for discerning fine details. An example of such truncation is apparent in the limitation imposed by lens aperture size in optical microscopy.

Another important example of the convolution theorem at work is provided by the shift operation. A multiplication in the time domain by the oscillation $\exp(\mathrm{i}2\pi f_0 t)$ implies a convolution with $\delta(f - f_0)$ in conjugate space. This particular convolution has the effect of shifting the frequency domain function along the frequency axis by f_0. The shift effect is illustrated in Fig. 1.4(b). This particular example corresponds to the relationship between an amplitude-modulated oscillation and its spectrum in frequency space, a key element in the process of selective excitation in MRI.

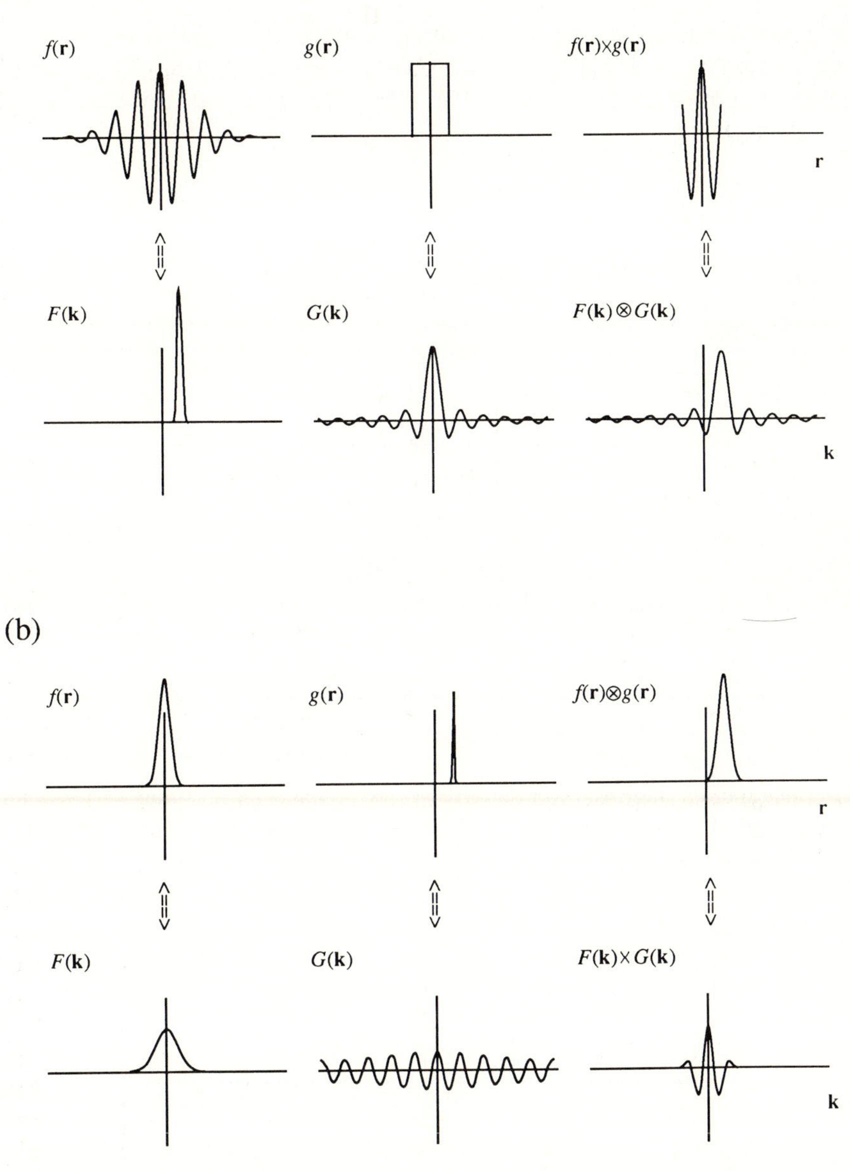

Fig. 1.4 The convolution theorem at work. (a) The 'window' effect. The finite window used to observe the signal may be represented by hat function multiplication. In the Fourier domain this leads to a convolution with the hat transform, the sinc function. (b) The 'shift' effect. Shifting the signal in the original space is equivalent to convolution with a displaced spike. This leads to multiplication by the spike transform (an oscillation) in Fourier space. Note that only the real parts of the functions are shown in (a) and (b).

1.2.4 *Digital Fourier transformation*

In examples pertaining to the use of computers, the signal is digitized and represented in a finite space of data points. Consider N time domain points sampled at interval T as shown in Fig. 1.5(a). The conjugate transformations are given by

$$G(n/NT) = \sum_{m=-N/2}^{N/2-1} g(mT)\,e^{-i2\pi mn/N} \qquad n = -N/2, \ldots, N/2 - 1 \quad (1.14)$$

(a)

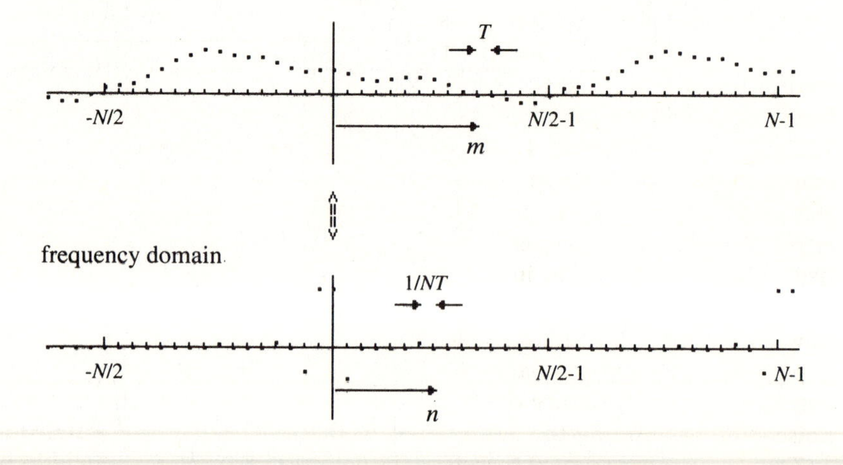

(b)

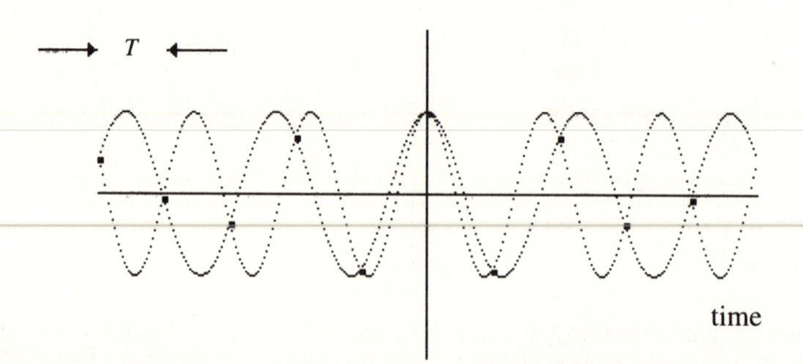

Fig. 1.5 (a) Digital space. The conjugate time and frequency domains shown are those defined in eqns (1.14) and (1.15). Note the periodicity of digital Fourier space. (b) The sampling theorem at work. A spectral component at $1/2T+\delta$ will appear identical to one with frequency $1/2T-\delta$.

$$g(mT) \quad = \frac{1}{N} \sum_{n=-N/2}^{N/2-1} G(n/NT)\, e^{i2\pi mn/N} \quad m = -N/2, \ldots, N/2 - 1. \quad (1.15)$$

As a consequence of sampling $g(mT)$ at intervals of T, the transformed function $G(n/NT)$ is periodic as shown in Fig. 1.5(b), with

$$G\{(N - n)/NT\} = G(-n/NT). \quad (1.16)$$

Furthermore, the inverse relationship imposes the same (presumed) periodicity on the initial signal $g(mT)$ where the repetition period is now NT.

When oscillatory data are digitized in this manner there is a limit to the frequencies which can be distinguished. Fig. 1.5(a) shows an example of a sinusoidal wave-form which is sampled periodically at intervals of T. It is clear that wave-forms with periods less than $2T$ will appear to oscillate with a period longer than $2T$, an effect known as 'aliasing'. The theorem which governs this effect, due to H. Nyquist,[9] states that $1/2T$ is the maximum frequency which can be measured. A spectral component with frequency $1/2T + \delta$ will appear at $1/2T - \delta$. Of course if quadrature detection is employed as described in the subsequent section, then the *sign* of the frequency may be distinguished. This means in effect that we can discriminate spectral components with frequencies between $\pm 1/2T$ so that the detection bandwidth is the inverse of the sampling interval, namely $1/T$. One important consequence of this bandwidth limitation is the need to prevent unwanted aliasing of signals outside the desired observation bandwidth. This is performed by using a bandpass filter. Such a filter plays a vital role in determining the sensitivity of the measurement since it also acts to prevent noise aliasing, limiting the noise power to that of the detection bandwidth, $1/T$. In an ideal experiment therefore the detection and filter bandwidths will be identical.

Computation of the discrete Fourier transformation according to eqns (1.14) and (1.15) involves the product of the two functions leading to N^2 multiplications. In a typical domain of 256 points this would require excessive computation time. The problem of how to shorten this time was solved by Cooley and Tukey in 1965[10] who devised an algorithm which reduces the number of multiplications to around $N \log N$. While the method does restrict the sample dimension, N, to a power of 2, this is a small price to pay for the gain in speed.

1.2.5 *Real and imaginary parts of the transform*

In practice the complex space in which the Fourier transformation is calculated is divided into real and imaginary parts. The complex transformation can then be written

$$F(n/N) = \sum_{m=-N/2}^{N/2-1} [\mathrm{Re}\,\{f(m)\}\cos(2\pi nm/N) + \mathrm{Im}\,\{f(m)\}\sin(2\pi nm/N)]$$
$$+\,\mathrm{i}\sum_{m=-N/2}^{N/2-1} [\mathrm{Im}\{f(m)\}\cos(2\pi nm/N) - \mathrm{Re}\,\{f(m)\}\sin(2\pi nm/N)] \tag{1.17}$$

where the sampling period, T, is set to 1 to reflect the dimensionless character of digital space. The real and imaginary parts of the function $f(m)$ might, for example, refer to the in-phase and quadrature signals in an NMR experiment. The computed transform $F(n/N)$ also has both real and imaginary parts, and contributions will be made to each according to the character of the initial signal $f(m)$. If, for example, $\mathrm{Re}\{f(m)\}$ is symmetric and $\mathrm{Im}\{f(m)\}$ is antisymmetric then the transformed function is entirely real. Examples of transform pairs for simple oscillatory behaviour are shown in Fig. 1.3. It is important to emphasize that acquisition of a single-phase signal alone implies that positive and negative frequencies cannot be distinguished.

1.3 Fourier transformation and optical microscopy

The role of the conjugate space in reconstruction tomography is illustrated in the model of Fig. 1.2. While that role is explicit in the case of X-ray and magnetic resonance tomography, the significance of the Fourier transformation in optical microscopy seems, at first sight, somewhat obscure. Consider the simple two-lens arrangement shown in Fig. 1.6 where the object is a point source. This example uses lenses of equal focal length with object and image at the focal point of each lens.

Of course the optical microscope utilizes a somewhat different and more complicated arrangement of lenses but the details of the geometric optics need not concern us here. What our simple example illustrates is the role of the conjugate space.

Suppose we ignore for the moment the action of lens 1 and consider a plane wave incident on lens 2. In our first example the incident wave of wavelength λ travels along the lens axis and the effect of the lens medium is to slow the wave more at its wide centre than at its thinner extremes. The result is to cause points of equal phase (i.e., the wavefront) to form a spherical surface centred on the focal point so that each part converges on this point with identical phase despite having travelled different distances. To account for this effect we ascribe to all parts of the wavefront on the image side of the lens, the wavevector belonging to the wavefront centre which passes through the lens pole.

In the example, shown in Fig. 1.6, the wave is slightly inclined to the axis

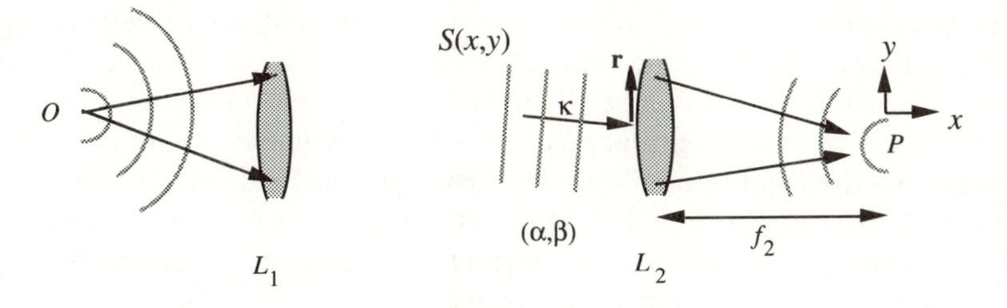

Fig. 1.6 The formation of an optical image P of a source O by the use of lenses. (α, β) are the direction cosines of the plane wave incident on lens 2.

such that its $\boldsymbol{\kappa}$-vector $\frac{2\pi}{\lambda}(\alpha\mathbf{i} + \beta\mathbf{j})$ has direction cosines α and β with respect to the axes x and y. This gives a phase variation $e^{-i\boldsymbol{\kappa}\cdot\mathbf{r}}$ along $\mathbf{r}$. From a knowledge of ray diagrams one can predict that the part of the wavefront passing through the pole of the lens (i.e., the ray) will continue in the same direction, so that $\boldsymbol{\kappa}$ defines the angular position of the image point P. The action of the lens is to cause all parts of the wavefront over the aperture of the lens to be co-added in the focal plane at a point with coordinates $x\mathbf{i} + y\mathbf{j} = (\alpha\mathbf{i} + \beta\mathbf{j})f_2$. Clearly the position of P is defined by the direction of $\boldsymbol{\kappa}$ and the focal length f_2. Again, the lens action causes all points of an incident plane wave front to converge at P as though each had the same wavevector $\boldsymbol{\kappa}$ and lay on the part of the wavefront which passes through the pole. This means that the coherent superposition of contributions from across the lens leads to an amplitude at P given by

$$\mathbf{E}_P(\boldsymbol{\kappa}) = \mathbf{E}_0\, e^{i\kappa f_2} \iint_{\text{aperture}} S(\mathbf{r})\, e^{-i\boldsymbol{\kappa}\cdot\mathbf{r}}\, d\mathbf{r} \tag{1.18}$$

where $S(\mathbf{r})$ represents the amplitude (and phase) distribution across the aperture on the incidence side and the integration is performed over the aperture of the lens. This is equivalent to a Fourier transformation in which a lens weighting function $g(\mathbf{r})$ is used where $g(\mathbf{r})$ is unity inside the aperture and zero outside. Thus

$$\mathbf{E}_P(\boldsymbol{\kappa}) = \mathbf{E}_0\, e^{i\kappa f_2} \iint S(\mathbf{r})\, g(\mathbf{r})\, e^{-i\boldsymbol{\kappa}\cdot\mathbf{r}}\, d\mathbf{r}. \tag{1.19}$$

Suppose that $S(\mathbf{r})$ is uniform, corresponding to a point source at O (and an 'ideal' lens 1). The image is then said to be a **diffraction pattern** of the aperture. Ignoring the factor in front of the integral, $\mathbf{E}_P(\boldsymbol{\kappa})$ is just the Fourier transform of $g(\mathbf{r})$ and is known as the point spread function. For a circular aperture of diameter d the spread function is a sort of 'Bessel sinc function' with a characteristic zero at a $\boldsymbol{\kappa}$-vector orientation angle of $\sin^{-1}(1.22\,\lambda/d)$ from the centre of the image.

The relationship in eqn (1.19) expresses the fact that the finite lens aperture introduces a point spread function which is simply the two-dimensional spatial FT of the aperture. The significance of the model is that amplitude modulation effects taking place between the two lenses occur in a space conjugate to the final image space. Suppose that instead of a point source we have an extended source and hence a distribution $O(\boldsymbol{\kappa})$. The role of the first lens can then be understood by virtue of reciprocity. It converts the density function for light emerging from the object, $O(\boldsymbol{\kappa})$ into a conjugate signal $S(\mathbf{r})$. Of course because we now have an extended object there will be a range of $\boldsymbol{\kappa}$ contributions to be summed at each lens. Two lenses therefore perform, in effect, two successive Fourier transformations.

Consider the case of lens 1. It generates a plane-wave amplitude function

$$S(\mathbf{r}) = \iint O(\boldsymbol{\kappa})\, e^{i\boldsymbol{\kappa}\cdot\mathbf{r}}\, d\boldsymbol{\kappa} \tag{1.20}$$

where common phase and amplitude factors have been ignored. Non-ideality of the first lens is accounted for by multiplying $S(\mathbf{r})$ by the appropriate spatial function $g_1(\mathbf{r})$. The second lens then reconstructs the image via the inverse transformation

$$I(\boldsymbol{\kappa}) = \iint S_A(\mathbf{r}) g_2(\mathbf{r})\, e^{-i\boldsymbol{\kappa}\cdot\mathbf{r}}\, d\mathbf{r} \tag{1.21}$$

where we have absorbed the multiplicative factor $g_1(\mathbf{r})$ into $S_A(\mathbf{r})$ in accordance with the scheme of Fig. 1.2. The net effect of both lenses is therefore given by $g(\mathbf{r}) = g_1(\mathbf{r})g_2(\mathbf{r})$. The consequence of multiplication of the 'ideal' conjugate signal by $g(\mathbf{r})$ can be understood by application of the convolution theorem. The image and object distributions are related by

$$I(\boldsymbol{\kappa}) = O(\boldsymbol{\kappa}) \otimes \mathfrak{F}\{g(\mathbf{r})\} \tag{1.22}$$

where $\mathfrak{F}\{g(\mathbf{r})\}$ is the point-spread function.†

In optical microscopy, magnification is achieved by appropriately adjusting the relative separation of the object and image planes for their respective lenses so that the wavevector $\boldsymbol{\kappa}$ translates into differing spatial separations. The point spread function due to the finite lens apertures introduces a convolution broadening in $\boldsymbol{\kappa}$ which corresponds to an image spatial separation of order $f_2\lambda/d$. Note that d/f_2 is, in effect, the maximum angular aperture subtended by the lens.

† Strictly speaking one measures intensity rather than amplitude in optical microscopy so that the spread function turns out to be $\mathfrak{F}\{g(x,y) \otimes g^*(-x,-y)\}$ if I and O refer to intensity distributions. Further discussion of this point, and of the role of coherence in image formation can be found in Stroke, 1969.[12]

The result can be achieved in another way, due originally to Abbe in 1873.[11] The resolution of optical microscopy is determined by the highest object spatial frequency which can be accommodated. We can think of this limiting Fourier component as comprising a sort of sinusoidal grating in the object. When this grating spacing is sufficiently fine that the first order maxima occur at angles wider than the angle θ_{max} subtended by the lens, then the limit is reached. The Abbe criterion therefore corresponds to a spatial period $\lambda/\sin\theta_{max}$. By this simple argument the resolution is given by

$$\delta \sim \lambda/\sin\theta_{max} \sim 2\pi/\kappa_{max}. \tag{1.23}$$

A more precise analysis applies a resolution criterion (for example the Rayleigh criterion[13]) to the image points and takes account of the refractive index n through which the light must travel to the objective lens. The resolution is in practice limited by the objective lens and is given by[12]

$$\delta = 0.61\,\lambda/n\sin\theta_{max}. \tag{1.24}$$

The highest value of $n \sin\theta_{max}$ which can be achieved in practice is around 1.5, giving optical microscopy a transverse resolution of around 0.2 μm in a standard slice thickness of 30 μm.

1.4 Photon intensity tomography

The principal disadvantage of optical microscopy is the need to prepare thin slices corresponding to the depth of field of the objective lens. In electron microscopy, where even greater resolution is possible, the sectioning and disturbance to the sample is even more severe. The purpose of a tomographic reconstruction is the provision of an image of a plane within the specimen without the need for sectioning. It is therefore necessary to employ radiations with sufficient penetrating power to traverse the object in question. For a few centimetres of biological tissue a schematic spectrum of relative absorption for both sound and electromagnetic radiation is shown in Fig. 1.7. This comparison is due to Ernst[14] who identified three 'windows' of transparency. These are low-frequency ultrasound, and X-rays and radio-waves in the case of the electromagnetic spectrum. For sound waves the penetration is poor below 100 μm. If we presume that resolution is, as in optical microscopy, wavelength limited, the potential for tomographic acoustic microscopy appears slight. As a surface technique however, the scanning acoustic microscope, operating in the micron wavelength regime, is a powerful tool in materials science.[15] For X-rays the wavelength is exceedingly short and the quantum energy high. In principle these features point to high resolution and high sensitivity. In fact the quantum of energy is sufficiently high to cause significant damage to biological specimens so

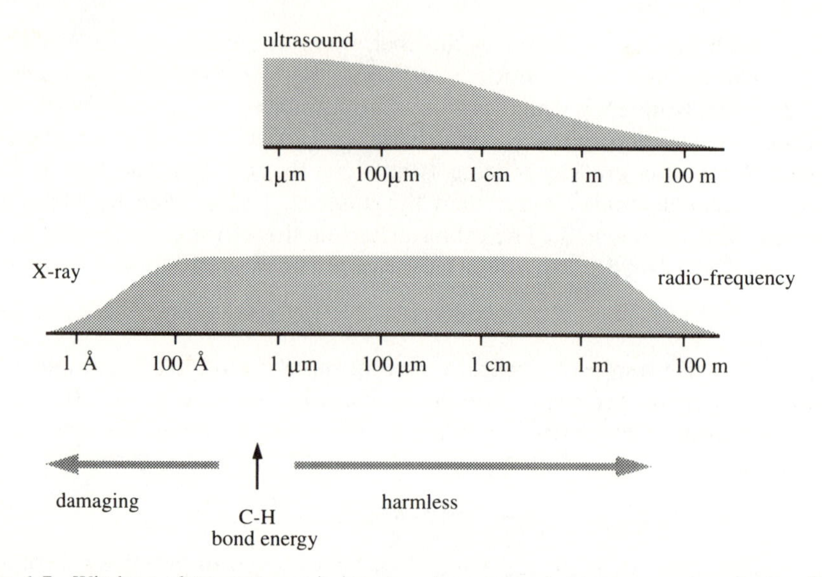

Fig. 1.7 Windows of transparency in imaging via sound and electromagnetic radiation. The vertical scale measures absorption in tissue (From Ernst[14]).

that X-ray penetration cannot strictly be termed non-invasive. Furthermore, as we shall see, the wavelength limit cannot be achieved for short wavelength (hard) X-rays because lenses do not exist in this part of the electromagnetic spectrum. However for soft X-rays, with wavelengths greater than 20 Å, it is possible to perform X-ray microscopy for thin samples using Fresnel lenses although such methods require intense radiation from a synchrotron or laser-produced plasma. These systems can give a resolution of around 0.1 μm.[16]

In the low-frequency window, the use of radio-waves appears hopeless because of the large wavelengths involved. In fact a technique is available which averts the diffraction limit. This is possible because a mechanism exists for imparting a frequency signature to the radiation which relates to its origin within the sample. It is the low photon energy which determines the ultimate spatial limit in NMR microscopy in the subtle 'trade-off' between sensitivity and resolution.

The basic scheme governing photon intensity tomography in the short wavelength electromagnetic window is shown in Fig. 1.8, adapted from Brooks and de Chiro.[17] The two examples are labelled absorption and emission imaging and refer, respectively, to X-ray CT and Positron Emission Tomography (PET). The photons are detected by a rotatable one-dimensional array of collimated detectors which define ray paths (r, ϕ). The coordinate r refers to the specific detector while ϕ refers to the orientation

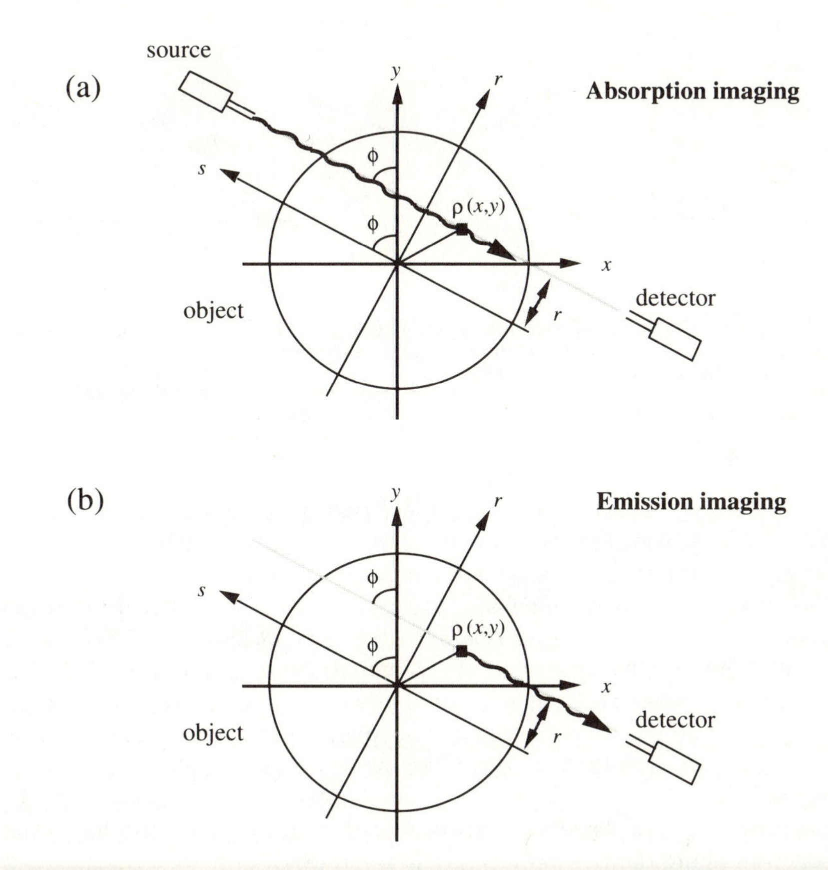

Fig. 1.8 Photon intensity tomography. (a) In absorption imaging the integrated absorption along a column through the object is measured. An array of detectors therefore measures a 'shadow profile'. (b) In emission imaging the integrated emitter density is measured.

of the array. In example (a) the planar beam of X-rays is generated by an array of collimated sources which in practice may be combined into a single source fan beam provided that the appropriate geometrical corrections are incorporated into the reconstruction.

Photon intensity measured by a detector at (r, ϕ) then gives a measure of the integrated absorption density along the ray path. Fig. 1.8(b) shows the arrangement for emission imaging. Here the radiation (i.e., γ-rays) is of shorter wavelength so that absorption effects are less important. The photon intensity at (r, ϕ) then gives a measure of the integrated emitter density along the ray path once suitable corrections are made for attenuation due to inverse square law and absorption effects.

Clearly the shadow profile cast by the object at various orientations of the detector array can be used to reconstruct the image density. At first sight,

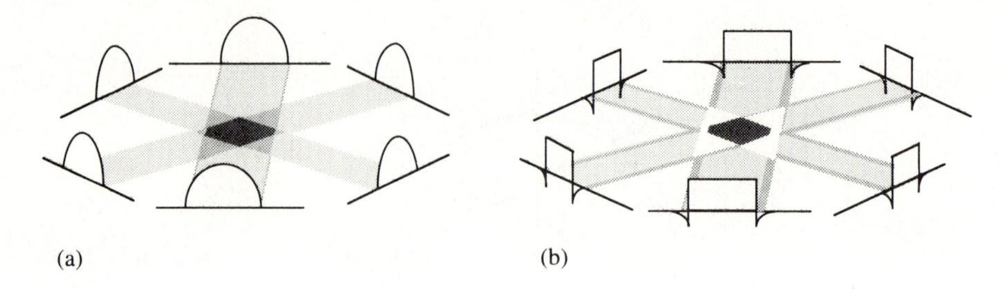

Fig. 1.9 Simple illustration of back-projection. In (a) the shadow profile of an object is back-projected, leading to 'star artefacts' in the image. In (b) the artefacts are removed by first 'filtering' the shadow profiles.

as shown in Fig. 1.9(a), it might apppear that a simple back-projection of intensity will provide such a reconstruction. In fact this method formed the basis of early attempts at X-ray CT. However it is clear in Fig. 1.9(a) that such a method yields so-called 'star artefacts'. The correct algorithm which is demonstrated in Fig. 1.9(b), is simply derived and illustrates how photon intensity tomography relates to the schematic system model of Fig. 1.2. This form of image reconstruction is known as 'filtered back-projection', 'Fourier filtration', or, more commonly, 'projection reconstruction'. In the present case we will look at the relationships which apply in absorption imaging.[17-21]

Suppose that the density of absorbers is $\rho(x, y)$. Then the integrated absorption along the ray path (r, ϕ) is given ideally by

$$P_\phi(r) = \int_{r,\phi} \rho(x, y)\, \mathrm{d}s \tag{1.25}$$

where s represents the distance measured along the ray. In practice, for X-ray CT, the integrated absorption projection $P_\phi(r)$ is proportional to the logarithm of the photon intensity because of the relationship

$$I = I_0 \exp\left\{-\int_{r,\phi} \mu(x, y)\, \mathrm{d}s\right\} \tag{1.26}$$

where μ is the absorption coefficient at that particular photon frequency. The 'signal' $P_\phi(r)$ is used to reconstruct the image and so obtain a representation of $\rho(x, y)$. In photon emission and absorption CT, therefore, the signal is obtained in real space although, as we shall see, the reconstruction process is most easily performed in conjugate space. The process may be visualized by starting from the Fourier transformation of $\rho(x, y)$ in Cartesian coordinates:

$$S(k_x, k_y) = \int_{-\infty}^{\infty} \int_{-\infty}^{\infty} \rho(x, y) \exp[\mathrm{i}\, 2\pi(k_x x + k_y y)] \, \mathrm{d}x \, \mathrm{d}y \tag{1.27}$$

$$\rho(x, y) = \int_{-\infty}^{\infty} \int_{-\infty}^{\infty} S(k_x, k_y) \exp[-\mathrm{i}\, 2\pi(k_x x + k_y y)] \, \mathrm{d}k_x \, \mathrm{d}k_y. \tag{1.28}$$

Suppose we transform from the (x, y) Cartesian basis to one based on (r, s) where

$$x = r \cos\phi - s \sin\phi \qquad \text{and} \qquad y = r \sin\phi + s \cos\phi. \tag{1.29}$$

If (k_x, k_y) is transformed to a polar coordinate frame where

$$k_x = k \cos\phi \qquad \text{and} \qquad k_y = k \sin\phi \tag{1.30}$$

then it is simple to show that

$$S(\mathbf{k}) = \int_{-\infty}^{\infty} \int_{-\infty}^{\infty} \rho(x, y) \, \mathrm{e}^{\mathrm{i}\, 2\pi k r} \, \mathrm{d}r \, \mathrm{d}s \tag{1.31}$$

where $k = |\mathbf{k}| = (k_x^2 + k_y^2)^{1/2}$. Changing the order of the integration, we can immediately recognize the absorption projection integral $P_\phi(r)$ so that

$$S(\mathbf{k}) = \int_{-\infty}^{\infty} P_\phi(r) \, \mathrm{e}^{\mathrm{i}\, 2\pi k r} \, \mathrm{d}r. \tag{1.32}$$

This expression has the form of a one-dimensional Fourier transformation of the 'integrated absorption profile measured transversely to the photon beam'. At this point it is instructive to rewrite eqn (1.28), which is expressed in conjugate Cartesian coordinates $(x, y) \Leftrightarrow (k_x, k_y)$, in terms of polar coordinates $(r, \phi) \Leftrightarrow (k, \phi)$ where we deliberately choose a common azimuthal angle. The area element $\mathrm{d}k_x \, \mathrm{d}k_y$ becomes $|k| \, \mathrm{d}k \, d\phi$. Thus

$$\begin{aligned} \rho(x, y) &= \int_0^{\pi} \int_{-\infty}^{\infty} S(\mathbf{k}) \, \mathrm{e}^{-\mathrm{i}\, 2\pi k r} \, |k| \, \mathrm{d}k \, \mathrm{d}\phi \\ &= \int_0^{\pi} \left\{ \int_{-\infty}^{\infty} \left[\int_{-\infty}^{\infty} P_\phi(r) \, \mathrm{e}^{\mathrm{i}\, 2\pi k r} \, \mathrm{d}r \right] |k| \, \mathrm{e}^{-\mathrm{i}\, 2\pi k r} \, \mathrm{d}k \right\} \mathrm{d}\phi \end{aligned} \tag{1.33}$$

where $r = x \cos\phi + y \sin\phi$.

The reconstruction process represented by eqn (1.33) may be viewed as follows. First (i) the absorption profile, P_ϕ, is measured transverse to the beam oriented at angle ϕ and a one-dimensional Fourier transformation is performed. This is the process illustrated inside the square bracket of eqn (1.33). Next (ii) the transformed signal is multiplied by a ramp function $|k|$ in conjugate space and the result subjected to an inverse transformation. This is illustrated inside the curly brackets. This process of transformation, multiplication, and inverse transformation is known as filtering. Fig. 1.9(a) shows the projection profiles of a uniform circular object while 1.9(b) shows the filtered profiles. The resulting signal applies to all points (x, y) which satisfy eqn (1.29). This corresponds to the back-projection step in which the filtered profile $\rho_\phi(r)$ is added to all points lying on rays normal to the profile. The filtered back-projection process is illustrated in Fig. 1.9(b). Finally (iii) the back-projection process is repeated for all angles $0 < \phi < \pi$, the final integration over ϕ in eqn (1.33). The negative lobes of the filtered profiles have the effect of cancelling the star artefacts. Because the original profiles $\rho_\phi(r)$ are independent of the sense of the rays, two quadrants of back-projection are sufficient to reconstruct the image.

Of course the process of multiplication in Fourier space is entirely equivalent to convolution in real space. The real space transform of $|k|$ is $-1/2\pi^2r^2$ so that the filtered profile can be obtained by means of the convolution integral

$$\rho_\phi = -\frac{1}{2\pi^2} \int_{-\infty}^{\infty} \frac{P_\phi(r')}{(r - r')^2}\, dr'. \tag{1.34}$$

This integral is known as a Radon filter after its discoverer who used it in 1917 in the context of gravitational theory.[22] The method requires more computation than Fourier filtration and suffers from problems associated with singularities in the integrand. Although Radon filtering has been applied to X-ray imaging the preferred reconstruction method is Fourier filtration.

The process of double Fourier transformation bears some resemblance to optical microscopy. Indeed in one early reconstruction method, which avoided the need for digital reconstruction, Bates and Peters[23] employed an optical system with lenses to carry out the Fourier transformation. This example serves to illustrate the common features represented by the schematic of Fig. 1.2. However there is an important difference between optical microscopy and photon intensity tomography which severely limits the resolution of the latter. In microscopy the emitted photons are 'collected' by a lens whose dimensions are conjugate to the object space. In short, the larger the lens, the better the resolution. By contrast, in photon intensity tomography the detection process is irrevocably confined to real space and the resolution is strictly limited by the thickness of the X-ray beam and the aperture of the collimators. In practice this restricts the resolution to of order 1 mm.

1.5 The radio-frequency window

A brief glance at Fig. 1.7 can convince us that radio-frequency radiation will be of little use in microscopy if we attempt to spatially localize photons in the manner of X-ray tomography. The problem is rather like attempting to use the intensity distribution of radio waves to decide from which of two very distant, but adjacent, radio antennas our signals originated. Of course the problem has a simple solution if the antennas broadcast at different frequencies and we happen to know which is which.

The radio-frequency spectroscopy of materials began in 1946 with the simultaneous discovery of nuclear magnetic resonance by Purcell, Torrey, and Pound[24] at Harvard and Bloch, Hansen, and Packard[25] at Stanford. Stable atomic nuclei possess angular momentum quantized in integral or half-integral units (I) of $\hbar = h/2\pi$ (where h is the Planck constant), depending on whether the number of constituent nucleons is even or odd. For $I \geq \frac{1}{2}$, the nucleus will possess an intrinsic magnetic dipole moment and so, when placed in a magnetic field, will experience a torque tending to align it. This torque however is applied to an already existing angular momentum and induces a precessional motion in which the nucleus reorients around the magnetic field at a characteristic rate known as the Larmor frequency. In typical laboratory magnetic fields this frequency is in the part of the electromagnetic spectrum occupied by HF and UHF radio transmission. Two aspects of this precessional motion are of particular significance in understanding the tomographic applications of nuclear magnetism. First, in the nuclear magnetic resonance (NMR) precession, the nuclei can be excited from their equilibrium state by applying a radio-frequency magnetic field at exactly the Larmor frequency. Subsequently the nuclei will produce a detectable radio-frequency signal in a sufficiently sensitive receiver system by using a coil antenna as the primary detection element. We thus have the potential to cause the abundant stable nuclei in the object to act as a source of photons for emission imaging. Second, the Larmor frequency which characterizes the nuclear radiation is proportional to the magnetic field strength and so it is possible to precisely encode the spatial origin of the emission by using a magnetic field which varies with position.

Providing that the field variation is sufficiently linear to permit accurate reconstruction and sufficiently large to dominate intrinsic spectral features in the NMR signal, the resolution, according to this simple argument, can be made arbitrarily fine. Nature however is not quite so kind. The radio-frequency photons, while abundant in number, have exceedingly low energies and must be detected in competition with the thermal noise which is present in the receiving coil. This noise power is distributed across the image spectrum comprising NMR emissions in an array of volume elements. As the volume element is reduced in size the contribution from the signal declines and the signal-to-noise ratio from that element eventually deteriorates beyond acceptable limits. Thus the resolution in an ideal NMR imaging system is inherently limited by sensitivity. A major factor in determining sensitivity is the available field strength. Using the largest laboratory magnetic fields in existence a voxel resolution of around $10\,\mu m^3$ is possible.

While the initial imaging experiments began in laboratory spectrometers intended for more traditional NMR spectroscopy, subsequent developments consisted principally of 'scaling up' for whole-body tomography. Today, because of this medical application known as magnetic resonance imaging (MRI), [26,27] the phenomenon of nuclear magnetic resonance is familiar to most physicians and to many of the non-scientific public. The literature of NMR imaging exhibited a phenomenal growth from 8 publications per annum in 1975 to 30 in 1980, 511 in 1985 and 687 in 1990. During the period up to 1985 however, the application of NMR imaging to microscopy had been largely neglected. While a growing number of non-medical applications had been reported, in all these the resolution had been of order 1 mm although some workers had reported applications in which resolutions of order 0.1 mm to 0.3 mm were achieved.

In 1986 four groups (Aguayo *et al.*,[28] Eccles and Callaghan,[29] Lauterbur and Hedges,[30] and Kamei and Katayama[31]) achieved a resolution with voxel elements smaller than $(100\,\mu m)^3$. Since then a large number of NMR microscopy applications in both biological and medical science have been reported. The ultimate resolution to be achieved with this method is uncertain. Calculations based on existing magnetic fields and receiver coil systems do not take account of the dramatic changes which are a likely consequence of recent advances in superconductor technology. However, to focus on resolution alone is to miss many remarkable features. While nuclear magnetic resonance microscopy has yet to achieve the resolving power of the optical microscope, it offers the following advantages. It is non-invasive, allowing slices of sample to be examined without physical cutting. It it molecular specific, allowing, in principle, the total density of a molecule of interest to be determined. It is sensitive to molecular dynamics through a host of contrast mechanisms previously developed for non-spatially resolved NMR spectroscopy. One of the more remarkable examples of such a contrast is the facility to encode in the signal a signature arising from molecular translation over a predetermined time interval. Applications of this method, including the imaging of diffusion and flow, are discussed in Chapter 8.

Over the last few years concurrent efforts have taken place to produce an imaging technique based on Electron Spin Resonance (ESR).[32-35] Despite the difficulty in finding suitable paramagnetic electron species capable of participating in the resonance process and various complications which exist in the interpretation of spectra, ESR is inherently more sensitive than NMR because of the much greater electron magnetic moment and the consequently higher Larmor frequencies. Between 1985 and 1990 some articles have appeared in which a transverse resolution below 1 mm has been reported.[36-39] These developments are potentially significant and ESR microscopy is a subject sufficient for a separate book. None the less there is no doubt that imaging using nuclear precession is considerably more advanced and, following the medical example, we shall employ the terms magnetic resonance microscopy and NMR microscopy interchangeably and set out the principles and applications in the pages to follow.

1.6 References

1. Hounsfield, G. N. (1973). *British Patent No.* 1283915 (1972) and Br. J. Radiol. **46,** 1016.
2. Bracewell, R. N. (1956). *Austr. J. Phys.* **9,** 109–217.
3. Cormack, A. M. (1963). *J. Appl. Phys.* **34,** 2722–7.
4. Lauterbur, P. C. (1973). *Nature* **242,** 190.
5. Mansfield, P. and Grannell, P. K. (1973). *J. Phys. C* **6,** L422.
6. Brigham, E. O. (1974). *The fast Fourier transform*, Prentice Hall, Englewood Cliffs.
7. Champeney, D. C. (1973). *Fourier transforms and their physical applications*, Academic Press, New York and London.
8. Titchmarsh, E. C. (1962). *Introduction to the theory of Fourier integrals*, Clarendon Press, Oxford.
9. Black, H. S. (1953). *Modulation theory*, Van Nostrand, Princeton.
10. Cooley, J. W. and Tukey, J. W. (1965). *Math. Computation* **19,** 297.
11. Abbe, E. (1873). *Arch. Mikrosk. Anat.* **9.**
12. Stroke, G. W. (1969). *An introduction to coherent optics and holography*, Academic Press, New York.
13. Lord Rayleigh, (1879). *Phil. Mag.* **8,** 261.
14. Ernst, R. R. (1987). *Q. Rev. Biophysics* **19,** 183.
15. Briggs, A. (1985). *An introduction to scanning acoustic microscopy*, Oxford University Press.
16. Kirz, J. and Rarback, H. (1985). *Rev. Sci. Instr.* **56,** 1.
17. Brooks, R. A. and di Chiro, G. (1976). *Phys. Med. Biol.* **21,** 689.
18. Ledley, R. S., di Chiro, G., Luessenhop, A. J., and Twigg, H. L. (1974). *Science* **186,** 207.
19. Erikson, H. R., Fry, F. J., and Jones, J. P. (1974). *IEEE Trans. Sonics Ultrasonics*, **SU21,** 144.
20. Robinson, A. L. (1975). *Science* **190,** 542, 647.
21. Ter-Pogossian, M. (1977). *Sem. Nucl. Med.* **7,** 109.
22. Radon, J. (1917). *Ber. Verh. Sachs. Akad. Wiss.* **69,** 262.
23. Bates, R. H. T. and Peters, T. M. (1971). *N.Z. J. Sci.* **14,** 883.
24. Purcell, E. M., Torrey, H. C., and Pound, R. V. (1946). *Phys. Rev.* **69,** 37.
25. Bloch, F., Hansen, W. W., and Packard, M. (1946). *Phys. Rev.* **70,** 474.

26. Mansfield, P. and Morris, P. G. (1982). *NMR imaging in bio-medicine*, Academic Press, New York.
27. Morris, P. G. (1986). *NMR imaging in biology and medicine*, Clarendon Press, Oxford.
28. Aguayo, J. B., Blackband, S. J., Shoeniger, J., Mattingley, M. A., and Hinterman, M. (1986). *Nature* **322,** 198.
29. Eccles, C. D. and Callaghan, P. T. (1986). *J. Magn. Reson.* **68,** 393.
30. Lauterbur, P. C. (1986). In *NMR in biology and medicine*, (ed. S. Chien and C. Ho), Raven Press, New York.
31. Kamei, H. and Katayama, Y. (1986). *IEEE/Eighth Annual Conference of the Engineering in Medicine and Biology Society*, Abstracts, p. 1159.
32. Ohno, K. (1986). *Appl. Spectr. Rev.* **22,** 1.
33. Ohno, K. (1987). *Magn. Reson. Rev.* **11,** 275.
34. Eaton, S. S. and Eaton, G. R. (1986). *Spectroscopy* **1,** 32.
35. Eaton, G. R. and Eaton, S. S., (1988). *Bull. Magn. Reson.* **10,** 22.
36. Ohno, K. and Murakami, T. (1988). *J. Magn. Reson.* **79,** 343.
37. Cleary, D. A., Shin, Y. K., Schneider, D. J., and Freed, J. H. (1988). *J. Magn. Reson.* **79,** 474.
38. Ohno, K. and Murakami, T. (1988). *J. Magn. Reson.* **79,** 343.
39. Woods, R. K., Bacic, G. G., Lauterbur, P. C., and Swartz, H. M. (1989). *J. Magn. Reson.* **84,** 247.

2
INTRODUCTORY NUCLEAR MAGNETIC RESONANCE

2.1 Elementary quantum mechanics and nuclear magnetization

Atomic nuclei are characterized by states which are inherently quantum mechanical in their behaviour. This means in effect that the properties we observe for a single nucleus belong to one of a discrete set of possibilities. When performing nuclear magnetic resonance, however, we deal with exceedingly large numbers of nuclei acting largely independently, so that at the macroscopic level the collection of particles appears continuous, the so-called 'ensemble' behaviour. We shall be mainly concerned with spin-$\frac{1}{2}$ nuclei. For these, all states of the ensemble may be characterized by a simple vector quantity which is referred to as the nuclear magnetization. Having established this point we may embark on a classical explanation of the nuclear magnetic resonance phenomenon. But the quantum mechanics is never far below the surface and will emerge explicitly whenever spin-$\frac{1}{2}$ (or any higher spin) nuclei experience mutual interactions. It will also be apparent when we perform NMR on independent nuclei of spin $> \frac{1}{2}$ in the presence of electric quadrupole interactions. Many of the vast panoply of tricks with which nuclear spins can be manipulated involve quantum behaviour. An understanding of these tricks and of the quantum mechanical principles underlying them is therefore central to an appreciation of some of the important contrast phenomena available in NMR microscopy. There are many excellent texts on this subject[1-6] but it is appropriate to review briefly the salient points relating to the nuclear magnetization. A knowledge of simple linear algebra and matrix multiplication is presumed in this approach. A detailed discussion of the principles of quantum mechanics can be found elsewhere[7] but a brief sketch is given here. Readers preferring to adopt a semi-classical view should take up the description from Section 2.3.

Four precepts are essential in understanding the quantum description of nuclear spin behaviour. The first concerns the concept of a spin **state** which may be described in terms of a set of discrete possibilities known as the basis set. We will deal exclusively with a basis set described by the component of angular momentum, m, as measured along some axis, z. For the moment we note that m may be any one of a discrete set of integer or half-integer values in the range $-I, -I+1, \ldots, I-1, I$ where I is an integer or half-integer number called the angular momentum quantum number. The origin of this description lies in the symmetry properties of the nuclear particles and need not concern us here. It is important to note that the spin quantum number, I, (sometimes called the 'spin') is a fixed quantity characterizing a nucleus in its stable ground state. For example, a ^{1}H nucleus (proton) or ^{13}C nucleus has $I = \frac{1}{2}$ while the deuteron has $I = 1$. Note that the angular momentum is actually measured in units of $\hbar$ (Planck's constant/2π). We will only introduce this unit when it is specifically required.

In general an arbitrary state, $|\Psi\rangle$, of a nucleus may be written in terms of a linear combination of basis states labelled by m. In other words

$$|\Psi\rangle = \sum_m a_m |m\rangle. \tag{2.1}$$

It is important to note that in quantum mechanics the amplitudes a_m have phase as well as magnitude so that it is helpful to represent these by complex numbers.

The second precept concerns the so-called **eigenvalue equation**. We will consider this equation when the angular momentum is the observable of interest. The process of observation of the angular momentum component along the z-axis for a single nucleus in a basis state $|m\rangle$ is described by the eigenvalue equation,

$$I_z|m\rangle = m|m\rangle. \tag{2.2}$$

The eigenvalue m is the result of an observation while I_z is the 'operator' for the angular momentum along the z-axis. Since we have a discrete set of $|m\rangle$ states we can represent these states using column vectors. For example, where $I = \frac{1}{2}$ we have the states

$$|+\tfrac{1}{2}\rangle = \begin{bmatrix} 1 \\ 0 \end{bmatrix} \qquad |-\tfrac{1}{2}\rangle = \begin{bmatrix} 0 \\ 1 \end{bmatrix}. \tag{2.3}$$

It is therefore obvious that the operator I_z is a matrix, namely

$$I_z = \frac{1}{2}\begin{bmatrix} 1 & 0 \\ 0 & -1 \end{bmatrix}. \tag{2.4}$$

Continuing this approach it can be seen that the result of a measurement, namely $m = \pm\frac{1}{2}$, is given by calculating $\langle m|I_z|m\rangle$ where $\langle m|$ is the row vector conjugate to $|m\rangle$. (More formally, $\langle\Psi|$ and $|\Psi\rangle$ are known, respectively, as bra and ket vectors. They are each the complex conjugate transpose of the other.) Thus, for example, a measurement of the I_z observable when the spin system is in the definite 'spin-up' state yields

$$\langle\tfrac{1}{2}|I_z|\tfrac{1}{2}\rangle = \begin{bmatrix} 1 & 0 \end{bmatrix} \frac{1}{2}\begin{bmatrix} 1 & 0 \\ 0 & -1 \end{bmatrix}\begin{bmatrix} 1 \\ 0 \end{bmatrix}$$

$$= \frac{1}{2} \tag{2.5}$$

Suppose our nucleus is not in an eigenstate $|m\rangle$ but in an admixed state given by the superposition, $|\Psi\rangle = \Sigma_m a_m |m\rangle$. Then the result of a measurement, the so-called 'expectation value', is given by

$$\langle\Psi|I_z|\Psi\rangle = \sum_{m,\,m'} a_m a^*_{m'} \langle m'|I_z|m\rangle$$

$$= \sum_{m,\,m'} a_m a^*_{m'}\, m\langle m'|m\rangle. \tag{2.6}$$

Because the basis vectors are orthogonal we are left with the result

$$\langle\Psi|I_z|\Psi\rangle = \sum_m |a_m|^2 m. \tag{2.7}$$

The meaning of this equation needs some explanation. If we have a large number of identically prepared nuclei then the sum on the right represents a mean of eigenvalue results weighted by the normalized probabilities $|a_m|^2$. For a single nucleus the interpretation is that the measurement has a probability $|a_m|^2$ of returning the result m.

Of course we have been dealing only with the operator I_z which is, incidentally, a diagonal matrix. In other words we have chosen to represent our states as eigenvalues of I_z. If, for example, we consider the measurement of angular momentum about other axes we find that the corresponding operators cannot be diagonal. This reflects the Heisenberg uncertainty principle whereby systems cannot simultaneously exist in definite states of certain pairs of observables. These observables are represented by operators which do not commute. This brings us to the third precept which concerns the algebra of angular momentum operators, I_x, I_y, I_z. This algebra is contained in the famous **commutation relationship**

$$[I_x, I_y] = I_x I_y - I_y I_x = \mathrm{i} I_z. \tag{2.8}$$

A full discussion of this relationship is beyond the scope of this book and can be found elsewhere.[8,9] The relationship arises from the fact that rotations about different axes do not commute and from the fact that the operator for rotation (for example about the z-axis) is related to the operator for angular momentum (I_z) by the rule

$$R_z(\phi) = \exp(\mathrm{i}\phi I_z) \tag{2.9}$$

where the exponential function of the operator I_z is interpreted as the usual power series. The axiom represented by eqn (2.8) is a starting point for understanding many of the properties of spin systems.

A particularly useful pair of operators, formed by taking linear combinations of I_x and I_y are the raising and lowering operators defined by

$$I_+ = I_x + \mathrm{i}I_y \qquad \text{and} \qquad I_- = I_x - \mathrm{i}I_y. \tag{2.10}$$

These operators have the effect of converting a $|-\frac{1}{2}\rangle$ state into a $|+\frac{1}{2}\rangle$ state, and vice versa, according to

$$I_+|-\tfrac{1}{2}\rangle = |+\tfrac{1}{2}\rangle \quad \text{and} \quad I_-|+\tfrac{1}{2}\rangle = |-\tfrac{1}{2}\rangle. \tag{2.11}$$

The final precept concerns the dynamics of the system which is described by the **Schrödinger equation**

$$\mathrm{i}\hbar\frac{\partial}{\partial t}|\Psi(t)\rangle = H|\Psi(t)\rangle \tag{2.12}$$

where H is the Hamiltonian (or energy) operator. Note that this formulation (the Schrödinger picture) has the states as functions of time but the operators stationary. However H may have some explicit time dependence if our quantum system is subjected to some fluctuating disturbance. If H is constant with time, then the Schrödinger equation yields the result

$$|\Psi(t)\rangle = U(t)|\Psi(0)\rangle \tag{2.13}$$

where

$$U(t) = \exp(-\mathrm{i}Ht/\hbar). \tag{2.14}$$

$U(t)$ is known as the evolution operator.

Atomic nuclei have a magnetic dipole moment proportional to the angular momentum, the constant of proportionality being known as the gyromagnetic ratio, γ. The interaction energy of a magnetic dipole $\boldsymbol{\mu}$ in a magnetic field $\mathbf{B}_0$ is written classically and quantum mechanically as $-\boldsymbol{\mu}\cdot\mathbf{B}_0$ so that the Hamiltonian operator for the case of B_0 oriented along the z-axis is just

$$H = -\gamma\hbar B_0 I_z. \tag{2.15}$$

This form of magnetic interaction is known as a Zeeman interaction. Since the eigenvalues of I_z range from $-I$ to I, the Zeeman Hamiltonian gives rise to the energy level structure shown in Fig. 2.1. Whatever the quantum number I, the energy separation between the levels is always $\hbar\gamma B_0$. Except where energy units are specifically required we shall, in future, drop the factor $\hbar$ and write our Hamiltonian in units of angular frequency.

Consider the evolution of a nuclear spin quantum state $|\Psi\rangle$ under the influence of the B_0 field. The evolution operator, $U(t) = \exp(\mathrm{i}\gamma B_0 I_z t)$, is identical to a clockwise rotation of the state about the z-axis by an angle $\gamma B_0 t$. The existence of the field causes all states to precess at the Larmor frequency, ω_0, given by

$$\omega_0 = \gamma B_0. \tag{2.16}$$

Taking the example of protons ($\gamma = 2.675 \times 10^8$ rad s^{-1} T^{-1}) in a large laboratory electromagnet, $B_0 = 1.4$ Tesla, we find a cyclic frequency f_0,

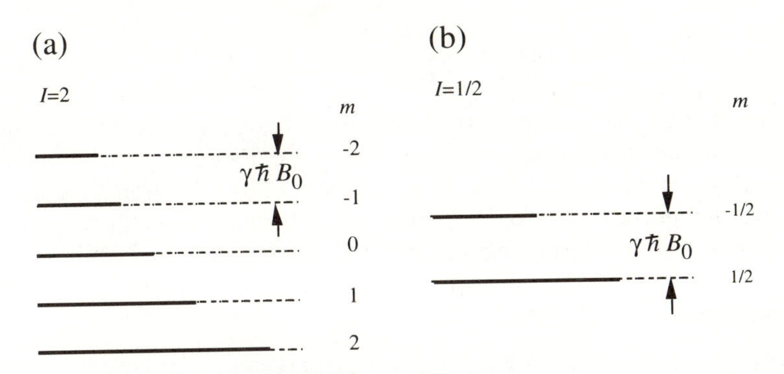

Fig. 2.1 Energy level diagram for spins experiencing a Zeeman Hamiltonian. In (a) $I=2$ while in (b) $I=1/2$. The energy separation for levels separated by $m=\pm 1$ is $\Delta E=\gamma\hbar B_0$. The bold line schematically represents the relative population in each state for an ensemble of systems in thermal equilibrium.

($f_0 = \omega_0/2\pi$) of 60 MHz. In a large superconductive magnet where B_0 may be 9.4 Tesla, the proton precession frequency will be 400 MHz. These frequencies correspond to the VHF and UHF parts of the radio-frequency spectrum, respectively.

Quantum mechanics has therefore given us a simple explanation for precession and in particular describes in detail how a magnetic field influences the state of a nuclear spin. Of course the description we are using refers to a single nucleus in isolation. In fact we always deal with large ensembles in which different nuclei may occupy different states $|\Psi\rangle$. Our description must therefore account for the ensemble averages which will result. This is done by representing the average by a sum over all subensembles, each with classical probability p_Ψ. In each subensemble all nuclei are in identical states $|\Psi\rangle$. The averaged expectation value of eqn (2.7) becomes

$$\overline{\langle\Psi|I_z|\Psi\rangle} = \sum_{\Psi} p_\Psi \langle\Psi|I_z|\Psi\rangle. \tag{2.17}$$

The bar over a quantity is taken to represent the averaging between the subensembles.

Suppose $I = \frac{1}{2}$, the simplest possible non-trivial example of a system in quantum mechanics. In nuclear magnetic resonance the dominant interaction of a spin with its environment is always via the Zeeman interaction of eqn (2.15). This means that the 'natural' eigenstates are those whose quantum numbers are eigenvalues of I_z, namely $|+\frac{1}{2}\rangle$ and $|-\frac{1}{2}\rangle$. In general therefore we express any state in this basis and write $|\Psi\rangle = a_{\frac{1}{2}}\begin{bmatrix}1\\0\end{bmatrix} + a_{-\frac{1}{2}}\begin{bmatrix}0\\1\end{bmatrix}$. Given the operator for I_z of eqn (2.4),

$$\overline{\langle \Psi | I_z | \Psi \rangle} = \frac{1}{2} \left(\overline{|a_{\frac{1}{2}}|^2} - \overline{|a_{-\frac{1}{2}}|^2} \right). \tag{2.18}$$

Eqn (2.18) may be interpreted by saying that the ensemble averaged expectation value of I_z is determined by the difference in population between the upper and lower energy levels. This difference is said to describe the polarization of the ensemble. A population difference will arise in thermal equilibrium according to the Boltzmann probability factor. In thermal equilibrium the two levels, separated by $\hbar \gamma B_0$, will have populations

$$\overline{|a_{\pm\frac{1}{2}}|^2} = \frac{\exp(\pm \hbar \gamma B_0 / 2k_\mathrm{B} T)}{\exp(-\hbar \gamma B_0 / 2k_\mathrm{B} T) + \exp(\hbar \gamma B_0 / 2k_\mathrm{B} T)}. \tag{2.19}$$

At room temperature in any laboratory magnet the energy difference $\gamma \hbar B_0$ is over five orders of magnitude smaller than the Boltzmann energy $k_\mathrm{B} T$. In consequence the expression for the populations, with a high degree of accuracy, may be written

$$\overline{|a_{\pm\frac{1}{2}}|^2} = \frac{1}{2} [1 \pm \hbar \gamma B_0 / 2k_\mathrm{B} T] \tag{2.20}$$

and in general, for any I, may be written

$$\overline{|a_m|^2} = \frac{1}{(2I+1)} [1 + m \hbar \gamma B_0 / k_\mathrm{B} T]. \tag{2.21}$$

Measurement of the x-component of angular momentum provides an interesting example by way of contrast. This particular observable provides the signal in the NMR experiment. I_x is the off-diagonal operator

$$I_x = \frac{1}{2} \begin{bmatrix} 0 & 1 \\ 1 & 0 \end{bmatrix} \tag{2.22}$$

so that

$$\overline{\langle \Psi | I_x | \Psi \rangle} = \frac{1}{2} \left[\overline{a^*_{\frac{1}{2}} a_{-\frac{1}{2}}} + \overline{a_{\frac{1}{2}} a^*_{-\frac{1}{2}}} \right]. \tag{2.23}$$

This is a quite different average from that represented by eqn (2.18). It reflects the phase coherence between the $+\frac{1}{2}$ and $-\frac{1}{2}$ states. The term in the bracket is said to describe the degree of 'single quantum coherence' of the ensemble.

The phenomena represented by both eqns (2.18) and (2.23) have a simple semi-classical interpretation. Consider, for example, Fig 2.1 in the case of thermal equilibrium. The ensemble has a slightly higher population in the lower energy state so that a net positive angular momentum z-component (*i.e.*, longitudinal magnetization) exists. On the other hand there is no a priori phase coherence in thermal equilibrium so that the the terms $\overline{a_{\frac{1}{2}}^* a_{-\frac{1}{2}}}$ are zero and no transverse magnetization exists.

At this point we are sufficiently well equipped to embark on a description of the nuclear magnetic resonance phenomenon. However, before leaving the discussion of quantum mechanics, it is useful to define one more operator which can serve us well in providing a simple description of a whole range of NMR phenomena not amenable to a semi-classical magnetization picture. This operator ρ, known as the density matrix,[1] generalizes the method of ensemble averaging. We define

$$\rho = \sum_{\Psi} p_{\Psi} |\Psi\rangle\langle\Psi|. \tag{2.24}$$

Suppose ρ is written as a matrix in the chosen $|m\rangle$ representation. Then the matrix elements of ρ are

$$\begin{aligned} \rho_{mn} = \langle m|\rho|n\rangle &= \sum_{\Psi} p_{\Psi} \langle m|\Psi\rangle\langle\Psi|n\rangle \\ &= \overline{a_m a_n^*}. \end{aligned} \tag{2.25}$$

The bar above a quantity is a shorthand notation for the ensemble average sum, $\Sigma_{\Psi} p_{\Psi}$. The usefulness of ρ lies in the fact that the expectation value of any operator O may be written

$$\overline{\langle\Psi|O|\Psi\rangle} = \mathrm{Tr}(O\rho) \tag{2.26}$$

where the right-hand side represents the trace or diagonal sum over the matrix product $O\rho$. It can easily be seen that eqn (2.26) gives results identical to (2.18) and (2.23) for the previously chosen examples. Three properties of ρ should be noted. First, it contains all the information we need to calculate the value of any observable quantity of interest. Second, the evolution of ρ with time may be simply deduced from the Schrödinger equation and may be written

$$\mathrm{i}\,\frac{\partial\rho}{\partial t} = [H, \rho]. \tag{2.27}$$

This relation is known as the Liouville equation and yields, for the case of constant H,

$$\begin{aligned} \rho(t) &= U(t)\rho(0)U^{-1}(t) \\ &= \exp(-\mathrm{i}Ht)\rho(0)\exp(\mathrm{i}Ht). \end{aligned} \tag{2.28}$$

Finally, the sum of the diagonal elements of ρ is just the sum of state probabilities and therefore equals unity.

For spin-$\frac{1}{2}$ therefore, only three numbers are needed to specify all the four matrix elements of ρ and therefore the results of any measurement. Using eqn (2.26) and the matrices for the operators I_x, I_y, and I_z it is straightforward to show that for $I = \frac{1}{2}$

$$\rho = \begin{bmatrix} \frac{1}{2} + \overline{\langle I_z \rangle} & \overline{\langle I_x - \mathrm{i}I_y \rangle} \\ \overline{\langle I_x + \mathrm{i}I_y \rangle} & \frac{1}{2} - \overline{\langle I_z \rangle} \end{bmatrix}. \tag{2.29}$$

In other words all states of the ensemble for independent spin-$\frac{1}{2}$ nuclei may be described by specifying the components of the vector $(\overline{\langle I_x \rangle}\,\mathbf{i} + \overline{\langle I_y \rangle}\,\mathbf{j} + \overline{\langle I_z \rangle}\,\mathbf{k})$. In macroscopic terms this is equivalent to specifying the magnetization vector, $\mathbf{M} = N\gamma\hbar[\overline{\langle I_x \rangle}\,\mathbf{i} + \overline{\langle I_y \rangle}\,\mathbf{j} + \overline{\langle I_z \rangle}\,\mathbf{k}]$ where N is the number of spins per unit volume. It may be shown that the influence of any Hamiltonian on such an ensemble is always equivalent to a precessional motion. Clearly ρ can be expressed as a linear combination of I_x, I_y, I_z, and the identity operator. Note that the I_z operator is diagonal and therefore gives information about the population terms $\overline{|a_{+\frac{1}{2}}|^2}$, $\overline{|a_{-\frac{1}{2}}|^2}$ whereas I_x and I_y are off-diagonal operators and tell us about the coherences, $\overline{a_{+\frac{1}{2}}a^*_{-\frac{1}{2}}}$.

These important ideas can be extended to higher dimensional systems. For example, a nuclear spin system with three or more energy levels can no longer be described by a density matrix expressed in a basis of the four operators, I_x, I_y, I_z and the identity matrix. What is required is a basis of the appropriate tensor rank. There are a variety of representations available to derive an algebra for ρ but that used most often in high-resolution NMR is the so-called product operator formalism[6,10] which is briefly described here.

Consider, for example, a system of spins in which pairs of spins are coupled. As a consequence there will be terms in the Hamiltonian which operate in the space of both spins, such as the scalar coupling interaction considered in Section 2.4.5. The eigenkets will involve the quantum numbers of two spins and the energy spectrum will consist of four levels, $|+\frac{1}{2}, +\frac{1}{2}\rangle$, $|+\frac{1}{2}, -\frac{1}{2}\rangle$, $|-\frac{1}{2}, +\frac{1}{2}\rangle$, and $|-\frac{1}{2}, -\frac{1}{2}\rangle$. The density operator for a four-level system is a 4×4 matrix with sixteen independent terms. This means that, whereas the isolated spin system operators could be expressed in a basis of four operators, the coupled two-spin system requires sixteen. In the product operator formalism the fifteen in addition to the identity matrix are shown in Table 2.1.

Table 2.1. *Basis operators for a coupled two-spin-$\frac{1}{2}$ system.*

one-spin operators[a]	*name*
I_{1z}, I_{2z}	polarization of spins 1 and 2
I_{1x}, I_{2x}	in-phase x-coherence of spins 1 and 2
I_{1y}, I_{2y}	in-phase y-coherence of spins 1 and 2
two-spin operators[b]	*name*
$2I_{1x}I_{2z}$, $2I_{1z}I_{2x}$	x-coherence of spin 1 in antiphase with respect to spin 2 (and vice versa)
$2I_{1y}I_{2z}$, $2I_{1z}I_{2y}$	y-coherence of spin 1 in antiphase with respect to spin 2 (and vice versa)
$2I_{1x}I_{2x}$, $2I_{1y}I_{2y}$, $2I_{1x}I_{2y}$, $2I_{1y}I_{2x}$	two-spin coherence of spins 1 and 2
$2I_{1z}I_{2z}$	longitudinal two-spin order of spins 1 and 2

[a] Note that each operator is implicitly a product of a spin 1 and a spin 2 operator. For these superficially one-spin operators the second spin operator is the identity matrix.

[b] Note that the factors of 2 in front of the two-spin operators are arbitrarily chosen to simplify the commutation algebra analogous to eqn. (2.8).

2.2 Resonant excitation and the rotating frame

We now consider the disturbance which results when an oscillating magnetic field of amplitude B_1 is applied transverse to a set of equivalent spins in thermal equilibrium in a longitudinal field $B_0\mathbf{k}$. The Hamiltonian in the laboratory frame of reference may be written

$$H_{\text{lab}} = -\gamma B_0 I_z - 2\gamma B_1 \cos\omega t\, I_x. \tag{2.30}$$

It is conventional to represent the linearly polarized oscillatory field as a sum of two counter-rotating circularly polarized components each of amplitude B_1 as shown in Fig. 2.2. One of these components will rotate in the same sense as the nuclear spin precession and will be responsible for the resonance phenomenon when $\omega = \omega_0$. The counter-rotating component can be ignored provided $B_1 \ll B_0$, which is invariably the case in the examples considered here. Now we make a coordinate transformation to the frame where the rotating field B_1 is stationary. This transformation has the great advantage that the Hamiltonian is made time independent and the simple evolution operator of eqn (2.14) applies. In future we will employ this frame of reference almost exclusively.

In the rotating frame the Hamiltonian of eqn (2.30) becomes

$$H_{\text{rot}} = -\gamma(B_0 - \omega/\gamma)I_z - \gamma B_1 I_x. \tag{2.31}$$

The reduction in the longitudinal field is simple to understand via a

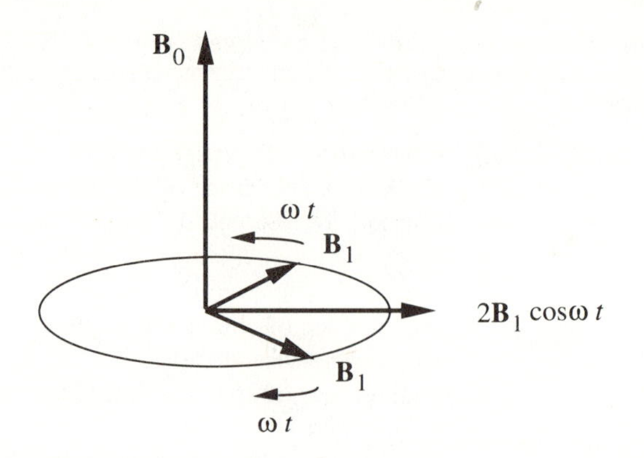

Fig. 2.2 Linearly polarized, oscillatory transverse field $2B_1 \cos \omega t$, represented as two counter-rotating circularly polarized fields.

semi-classical argument. Since the frame of reference is now rotating in the same sense as the spin precession, the apparent precessional speed is reduced by ω and so, if we are to preserve the Schrödinger equation, we must interpret the magnetic field as having been reduced by ω/γ. At resonance, $\omega = \omega_0$ and the apparent longitudinal field vanishes, thus leaving the effective field along the rotating frame x-axis. As a consequence the spins precess about this axis and the magnetization orientation is disturbed.

The quantum mechanical operator concept can be used to give another perspective on this process. In the rotating frame of reference the perturbing field has the form $\gamma B_1 I_x$ as shown in eqn (2.31). The I_x operator is just the linear combination of the raising and lowering operators, $\frac{1}{2}(I_+ + I_-)$. Application of eqns (2.13) and (2.14) tells us therefore that the evolution of the system with time corresponds to an inter-conversion of each spin between $|+\frac{1}{2}\rangle$ and $|-\frac{1}{2}\rangle$ at a rate γB_1.

So far the spin Hamiltonian has comprised only the Zeeman interaction arising from the uniform polarizing field, B_0, and the r.f. field B_1. In Section 2.4 we discuss the influence of other Hamiltonian terms. In doing so it is most convenient to use the rotating reference frame since this is the frame where description of the spin manipulation by the r.f. field and the subsequent phase-sensitive detection of the precessing transverse magnetization is particularly simple. The effect of transforming to the rotating frame is to subtract the term ωI_z from the main Zeeman Hamiltonian, $\omega_0 I_z$ as shown in eqn (2.31). Any further diagonal Hamiltonian terms (such as those involving the operator I_z) are unchanged while static laboratory terms in I_x or I_y will oscillate at the frequency ω, which at resonance is equal to ω_0.

Consider the effect of these additional oscillating components. The laboratory frame Hamiltonian is dominated overwhelmingly by B_0 which means that ω_0 is much greater than any possible precession about an additional Hamiltonian component. If this component fluctuates at ω_0 in the rotating frame then its own weaker induced spin precession will be unable to 'follow' this fluctuation and the effect will be as if the term were averaged to zero. This is the basis of our retaining only the diagonal or 'secular' terms in the laboratory frame Hamiltonian in calculating the spin evolution in the rotating frame. The off-diagonal terms which oscillate in the rotating frame have no effect on the energy levels and hence the position of resonances in the NMR spectrum. They are, however, important in determining the relaxation behaviour of the spins.

2.3 The semi-classical description

2.3.1 *Excitation*

In the case of independent spin-$\frac{1}{2}$ nuclei the motion of the ensemble of spins may always be described in terms of the precession of the spin magnetization vector. In such a model the macroscopic angular momentum vector is simply $\mathbf{M}/\gamma$ where $\mathbf{M}$ is the magnetization and γ the nuclear gyromagnetic ratio. By equating the torque to the rate of change of angular momentum we obtain

$$\frac{\mathrm{d}\mathbf{M}}{\mathrm{d}t} = \gamma \mathbf{M} \times \mathbf{B}. \tag{2.32}$$

The solution to eqn (2.32) when $\mathbf{B}$ is a magnetic field of amplitude B_0 corresponds to a precession of the magnetization about the field at rate $\omega_0 = \gamma B_0$, the Larmor frequency. The resonance phenomenon results on application of a transverse magnetic field oscillating at ω_0. To obtain the expression for the spin evolution we need retain only the circularly polarized component of the oscillating transverse field which is rotating in the same sense as the spin precession, namely

$$\mathbf{B}_1(t) = B_1 \cos \omega_0 t\, \mathbf{i} - B_1 \sin \omega_0 t\, \mathbf{j} \tag{2.33}$$

where $\mathbf{i}$, $\mathbf{j}$, $\mathbf{k}$ are unit vectors along the x, y, and z axes, respectively. Then eqn (2.32) gives

$$\frac{\mathrm{d}M_x}{\mathrm{d}t} = \gamma[M_y B_0 + M_z B_1 \sin \omega_0 t]$$

$$\frac{\mathrm{d}M_y}{\mathrm{d}t} = \gamma[M_z B_1 \cos \omega_0 t - M_x B_0]$$

$$\frac{\mathrm{d}M_z}{\mathrm{d}t} = \gamma[-M_x B_1 \sin \omega_0 t - M_y B_1 \cos \omega_0 t]. \tag{2.34}$$

It is easy to show by inspection that the solution, under the initial condition $\mathbf{M}(t) = M_0\mathbf{k}$, is

$$M_x = M_0 \sin \omega_1 t \sin \omega_0 t$$

$$M_y = M_0 \sin \omega_1 t \cos \omega_0 t$$

$$M_z = M_0 \cos \omega_1 t \tag{2.35}$$

where $\omega_1 = \gamma B_1$. Eqn (2.35) implies that on application of a rotating magnetic field of frequency ω_0, the magnetization simultaneously precesses about the longitudinal polarizing field B_0 at ω_0 and about the r.f. field B_1 at ω_1. The phenomenon is illustrated in Fig. 2.3(a). Of course this means that in the frame of reference rotating with B_1 about B_0 (shown in Fig. 2.3(b)),

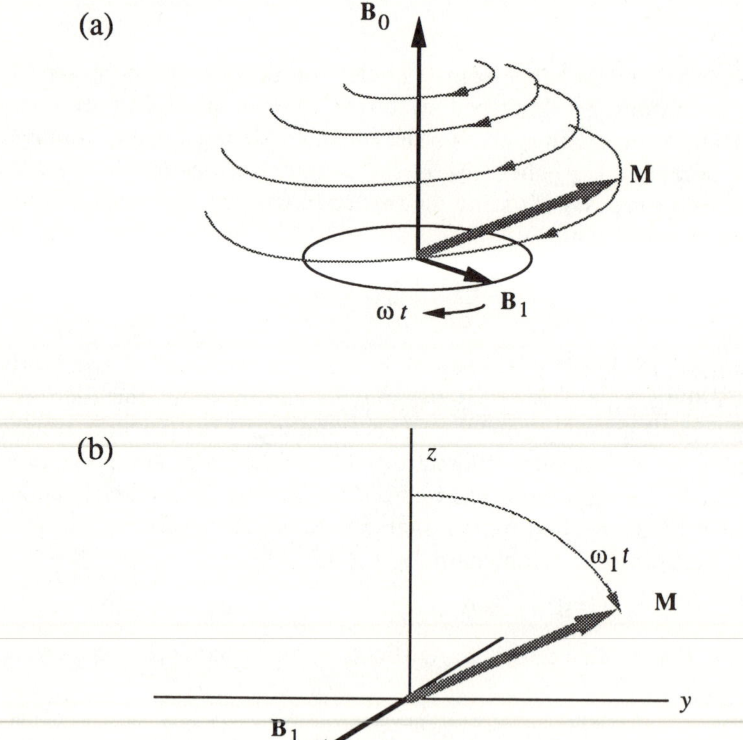

Fig. 2.3 (a) Evolution of the nuclear spin magnetization, in the laboratory frame, in the presence of a longitudinal field, $\mathbf{B}_0$, and a transverse rotating field, $\mathbf{B}_1$. When $\omega = \omega_0$ the magnetization vector simultaneously precesses about $\mathbf{B}_0$ at ω_0 and about $\mathbf{B}_1$ at ω_1. (b) As for (a) but in the rotating frame where $\mathbf{B}_1$ is stationary. The effective longitudinal field is zero on resonance and only the precession about $\mathbf{B}_1$ is apparent.

the motion is simply a precession about B_1. B_1 values are typically a few 10^{-4} T so that the precession about the r.f. field, using the proton example again, is typically at a cyclic frequency of 10 kHz.

Off-resonant behaviour is of importance in NMR microscopy because of its role in selective excitation. When $\omega \neq \omega_0$ the effective longitudinal field in the rotating frame is no longer zero but takes the value $(B_0 - \omega/\gamma)$. The net effective magnetic field, $\mathbf{B}_{\text{eff}}$, is shown in Fig. 2.4 and under these circumstances the spin magnetization precesses about B_{eff} in the rotating frame of reference. This phenomenon is discussed in detail in Chapter 3.

It is now easy to visualize the influence of a short burst of resonant r.f. field, B_1, known in NMR as the r.f. pulse. If the duration of the pulse is t, the magnetization will rotate by an angle $\omega_1 t$ about the direction of $\mathbf{B}_1$ in the rotating frame. This is the effect illustrated in fig. 2.3. At this point in the discussion an obvious question arises. What is the meaning of the *direction* of $\mathbf{B}_1$ in the *laboratory* frame where our experiment is performed? Clearly the instantaneous direction will depend on the instant that we start the r.f. pulse, and that is quite arbitrary. The significance of this direction becomes apparent, however, when a second pulse is applied at some later time. The first pulse establishes a rotating reference axis and, provided that the second pulse oscillates in phase with the first, this axis system will be maintained. Labelling the rotating frame direction of the first pulse as X then the second pulse will also correspond to a field along the X-axis. This immediately suggests that the rotating frame r.f. field direction can be oriented at will simply by changing the phase of the pulses. In most imaging experiments these phases are shifted in multiples of 90°. We denote an r.f. pulse with duration t as $\Theta_{\pm X}$ or $\Theta_{\pm Y}$, where $\Theta = \omega_1 t$ (expressed in degrees) and the subscript indicates the phase according to the chosen direction in the

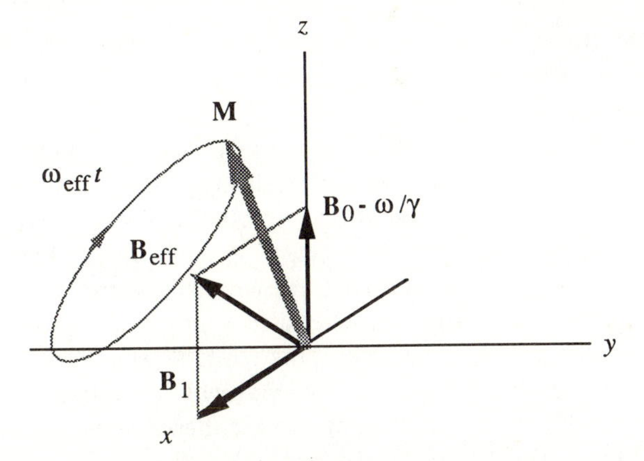

Fig. 2.4 Precession of the nuclear magnetization vector in the rotating frame when the r.f. field is off-resonant. The direction of the effective field is determined by the amplitude γB_1 and by the offset frequency, $\Delta\omega = \gamma B_0 - \omega$.

rotating frame. To emphasize the distinction between the rotating (X, Y, Z) and laboratory (x, y, z) frames, different fonts are used here. Subsequently, however, identical fonts will be used with the understanding that the r.f. field axes will always correspond to the rotating frame of reference.

NMR can be performed with an almost infinite array of pulse train possibilities. The capacity of the experimenter to change both the duration and phase of r.f. pulses, coupled with the existence of a host of environmentally sensitive terms in the nuclear spin interactions, leads to the essential richness of this branch of spectroscopy.

2.3.2 *Relaxation*

The effect of a resonant r.f. pulse is to disturb the spin system from its thermal equilibrium state. In due course that equilibrium will be restored by a process known as spin–lattice relaxation. As the name implies, the process involves an exchange of energy between the spin system and the surrounding thermal reservoir, known as the 'lattice', with which it is in equilibrium. The equilibrium is characterized by a state of polarization with magnetization M_0 directed along the longitudinal magnetic field, B_0. The restoration of this equilibrium is therefore alternatively named longitudinal relaxation. The phenomenological description of this process is given by the equation

$$\frac{\mathrm{d}M_z}{\mathrm{d}t} = -\,(M_z - M_0)/T_1 \tag{2.36}$$

with solution

$$M_z(t) = M_z(0)\exp(-t/T_1) + M_0\,(1 - \exp(-t/T_1)). \tag{2.37}$$

T_1 is known as the spin–lattice or longitudinal relaxation time. At room temperature it is typically in the range 0.1 to 10 seconds for protons in dielectric materials.

At first sight it may appear that the time constant T_1 will also describe the lifetime of transverse magnetization resulting from the application of an r.f. pulse. In fact transverse relaxation, which is characterized by the time constant T_2, is the process whereby nuclear spins come to thermal equilibrium among themselves. It is therefore known also as spin–spin relaxation. While indirect energy exchange via the lattice may play a role, additional direct processes are also responsible. This leads to the result $T_2 \leq T_1$. As indicated in the earlier quantum mechanical description, transverse magnetization corresponds to a state of phase coherence between the nuclear spin states. This means that transverse relaxation, unlike longitudinal relaxation, is sensitive to interaction terms which cause the nuclear spins to dephase. As we shall see later in Section 2.5 this may lead to T_2 relaxation being exceedingly rapid in comparison with T_1 where the interaction between the nuclear spins fluctuates very slowly, as in the case of solids or rigid macromolecules. T_2 values are usually in the range 10 μs to 10s.

The phenomenological description for transverse relaxation is written

$$\frac{\mathrm{d}M_{x,y}}{\mathrm{d}t} = \frac{-M_{x,y}}{T_2} \tag{2.38}$$

with solution

$$M_{x,y}(t) = M_{x,y}(0)\exp(-t/T_2). \tag{2.39}$$

It should be emphasized that the exponential description applies in the case where the interaction terms responsible for transverse relaxation are weak. This is the regime of the Bloembergen, Purcell, and Pound (BPP) theory,[11] an approach which works well for spins residing in liquid state molecules. However, for solids and macromolecules undergoing very slow motions, the decay is more complicated than that represented by eqn (2.39). For many NMR imaging applications, however, we are confined to observe slowly relaxing spins for which the phenomenological approach is entirely appropriate.

Combining (2.32), (2.36), and (2.38) in the rotating frame applicable in Fig. 2.4 yields a set of relationships known as the Bloch equations.[1]

$$\frac{\mathrm{d}M_x}{\mathrm{d}t} = \gamma M_y(B_0 - \omega/\gamma) - \frac{M_x}{T_2}$$

$$\frac{\mathrm{d}M_y}{\mathrm{d}t} = \gamma M_z B_1 - \gamma M_x(B_0 - \omega/\gamma) - \frac{M_y}{T_2}$$

$$\frac{\mathrm{d}M_z}{\mathrm{d}t} = -\gamma M_y B_1 - \frac{(M_z - M_0)}{T_1}. \tag{2.40}$$

These provide a valuable reference in describing many phenomena important in NMR imaging.

2.3.3 *Signal detection*

The excitation of the nuclear spins by an r.f. pulse may be viewed in terms of transitions between the energy levels as shown in Fig. 2.5(a). Such a picture, which uses only populations to describe the resonant excitation, is only partially complete since it fails to take account of coherence effects so important in nuclear magnetic resonance. The same defect applies to depiction of the detection process as an emission of photons as shown in Fig. 2.5(b). Furthermore, it is important to realize that the independent radio-frequency photon is too small in energy to be detected in isolation and

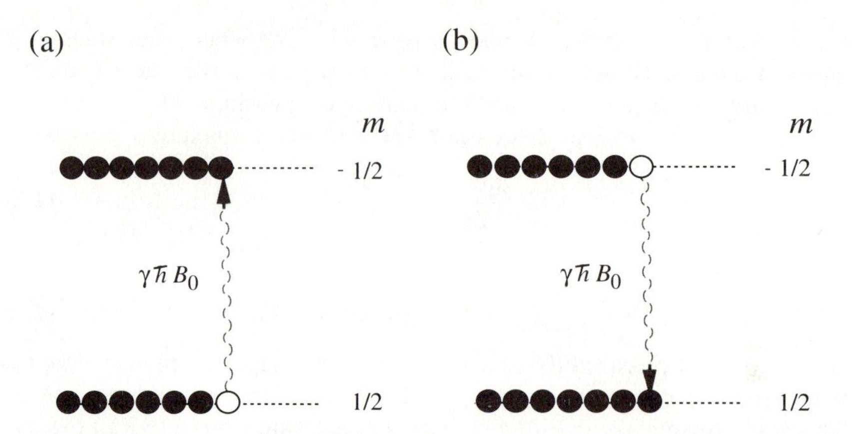

Fig. 2.5 Depiction of resonant excitation and emission in terms of a simple population transfer. While this picture is useful for transitions involving quantum states separated by large energies, (for example, principal atomic and nuclear levels) it is not helpful in the case of the delicate nuclear Zeeman levels. (b) fails because the Larmor frequency is so low that spontaneous emission has vanishingly small probability. Once the Zeeman system is excited, de-excitation will only occur via a resonant perturbation, such as that due to an externally applied r.f. field or to the relative motion of the nuclear dipoles, an effect known as relaxation. Nuclear Zeeman systems are remarkable for the long lifetime (μs to s) of excited coherences. The process shown in (a) represents a single particle transition, undetectable because of the low r.f. photon energy. As a multiparticle description it represents only the populations and fails to provide a picture for coherences between particles.

that the excitation and detection processes can only be adequately explained using the ensemble picture. The correct quantum mechanical description is that given by eqn (2.26) using the appropriate detection operator. This description will be required in dealing with those specific aspects of NMR microscopy where non-Zeeman interactions play a role. For most of what follows, however, the semi-classical magnetization approach will serve us very well.

Suppose a coil is placed around the sample with its symmetry axis transverse to the polarizing field $\mathbf{B}_0$. In the laboratory frame any transverse magnetization precessing at the Larmor frequency will induce an oscillatory e.m.f. at frequency ω_0. A quantitative analysis of this signal amplitude follows in Chapter 4. For the moment we need only note that this amplitude is typically of order μV and will be proportional to the transverse magnetization which exists in the coil. Of course the detection process is governed by Faraday's law and depends on the motion of the magnetization vector. The existence of precessing transverse magnetization is therefore essential. Longitudinal magnetization will induce no signal.

The geometry of the receiver coil depends on the geometry of the magnet as illustrated in Fig. 2.6. For superconducting magnets, easy entry of samples down the main field solenoid axis requires that the r.f. receiver coil

(a)

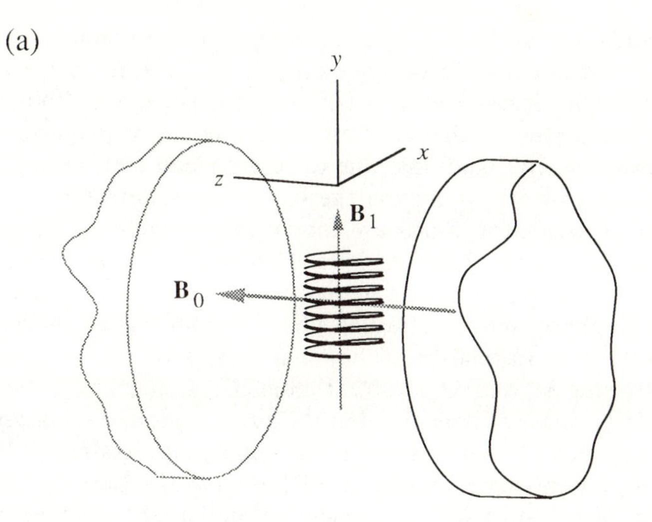

(b)

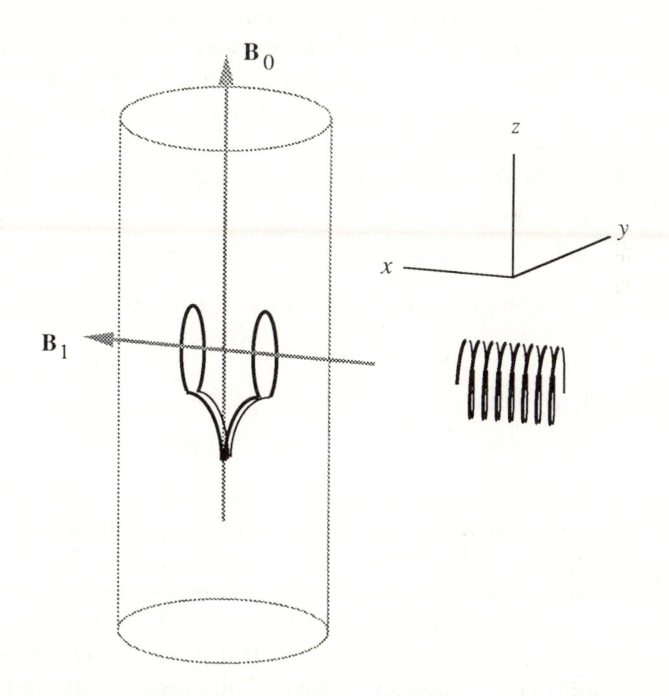

Fig. 2.6 (a) Solenoidal r.f. coil in electromagnet geometry. Sample access is down the y-axis. (b) Saddle r.f. coil in superconducting solenoid geometry with sample access down z. Note that the use of a more efficient solenoidal r.f. coil requires transverse sample loading.

have a saddle[12] or birdcage[13] configuration in order to establish a transverse r.f. field. For small-scale samples only the saddle geometry is practical and such a shape is never as efficient as an r.f. solenoid, although the high Larmor frequencies available from superconductive magnets more than compensate for any signal loss due to receiver coil inefficiency.

R.f. receivers work by mixing the signal e.m.f. with the output from a reference r.f. oscillator, a process known as heterodyning. This method of detection is inherently phase sensitive. This means that by separately mixing the e.m.f. with two heterodyne references each 90° out of phase, we obtain separate in-phase and quadrature phase output signals which are each respectively proportional to orthogonal phases of the magnetization, in effect detecting M_x and M_y. Nuclear magnetic resonance can be performed using a single phase of detection but the advantages of dual phase detection are so numerous[14–16] that this method is almost universally employed. Note that where the mixing reference is oscillating at the Larmor frequency we obtain d.c. quadrature output signals. At any other 'reference' mixing frequency ω_r, the signal will oscillate at the offset frequency $\Delta\omega = \omega_0 - \omega_r$.

The mathematics of the detection process is conveniently handled using complex numbers where the real part is used to represent the x-direction in the rotating frame (**i**) and the imaginary part the y-direction (**j**). Such an approach lends itself to simple interpretation using the Fourier theory outlined in Chapter 1. Consider, for example, the simple experiment depicted in Fig. 2.7 where a 90° r.f. pulse is applied to the equilibrium spin magnetization, $M_0\mathbf{k}$.

The laboratory frame magnetization at time t following the pulse is, in Cartesian notation,

$$\mathbf{M}(t) = [M_0 \cos \omega_0 t\, \mathbf{i} + M_0 \sin \omega_0 t\, \mathbf{j}] \exp(-t/T_2). \tag{2.41}$$

By contrast, in complex number notation this becomes

$$M_+(t) = M_0 \exp(\mathrm{i}\, \omega_0 t) \exp(-t/T_2). \tag{2.42}$$

The heterodyne signal at offset $\Delta\omega$ is therefore

$$S(t) = S_0 \exp(\mathrm{i}\, \phi) \exp(\mathrm{i}\Delta\omega t) \exp(-t/T_2), \tag{2.43}$$

where ϕ is the absolute receiver phase (an arbitrary but adjustable parameter for the apparatus) and S_0 is the signal amplitude immediately following the pulse, a number which is simply proportional to M_0. The primary NMR signal is measured in the time domain as an oscillating, decaying e.m.f. induced by the magnetization in free precession. It is therefore known as the Free Induction Decay (FID). By Fourier transformation the same signal may be represented in the frequency domain. This process is shown in Fig. 2.7 for the case $\phi = 0$. The result in the real part of the domain is a Lorentzian lineshape of cyclic frequency offset $\Delta\omega/2\pi$ and Full-Width-Half-Maximum

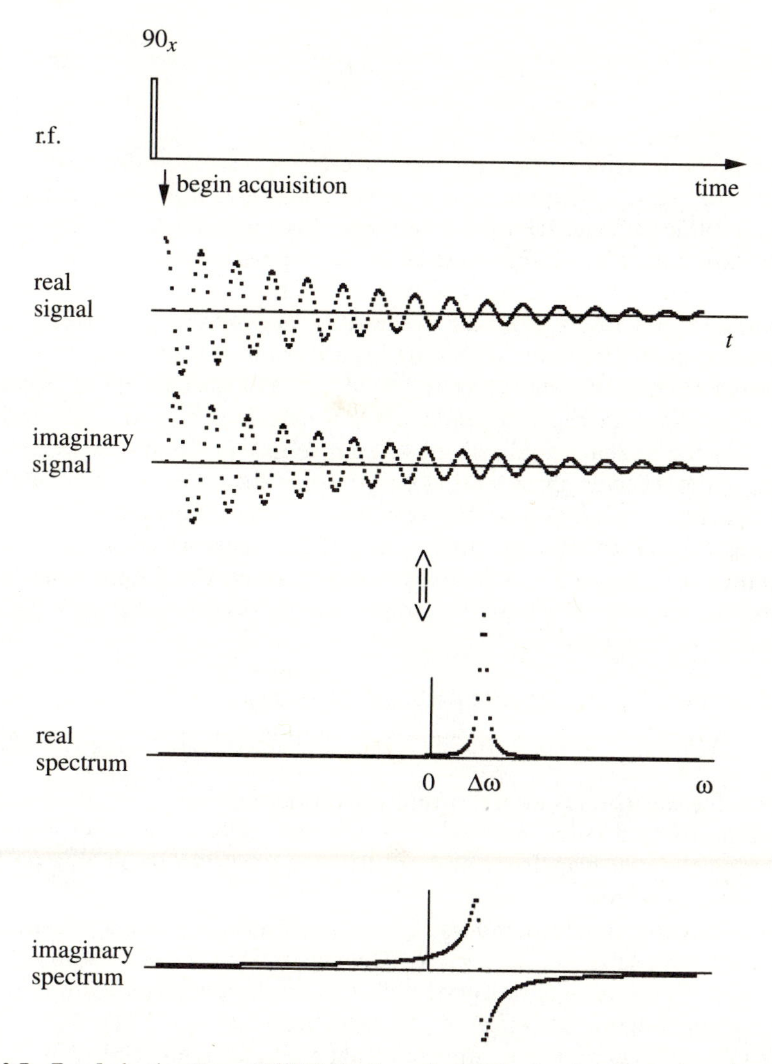

Fig. 2.7 Free Induction Decay (FID) following a single 90° r.f. pulse. The real and imaginary parts of the signal correspond to the in-phase and quadrature receiver outputs. The signal is depicted with receiver phase $\phi = 0$ and, on complex Fourier transformation, gives real absorption and imaginary dispersion spectra at the offset frequency, $\Delta\omega = \omega_0 - \omega$.

(FWHM) $1/\pi T_2$, the so-called absorption spectrum. The imaginary lineshape is known as the dispersion spectrum and for this function the width is somewhat greater. It is an elementary exercise in calculus to show that the real and imaginary parts of $\mathfrak{F}(S(t))$ are given by

$$\mathrm{Re}\{\mathfrak{F}(S(t))\} = \cos\phi \frac{T_2}{1 + (\omega - \Delta\omega)^2 T_2^2} + \sin\phi \frac{(\omega - \Delta\omega) T_2^2}{1 + (\omega - \Delta\omega)^2 T_2^2}$$

$$\mathrm{Im}\,\{\mathfrak{F}(S(t))\} = \sin\phi\,\frac{T_2}{1+(\omega-\Delta\omega)^2T_2^2} - \cos\phi\,\frac{(\omega-\Delta\omega)T_2^2}{1+(\omega-\Delta\omega)^2T_2^2}. \tag{2.44}$$

Note that the imaginary dispersion spectrum only exists, in the case $\phi = 0$, because the signal is acquired for positive time only. It is hard to imagine what sampling at 'negative time' could possibly mean in the context of the simple experiment being discussed here, so the existence of the dispersion spectrum seems quite natural. However, where the dephasing of the magnetization does not result from an irreversible T_2 process, but instead from some coherent interaction, it is possible to produce rephasing as well as dephasing effects. We will return to this point, and the concept of negative time acquisition, in the discussion of spin echoes in Section 2.6, and in discussing the imaging applications of spin echoes in Chapter 3.

Setting the receiver phase so that $\phi = 0$ is not always easy. Fortunately it is not necessary since the appropriate phase correction can be made later on. Where $\phi \neq 0$ the real and imaginary parts of the transformed signal contain admixtures of absorption and dispersion parts and the spectrum may be restored to the 'correct phase' by simply multiplying by $\exp(-\mathrm{i}\phi)$. This is equivalent to the operation

$$\begin{aligned}\mathrm{Re}\,\{\mathfrak{F}(S(t,0))\} &= \cos\phi\,\mathrm{Re}\,\{\mathfrak{F}(S(t,\phi))\} + \sin\phi\,\mathrm{Im}\,\{\mathfrak{F}(S(t,\phi))\}\\ \mathrm{Im}\,\{\mathfrak{F}(S(t,0))\} &= \cos\phi\,\mathrm{Im}\,\{\mathfrak{F}(S(t,\phi))\} - \sin\phi\,\mathrm{Re}\,\{\mathfrak{F}(S(t,\phi))\}.\end{aligned} \tag{2.45}$$

Phasing the spectrum is an important operation in NMR spectroscopy and can be performed either manually, or automatically in the computer, by inspecting the real transform and adjusting ϕ until the result is a perfect absorption spectrum.

In practice the signal is sampled digitally using a finite number of sampling points, N, with sampling interval T. This means that the acquisition bandwidth is $1/T$ and the separation of points in the frequency domain is $1/NT$ with values ranging between $-\frac{1}{2}N(1/NT)$ and $(\frac{1}{2}N-1)\,(1/NT)$. This finite sampling range means in effect that the data is multiplied by a hat function given by

$$H(t) = \begin{cases} 1 & 0 \le t \le NT \\ 0 & NT < t \end{cases} \tag{2.46}$$

which, in turn, means that the frequency domain spectrum is convoluted with the digital Fourier transform of $H(t)$, namely, the point spread function

$$\mathfrak{F}(H(t)) = NT\,\frac{\sin 2\pi fNT}{2\pi fNT} + \mathrm{i}\,\frac{NT(1-\cos 2\pi fNT)}{2\pi fNT} \tag{2.47}$$

The point spread function tells us the contribution to each point in digital frequency space arising from NMR signals displaced by frequency f.

2.4 Other nuclear interactions

2.4.1 *The visibility of fine details*

The longitudinal field Zeeman interaction dominates the energy of the nuclear spins. It is a feature of nuclear magnetic resonance that many other much weaker interaction terms can still be observed, a consequence of the remarkable coherence times which spin ensembles exhibit. For example, because the T_2 times for nuclear spins in liquids may be of order seconds, the fundamental linewidth $(1/\pi T_2)$ may be eight or nine orders of magnitude smaller than the Larmor frequency. This means that very fine features in the Hamiltonian may be detected. The situation is not so favourable in solids where the strong dipolar interactions between the spins serve to broaden the linewidth considerably, so destroying the transverse magnetization coherence a few tens of microseconds after its formation.

The influence of interactions superposed on the applied longitudinal Zeeman field are of considerable importance in NMR microscopy. A more complete discussion of these effects may be found in standard texts on NMR spectroscopy but the principles underlying the interactions may be summarized as follows.

2.4.2 *Magnetic field inhomogeneity*

The manufacturers of NMR magnets go to great lengths to provide a homogeneous magnetic field in the region of the sample space, incorporating first-, second-, and higher-order shim coils for fine correction. However it is inevitable that some variation in B_0 across the sample will occur, although typically this leads to a broadening of no more than a few Hz in the proton NMR spectrum. Such a broadening causes severe distortion in NMR microscopy when the acquisition bandwidth is very narrow. These effects are discussed further in Chapter 4. For the moment we note that the additional broadening caused by field inhomogeneity can result in a decay of the FID more rapid than that due to T_2 effects alone. The resultant time constant is often labelled T_2^* but it is important to appreciate that coherence loss due to relaxation is inherently random and irreversible whereas that caused by field imperfection is ordered and, given an appropriate pulse sequence, can be 'undone'. This leads to the important distinction between inhomogeneous and homogeneous broadening of the NMR spectral line. In practice, coherence loss which can be refocused by an appropriate pulse sequence is said to arise from inhomogeneous broadening.

This broadening is typically caused by local differences in the Hamiltonian between spin packets (isochromats). Homogeneous broadening, by contrast, is common to each spin packet and is essentially irreversible; it arises from random motion of the spins.

2.4.3 *Dipolar interactions*

In comparison with their dimensions, the separation of atomic nuclei in condensed matter is enormous. Despite this, the intrinsic magnetic moment associated with each nuclear spin dipole exerts a severe influence on its neighbours via the magnetic field produced by this dipole acting on the dipole moments of remote spins. The geometry relating to the magnetic dipole–dipole interaction for a pair of spins is shown in Fig. 2.8.

The associated Hamiltonian of an ensemble of spin pairs may be written[1]

$$H_{\mathrm{D}} = \frac{\mu_0}{4\pi}\sum_{i<j}\gamma_i\gamma_j\hbar\, r_{ij}^{-3}[\mathbf{I}_i\cdot\mathbf{I}_j - 3(\mathbf{I}_i\cdot\mathbf{r}_{ij})(\mathbf{I}_j\cdot\mathbf{r}_{ij})r_{ij}^{-2}] \qquad (2.48)$$

where $\mathbf{I}$ is the vector operator given by $I_x\mathbf{i} + I_y\mathbf{j} + I_z\mathbf{k}$. The ensemble sum is written in such a way that each pair of spins is accounted for once only.

Suppose we focus our attention on a single spin pair i, j separated by distance r_{ij}. The interaction represented by eqn (2.48) can be written

$$H_{\mathrm{D}_{ij}} = \mathbf{I}_i\cdot\mathbf{D}\cdot\mathbf{I}_j$$

$$= [I_{ix} \quad I_{iy} \quad I_{iz}]\begin{bmatrix} D_{xx} & D_{xy} & D_{xz} \\ D_{yx} & D_{yy} & D_{yz} \\ D_{zx} & D_{zy} & D_{zz} \end{bmatrix}\begin{bmatrix} I_{jx} \\ I_{jy} \\ I_{jz} \end{bmatrix}. \qquad (2.49)$$

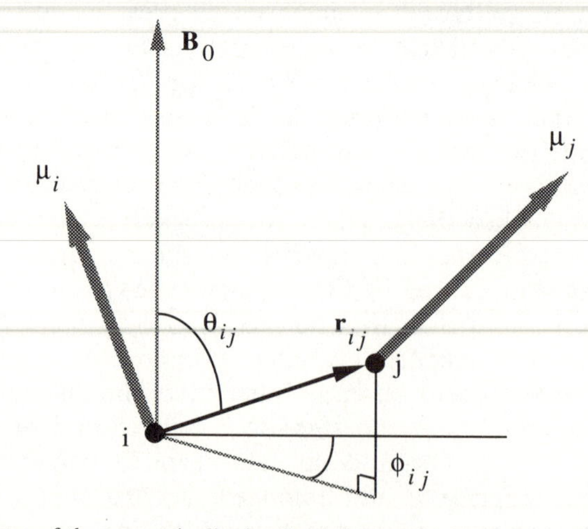

Fig. 2.8 Geometry of the magnetic dipole–dipole interaction. The inter-nuclear vector, $\mathbf{r}_{ij}$, is oriented at (θ_{ij}, ϕ_{ij}) to the external field. Note that the operator for μ_i is $\hbar\gamma\mathbf{I}_i$.

The matrix **D** is a rank-two tensor whose elements can be simply deduced by inspection of eqn (2.48). The Cartesian representation in which it is written is, however, quite clumsy, especially where we wish to describe the effect of rotations. Here the spherical tensor representation is much more convenient.[8,9] Suppose that the internuclear vector has polar coordinates $(r_{ij}, \theta_{ij}, \phi_{ij})$. Since the Zeeman field is very much larger than the dipolar field, we use the same zeroth-order representation in which the operator I_z is diagonal. The perturbing dipolar Hamiltonian for a single pair of identical spins can be written in the spherical tensor formalism as

$$H_{D_{ij}} = \sum_m (-1)^m F_2^m(\theta_{ij}, \phi_{ij}) T_2^{-m} \tag{2.50a}$$

where

$$F_2^m = -\frac{\mu_0 \gamma^2 \hbar}{4\pi r_{ij}^3} (24\pi/5)^{\frac{1}{2}} Y_2^m \tag{2.50b}$$

the Y_2^m being spherical harmonics of order 2 and component m, while the T_2^{-m} are products of spin operators.[17] F_2^m and T_2^{-m} transform under rotations as tensors of rank 2 and their component indices, m, range between -2 and 2 in integer steps.The products of spatial and spin tensors are shown in detail in Table 2.2. In general a rank 2 tensor is written as a superposition of spherical tensors[8,9] of order 0, 1, and 2. In the particular case of the dipolar interaction the order 0 and 1 parts of this superposition, T_0^0, T_1^0, and $T_1^{\pm 1}$, are absent. Since the product sum H_D represents the energy it is, of course, a scalar.

This complicated dipole–dipole Hamiltonian has enormous significance in nuclear spin systems. Note that its magnitude, in frequency units, is of order $\mu_0\gamma^2\hbar/4\pi r_{ij}^3$. Taking $r_{ij} \approx 1.5$ Å we obtain, for protons, an interac-

Table 2.2. *Tensor product components of the rank 2 tensor Hamiltonian. In each case the product term of the spatial and spin tensors transforms under rotation as a scalar.*

component	$Y_2^m(\theta_{ij}, \phi_{ij})$	T_2^{-m}
$m = 0$	$(5/4\pi)^{1/2} \frac{1}{2} (3\cos^2\theta_{ij} - 1)$	$\frac{1}{\sqrt{6}} (3I_{iz}I_{jz} - \mathbf{I}_i \cdot \mathbf{I}_j)$
$m = \pm 1$	$-(15/8\pi)^{1/2} \sin\theta_{ij} \cos\theta_{ij} \exp(\pm i\phi_{ij})$	$\pm\frac{1}{2}(I_{iz}I_{j\mp} + I_{i\mp}I_{jz})$
$m = \pm 2$	$(15/2\pi)^{1/2} \frac{1}{4} \sin^2\theta_{ij} \exp(\pm i2\phi_{ij})$	$\frac{1}{2} I_{i\mp} I_{j\mp}$

tion strength of around 100 kHz. In the case of solids where the inter-nuclear vectors have fixed orientations, the lineshape is dominated by dipolar interactions. A detailed discussion of these effects is beyond the scope of this book and well covered elsewhere.[17]

The method of calculating the influence of solid state dipolar interactions involves retention of only the secular (diagonal matrix) T_2^0 terms in eqn (2.47) and their summation over all spin pairs. The dipolar interaction between spins is weak in comparison with the dominant Zeeman interaction of the spins with the polarizing field. Only diagonal terms can influence the energy levels to first order, giving shifts on the scale of $\mu_0\gamma^2\hbar/4\pi r_{ij}^3$, whereas the off-diagonal terms produce second-order shifts on the scale of $(\mu_0\gamma^2\hbar/4\pi r_{ij}^3)^2/\gamma B_0$, a factor of 10^3 smaller. This secular term for the ensemble of spin pairs is therefore

$$H_{D_0} = \frac{\mu_0\gamma^2\hbar}{4\pi}\sum_{i<j}\frac{1}{r_{ij}^3}\frac{1}{2}(1 - 3\cos^2\theta_{ij})[3I_{iz}I_{jz} - \mathbf{I}_i\cdot\mathbf{I}_j]. \tag{2.51}$$

It is helpful to define a normalized free induction decay, $G(t)$, in which the laboratory frame Larmor frequency and arbitrary detection phase factors shown in eqns (2.42) and (2.43) are ignored. This function represents the decay of a 'correctly phased' signal 'on-resonance'. For a spin system experiencing the dipolar interaction of eqn (2.51), $G(t)$ may be written, in the short-time approximation, as

$$G(t) = 1 - M_2t^2/2! + M_4t^4/4! \tag{2.52}$$

where M_2 and M_4 are the first and second moments of the linewidth and may be computed, for a given lattice, as

$$M_2 = -\mathrm{Tr}([H_{D_0}, I_x]^2)/\mathrm{Tr}(I_x^2)$$

$$M_4 = \mathrm{Tr}([H_{D_0},[H_{D_0},I_x]]^2)/\mathrm{Tr}(I_x^2). \tag{2.53}$$

The significance of eqn (2.52) is threefold. First, the linewidth for many solids is of order 100 kHz with corresponding FIDs (known as the Bloch decay) of a few microseconds. This has severe consequences in NMR imaging since the broadening is so profound as to render the solid state proton signal invisible under normal circumstances. Second, the FID approximates a Gaussian in its time dependence. In consequence the spectral lineshape is also approximately Gaussian in contrast with the Lorentzian line associated with liquid state exponential T_2 relaxation (eqn (2.40)). Finally this Gaussian lineshape arises from ordered interactions and represents a form of inhomogeneous broadening. In principle, and given the right pulse sequence, it can be refocused. An example of a pulse sequence which causes such refocusing is the solid echo discussed in Section 2.6.9.

The situation for liquids is dramatically different. Motion causes (θ_{ij}, ϕ_{ij}) and hence H_{D_0} to fluctuate. Provided that the molecular tumbling motion is more rapid than the dipolar interaction strength, which is true for all but the most viscous liquids, the dipolar Hamiltonian is averaged to zero. Static dipolar broadening therefore makes no contribution to the linewidth in liquids. By contrast, the rapidly fluctuating non-secular (off-diagonal) terms in H_D can induce transitions between the nuclear energy levels and are therefore responsible both for spin–lattice and spin–spin relaxation. Provided field inhomogeneity is made sufficiently small it is the T_2 relaxation which determines the linewidth in liquids. These effects are discussed in greater detail in Section 2.5.

Normally the signal employed in NMR microscopy arises overwhelmingly from those nuclei residing in liquid state molecules. The intrinsic linewidth for these spins is in the range 1 to 100 Hz, sufficiently narrow to be influenced by fine structure in the Hamiltonian arising from the local chemical environment of the nucleus.

2.4.4 *Chemical shift*

In condensed matter nuclei are surrounded by atomic or molecular electron clouds which interact with the nuclear spin angular momentum. These interactions are characteristic of the local electronic environment and the discovery of these chemical fingerprints in the 1950s revolutionized chemistry. NMR is an essential tool in structural organic chemistry. The principal influence of the surrounding electrons is the magnetic shielding which results when electronic orbitals are perturbed by the applied magnetic field. This effect, known as the chemical shift, causes Larmor precession frequencies to be slightly displaced in a manner which is characteristic of the chemical environment. The Hamiltonian term associated with this chemical shift is simply a Zeeman operator.

Chemical shifts depend strongly on the atomic number and are of order a few p.p.m. in ^1H but several 100 p.p.m. in ^{13}C and ^{31}P. Table 4.1 lists the range of shifts obtaining in a variety of local environments for these nuclei. The offset of about 3.5 p.p.m. between methylene and water protons resonances[18] is of particular interest in NMR imaging. Their separation represents 210 Hz and 1400 Hz, respectively, at 60 and 400 MHz. In samples where both water and fat signals contribute to the image, such offsets can lead to severe distortion. However, this separation in frequency can, given the appropriate pulse tricks, enable separate images from the different molecular species to be obtained. This facility has been termed chemical microscopy.

In ordered environments, such as in solids and liquid crystals, the molecular orbital can have rotational anisotropy. This means that the shielding effect of the electron cloud has a tensorial character, reflecting the possibility that the field applied in one direction, say z, can result in an induced field along some other axis, say x. The spin Hamiltonian is therefore

$$H_{CS} = -\mathbf{I}\cdot\mathbf{S}\cdot\mathbf{B}_0 \tag{2.54}$$

where

$$\mathbf{S} = \gamma\begin{bmatrix} \sigma_{xx} & \sigma_{xy} & \sigma_{xz} \\ \sigma_{yx} & \sigma_{yy} & \sigma_{yz} \\ \sigma_{zx} & \sigma_{zy} & \sigma_{zz} \end{bmatrix} \tag{2.55}$$

As with the dipolar interaction tensor, the chemical shift tensor, $\mathbf{S}$, can be written in spherical form as a superposition of spin tensors. However, unlike the dipolar interaction, the coefficient of T_1^0 in the chemical shift is non-zero. Under rapid isotropic rotation only the T_1^0 component of a spin tensor remains. This is the basis for the disappearance of the dipolar interaction for spins in rapidly tumbling molecules, but in the case of the chemical shift, isotropic rotation leaves a residual Hamiltonian

$$H_{CS} = -\sigma_i \omega_0 I_z \tag{2.56}$$

where σ_i is the isotropic chemical shift given by the diagonal sum $\frac{1}{3}(\sigma_{xx} + \sigma_{yy} + \sigma_{zz})$. For solid state environments, in which the secular part of the chemical shift interaction affects the energy levels, the full Hamiltonian is[17]

$$H_{CS} = -\sigma_i \omega_0 I_z - \frac{1}{2}(3\cos^2\beta - 1)(\sigma_{zz} - \sigma_i)\,\omega_0 I_z \tag{2.57}$$

where β is the polar angle between the polarizing field direction and the principal axis system of the shift tensor. The additional term is known as the anisotropic chemical shift and, like the secular part of the dipolar interaction, transforms under rotation as the second-rank tensor Y_2^0.

2.4.5 *Scalar coupling*

The finest structural detail in the liquid state nuclear spin Hamiltonian is the scalar spin–spin coupling. This indirect interaction between nuclei arises via the mediation of electrons in the molecular orbital. The nuclear spin causes a slight electron polarization which, because of electron delocalization, is transmitted to neighbouring nuclei. The indirect nature of the spin–spin interaction leads to rotational invariance so that the scalar spin–spin Hamiltonian between nuclei is written

$$H_{\text{scalar}} = \pi J 2\mathbf{I}_1\cdot\mathbf{I}_2. \tag{2.58}$$

J is the coupling constant and has a magnitude typically between 0 and 18 Hz for protons. This spin–spin interaction, acting in conjunction with the chemical shift, imparts a characteristic signature to the high-resolution spectrum. It is important to note that because the spin–spin interaction requires a molecular orbital, it acts only through the medium of covalent bonds. For this reason the scalar coupling is always intramolecular. This is to be contrasted with direct dipole–dipole interactions which act through space.

Important differences in the spectral features associated with scalar couplings arise according to the relative size of the coupling strength and the chemical shift separation between interacting nuclei. Spectra are labelled by a scheme which describes the number of nuclei and their chemical shift proximity which is indicated by alphabetic proximity of the letters used for nuclear labelling. For example, A_2X represents two identical nuclei A in identical chemical locations coupled to a nucleus X with chemical shift from A much larger than the AX scalar coupling strength. AB, by contrast, represents two nuclei with chemical shift difference of the same order as their spin–spin interaction, the so-called strong coupling case. When the chemical shift difference is zero, scalar coupling effects vanish. This leads to singlet resonances for the protons in water molecules and for methylene protons in $(CH_2)_n$ groups distant from the ends of a fatty acid chain.

Fig. 2.9 illustrates spectral characteristics for protons in three different chemical environments in ethyl alchohol, $CH_3{-}CH_2{-}OH$, an A_3M_2X spectrum. Both the methyl and hydroxyl protons are J-coupled to the chemically identical two-spin (CH_2) group. This group of spins internally couple[8] to give total angular momentum 1 and 0 and, in consequence, the methyl and hydroxyl resonances exhibit binomially-weighted triplet splittings. By contrast the methylene protons are J-coupled to the single hydroxyl proton (total angular momentum $\frac{1}{2}$) and the chemically identical three-spin methyl group (total angular momenta $\frac{3}{2}, \frac{1}{2}, \frac{1}{2}$). Thus the methylene subresonance is a doublet of binomially-weighted quartets.

A simple analysis of the spectral features of an AX spin system is given in Fig. 2.10. Because the chemical shift difference, δ is much larger than the coupling frequency, J, the Zeeman terms dominate in the rotating frame Hamiltonian. This allows the use of an $|m_1\ m_2\rangle$ product representation where m_1 and m_2 are the respective eigenvalues of the Zeeman operators, I_{1z} and I_{2z}.

Spectral effects due to scalar couplings are delicate, but, as with chemical shift differences, they offer the prospect of identifying molecular species within the image. It should be noted, however, that the transformation properties of the scalar spin–spin Hamiltonian, the chemical shift Hamiltonian, and the dipole–dipole Hamiltonian are all different so that the

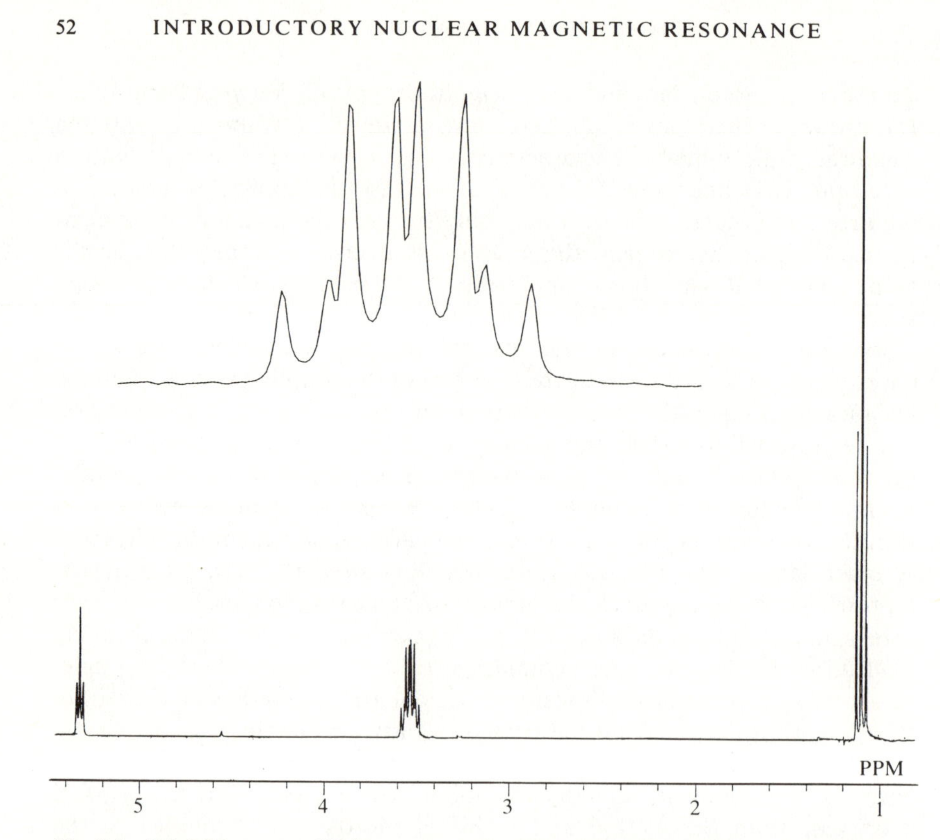

Fig. 2.9 270 MHz proton NMR spectrum for CH_3–CH_2–OH (A_3M_2X). The methyl and hydroxyl resonances at 1.1 ppm and 5.2 ppm are split as triplets while the methylene proton exhibits a doublet of quartets (shown expanded) because of *J*-coupling to both the single hydroxyl spin and the three identical methyl spins.

radio-frequency pulse trains required to distinguish their effects will necessarily differ.

2.4.6 $I > \frac{1}{2}$: *The quadrupole interaction*

The interaction of a single spin-$\frac{1}{2}$ nucleus with its environment can be described in terms of two basis states, spin-up and spin-down, and the Hamiltonian for the spin-$\frac{1}{2}$ nucleus is inherently rank 1 in the spin operator, involving only Zeeman-like interactions with its surroundings. Once we consider a system of two or more such spins, however, in which bilinear interactions are possible, higher rank terms will be present. However the restricted rank-one formalism will work for a large ensemble of spin-$\frac{1}{2}$ nuclei provided that these are well separated and no bilinear terms are present. This provides the theoretical justification for using a vector magne-

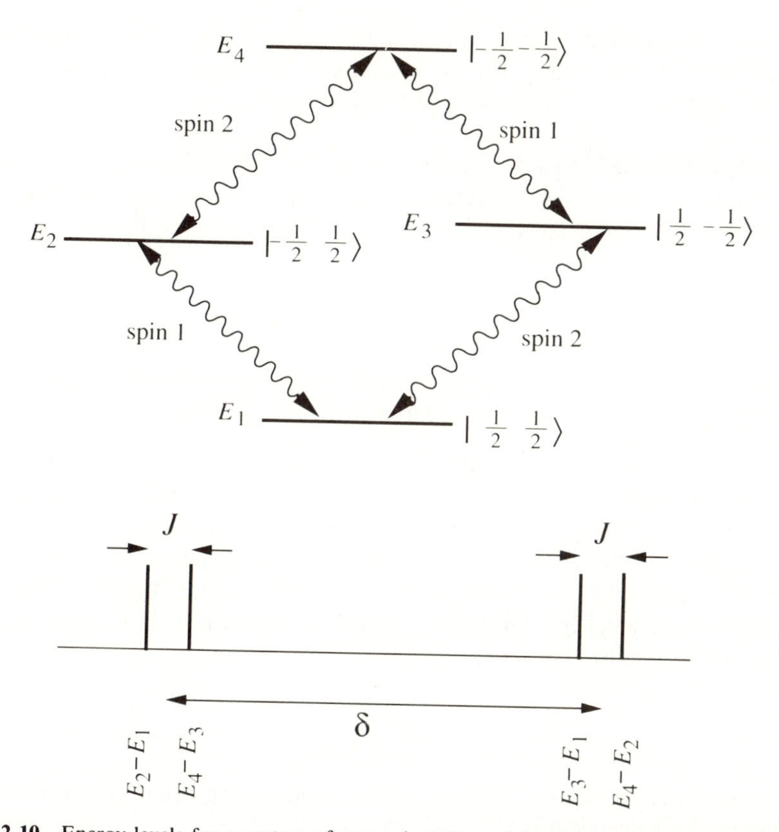

Fig. 2.10 Energy levels for a system of two spin-1/2 particles weakly coupled by the scalar interaction, $2\pi J\ \mathbf{I}_1 \cdot \mathbf{I}_2$ and experiencing a dominant (rotating frame) Zeeman interaction, $-(\Delta\omega_1 I_{1z} + \Delta\omega_2 I_{2z})$, where $\Delta\omega_1$ and $\Delta\omega_2$ are the Larmor frequencies in the rotating frame. The energy levels (in the product representation $|m_1\ m_2\rangle$) are: $E_1 = -(\Delta\omega_1 + \Delta\omega_2)/2 + \pi J/2$; $E_2 = -(\Delta\omega_2 - \Delta\omega_1)/2 - \pi J/2$; $E_3 = (\Delta\omega_2 - \Delta\omega_1)/2 - \pi J/2$; $E_4 = (\Delta\omega_1 + \Delta\omega_2)/2 + \pi J/2$; The single spin transitions shown correspond to the I_{1x} and I_{2x} operators of the resonant r.f. field and lead to the spectrum (in cyclic frequency units) shown in the lower part of the diagram, where $\delta = (1/2\pi)(\Delta\omega_2 - \Delta\omega_1)$.

tization (the Bloch vector) to describe the various evolutions of spin-$\frac{1}{2}$ ensembles.

For nuclei with $I > \frac{1}{2}$, even where interactions between spins are specifically excluded, the basis states which represent the spin are greater than 2 so that a simple vector description in general is no longer possible and higher rank tensors must be used. For a spin-1 nucleus the Hamiltonian describing the interaction of that spin with its environment will contain not only Zeeman terms linear in I_x, I_y, and I_z but also quadrupolar terms which are quadratic in the spin operators. Quadrupole terms arise from an interaction between the nuclear quadrupole moment and the surrounding electric field gradient. These interactions are of considerable significance in deuterium NMR but are non-existent for protons since symmetry considerations dictate that the quadrupole moment of a spin-$\frac{1}{2}$ nucleus is identically zero.

The quadrupole interaction Hamiltonian for a single spin bears a resemblance to that shown for the dipolar interaction of two spins in eqn (2.47), except that operators for a single spin only are employed.

$$H_Q = \mathbf{I}_i \cdot \mathbf{Q} \cdot \mathbf{I}_i \tag{2.59}$$

where

$$\mathbf{Q} = \frac{eQ}{2I(2I-1)\hbar} V \tag{2.60}$$

with Q being the nuclear quadrupole moment of spin-I and $\mathbf{V}$ being the electric field gradient tensor. For example, the component V_{xy} represents the second derivative of the electric potential, $\partial^2 V/\partial x \partial y$. It is convenient to express quadrupole interactions in terms of a principal (z) axis system with an asymmetry parameter η given by $(V_{xx} - V_{yy})/V_{zz}$.[1,17,19] We shall only be concerned here with axial symmetry where $\eta = 0$. In this case the Hamiltonian has the form

$$H_Q = \sum_m (-1)^m F_2^m(\theta, \phi) T_2^{-m} \tag{2.61}$$

where (θ, ϕ) describes the relative orientation of the quadrupole and spin quantization axis systems.

$$F_2^m = \frac{3eV_{zz}Q}{4I(2I-1)\hbar}(24\pi/5)^{\frac{1}{2}} Y_2^m \tag{2.62}$$

with the Y_2^m and T_2^{-m} as given in Table 2.2, the latter tensor being written in terms of operators of a single spin.

2.4.7 *Multiple quantum coherence*

The energy level diagram for two weakly J-coupled nuclei (AX system) shown in Fig. 2.10 provides a useful representation of the various states of the coupled spin-$\frac{1}{2}$ system. In Section 2.1 the operators for such a two-spin system were written in a basis of products of operators I_{1x}, I_{2y}, etc. A graphical description of these states of the density matrix can be given by representing the I_z operator in terms of a difference in populations and the I_x and I_y operators as coherences between states. In fact it is more convenient to choose a slightly different basis set using I_+ and I_- instead of I_x

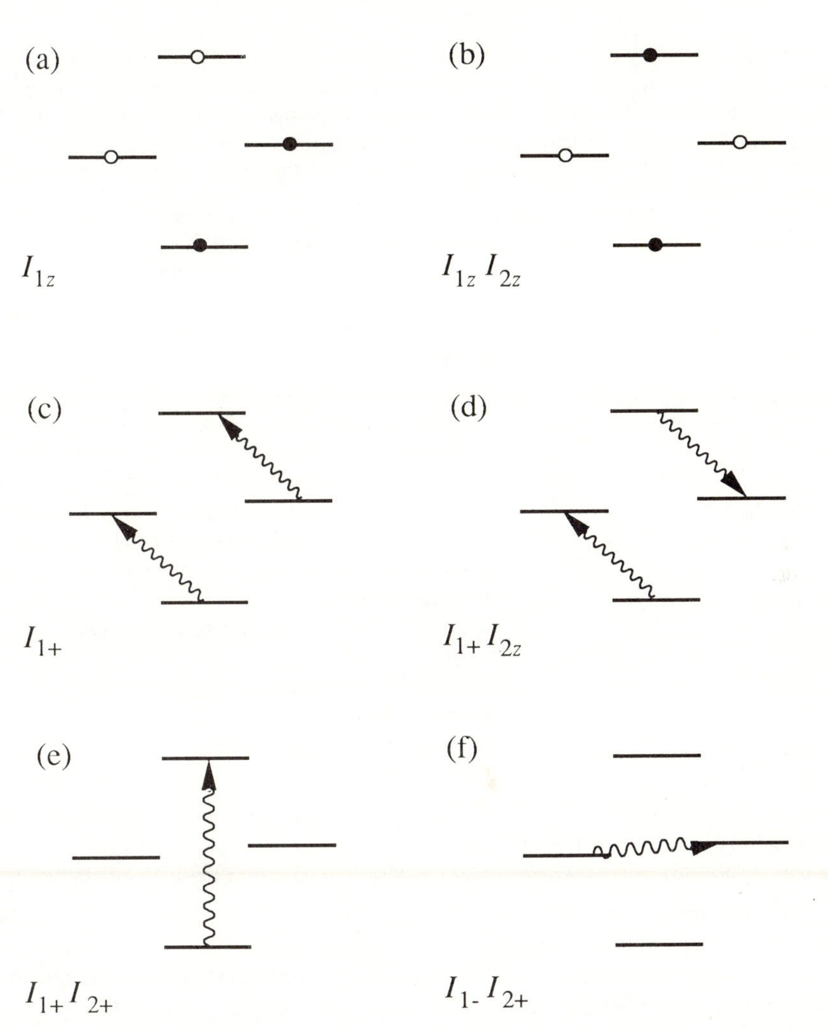

Fig. 2.11 Pictorial representation of basis operators for a system of coupled spin-1/2 pairs. States of the density matrix which are represented by I_+ or I_- operators are shown as wavy lines in the energy level diagrams with the arrow indicating raising or lowering. States represented by I_z operators are shown as population differences, indicated by filled or empty circles. The states shown are (a) polarization of spin 1, (b) longitudinal two-spin order of spins 1 and 2, (c) +1QC of spin 1, (d) +1QC of spin 1 in antiphase with spin 2, (e) +2QC of spins 1 and 2, and (f) ZQC of spins 1 and 2.

and I_y since the raising and lowering operators represent single quantum transitions in a specific direction. These operators are illustrated schematically in Fig. 2.11 using a formalism developed by Sorensen *et al.*[6,10]

Systems in which the spins do not interact with each other are incapable of exhibiting coherences other than ±1QC. Systems involving *p* scalar

Table 2.3. *One-spin and two-spin basis operators for a system of coupled spin-$\frac{1}{2}$ pairs. Note that the non-diagonal coherences are written in terms of raising and lowering operators, I_+ and I_-. Since I_+ and I_- are linear combinations of I_x and I_y, these coherences are, in turn, admixtures of those shown in Table 2.1.*

one-spin operator states	*name*
I_{1z}, I_{2z}	polarization of spins 1 and 2
I_{1+}, I_{2+}	in-phase +1-quantum coherence (+1QC) of spins 1 and 2
I_{1-}, I_{2-}	in-phase −1-quantum coherence (−1QC) of spins 1 and 2
two-spin operator states	*name*
$I_{1+}I_{2z}$, $I_{1z}I_{2+}$	+1-quantum coherence of spin 1 in antiphase with respect to spin 2 (and vice versa)
$I_{1-}I_{2z}$, $I_{1z}I_{2-}$	−1-quantum coherence of spin 1 in antiphase with respect to spin 2 (and vice versa)
$I_{1+}I_{2+}$	in-phase +2-quantum coherence (+2QC) of spins 1 and 2
$I_{1-}I_{2-}$	in-phase −2-quantum coherence (−2QC) of spins 1 and 2
$I_{1-}I_{2+}$, $I_{1+}I_{2-}$	in-phase zero-quantum coherence (ZQC) of spins 1 and 2
$I_{1z}I_{2z}$	longitudinal two-spin order of spins 1 and 2

coupled spins can exhibit all multiple quantum coherences up to p. These two features are important in simplifying complex NMR spectra since they enable the design of filter pathways in which all resonances due to spin systems with less than p-coupled spins are suppressed.

The existence of states of multiple quantum coherence is a consequence of there being spin operators in the Hamiltonian which do not transform as simple vectors (as in the case of the Zeeman spin operator) but as higher rank tensors. In consequence the existence of a quadrupole interaction for spin-1 (or higher spin) nuclei can result in the production of multiple quantum coherences. The preferred algebra used in describing the states of quadrupolar nuclei is the spherical tensor formalism[20,21] of Table 2.2 since the product operator method, while simple for weak scalar couplings where the Hamiltonian is approximately $\pi J\, 2I_{1z}I_{2z}$, is excessively cumbersome for the more complex quadrupolar interaction. A detailed description of the various evolutions possible for spin > $\frac{1}{2}$ systems in terms of irreducible tensor operators has been given by Bowden and co-workers.[22-25] The states of the density matrix represented in terms of spherical spin tensor operators are shown in Table 2.4.

Table 2.4. *Spherical tensor basis operators appropriate to a system of spin-1 particles or coupled spin-½ pairs.*

spherical tensor operator states	*name*
T_1^0	Zeeman polarization $(\sim I_z)$
$T_1^{\pm 1}$	single quantum coherence $(\sim I_{\pm})$
T_2^0	quadrupolar polarization $(\sim I_z^2)$
$T_2^{\pm 1}$	single quantum combination state $(\sim I_{\pm} I_z)$
$T_2^{\pm 2}$	two-quantum coherence $(\sim I_{\pm}^2)$

Note that the spherical tensor operators can be expressed in a basis of ordinary spin operators, examples of which are given in brackets. For complete expressions see Buckmaster *et al.*[20]

2.4.8 *Bandwidth of the r.f. pulse*

Given the spread in NMR spectral features resulting from these three interactions it is reasonable to enquire about the meaning of the condition 'on-resonance' as it applies to the r.f. excitation pulse. It is apparent from Fig. 2.4 that the nuclear magnetization reorientation resulting from the application of an off-resonant r.f. field of duration t and amplitude B_1 will depend on the orientation of B_{eff}. Clearly the criterion for resonant reorientation in the yz plane is that B_{eff} is nearly transverse and hence that $B_1 \gg \Delta\omega$. This criterion can be conveniently viewed in the frequency domain if we assume, to a first approximation, that the spins behave as a linear system. The bandwidth of the r.f. pulse is inversely related to the pulse duration and is of order t_p^{-1}. For a significant turn angle we require that $\gamma B_1 t_p \sim 1$ and so we can see that the pulse bandwidth is of order the r.f. Larmor frequency γB_1. The highest bandwidth is achieved by obtaining the largest possible B_1 from the pulse transmitted to the excitation coil and this is largely determined by the available transmitter power. Typical r.f. bandwidths for non-selective pulses are in excess of 20 kHz and may be as high as 200 kHz. This magnitude is much larger than the spectral features considered in this section with the exception of solid state dipolar interactions. Except in a few cases pertaining to solid state NMR, such non-selective r.f. pulses are capable of exciting all nuclei for a given spin species.

2.5 Relaxation and molecular motion

Nuclear spin relaxation is caused by the distribution of local interactions experienced by the nuclear spins. The rate of relaxation is sensitive not only to the magnitude of those interactions but also to their rate of fluctuation.

In Section 2.3.2 a semi-classical description was given in which the Bloch equations contained a phenomenological term leading to exponential relaxation. In practice this description is quite accurate for spins in rapidly tumbling molecules but breaks down when the molecular motion becomes slow. However, the phenomenological approach cannot indicate how the relaxation times will depend upon the magnitude and fluctuation rates of spin interactions, nor does it give a clue as to when such exponential behaviour is applicable. Here we address that question and in the process define the relevant 'NMR time-scales' which enable us to appreciate the various behavioural regimes.

Relaxation theory is complicated, requiring a deep understanding of quantum statistics. A rigorous analysis is not appropriate here and readers can find this elsewhere,[1,2,26-28] but it is useful to outline the overall approach used to obtain some of the well-known formulae. The behavioural regimes may be broadly divided into what is termed rapid and slow motion. The rapid-motion regime is applicable to most molecules in the liquid state. We shall not be concerned here with the problem of relaxation in the case of physically distinct phases of molecules in which the total NMR signal is a superposition from spins in these different regions, nor with the effect on relaxation of exchange between phases. These topics, as well as the effect of motion in an inhomogeneous field, are dealt with in Sections 5.1 and 7.6.

In spin-$\frac{1}{2}$ nuclei, the dominant interaction causing spin relaxation arises from the dipolar Hamiltonian, whereas in higher spin systems quadrupolar interactions are significant. Because of the pre-eminent role of the proton in NMR microscopy, our discussion will concentrate on the effect of H_D.

2.5.1 *Liquids*

In liquids H_D fluctuates because of molecular tumbling, in all but exceptional cases leading to a time-averaged H_D of zero. The exceptions include ordered liquids such as liquid crystals or polymer melts, and the residual time-averaged dipolar (or quadrupolar) Hamiltonian in these cases engenders a spectral character akin to that of a solid, albeit with reduced effective NMR linewidth. The influence of such static Hamiltonian terms was treated in Section 2.4.3. Here we consider the effect of the fluctuation in what is generally called the rapid-motion case. The original description was given by Bloembergen, Purcell, and Pound[11] and as a consequence this theory of relaxation is often called the BPP theory.

The easiest case to treat is that of spin–lattice relaxation. This is because we may consider this 'z-axis' relaxation process in the laboratory frame where the difference in energy levels is dominated by the longitudinal Zeeman field and the dipolar interaction can be regarded as weak by comparison. This permits the use of time-dependent perturbation theory. By a simple argument Hebel and Slichter[29] showed that the spin–lattice relaxation rate could be written

$$\frac{1}{T_1} = \frac{1}{2}\,\frac{\Sigma_{nm} W_{nm}(E_n - E_m)^2}{\Sigma_n E_n^2} \tag{2.63}$$

where the transition rate between the two states n and m of energies E_n and E_m is given by perturbation theory as

$$W_{nm} = \frac{1}{\hbar^2}\int_0^t \exp[\mathrm{i}(E_n - E_m)(t' - t)/\hbar]\,\overline{\langle n|\hbar H_D(t')|m\rangle\langle m|\hbar H_D(t)|n\rangle}\,\mathrm{d}t'. + \text{complex conjugate} \tag{2.64}$$

Using the decomposition of $H_D(t)$ into spatial and spin operators as shown in eqn (2.50) W_{nm} may be written in the limit $t \gg \hbar/(E_n - E_m)$

$$W_{nm} = \sum_{qq'}\left\{\int_{-\infty}^{\infty} \exp[\mathrm{i}(E_n - E_m)\tau/\hbar]\,\overline{F_2^q(0)F_2^{-q'}(\tau)}\,\mathrm{d}\tau\right\}\langle n|T_2^q|m\rangle\langle m|T_2^{-q'}|n\rangle. \tag{2.65}$$

For isotropic random motion, $\overline{F_2^q(0)F_2^{-q'}(\tau)} = \delta_{qq'}G^q(\tau)$, where $G^q(\tau)$ is the autocorrelation function of the spatial tensor component q. The term in curly brackets in eqn (2.65) is the Fourier transform of $G^q(\tau)$, termed the spectral density function, $J^{(q)}(\omega)$. (Note that $G^q(\tau)$ should not be confused with the normalized FID signal, $G(t)$.) $J^{(q)}(\omega)$ represents the intensity of fluctuations in F_2^q at frequency ω. $G^q(\tau)$ will have some characteristic time known as the correlation time, τ_c, where $G^q(t) \to 0$ as $t \gg \tau_c$. Similarly $J^{(q)}(\omega)$ has a characteristic frequency, τ_c^{-1}, with $J^{(q)}(\omega) \to 0$ as $\omega \gg \tau_c^{-1}$. (The idea of a correlation function is very important in NMR theory and will be referred to later in Chapter 6 in the context of fluctuating translational motion.)

Because of the selection rule inherent in the matrix elements $\langle n|T_2^q|m\rangle$ (i.e., $n - m = \pm 1$ and $n - m = \pm 2$) the relevant frequencies for transitions are the Larmor frequency and twice the Larmor frequency. Following some algebraic manipulation[20] eqns (2.63) and (2.65) give

$$\frac{1}{T_1} = \left(\frac{\mu_0}{4\pi}\right)^2 \gamma^4\hbar^2\,\frac{3}{2}\,I(I+1)\left[J^{(1)}(\omega_0) + J^{(2)}(2\omega_0)\right]. \tag{2.66}$$

The case of transverse relaxation is not amenable to such an approach since here we are dealing with a process which is naturally described in the rotating frame where the transverse magnetization is stationary. In this frame of reference there is no large zeroth-order Hamiltonian which will dominate the dipolar interaction. Instead the behaviour of the magnetization is best handled using the density operator formalism.[1] Following eqn (2.27), the density matrix in the rotating frame, $\rho^*(t)$, obeys

$$\mathrm{i}\,\frac{\mathrm{d}\rho^*(t)}{\mathrm{d}t} = [H_{\mathrm{D}}^*(t),\, \rho^*(t)] \tag{2.67}$$

where H_{D}^* is the transformed dipolar Hamiltonian, $\exp[\mathrm{i}\omega_0 t I_z]\, H_{\mathrm{D}}(t)$ $\exp[-\mathrm{i}\omega_0 t I_z]$. Once the evolution of $\rho^*(t)$ is calculated then all relevant spin properties, such as the decay of the transverse magnetization, can be determined. (This approach can, of course, be used as an alternative means of calculating the spin–lattice relaxation.) The problem with an equation such as (2.67) is that it cannot easily be integrated. To first order however[1]

$$\frac{\mathrm{d}\rho^*(t)}{\mathrm{d}t} = -\mathrm{i}[H_{\mathrm{D}}^*(t),\, \rho^*(0)] - \int_0^t [H_{\mathrm{D}}^*(t),\, [H_{\mathrm{D}}^*(t'), \rho^*(0)]]\, \mathrm{d}t'. \tag{2.68}$$

Taking the ensemble average, noting that $\overline{H_{\mathrm{D}}^*(t)} = 0$, replacing $t' = t - \tau$, and extending the integration limit to ∞, one obtains[1] the vital result from which the decay of M_y (i.e., Trace$[\rho^* I_y]$) can be calculated, namely,

$$\frac{\mathrm{d}\rho^*(t)}{\mathrm{d}t} = -\int_0^\infty \overline{[H_{\mathrm{D}}^*(t)\, [H_{\mathrm{D}}^*(t-\tau)\, \rho^*(0)]]}\, \mathrm{d}\tau. \tag{2.69}$$

This linear differential equation leads to exponential relaxation, a key property of BPP theory. Eqn (2.69) contains oscillatory phase factors due to the transformation of H_{D} to the rotating frame and so, like eqn (2.65), can be shown to comprise the Fourier spectra of dipolar correlation functions, $G^q(\tau)$. The various matrix element evaluations in the calculation of $\mathrm{d}M_y/\mathrm{d}t$ lead to

$$\frac{1}{T_2} = \left(\frac{\mu_0}{4\pi}\right)^2 \gamma^4 \hbar^2 \frac{3}{2} I(I+1) \left[\frac{1}{4} J^{(0)}(0) + \frac{5}{2} J^{(1)}(\omega_0) + \frac{1}{4} J^{(2)}(2\omega_0)\right]. \tag{2.70}$$

The assumptions involved in the various steps leading to eqn (2.69) are delicate but the key assumption underpining BPP theory is $\overline{\langle H_{\mathrm{D}}^2 \rangle}\, \tau_c^2 \ll 1$ (or $M_2 \tau_c^2 \ll 1$). Thus the relevant time-scale for fast motion in the BPP sense is the 'precession period in the dipolar field'. This is simply the inverse of the dipolar linewidth in the absence of motion, $M_2^{-\frac{1}{2}}$. The regime of motion in which $M_2 \tau_c^2 \ll 1$ is termed 'motionally narrowed'.

One other relaxation time can be calculated using the spectral density approach. This is the rotating frame relaxation time, $T_{1\rho}$, which describes the rate at which transverse magnetization decays in the presence of an r.f. field, B_1. Provided the rotating frame Zeeman energy of the spins in the r.f. field, γB_1, exceeds the residual dipolar interaction (i.e., the NMR linewidth) the magnetization is said to be 'spin-locked' and the perturbative treatment of relaxation which applies in the case of the T_1 process along the B_0 Zeeman field in the laboratory frame, also applies for the $T_{1\rho}$ process along the B_1 Zeeman field in the rotating frame.[27] The result is derived from an expression akin to eqn (2.65) but in this case m and n refer to eigenstates along I_x, and $H_D(t)$ is now $H_D^*(t)$, the dipolar interaction in the rotating frame. The relevant matrix elements $\langle n|T_2^q|m\rangle$ cause selection rules $n - m = 0, \pm 1, \pm 2$ and their evaluation yields

$$\frac{1}{T_{1\rho}} = \left(\frac{\mu_0}{4\pi}\right)^2 \gamma^4 \hbar^2 \frac{3}{2} I(1+1) \left[\frac{1}{4} J^{(0)}(\omega_1) + \frac{5}{2} J^{(1)}(\omega_0) + \frac{1}{4} J^{(2)}(2\omega_0)\right]. \tag{2.71}$$

As expected, the expression for $T_{1\rho}$ equals that for T_2 in the limit as $\omega_1 \to 0$ although in practice the spin-locking condition breaks down as the r.f. field amplitude is decreased.

In order to appreciate the physical significance of eqns (2.66), (2.70), and (2.71), it is helpful to evaluate them for a simple isotropic rotational diffusion model, an excellent representation of the fluctuations in dipolar interactions which occur in most liquids. Here the $J(q)$ are given by[1]

$$J^{(0)}(\omega) = \frac{24}{15r_{ij}^6} \frac{\tau_c}{1 + \omega^2\tau_c^2}$$

$$J^{(1)}(\omega) = \frac{4}{15r_{ij}^6} \frac{\tau_c}{1 + \omega^2\tau_c^2}$$

$$J^{(2)}(\omega) = \frac{16}{15r_{ij}^6} \frac{\tau_c}{1 + \omega^2\tau_c^2} \tag{2.72}$$

where τ_c is the rotational correlation time (for reorientation of the rank 2 spatial tensor). The result of substituting these spectral density functions in the expressions for T_1 and T_2 is shown in Fig. 2.12. The most obvious feature is the existence of two distinct motional regimes separated by a minimum in T_1 when the correlation time is of order the Larmor period. The regime corresponding to $\tau_c^{-1} \gg \omega_0$ is characterized by identity of T_1 and T_2 and, in accordance with the reduction in the homogeneous linewidth, $1/\pi T_2$, as T_2 increases, is termed 'extreme narrowed'. Such a regime applies typically for small molecules in the liquid state (such as water at room temperature) where correlation times are of order 10^{-12} s to 10^{-14} s.

The divergence of T_1 and T_2 below the T_1 minimum is characteristic of highly viscous liquids, concentrated flexible polymers, and semi-rigid polymers where rotational tumbling of the inter-nuclear vector $\mathbf{r}_{ij}$ is slowed

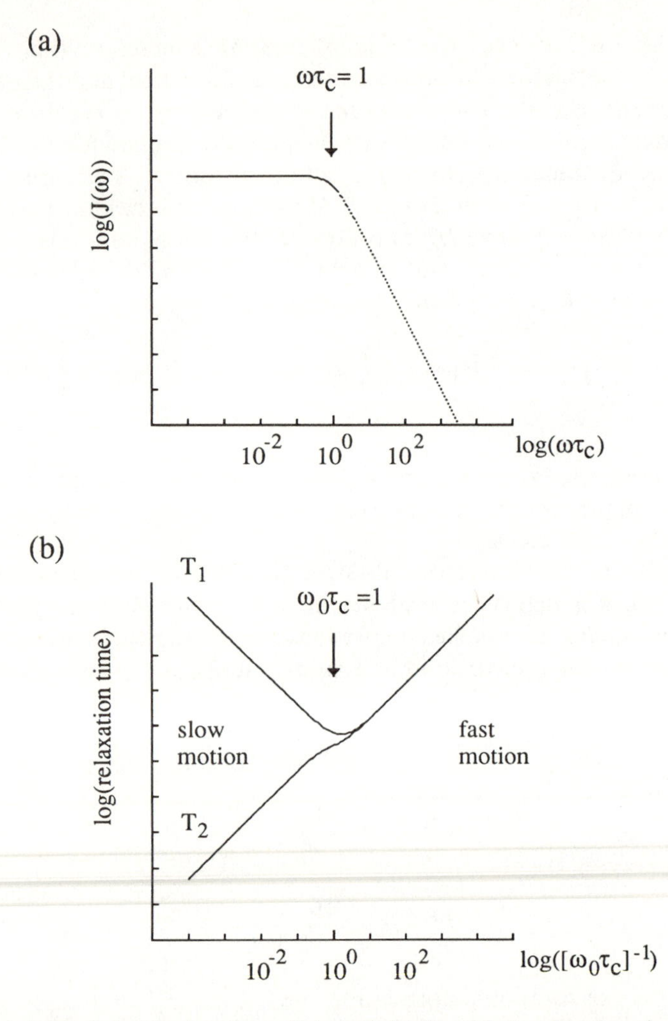

Fig. 2.12 (a) Spectral density function, $J(\omega)$, and (b) T_1 and T_2 relaxation times for a spin-1/2 system undergoing like-spin dipolar relaxation. In (a) the spectral density is plotted as a function of frequency ω for a fixed correlation time, τ_c. In (b) the T_1 and T_2 relaxation times are plotted as a function of rotational correlation time, τ_c, at a fixed Larmor frequency. The T_1 minimum divides the regimes for τ_c^{-1} into slow and fast in comparison with the Larmor frequency.

or restricted. Molecular motion in which $M_2^{1/2} \ll \tau_c^{-1} \ll \omega_0$ is identified by $T_2 \ll T_1$. This turns out to be a difficult regime for NMR microscopy as discussed in Chapters 4 and 5. In cases where T_2 is very short and close to the solid limit, the free induction decay vanishes a few μ s after the initial r.f. pulse and, unless very special pulse sequences are employed (see Section 2.6.8), the signal will be entirely invisible in the NMR image.

All three expressions, (2.66), (2.70), and (2.71) correspond to a dipolar interaction between a pair of like spins. Where a spin experiences a dipolar interaction with more than one other spin these expressions can be scaled up proportionately. Similarly if the interaction arises between spins in different molecules in relative translational motion some effective value for $\overline{(1/r_{ij}^6)}$ is required. Furthermore the expressions can be simply modified for dipolar interactions between unlike spins (such as ^{13}C and ^{1}H) or to allow for different types of perturbative interactions such as the quadrupole interaction. Such formulae are covered in detail elsewhere.[1,6] In all such analyses the motional features apparent in Fig. 2.12 remain the same.

2.5.2 *Solids and slow motions*

When the reorientation of the dipolar interaction slows sufficiently that $M_2\tau_c^2 \geq 1$, the line-narrowing assumption inherent in the BPP theory breaks down. However the perturbation theory approach which yields the spin-lattice relaxation behaviour is still quite good since $\gamma B_0 \gg M_2^{1/2}$. Consequently the T_1 formula, eqn (2.66), is applicable in the slow-motion limit. In the case of solids it is clear that T_1 will be long (usually many seconds) and in insulators will arise from the spectrum of lattice vibrations.

For T_2 (and $T_{1\rho}$) in solids and semi-solids a different approach is required. A nice treatment, due to Anderson and Weiss[30] represents the fluctuating dipolar fields as a time-dependent Larmor frequency, $\Delta\omega(t)$, which varies randomly such that the time-averaged and ensemble-averaged mean square values are the same, namely, $\overline{\Delta\omega^2}$. The method is also applicable to any other randomly varying local field. Note that $\Delta\omega(t)$ represents the offset from the average longitudinal field precession frequency, ω_0.

The normalized free induction decay is therefore,

$$G(t) = \overline{\exp\left(\mathrm{i}\int_0^t \Delta\omega(t')\,\mathrm{d}t'\right)}. \tag{2.73}$$

Two assumptions make the problem tractable. The first is that the distribution of $\Delta\omega(t)$ is Gaussian, in which case the distribution of $X(t) = \int_0^t \Delta\omega(t')\,\mathrm{d}t'$ is also Gaussian with probability distribution $P(X) = [2\pi\overline{X^2}]^{-\frac{1}{2}} \exp[-\frac{1}{2}X^2/\overline{X^2}]$. This leads to the result

$$\overline{\exp\left(\mathrm{i}\int_0^t \Delta\omega(t')\,\mathrm{d}t'\right)} = \int P(X)\,\mathrm{e}^{\mathrm{i}X}\,\mathrm{d}X$$

$$= \exp\left(-\frac{1}{2}\overline{X^2}\right). \tag{2.74}$$

The second assumption is that the local field fluctuation is described in the usual manner by a correlation function

$$\overline{\Delta\omega(t)\,\Delta\omega(t-\tau)} = \overline{\Delta\omega^2}\,g(\tau). \tag{2.75}$$

(Note that $g(\tau)$ is a sort of 'normalized, q-averaged' $G^q(\tau)$.) This allows one to calculate $\overline{X^2}$ as $2\overline{\Delta\omega^2}\int_0^t (t-\tau)g(\tau)\,d\tau$ so that

$$G(t) = \exp\left[-\overline{\Delta\omega^2}\int_0^t (t-\tau)g(\tau)\,d\tau\right]. \tag{2.76}$$

This Anderson–Weiss expression is an extremely useful result, applicable in a wide range of dynamical situations. For example, we shall use this result in Chapter 7 to deal with spin relaxation in microscopically inhomogeneous media. Before leaving the model we note that it reproduces two familiar results in the slow-and fast-motion limits of the dipolar interaction. For this Hamiltonian $\overline{\Delta\omega^2} = M_2$. In the slow-motion limit, $M_2\tau_c^2 \gg 1$, $g(\tau)$ in eqn (2.76) is approximately unity and the solid state behaviour of eqn (2.52) is reproduced as $t \to 0$. Strictly speaking such a decay is due to inhomogeneous broadening and cannot strictly be termed 'relaxation', since the signal may be recovered by using appropriate pulse sequences as discussed in Section 2.6.8.

In the fast-motion limit $g(\tau)$ decays rapidly in eqn (2.76) and the integral approximates to $t\int_0^\infty g(\tau)\,d\tau$, leading to a free induction decay given by $\exp(-M_2\tau_c t)$. Such an exponential decay corresponds to a transverse relaxation rate $1/T_2 = M_2\tau_c$, and, using the definition of the second moment in eqn (2.53), this rate turns out to be identical to the dominant fast-motion term of the BPP theory (eqn (2.70)), namely

$$\frac{1}{T_2} = \left(\frac{\mu_0}{4\pi}\right)^2 \gamma^4\hbar^2\frac{3}{2}I(I+1)\left[\frac{1}{4}J^{(0)}(0)\right]$$

$$= \left(\frac{\mu_0}{4\pi}\right)^2 \gamma^4\hbar^2\frac{3}{5}I(I+1)\frac{1}{r_{ij}^6}\tau_c$$

$$= M_2\tau_c \tag{2.77}$$

2.6 Introductory spin manipulation

Among the enormous number of pulse sequences which are used in modern NMR are four whose origins date from the birth of the subject but whose usefulness makes them starting points for newcomers to spectroscopy. These are the inversion recovery sequence, the simple Hahn echo, the Carr–Purcell echo train, and the stimulated echo. Each is concerned with the manipulation of the spin system under the influence of the Zeeman Hamiltonian and the T_1 and T_2 relaxation processes, and each can be understood using a simple vector description of the nuclear magnetization. The final sequences considered in this section involve coherence transfer and are much more recent. As with many of the more sophisticated spin manipulations, these cannot be understood without recourse to quantum mechanics.

Usually, each application of a particular pulse sequence to the NMR spin system is not performed in a single shot but in rapid succession, the resulting signal from each experiment being successively added. It is this aspect of NMR methodology which we consider first.

2.6.1 *Signal averaging*

Because of the low sensitivity of NMR, it is customary to co-add signals from N successive experiments in order to enhance the signal-to-noise ratio. Successive addition has the effect that the signals add coherently while the noise adds in random phase. Because the noise power is additive, the r.m.s. noise amplitude is proportional to $N^{\frac{1}{2}}$. As a result the signal-to-noise ratio improves as $N/N^{\frac{1}{2}}$ or $N^{\frac{1}{2}}$. There is, however, a limitation to the rate at which successive additions can be performed. To retain full signal strength the nuclear spin system must be allowed to recover its z-axis magnetization between experiments and this recovery requires a time delay of several T_1 relaxation times. This recovery imposes a limit on the number of co-additions which are possible in a given time.

Of course it is possible to repeat excitation pulses at intervals less than the time taken for full recovery, in which case the spins establish an equilibrium longitudinal magnetization before each excitation pulse which is less than the thermal equilibrium value. This effect is known as partial saturation.[31] The time over which the longitudinal magnetization recovers before the next excitation pulse is simply the pulse repetition time. Given a repetition time T_R and longitudinal relaxation time T_1 the equilibrium magnetization immediately before the r.f. pulse for a repetitive single θ_x pulse experiment is

$$M_z = M_0 \frac{1 - \exp[-T_R/T_1]}{1 - \cos\theta \exp[-T_R/T_1]} \,. \tag{2.78}$$

The method by which this result is obtained is instructive since it is generally applicable to more complicated pulse sequences. The dynamic equilibrium is established by equating the z-magnetization just before the r.f. pulse, $M_z(0_-)$ to the z-magnetization, $M_z(T_R)$, which remains from the preceding sequence, following relaxation. In the derivation of eqn (2.78) a simplifying assumption is made, namely that the role of transverse magnetization can be neglected because of irreversible decay after each excitation and signal acquisition. Therefore the partial saturation condition represented by equation 2.78 will apply when $T_R \gg T_2$ or when the transverse magnetization is deliberately destroyed, for example by using magnetic field gradient pulses. Using the symbol $M_z(0_+)$ to represent the z-magnetization immediately following the r.f. pulse, we can write[6]

$$M_z(0_+) = M_z(0_-)\cos\theta \tag{2.79}$$

$$M_z(T_R) = M_z(0_+)\exp(-T_R/T_1) + M_0(1 - \exp(-T_R/T_1)). \tag{2.80}$$

Equating $M_z(T_R)$ and $M_z(0_-)$ eqn (2.78) follows directly.

The initial amplitude of the transverse magnetization just after the θ pulse is

$$M_y = M_0 \frac{1 - \exp[-T_R/T_1]}{1 - \cos\theta \exp[-T_R/T_1]} \sin\theta. \tag{2.81}$$

The maximum partial saturation signal amplitude is therefore not obtained for $\theta = 90°$ except where T_R/T_1 is large. An optimum condition can be found by adjusting the nuclear turn angle θ to a value known as the Ernst angle, θ_E, for which

$$\cos\theta_E = \exp(-T_R/T_1). \tag{2.82}$$

For T_R/T_1 of around 3 the relaxation between pulses is fairly complete and the optimum turn angle is close to 90°. The signal-to-noise advantage in a repetitive pulse experiment in which T_R is chosen to be much shorter than this and θ is set to θ_E is around $2^{\frac{1}{2}}$.

It should, however, be noted that use of the Ernst angle requires care. Incomplete transverse relaxation between successive experiments can lead to unexpected signals appearing in multiple pulse sequences.[6,32] Imaging experiments however are not necessarily subject to this effect because the magnetic field gradient pulses used during acquisition of the signal cause a destruction of magnetization coherence. This feature facilitates the use of rapidly repeated low turn angle pulse experiments, an example of which (FLASH) is discussed in the next chapter.

Whatever the repetition delay and nuclear turn angle employed, the T_1 relaxation time provides a fundamental limit to the available signal-to-noise ratio. This in turn limits the spatial resolution which is possible in NMR microscopy.

2.6.2 *Phase cycling*

Because the phase of the signal depends on the r.f. pulse phase, it becomes possible to distinguish the NMR FID signal from any background interference by r.f. phase alternation. For example, incrementing the r.f. phase by 180° will lead to signal inversion. Thus a successive phase alternation in 180° steps linked to successive addition and subtraction of the incoming signal from the data sum will lead to coherent superposition of the FID while the background interference is nullified. The process of addition and subtraction can be thought of as an alternation of the receiver phase so that the phase cycle can be written (0°, 0°)–(180°, 180°), where the r.f. transmitter and receiver phases are given in brackets. This particular cycle is called coherent noise cancellation.

The (0°, 0°)–(180°, 180°) sequence represents the simplest possible form of phase cycling. Other more sophisticated schemes enable the spectroscopist to correct for phase and amplitude anomalies in quadrature detection,[15] transverse magnetization interference due to rapid pulse repetition,[33] and echo artefacts due to pulse amplitude errors.[34] Among these, phase and amplitude errors in quadrature detection are the primary cause of artefacts in NMR imaging and can be nicely corrected in the four-pulse CYCLOPS sequence of Hoult and Richards,[15] a sequence which is so useful that it is generally incorporated as a subcycle of all other phase cycles.

When the amplitudes of the quadrature signal channels are unmatched or the relative phases are not precisely 90° apart, the complex Fourier transformation results in foldback artefacts. By successively swapping the channels used to acquire the real and imaginary signals, equivalent to a 90° transmitter and receiver phase shift, the error is compensated to first order as shown in Fig. 2.13. The CYCLOPS sequence also incorporates the usual 0°/180° alternation of coherent noise cancellation, giving a four-pulse cycle (0°, 0°)–(90, 90°)–(180°, 180°)–(270°, 270°).

Generally phase cycling procedures are performed via the data acquisition software. Most modern NMR spectrometers select the phase of the r.f. transmitter by digital means which means that the r.f. pulse phase can be simply manipulated through the software, usually in steps of 90°, although smaller steps are also available for some experiments involving multiple quantum filters. Similarly, it is possible to 'adjust the receiver phase' in 90° steps. In practice this is usually performed not in the phase-sensitive detector but after digitization by a software 'trick'. Whereas a shift of 180° corresponds to negation of both the incoming real and imaginary signals, a shift of 90° corresponds to negation of the incoming real followed by an interchange of the real and imaginary signals. It is, of course, this interchange which is so essential to the success of the CYCLOPS cycle.

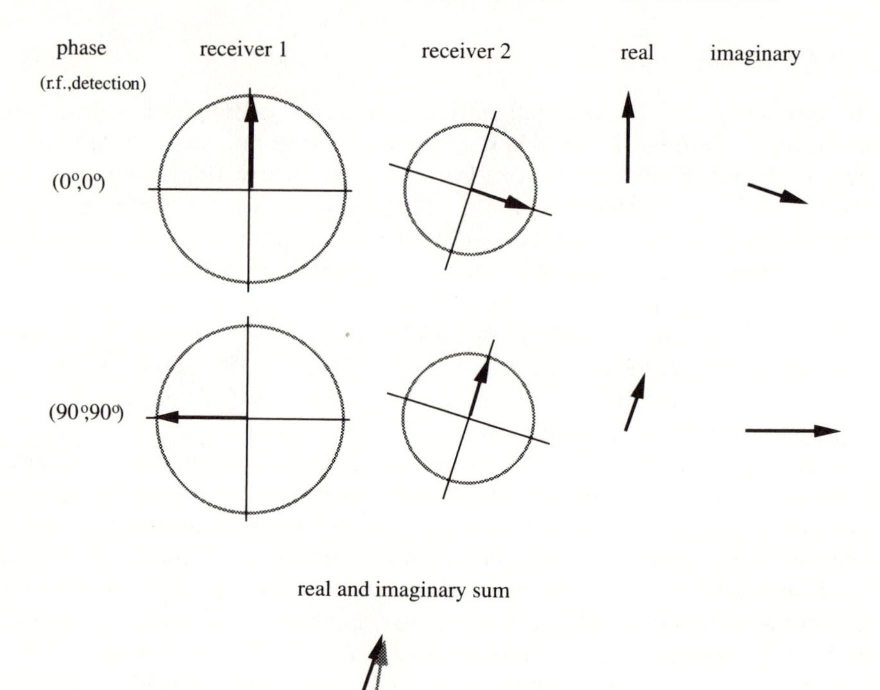

Fig. 2.13 The CYCLOPS method of phase cycling. The two receivers are shown as having different gains and with phase settings not precisely in quadrature. By alternating both the transmitter and receiver phases between 0° and 90° the real and imaginary signals are routed through both receivers and the summed resultants have matched amplitudes and quadrature phases to first order.

2.6.3 *Inversion recovery*

Inversion recovery[35] is used for the measurement of T_1 relaxation times as well as for the selective suppression of unwanted spin signals. The r.f. pulse sequence and resulting magnetization trajectories are shown in Fig. 2.14. The first 180° r.f. pulse inverts the magnetization vector, so subjecting the system to the most severe disturbance from equilibrium. Spin–lattice relaxation proceeds for a time t following which a 90 pulse is used to inspect the remaining longitudinal magnetization. The signal amplitude is described by

$$M(t) = M_0[1 - 2\exp(-t/T_1)]. \tag{2.83}$$

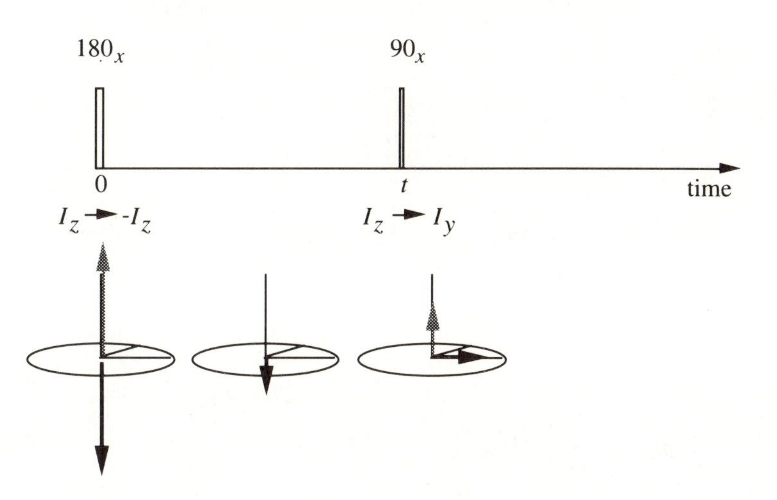

Fig. 2.14 Inversion recovery pulse sequence and magnetization trajectories. At each r.f. pulse the grey arrow represents the magnetization vector prior to the pulse while the black vector represents the magnetization after the pulse. Also shown is the evolution of the density matrix under the action of each pulse.

A noteworthy feature of this method is the change from a negative signal (proportional to $-M_0$) at $t = 0$ to a positive signal (proportional to M_0) as t becomes large. The cross-over through zero magnetization occurs at $t = 0.6931T_1$ and can be exploited to good effect. First, it is clear that a measurement of time for the cross-over null can yield T_1 in a single measurement. Second, the method may be used to remove the contribution from a spin with specific T_1 value by applying a 180_x pulse at a time interval $0.6931T_1$ before the imaging pulse sequence. Such signal suppression is particularly important in biological experiments where water suppression is desired. While the inversion recovery method of signal suppression is simple it can lead to distortions in the NMR spectrum. Other more sophisticated and effective methods for selective signal suppression are discussed in Chapter 5.

2.6.4 *Simple (Hahn) spin echo*

Magnetic field inhomogeneity causes nuclear spins to precess at differing Larmor frequencies according to their location in the sample, an effect which forms the basis of the NMR imaging method. Even in conventional NMR experiments where no magnetic field gradient is deliberately applied, the inhomogeneity of the polarizing magnet will result in a field spread across the sample of ΔB_0. This spread causes a dephasing of transverse magnetization following a 90_x r.f. pulse. Transverse magnetization phase

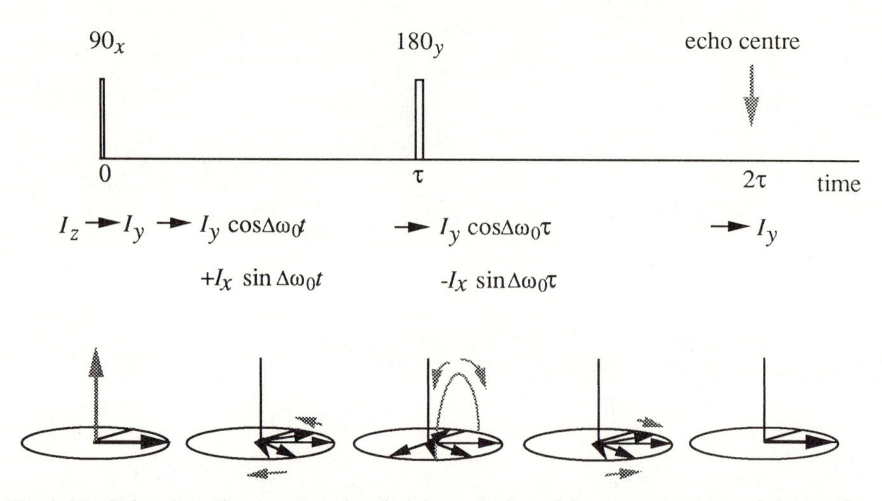

Fig. 2.15 Spin echo pulse sequence showing the evolution of the magnetization and the density matrix. Note that the 180_y pulse inverts the phase of each spin isochromat (i.e., $\Delta\omega_0\tau \rightarrow -\Delta\omega_0\tau$) so that perfect refocusing occurs at time $t=2\tau$. Following the 180_y pulse $I_y \cos\Delta\omega_0\tau$ evolves to $I_y \cos^2\Delta\omega_0\tau + I_x \cos\Delta\omega_0\tau \sin\Delta\omega_0\tau$ and $-I_x \sin\Delta\omega_0\tau$ evolves to $-I_x \sin\Delta\omega_0\tau \cos\Delta\omega_0\tau + I_y \sin^2\Delta\omega_0\tau$ giving a sum (averaged over all $\Delta\omega_0$) precisely equal to I_y.

coherence therefore lasts only for a time of order $(\gamma\Delta B_0)^{-1}$, a transience which apparently constrains the length of time over which this magnetization can be manipulated or detected. Many years ago E. Hahn[36] recognized that this loss of phase coherence was inherently reversible. Application of a second 180° r.f. pulse after a time delay τ will cause refocusing at 2τ as shown in Fig. 2.15 and the resulting phase coincidence is known as a spin echo with the phase of the 180° pulse determining the sign of the echo signal.

At this point it is useful to introduce the idea of precession using the density matrix picture. Eqn (2.27) describes the evolution of the density matrix in terms of operators involving the Hamiltonian. We will not solve these equations explicitly in the case of the spin echo but we will describe the successive evolution stages. To help visualize this process we can use rotating frame 'precession diagrams' like those shown in Fig. 2.16. Here the Hamiltonian term is, in each case, a Zeeman interaction, in (a) caused by an offset ΔB_0 in magnetic field applied along the z-axis giving a Hamiltonian $\Delta\omega_0 I_z$, in (b) caused by an r.f. pulse θ_x, and in (c) caused by an r.f. pulse θ_y.

In thermal equilibrium the spin system density matrix has a state of longitudinal polarization so that the density matrix is proportional to I_z. The first r.f. pulse, applied along the x-axis in the rotating frame, corresponds to a Hamiltonian $-\gamma B_1 I_x$ which causes a cyclic precession of I_z to I_y and back to I_z at a rate γB_1, a process which is illustrated in Fig. 2.16(b). 90° rotation is achieved by terminating the pulses at a time when ρ has

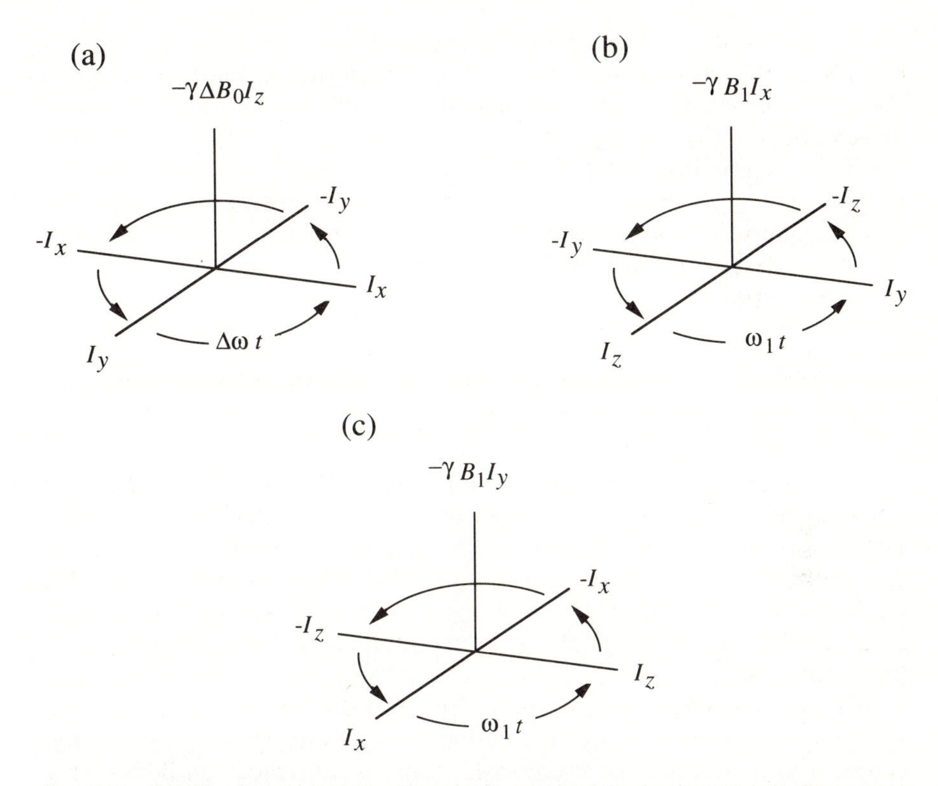

Fig. 2.16 Rotating-frame precession diagrams for states of the density matrix in the presence of a Zeeman interaction: (a) evolution of I_y under the off-resonant Zeeman interaction, $-\Delta\omega_0 I_z$; (b) I_z under the influence of the r.f. field $-\Delta\omega_1 I_x$; and (c) I_x under $-\Delta\omega_1 I_y$. Note that in each of these precessions the direction of rotation is shown as positive in a right-handed sense around the vertical axis. In Section 2.1, we saw that when the gyromagnetic ratio is positive the magnetic field $\Delta\mathbf{B}_0$ causes clockwise (i.e., negative) rotations about $\Delta\mathbf{B}_0$. In each of the precession diagrams the implied magnetic field direction (i.e., z, x, and y) is opposite to the 'Hamiltonian axis' (i.e., $-\gamma\Delta B_0 I_z$, $-\gamma B_1 I_x$, and $-\gamma B_1 I_y$) about which precession takes place.

evolved into I_y. Following the r.f. pulse the new density matrix is now subject to the local Zeeman field offset. (Of course this field was also present during the r.f. pulse but we are able to neglect its effect by ensuring that the r.f. field strength is very much greater than ΔB_0, a point which is discussed further in Section 2.4.7.) This next stage in the evolution, the precession due to the Hamiltonian $\gamma\Delta B_0 I_z$, is shown in Fig. 2.16(a). The evolution diagram for the second r.f. pulse can be easily worked out by cyclically permuting the vectors of Fig. 2.16(b). A 180_y pulse inverts the I_x terms in the density matrix but leaves the I_y terms unchanged. If the time of initial evolution in the field offset ΔB_0 was τ, the final evolution for an equal time τ in the same field offset ΔB_0 will therefore cause a perfect restoration of the I_y

density matrix state which existed immediately after the first r.f. pulse, a process which is also shown in Fig. 2.15. It is important to note that the phase of the 180° inversion pulse is relevant. A 180_x pulse, for example, results in an echo of negative sign.

We can follow the evolution of the density matrix polarization state by using a succession of arrows to represent the time progression where the Hamiltonian term under which ρ evolves is written above the arrow. For example, the evolution progression for the Hahn echo would have the following appearance:

$$I_z \xrightarrow{-(\pi/2)\,I_x} I_y \xrightarrow{-\Delta\omega_0\tau I_z} I_y\cos\phi + I_x\sin\phi \xrightarrow{-(\pi)\,I_y} I_y\cos\phi - I_x\sin\phi \xrightarrow{-\Delta\omega_0\tau I_z}$$

$$\overline{[\mathrm{I}_y\cos^2\phi + I_x\cos\phi\sin\phi - I_x\sin\phi\cos\phi + I_y\sin^2\phi]} = I_y \qquad (2.84)$$

where ϕ is the precessional phase shift, $\Delta\omega_0\tau$. Note that magnetic field inhomogeneity leads to a distribution of offsets, $\Delta\omega_0$, and hence a dephasing of the transverse magnetization. Upon averaging over the distribution of $\Delta\omega_0$ the final density matrix term in square brackets is exactly I_y. The 180_y r.f. pulse (denoted $-(\pi)I_y$) causes the phases to be inverted (i.e., angle $\Delta\omega_0\tau$) so that the effect of the next evolution period τ is to restore the polarization to I_y.

In spin systems experiencing only Zeeman Hamiltonians, the spin echo sequence refocuses all dephasing due to inhomogeneous broadening, chemical shift, and heteronuclear scalar spin–spin interactions. Because the 180° r.f. pulse inverts all spins and so leaves the interaction term $\pi J\, 2\mathbf{I}_1 \cdot \mathbf{I}_2$ invariant, homonuclear scalar spin–spin interactions are not refocused and therefore remain to modulate the echo. Over and above this modulation, residual attenuation of the echo is due to spin–spin relaxation alone and in principle, a plot of echo amplitudes $M_y(2\tau)$ obtained from differing τ values can be used to yield T_2 according to

$$M_y(2\tau) = M_0\exp(-2\tau/T_2). \qquad (2.85)$$

Earlier we saw that sampling of the decaying NMR signal at positive times only, leads to the existence of a dispersion image in the imaginary domain following Fourier transformation. Fig. 2.17 suggests a mechanism by which 'negative' time may be sampled. If we take the centre of the echo as our time origin and begin sampling the signal at this point, the resulting spectrum is the same as that shown earlier for the single pulse experiment, with the exception that T_2 relaxation will have attenuated the signal at $t = 0$ and hence reduced the area under the absorption peak by the same factor. (For the purpose of the present argument we will neglect the role of interactions such as homonuclear couplings which are non-refocusable via the Hahn echo.) Furthermore the dispersion spectrum will still exist.

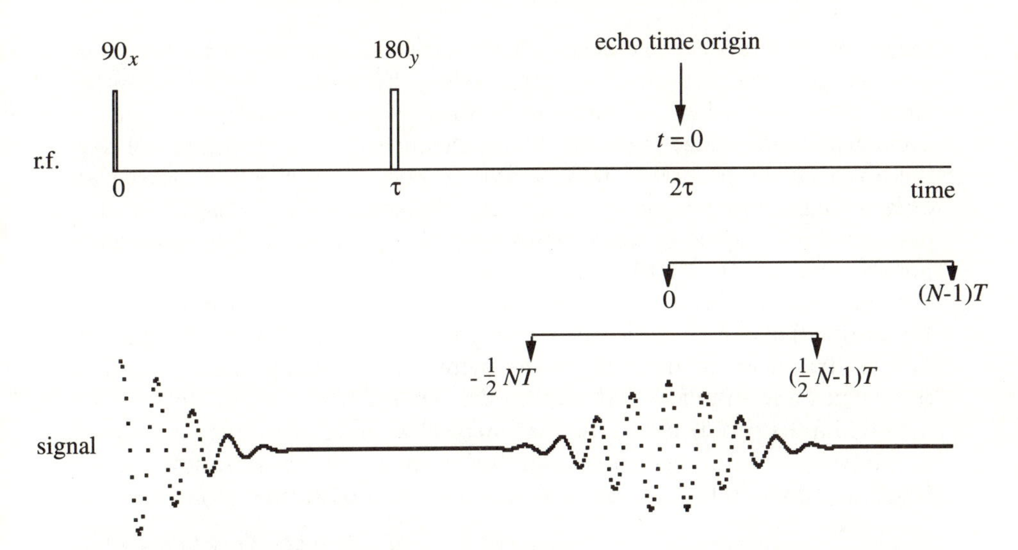

Fig. 2.17 Echo signal, corresponding to magnetization states I_y (real part) and I_x (imaginary part), in a simple spin echo. Note that the echo centre represents a local time origin. Following this the signal is akin to the FID resulting from the 90_x r.f. pulse, apart from an additional attenuation due to T_2 relaxation. Because of the echo symmetry, sampling of the full echo leads to a spectrum without dispersion, as illustrated in Fig. 1.3(a).

Suppose, however, that we were to sample the entire echo, dividing our sampling time period equally before and after the echo with $t = 0$ at the echo centre. In this case the digitized time values run between $-\frac{1}{2}NT$ and $(\frac{1}{2}N - 1)T$, and, as before, the Fourier domain points run between $-\frac{1}{2}N(1/NT)$ and $(\frac{1}{2}N - 1)(1/NT)$. Neglecting T_2 effects and presuming $\phi = 0$, it is clear that the in-phase and quadrature signals are respectively symmetric and antisymmetric, leading to an entirely real spectrum as shown in Fig. 1.3(a). If $\phi \neq 0$ then the spectrum is multiplied by $\exp(i\phi)$ and will contain both real and imaginary parts, although the real spectrum can be recovered by the usual 'phasing' process as discussed in Section 2.3.3. There is, however, a crucial difference between this situation and the case of $t > 0$ sampling. When full echo sampling is used the resulting **modulus** spectrum for $\phi \neq 0$ exactly reproduces the real spectrum for $\phi = 0$, a restoration process which is impossible in the case of data sampled only at positive times because of the existence of the dispersion component in the spectrum.

The symmetry relations imply that for ideal data detected 'in-phase', no additional information is present in the period before the echo. Furthermore the identity of the real spectra which result from 0 to $(N - 1)T$ and $-\frac{1}{2}NT$ to $(\frac{1}{2}N - 1)T$ sampling raises the interesting question as to what has happened to the additional information which is clearly contained in the first method. The answer would appear to be that it is contained in the dispersion spectrum. That being the case and, given that we may only be interested in the real spectrum, the question arises as to why we do not simply sample from 0 to $(\frac{1}{2}N - 1)T$. If this yields the desired spectrum then such a limited sampling range is clearly advantageous since it introduces one half the noise power compared with sampling from 0 to $(N - 1)T$, giving an overall signal-to-noise gain of $\sqrt{2}$. As we shall see in Chapter 4, the optimization of signal-to-noise ratios is vitally important in determining the possible resolution in NMR microscopy.

In practice this limited sampling approach can be handled in two ways. Either the data can be zero-filled from $\frac{1}{2}NT$ to $(N - 1)T$ and subsequently Fourier transformed directly or the symmetry relations can be used to reconstruct the in-phase and quadrature signals from $-\frac{1}{2}NT$ to 0. Each method has its problems. In the first method we have effectively multiplied the data by a hat function of half the normal window duration. In the frequency domain this results in a convolution broadening

$$\mathfrak{F}(H(t)) = NT\frac{\sin\pi fNT}{\pi fNT} + \mathrm{i}\,\frac{NT(1 - \cos\pi fNT)}{\pi fNT} \qquad (2.86)$$

for which the imaginary part has non-zero values when $f \neq 0$ (i.e., when k is odd). This means that some additional broadening will be introduced into the real part of the spectrum from the dispersion component. For the second method the symmetrization can only be performed when the detection phase, ϕ, is precisely zero although in principle a post-detection adjustment process, similar to that discussed in Section 2.3.3, is possible, by admixing in-phase and quadrature components until the imaginary signal is zero at $t = 0$. Such a process is manifestly inferior to spectrum phasing since it depends on the accuracy with which a single point can be measured.

The various relationships between the approaches to sampling are summarized in Table 2.5.

Finally it should be noted that these symmetry relationships only apply when the spectrum of interest is implicitly real. While this is normally the case in NMR imaging there are, nevertheless, important examples where the image is inherently phase modulated, such as when the nuclear spins are in net translational motion. In these examples the echo has no symmetry and both negative and positive times must be sampled.

2.6.5 The *Carr–Purcell–Meiboom–Gill echo train*

The phase coherence recovered in the nuclear spin echo is subsequently lost for $t > 2\tau$. Successive recoveries are however possible if a train of additional 180° r.f. pulses is used as suggested by Carr and Purcell in 1954.[37] The choice of phase with which these pulses are applied is important if the cumulative effects of small turn angle errors are to be avoided. In the

Table 2.5. *Properties of spin echo sampling schemes.*

	relative signal amplitude	relative noise amplitude	convolution broadening	detection phase constraints	when $\phi \neq 0$ modulus $\Rightarrow$ real ($\phi = 0$)
full echo acquisition $-\frac{1}{2}NT$ to $(\frac{1}{2}N - 1)T$	1	1	none	none	yes
full positive time acquisition 0 to $(N - 1)T$	1/2	1	none	none	no
positive time acquisition 0 to $(\frac{1}{2}N - 1)T$, zero-filling $\frac{1}{2}NT$ to $(N - 1)T$	1/2	$1/\sqrt{2}$	yes	none	no
positive time acquisition 0 to $(\frac{1}{2}N - 1)T$, symmetrization $-\frac{1}{2}NT$ to 0	1	$\sqrt{2}$	none	$\phi = 0$	yes

Meiboom–Gill modification[38] to the Carr–Purcell train the use of quadrature 180_y pulses provides the appropriate compensation. The envelope of the echoes in a Carr–Purcell–Meiboom–Gill (CPMG) sequence is determined by T_2 decay alone and so it is possible to determine T_2 in a single experiment. The method is illustrated in Fig. 2.18.

The production of multiple echoes in the CPMG pulse sequence suggests an obvious application in signal averaging. Co-addition of echoes within a train leads to signal-to-noise enhancement in addition to that obtained by addition of the independent experiments separated by the T_1 recovery period, T_R. This has important consequences for NMR microscopy which will be examined in Chapter 4.

2.6.6 *The stimulated echo*

In many materials, especially those with molecules undergoing motion which is slow compared with the period of the nuclear Larmor precession, the transverse relaxation time T_2 is considerably shorter than T_1. The fact that the 'lifetime of spin polarization' exceeds the 'lifetime of first-order spin coherence' can be turned to advantage where one wishes to 'store' coherence over a long time interval. Suppose that the transverse magnetization existing at some point of time τ is required to be stored for later recall. An example

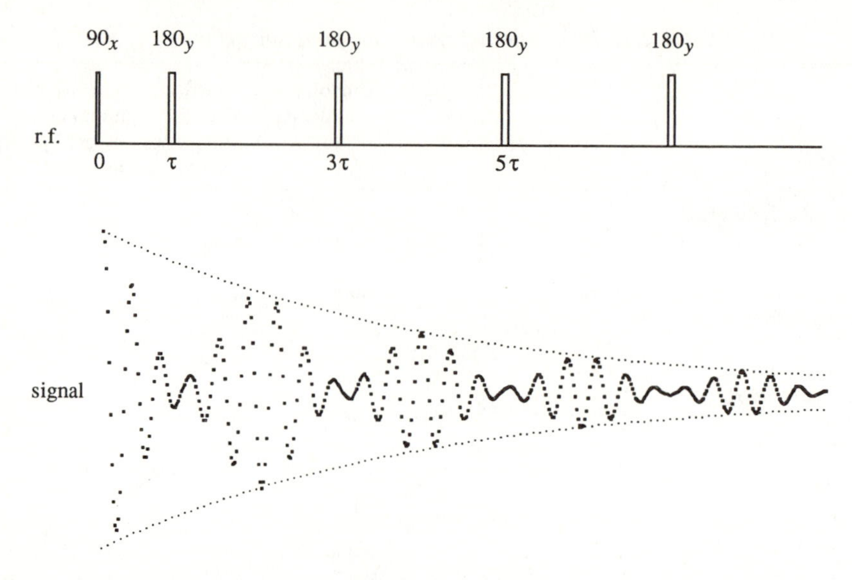

Fig. 2.18 CPMG pulse sequence exhibiting multiple spin echoes at $2n\tau$ modulated by a T_2 relaxation envelope.

might be where one wishes to use this magnetization at a later time to see how far the molecules containing the NMR nuclei have moved. The method used is shown in Fig. 2.19 where a single 90_x pulse applied after a time τ has the effect of rotating the y-component of magnetization into longitudinal polarization along the z-axis, a state in which only T_1 relaxation will occur. Of course any x-magnetization will be unaffected so that only half the transverse magnetization can be stored in this way. Recall is made at some later stage using another 90_x pulse. As shown, this leads to an echo at time τ after the last r.f. pulse. This stimulated echo[36] is of particular importance in a number of NMR imaging applications[39], especially where the translational motions of molecules are being measured using pulsed field gradient spin echo methods.

It should be noted that the stimulated echo r.f. pulse sequence also generates two additional spin echoes. These are, respectively, the echo of the initial pulse FID caused by the second pulse, and the echo of the second pulse FID caused by the third pulse. Special care is needed to avoid interference between the stimulated echo and the two spin echoes. One effective approach is the use of a homogeneity-spoiling (homospoil) magnetic field gradient pulse applied during the 'z-storage' period between the second and third r.f. pulses. This has the effect of destroying the unwanted transverse magnetization without influencing the magnetization which has been stored along the z-axis.

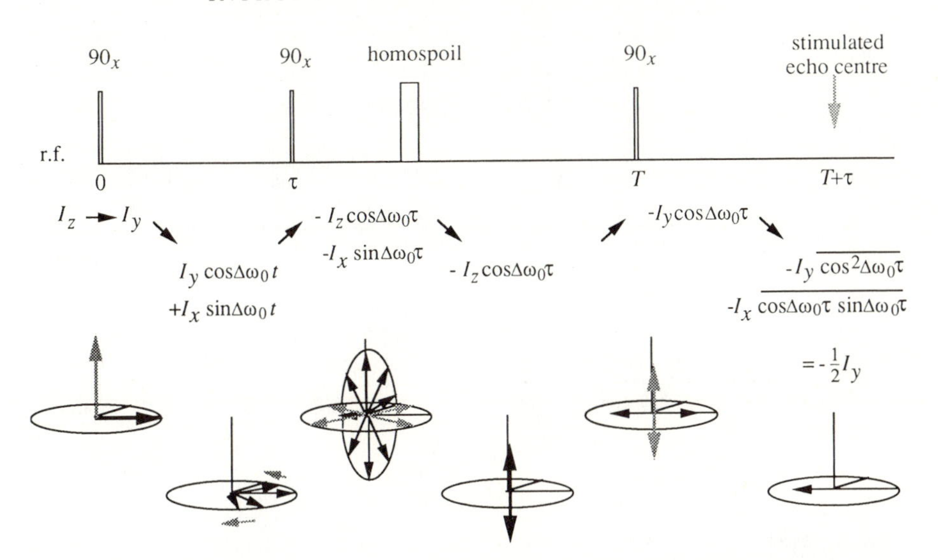

Fig. 2.19 Stimulated echo pulse sequence using a homospoil pulse during the 'z-storage' period. The representation is similar to that used for the simple spin echo in Fig. 2.15. Note that the final density matrix state is an ensemble average over the isochromats labelled by $\Delta\omega_0$.

2.6.7 *Steady-state free precession and driven equilibrium*

In the CPMG train the r.f. pulse spacing, 2τ, is less than the transverse relaxation time T_2 so that transverse coherence is maintained. Eventually, however, this coherence is lost and the spin system is allowed to come to thermal equilibrium in a time of order T_1 so that a new pulse train can again be applied. An alternative approach to the retention of transverse coherence was suggested by Carr[32] in which a train of equally spaced r.f. pulses is applied continuously and the spin magnetization is allowed to reach a dynamic equilibrium. This experiment is known as Steady-State Free Precession (SSFP). In principle the turn angle of the repeated r.f. pulse may be chosen at will but it is usual[40] to alternate the r.f. phase by 180° as shown in Fig. 2.20(a) so that the double cycle time is 2τ. This procedure causes phase coherence for spins whose isochromats are either on resonance or at an integral multiple of the offset $\Delta\omega = 2\pi/\tau$.

The essential difference between steady-state free precession and the partial saturation condition discussed in Section 2.6.1 concerns the role of the transverse coherence. In SSFP the repetition time between the r.f. pulses is allowed to be sufficiently short that both the transverse and longitudinal magnetizations must be accounted for in determining the steady state condition. Calculation of the resulting equilibrium values of M_y following each r.f. pulse must account for T_1 and T_2 relaxation as well as

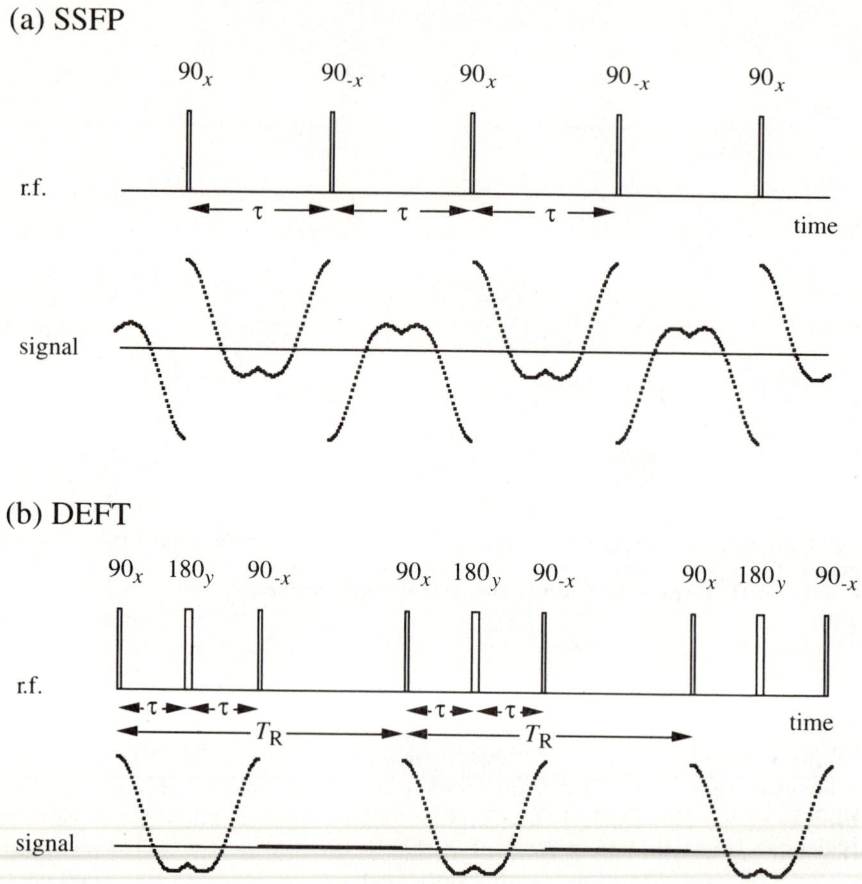

Fig. 2.20 (a) Steady-state free precession in which the equilibrium magnetization value is determined by the interval τ. The alternation of r.f. pulse phase helps avoid cumulative errors due to r.f. inhomogeneity. (b) Driven equilibrium. The magnetization is stored along the z-axis to allow some T_1 recovery over the period $T_R - 2\tau$, where T_R is the repetition interval.

the precessions which take place due to the r.f. pulses and the intervening periods of phase evolution. The details of this calculation are available elsewhere[41] and only the final result for the equilibrium value of M_y immediately following the pulse in a train of identical θ_x pulses is given here, namely,

$$M_y = M_0 \frac{(1-\exp(-\tau/T_1))\sin\theta}{1+\exp(-\tau/T_1)\exp(-\tau/T_2)-\cos\theta(\exp(-\tau/T_1)+\exp(-\tau/T_2))}. \tag{2.87}$$

Eqn (2.87) is, as required, identical to eqn (2.81) when $\tau \gg T_2$. For a 90° pulse the expression reduces to

$$M_y = M_0 \frac{1 - \exp(-\tau/T_1)}{1 + \exp(-\tau/T_1)\exp(-\tau/T_2)}. \tag{2.88}$$

An alternative approach to generating equilibrium magnetization is the driven equilibrium cycle of Becker *et al.*,[42] usually called the Driven Equilibrium Fourier Transform (DEFT) method. In this cycle the transverse magnetization is refocused in a spin echo and then nutated back to the z-axis where it can be briefly stored, allowing partial recovery towards equilibrium, before recall to the following cycle. This process is shown in Fig. 2.20(b). Clearly DEFT requires specific pulse turn angles, namely 90_x, 180_y, 90_{-x}. The equilibrium transverse magnetization immediately following the 90_x pulse has been derived by Shoup *et al.*[43] and is given by

$$M_y = M_0 \frac{1 - \exp(-(T_R - 2\tau)/T_1)}{(1 - \exp(-2\tau/T_1)\exp(-(T_R - 2\tau)/T_1))}. \tag{2.89}$$

Both eqns (2.88) and (2.89) represent a value of M_y which is comparable with the partial saturation value for 90° pulses, $M_0(1 - \exp[-T_R/T_1])$. The relative advantages of these approaches, along with those of the multiple echo CPMG method, will be discussed in Chapter 4. Rapid acquisition methods are of particular interest in NMR imaging, both because of the possibility of obtaining rapid image rasters and hence images of moving objects, and also because the co-addition of a rapidly repeated train of signals offers the possibility of signal-to-noise enhancement. This latter feature is of particular interest in NMR microscopy.

2.6.8 *Coherence transfer via J-couplings, dipolar interactions, and quadrupole interactions*

The equilibrium state of the spin ensemble is one of simple longitudinal polarization in the external magnetic field. Usually, as in the above sequences, we start our experiment with a 90_x pulse, so transforming this polarization into a state of magnetization along the y-axis (in-phase y-coherence). In the echo experiments we saw how the existence of an off-resonant magnetic field term caused this coherence to precess from y to x magnetization. The spin echo occurs because a subsequent 180_y r.f. pulse inverts the phase of all x-coherences while leaving y-coherence unchanged.

Eqn. (2.27) describes how a given state of polarization, represented by ρ, evolves with time under the influence of a particular interaction, represented by a Hamiltonian H. Under the Zeeman interactions of the longitudinal polarizing field and the r.f. pulses, the initial equilibrium magnetization state evolves into other states of pure magnetization. In other words, the Zeeman Hamilton retains the rank-one character of the initial polarization, T_1^0 (i.e. I_z). The existence of terms in the Hamiltonian which are bilinear in the spin operators, such as J-couplings, dipolar interactions, or quadrupole interactions, causes 'precession' to higher rank polarization states. We illustrate this using the scalar coupling which is important in high-resolution proton NMR in the liquid state. Following the first r.f. pulse which turns I_z magnetization into I_y the scalar coupling interaction causes an unusual transformation of the resulting y-coherence and, by using this evolution along with a combination of r.f. pulses of various turn angles and phases, it is possible to generate most of the states available to the coupled spin system.[44-47] This remarkable property of being able to prepare the spin system in almost any desired state is the basis of a large number of coherence transfer and multiple quantum filtering experiments.[48-50] These are thoroughly described in the book by Ernst, Bodenhausen, and Wokaun and we shall focus on only two simple pulse sequences designed to generate two- and zero-quantum coherence, respectively. To understand these, however, it is necessary to appreciate the progression of evolutions. Some simple evolution diagrams, similar to those used to explain the spin echo, will be helpful and these are shown in Fig. 2.21. Note that in the case of a two-spin system we begin with a magnetization $(I_{1z} + I_{2z})$ which is transformed into $(I_{1y} + I_{2y})$ by the first non-selective 90_x r.f. pulse. Because such a pulse acts on both spin species it is represented by $-(\pi/2)(I_{1x} + I_{2x})$.

Fig. 2.21(b) shows the 'precession' of half of the polarization (corresponding to the spin labelled 1) under the influence of the secular part of the scalar interaction in which I_{1y} evolves between I_{1y} and $I_{1x}I_{2z}$ states at a rate πJ radians per second. For example, after a quarter cycle, at time $1/2J$, I_{1y} has evolved precisely into $-I_{1x}I_{2z}$, state of spin 1 x-coherence in anti-phase with spin 2. Consider the influence of a second r.f. pulse with x-polarization applied at this point in time. The evolution of $-I_{1x}I_{2z}$ under the Hamiltonian $-\theta I_x$ is shown in Fig. 2.21(c). This pulse causes precession between $-I_{1x}I_{2z}$ and $+I_{1x}I_{2y}$ at a rate γB_1 so that a '90_x' pulse of duration $1/4\gamma B_1$ will produce the state of two-spin coherence $I_{1x}I_{2y}$.

We can represent this transfer by:

$$I_{1z} \xrightarrow{-(\pi/2)(I_{1x}+I_{2x})} I_{1y} \xrightarrow{(\pi/2)\,2I_{1z}I_{2z}} -2I_{1x}I_{2z} \xrightarrow{-(\pi/2)(I_{1x}+I_{2x})} 2I_{1x}I_{2y}. \tag{2.90}$$

Eqn (2.90) follows the evolution of spin 1 under the influence of the r.f. pulses and scalar coupling. An equivalent process takes place for spin 2 which means that $2I_{1x}I_{2y}$ is only half the final density matrix and must be superposed on that which obtains from the other half of the initially excited polarization, I_{2y}, to produce a net result, $(I_{1x}I_{2y} + I_{1y}I_{2x})$, a state which

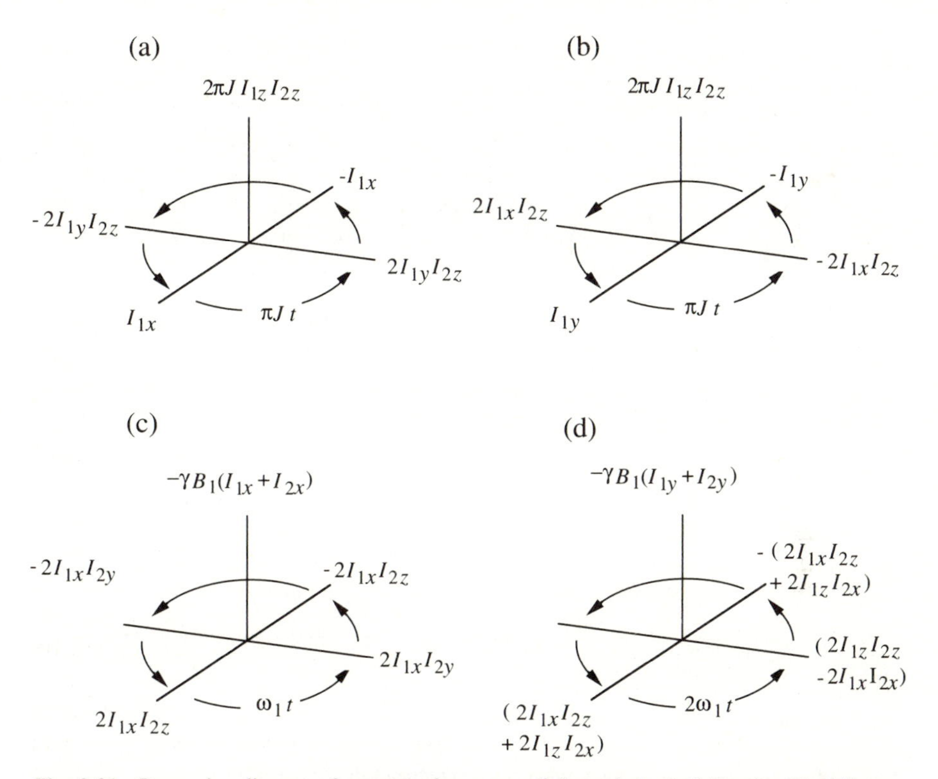

Fig. 2.21 Precession diagrams for states of a system of J-coupled spin-1/2 pairs. The bilinear interaction, $2\pi J I_{1z} I_{2z}$ causes the magnetization states to evolve into two-spin coherences. Note that the r.f. pulse consists of a sum of spin 1 and spin 2 operators if it is applied non-selectively (i.e., with broad bandwidth). (a) and (b) Evolution of I_x and I_y under the scalar coupling into ± 1QC of spin 1 in antiphase with spin 2. (c) and (d) Evolution of the ± 1QC antiphase state produced in (b) under the influence of a broadband r.f. pulse applied along the x and y directions of the rotating frame. Note that in (d) the precession rate is $2\omega_1 t$, meaning that a 45_y pulse causes a precession of the total spin density matrix state $(2I_{1x}I_{2z}+2I_{1z}I_{2x})$ into $(2I_{1z}I_{2z}-2I_{1x}I_{2x})$. $2I_{1x}I_{2x}$ is an equal admixture of 2QC and ZQC.

represents a superposition of $+2$QC and -2QC. These 2QC states are able to be 'recalled' at some later time by a 90° r.f. pulse. Such a pulse works by converting $I_{1x}I_{2y}$ back into $I_{1x}I_{2z}$ and $I_{1z}I_{2x}$ which, in turn, at a later time $1/2J$ return to the observable $-(I_{1y} + I_{2y})$ polarization under the action of the scalar coupling Hamiltonian.

A pulse sequence for the generation and recall of two-quantum coherence is shown in Fig. 2.22(a). Note the additional insertion of the 180_y pulses. These have the effect of refocusing all Zeeman influences such as those terms in the Hamiltonian due to chemical shift and B_0 field inhomogeneity. It is important to note that the pulse sequence of Fig. 2.22(a) also allows uncoupled spins to produce a signal in response to the final 90_x r.f. pulse,

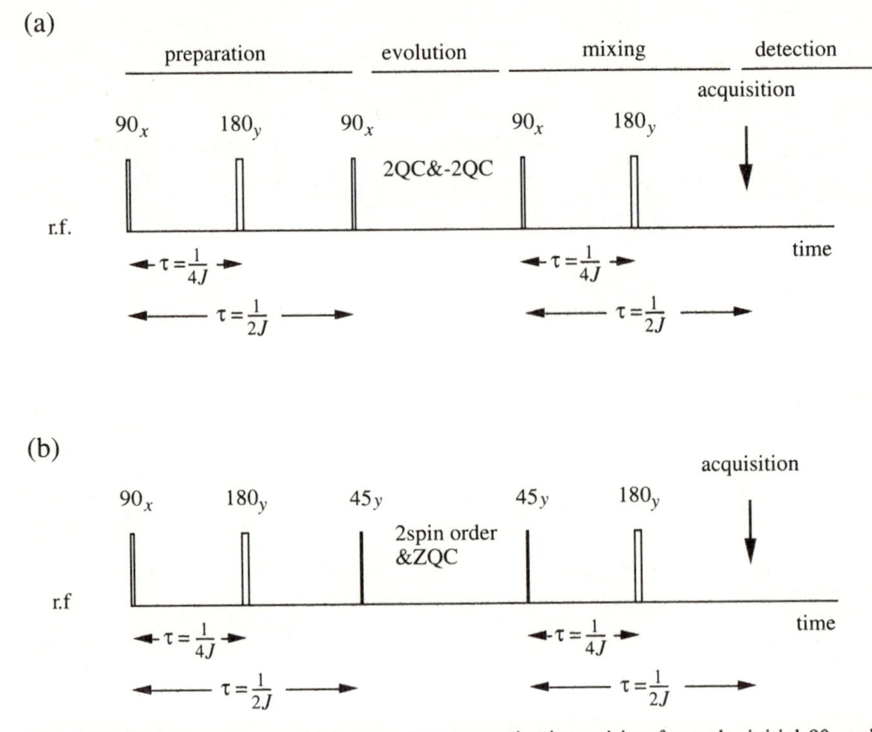

Fig. 2.22 Pulse sequences which convert y magnetization arising from the initial 90_x pulse into (a) double quantum coherence and (b) zero-quantum coherence. In each case the preparation period utilizes evolution under the bilinear scalar coupling interaction. Restoration of detectable magnetization is achieved by means of a mixing pulse sequence in which the stored ZQC or 2QC state is converted through a reversed evolution. The period labelled 'evolution' refers to NMR imaging applications in which a magnetic field gradient is applied to the prepared 2QC or ZQC states.

in addition to the signal from the coupled spins which have evolved through the two quantum coherence pathway. Prior to the mixing pulse segment these uncoupled spins have a magnetization along the z-axis caused by the preparation segment of two 90_x r.f. pulses. To filter only the signal from the coupled spins requires some appropriate phase cycling which will distinguish the various orders of coherence generated in the preparation sequence. In particular the transformation of the preparation sequence by a rotation operator $\exp(i\phi I_z)$, equivalent to a phase shift of all preparation sequence r.f. pulses by ϕ, results in a phase modulation of the final signal by $\exp(ip\phi)$ where p is the coherence order. Such a modulation can be distinguished by multiplying the final signal by $\exp(-ip\phi)$ and summing over all ϕ values in the phase cycling sequence. To illustrate how this works, consider the effect of alternating the phase of the preparation segment between 90_x and 90_y pulse pairs, corresponding to ϕ values of 0 and 90° respectively. This alternation results in a change of sign of the two quantum coherence state but no change in sign of the z-magnetization arising from the uncoupled spins. If successive acquisitions are multiplied by 1 and -1 respectively, the experiment will produce a null result for any spins which are not scalar coupled and therefore unable to pass the double quantum filtering process.

Fig. 2.22(b) shows an alternative pathway in which zero quantum coherence is produced in the intermediate state. This uses a 45_y r.f. pulse instead of the 90_x pulse, so converting $(2I_{1x}I_{2z} + 2I_{1z}I_{2x})$ into equal amounts of 2QC and ZQC each with half the amplitude of the 2QC polarization resulting from the filter in Fig. 2.22(a). Despite the poorer signal-to-noise ratio which results, the ZQC state offers one particular advantage. Because ZQC is invariant under a Zeeman Hamiltonian,[5,6] it does not suffer from phase spreading due to precession in an inhomogeneous Zeeman field. If a severe magnetic field gradient is deliberately applied in the form of a 'homospoil' pulse one- and two-quantum coherences are destroyed. This form of zero-quantum filter is particularly effective in removing any signals arising from non-scalar coupled spins.[51]

The scalar coupling interaction represents an interaction between the nuclear moments and is bilinear in the spin operators. On rotating the coordinate system the spin operator $\mathbf{I}_1 \cdot \mathbf{I}_2$ is invariant, and we say that it transforms under rotations as a scalar. By contrast, the dipolar and quadrupolar spin operators, which are also bilinear, transform under rotations as second-rank tensors. As with the scalar coupling interaction the evolution of the density matrix under these rank-two interactions can also be described in terms of a set of precession diagrams and it is possible to devise coherence transfer pathways which bear a considerable resemblance to those discussed previously. Two states of importance in the imaging of spin-1 nuclei undergoing quadrupolar interactions are two-quantum coherence, $T_2^{\pm 2}$, and the quadrupolar polarization, T_2^0. The 2QC state is invariant under the quadrupolar Hamiltonian (i.e. $[\omega_Q T_2^0, T_2^{\pm 2}] = 0$) but precesses at twice the Larmor frequency under the Zeeman interaction (i.e. $[\omega_0 T_1^0, T_2^{\pm 2}] = \pm 2\omega_0 T_2^{\pm 2}$. This means that, once created, the state of two-quantum coherence is protected from any dephasing which might result from inhomogeneities in H_Q but is available for spatial encoding using magnetic field gradient pulses. By contrast the quadrupole polarization is invariant under both the Zeeman and quadrupolar terms in the Hamiltonian, making it ideal for 'protective storage' of spin order. An example of an imaging pulse sequence using these properties is given in Chapter 5.

Fig. 2.23 shows precession diagrams and an r.f. pulse sequence relevant to production of the two-quantum coherent state via the state of quadrupolar

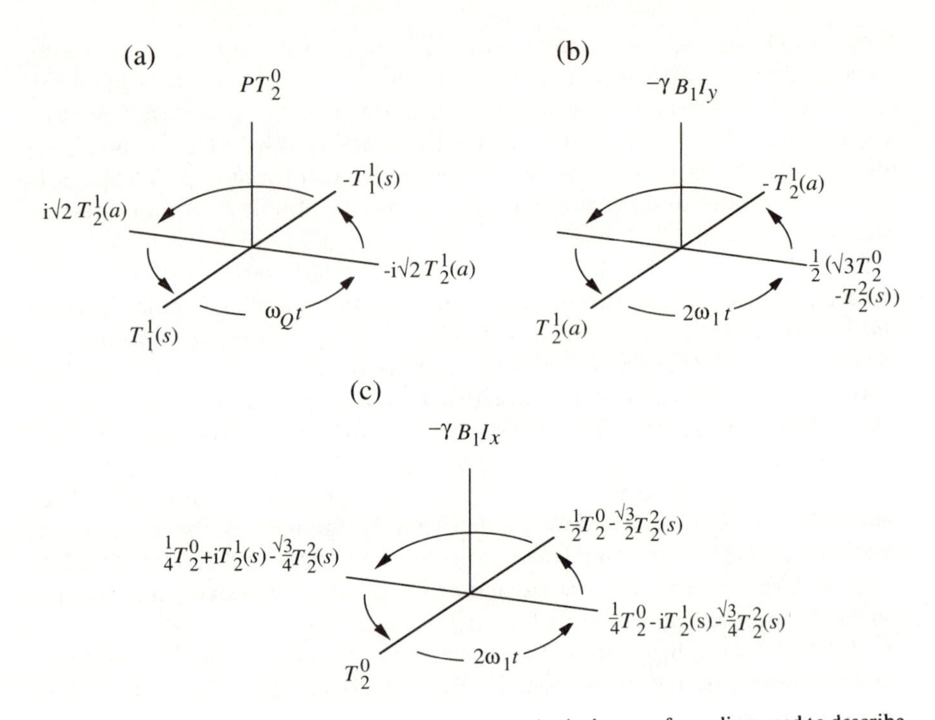

Fig. 2.23 Precession of density matrix states in the spherical tensor formalism used to describe quadrupolar interactions of higher spin systems. (*s*) and (*a*) refer to symmetric and antisymmetric combinations of tensor components. For example, $T_1^1(s) = (T_1^1 + T_1^{-1})/\sqrt{2}i \sim I_y$, while $T_1^1(a) = (T_1^1 - T_1^{-1})/\sqrt{2} \sim I_x$. (a) shows the evolution of the magnetization state I_y ($T_1^1(s)$) under the quadrupole interaction, PT_2^0. The precession frequency is $\omega_Q = \sqrt{3/2}\,P$. (b) shows the evolution of $T_2^1(a)$ into the quadrupole polarization state, T_2^0, under I_y. Note that the Larmor precession is doubled for this second-rank tensor. (c) shows the evolution of T_2^0 under I_x. Unlike all other evolutions represented in Figs. 2.16, 2.21, 2.23, this is not a simple vector rotation but a transformation between T_2^0, T_2^1, and T_2^2 states according to the irreducible representation of the rotation group, $D^{(2)}$. Further details may be found in references 22 to 25 and in reference 52.

polarization. Note that no NMR signal is observed until the spin system is returned to a rank-one state of single-quantum coherence, $T_1^{\pm 1}$.

2.6.9 *The solid echo*

Similar coherence pathways are possible for systems of spin-$\frac{1}{2}$ nuclei experiencing mutual dipolar interactions as for isolated spin-1 nuclei under the quadrupolar interaction, although the algebra is slightly more complicated. Fortunately, when dealing with the spin-1 deuteron in the solid state, we can neglect dipolar interactions because the deuteron gyromagnetic ratio is so much smaller than that of the spin-$\frac{1}{2}$ proton.

For both deuterium and proton NMR in the solid state the respective quadrupolar and dipolar interactions vary from nucleus to nucleus as a result of variations in the orientation of the electric field gradient in the case of the deuteron, and through variations in the orientation and length of internuclear vectors in the case of the proton. The dephasing caused by precession under the rank-two tensor interactions causes a rapid loss of signal as discussed in Section 2.4.3. However all precessions can be reversed and a spin echo formed by the appropriate coherence transfer, an effect known as the solid echo. The solid echo can be explained either in terms of the spherical tensor or product operator formalisms since the secular parts of the quadupole and dipolar interactions resemble T_2^0 and $I_{1z}I_{2z}$ in tensorial character. We will first consider the echo formation from the standpoint of the product operator formalism using the precession diagrams shown in Fig. 2.21.

Eqn (2.84) followed the evolution of the spin polarization in the ordinary spin echo as the magnetization was dephased by a spread in the Zeeman interaction, the 180_y pulse providing the refocusing process by inverting the phase angle $\Delta\omega\tau$. In the Bloch decay the dipolar interaction causes precession of the initial I_{1y} state of the density matrix into $I_{1x}I_{2z}$ so that the initial state I_{1y} dephases into an admixture $I_{1y}\cos(\Delta\omega_D\tau) - 2I_{1x}I_{2z}\sin(\Delta\omega_D\tau)$ where ω_D is some equivalent 'dipolar precession frequency'. Note that ω_D will be of order $M_2^{1/2}$.

To start the rephasing process we require an r.f. pulse which will convert $I_{1x}I_{2z}$ to $-I_{1x}I_{2z}$, thus in effect inverting the phase. This is achieved by a 90_y r.f. pulse in the following sequence:

$$I_{1z} \xrightarrow{-(\pi/2)(I_{1x}+I_{2x})} I_{1y} \xrightarrow{\omega_D I_{1z}I_{2z}} I_{1y}\cos(\Delta\omega_D\tau) - 2I_{1x}I_{2z}\sin(\Delta\omega_D\tau) \xrightarrow{-(\pi/2)(I_{1y}+I_{2y})}$$

$$I_{1y}\cos(\Delta\omega_D\tau) + 2I_{1x}I_{2z}\sin(\Delta\omega_D\tau) \xrightarrow{\omega_D I_{1z}I_{2z}} I_{1y}. \qquad (2.91)$$

For quadrupolar echoes the same 90_x–τ–90_y–τ sequence suffices to produce an echo. Here the dephasing over a time τ due to a spread in ω_Q following the 90_x pulse causes $T_1^1(s)$ (i.e., I_y) to precess into an admixture of $T_1^1(s)$ and $T_2^1(a)$ as illustrated in Fig. 2.23. The 90_y pulse has the effect of inverting the sign of $T_2^1(a)$ terms while leaving the $T_1^1(s)$ terms unchanged, a result equivalent to precisely reversing all the precession phases. Consequently an echo is formed at an equal time τ following the 90_y r.f. pulse.

2.6.10 *Multiple pulse line-narrowing and magic angle spinning*

Toggling the dipolar Hamiltonian

In the liquid state the rapid fluctuation of the dipolar interaction caused by molecular tumbling results in an average Hamiltonian in which the dipolar interaction is effictively zero. In order to produce this averaging it is necessary that the fluctuation rate, τ_c^{-1}, be greater than the dipolar linewidth, $M_2^{1/2}$. In 1966 Ostroff and Waugh[53], and Mansfield and Ware[54] discovered that by application of a suitable train of r.f. pulses, the effective dipolar spin Hamiltonian could be made to fluctuate in a controlled way thus leading to dipolar line narrowing. The ability to slow the decay of the transverse magnetization and hence to 'narrow the line' arises from the fact that, in the solid state, the dephasing of the FID is caused by static interactions which constitute an inhomogeneous broadening. Given the right pulse sequence these phase shifts can always be reversed, the solid echo being a simple example of such an approach.

A thorough description of multiple pulse line-narrowing is given in the book by Mehring[17] and the reader wishing to understand this subject in depth will find this text particularly helpful. Here we briefly review the essentials. The key element in the method is to introduce a time-dependence in the dipolar Hamiltonian which is periodic and then to strobe the signal acquisition synchronously with the period of this Hamiltonian. Time-dependence is caused by a sequence of r.f. pulses since these rotate the spin orientations and hence the vectors $\mathbf{I}_i$ and $\mathbf{I}_j$ in eqn (2.43). Of course these pulses not only affect the Hamiltonian but also cause a sudden transformation in the density matrix as well. We can think of this by saying that in the rotating frame the total evolution operator is a product $U_{rf}(t)\,U_D(t)$ where $U_{rf}(t)$ is the evolution operator representing the cumulative effect of the r.f. pulses and $U_D(t)$ is the evolution operator for the 'toggled' dipolar interaction. In order to separate the evolutions in this way, the appropriate time-dependent form for the dipolar Hamiltonian turns out to be

$$H_D^*(t) = U_{rf}^{-1}(t)\,H_D\,U_{rf}(t). \tag{2.92}$$

A key point to note is that, in comparison with density matrix evolution, the order of the U operators is reversed.

Before finding $U_D(t)$ we consider the role of $U_{rf}(t)$ in the product $U_{rf}(t)\,U_D(t)$. The longer-term influence of U_{rf} can be removed by making the r.f. pulse sequence cyclic, such that its associated evolution operator, $U_{rf}(t)$ is unity, where t is a multiple of the cycle time, t_c. Provided the signal is sampled stroboscopically at multiples of this period, the net effect on the density matrix is due only to the evolution caused by $U_D(t)$. The manner in which this fluctuation is introduced must have the symmetry necessary to cancel the dipole–dipole interaction and so make $U_D(t)$ equal to the identity operator.

Average Hamiltonian

Each time an r.f. pulse is applied, the Hamiltonian suddenly changes with the time dependence given by eqn (2.92). If we imagine that the r.f. pulses are infinitesimally narrow then $H_D^*(t)$ becomes 'piecewise constant' such that

$$H_D^*(t) = H_{D_k}^* \qquad \text{for } (\tau_1 + \tau_2 + \ldots + \tau_{k-1}) < t < (\tau_1 + \tau_2 + \ldots + \tau_k). \tag{2.93}$$

To understand how the evolution of the density matrix occurs we return to eqn (2.28). The relevant evolution operator is

$$U_D(t) = \exp(-iH_{D_n}^*\tau_n)\ldots\exp(-iH_{D_1}^*\tau_1) \tag{2.94}$$

where

$$t = \sum_{k=1}^{n} \tau_n. \tag{2.95}$$

Of course, the order of these successive evolution operators is crucial. Because a product of unitary transformations is itself unitary, the product in eqn (2.94) can be written as a single operator

$$U(t) = \exp[-i\overline{H_D}(t)t]. \tag{2.96}$$

$\overline{H_D}(t)$ is the average Hamiltonian.[55] In the case of the dipolar interaction we attempt to make this zero. $\overline{H_D}(t)$ is not simply the time average of the various $H_{D_k}^*$, although such an average is one of the leading terms. It can be shown that $\overline{H_D}(t)$ comprises a sum of terms, $\overline{H_D}^{(0)} + \overline{H_D}^{(1)} + \overline{H_D}^{(2)} + \ldots$, where, for example,

$$\overline{H_D}^{(0)} = \frac{1}{t}\{H_{D_1}^*\tau_1 + H_{D_2}^*\tau_2 \ldots + H_{D_n}^*\tau_n\}$$

$$\overline{H_D}^{(1)} = -\frac{i}{2t}\{[H_{D_2}^*\tau_2, H_{D_1}^*\tau_1] + [H_{D_3}^*\tau_3, H_{D_1}^*\tau_1] + [H_{D_3}^*\tau_3, H_{D_2}^*\tau_2] + \ldots\}. \tag{2.97}$$

The ideal r.f. pulse sequence is one which renders all terms $\overline{H_D}^{(k)}$ zero. Removal of the zeroth-order term is relatively straightforward, and can be achieved with a cycle of four pulses. Provided this cycle is symmetric in the sense that $H_D^*(t) = H_D^*(t_c - t)$ then all odd-order terms also vanish. Removal of higher even-order terms can, in principle, be achieved by including more pulses in the cycle, but often the cumulative effect of r.f. pulse imperfections leads to additional decay of the transverse magnetization.

WHH-4 and MREV-8 sequences

One of the first line-narrowing sequences proposed was the WHH cycle[56] shown in Fig. 2.24. The dipolar Hamiltonian starts in the form given by

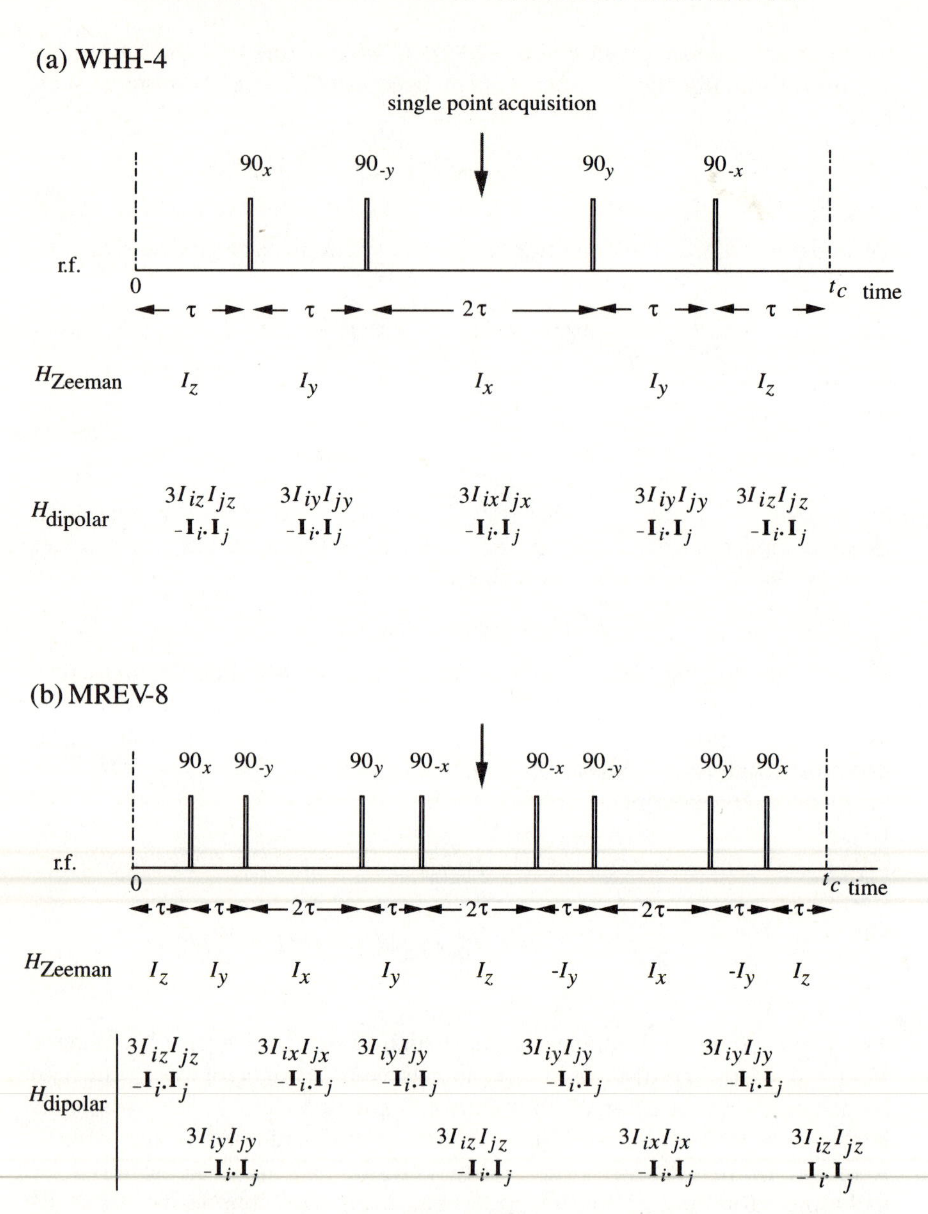

Fig. 2.24 Pulse sequences for (a) WHH-4 and MREV-8 dipolar line narrowing. The states of the Zeeman and dipolar Hamiltonians are shown below each respective precession frequency. The averaged H_{dipolar} is zero in both sequences while the averaged Zeeman Hamiltonian is $(1/3)\gamma B_0(I_x+I_y+I_z)=(1/\sqrt{3})\gamma B_0\ I_{z'}$ for WHH-4, and $(1/3)\gamma B_0(I_x+I_y)=(\sqrt{2}/3)\gamma B_0 I_{z''}$ for MREV-8.

eqn (2.50a). The trick involved in this cycle is to cause the $I_{iz}I_{jz}$ terms in H_D to progressively change to $I_{iy}I_{jy}$, then $I_{ix}I_{jx}$, then $I_{iy}I_{jy}$, and returning to $I_{iz}I_{jz}$, at the end of the cycle. Of course, the term $\mathbf{I}_i \cdot \mathbf{I}_j$ is invariant under the rotations caused by the pulses.

The zeroth-order average Hamiltonian term, $\overline{H_D}^{(0)}$ will vanish according to eqn (2.97) if the respective times spent in these states is $\tau, \tau, 2\tau, \tau$, and τ. Note that, because of the evolution matrix order reversal represented by eqn (2.92), the effect of the r.f. pulses on the vectors I_{iz}, etc, must also be calculated in reverse order. For the WHH sequence the r.f. pulses are antisymmetric such that pulse 4 is the inverse of pulse 1 and pulse 3 is the inverse of pulse 2. The piecewise constant $H^*_{D_k}$ terms are therefore

$$\begin{aligned}
H^*_{D_1} &= H_D \\
H^*_{D_2} &= U_{rf}(1)^{-1} H_D U_{rf}(1) \\
H^*_{D_3} &= U_{rf}(1)^{-1} U_{rf}(2)^{-1} H_D U_{rf}(2) U_{rf}(1) \\
H^*_{D_4} &= U_{rf}(1)^{-1} U_{rf}(2)^{-1} U_{rf}(3)^{-1} H_D U_{rf}(3) U_{rf}(2) U_{rf}(1) \\
&= U_{rf}(1)^{-1} H_D U_{rf}(1) \\
H^*_{D_5} &= U_{rf}(1)^{-1} U_{rf}(2)^{-1} U_{rf}(3)^{-1} U_{rf}(4)^{-1} H_D U_{rf}(4) U_{rf}(3) U_{rf}(2) U_{rf}(1) \\
&= H_D
\end{aligned} \tag{2.98}$$

So far we have considered the Hamiltonian in the rotating frame to be dominated by the secular part of the dipolar interaction. If this is largely removed by multiple pulse line narrowing then the weaker chemical shift terms can be revealed. Of course, these terms will also be affected by the r.f. pulse train, but, because the symmetry of the Zeeman and dipolar interactions differ, the average Hamiltonian is not necessarily zero. Writing the total (isotropic and anisotropic) chemical shift Hamiltonian, H_{CS}, of eqn (2.57) as $\delta_{CS} I_z$ and including the effect of any resonant offset term, $\Delta\omega_0 I_z$, the total rotating frame Zeeman Hamiltonian is,

$$H_Z = (\delta_{CS} + \Delta\omega_0) I_z. \tag{2.99}$$

It is a simple exercise to show that $\overline{H_Z}^{(0)}$ is $\frac{1}{3}(\delta_{CS} + \Delta\omega_0)(I_x + I_y + I_z)$ which results in a precession about a tilted axis I_z' in the rotating frame with effective frequency $1/\sqrt{3}(\delta_{CS} + \Delta\omega_0)$. The chemical shift and resonance offset do not vanish under the WHH-4 sequence but are reduced by a factor of $1/\sqrt{3}$. Normally, in solid state samples, the chemical shift is 'buried' in the much larger dipolar linewidth. Where information about H_{CS} is sought in these systems, retention of the isotropic and anisotropic chemical shift, albeit somewhat attenuated, is a valuable feature of pulse line-narrowing experiments.

The WHH-4 cycle does not remove $\overline{H_D}^{(2)}$. Furthermore it suffers from the effects of r.f. pulse imperfections. A noticeable improvement results in using an eight-pulse cycle such as the MREV-8 method proposed independently by Mansfield[57,58] and by Rhim, Elleman, and Vaughan.[59] Although MREV-8 also does not remove $\overline{H_{\mathrm{D}}}^{(2)}$ it has the added advantage of being less adversely affected by the effects of finite r.f. pulse widths, r.f. inhomogeneity, and pulse phase deviations. This particular sequence is shown in Fig. 2.24(b) In the MREV-8 cycle the chemical shift and resonance offsets are scaled by $\sqrt{2}/3$ in comparison with the $1/\sqrt{3}$ factor in WHH-4.

Sample rotation about the magic angle (MAS)

If the NMR sample is spun about an axis inclined to the polarizing field, the secular terms in the Hamiltonian are transformed from the sample frame into the laboratory frame in which the dominant Zeeman interaction is stationary. This produces an admixture of tensors, some of which are modulated at the spinning rate. According to the argument outlined at the end of Section 2.2, these oscillating terms can be neglected provided that the spinning rate is larger than the strength of the secular term expressed in frequency units. At lower rotation speeds these oscillating terms result in spectral sidebands which can be removed by stroboscopically sampling the NMR signal synchronously with the rotor frequency.

We will presume that the oscillating terms vanish, although it should be noted that the maximum achievable rotor speeds of around 20 kHz are insufficient to remove these in the case of proton–proton dipolar couplings in rigid solids. Such speeds are, however, quite sufficient in the case of anisotropic chemical shift effects and ^{1}H–^{13}C dipolar couplings for ^{13}C spins in solids. They can also be effective for proton dipolar interactions in non-rigid solids, such as polymeric elastomers.

The terms which remain following transformation from the sample frame to the laboratory frame are those which are independent of the rotor angle. These are 'zero-component' tensors, which, in the case of the secular anisotropic chemical shift and dipolar interactioin terms, correspond to T_2^0 and transform as $\frac{1}{2}(3\cos^2\theta - 1)$ where θ is the angle at which the rotor axis is inclined to the polarizing field, $\mathbf{B}_0$. Andrew and co-workers[60–62] have exploited this dependence, demonstrating than when the rotor axis is inclined at the 'magic angle' (about 54°), such that $3\cos^2\theta - 1$ is zero, then these remaining terms also disappear, leaving a high-resolution spectrum in which only the isotropic chemical shift and other Zeeman offsets (for example, due to $\mathbf{B}_0$ inhomogeneity) remain. Magic-Angle Spinning (MAS) has been widely applied to obtain ^{13}C high-resolution spectra in solids. While it cannot compete with multiple pulse methods in achieving dipolar line-narrowing for protons in rigid solids, it is very effective where the dipolar width has already been reduced by molecular motional narrowing. Both MAS and multiple pulse methods have been applied to obtain sufficient line-narrowing for the imaging of protons in solids. These applications will be discussed in Chapter 5.

2.7 References

1. Abragam, A. (1961). *Principles of nuclear magnetism*, Clarendon Press, Oxford.
2. Slichter, C. P. (1963). *Principles of magnetic resonance*, Harper and Row, New York.
3. Farrar, T. C. and Becker, E. D. (1971). *Pulse and Fourier transform NMR*, Academic Press, New York.
4. Shaw, D. (1984). *Fourier transform NMR spectroscopy*, (2nd edn) Elsevier, Amsterdam.
5. Bax, A. (1984). *Two-dimensional nuclear magnetic resonance in liquids*, Delft University Press, Dordrecht.
6. Ernst, R. R., Bodenhausen, G., and Wokaun, A. (1987). *Principles of nuclear magnetic resonance in one and two dimensions*, Clarendon Press, Oxford.
7. Schiff, L. I. (1968). *Quantum mechanics*, McGraw-Hill, New York.
8. Edmonds, A. R. (1974). *Angular momentum in quantum mechanics*, (2nd edn). Princeton University Press, Princeton.
9. Rose, M. E. (1957). *Elementary theory of angular momentum*, Wiley, New York.
10. Sorensen, O. W., Eich, G. W., Levitt, M. H., Bodenhausen, G., and Ernst, R. R. (1983). *Progr. NMR Spectroscopy* **16**, 163.
11. Bloembergen, N., Purcell, E. M., and Pound, R. V. (1948). *Phys. Rev*, **73**, 679.
12. Hoult, D. I. (1978). *Progr. NMR Spectr.* **12**, 41.
13. Hayes, C. E., Edelstein, W. A., Schenck, J. F., Mueller, O.M., and Eash, M. (1985). *J. Magn. Reson.* **63**, 622.
14. Ellett, J. D., Gibby, M. G., Haeberlen, U., Huber, L. M., Mehring, M., Pines, A., and Waugh, J. S. (1971). *Adv. Magn. Reson.* **5**, 117.
15. Hoult, D. I. and Richards, R. E. (1975). *Proc. Roy. Soc. (London)* **A344**, 311.
16. Stejskal, E. O. and Schaefer, J. (1974). *J. Magn. Reson.* **14**, 160.
17. Mehring, M. (1982). *High resolution NMR in solids*, Springer (2nd edn)., Berlin.
18. Mason, J. (1987). *Multinuclear NMR*, Plenum, New York.
19. Speiss, H. W. (1978). *NMR Basic Principles Prog.* **15**, 55.
20. Buckmaster, H. A., Chatterjee, R. and Shing, Y. H. (1972). *Phys. Status Solidi* **113**, 9.
21. Sanctuary, B. C. (1985). *Mol. Phys.* **55**, 1017.
22. Bowden, G. J. and Hutchison, W. D. (1986). *J. Magn, Reson.* **67**, 403.
23. Bowden, G. J., Hutchison, W. D. and Kachan, J. (1986). *J. Magn, Reson.* **67**, 415.
24. Bowden, G. J. and Hutchison, W. D. (1986). *J. Magn. Reson.* **70**, 341.
25. Bowden, G. J. and Hutchison, W. D. (1987). *J. Magn. Reson.* **72**, 61.
26 Wolf, D. (1979). *Spin-temperature and nuclear-spin relaxation in matter*, Clarendon Press, Oxford.
27. Look, D. C. and Lowe, I. J. (1966). *J. Chem. Phys.* **44**, 2995.
28. Pines, D. and Slichter, C. P. (1955). *Phys. Rev.* **100**, 1014.

29. Hebel, L. C. and Slichter, C. P. (1959). *Phys. Rev.* **116**, 583.
30. Anderson, P. W. and Weiss, P. R. (1953). *Rev. Mod. Phys.* **25**, 269.
31. Freeman, R. and Hill, H. (1971). *J. Chem. Phys.* **54**, 3367.
32. Carr, H. Y. (1958). *Phys. Rev.* **112**, 1693
33. Turner, C. J. and Patt. S. L. (1989). *J. Magn. Reson.* **85** 492.
34. Zur, Y. and Stokar, S. (1987). *J. Magn. Reson.* **71**, 212.
35. Vold, R. L., Waugh, J. S., Klein, M. P. and Phelps, D. E. (1968). *J. Chem. Phys.* **48**, 383.
36. Hahn, E. L. (1950). *Phys. Rev.* **80**, 580.
37. Carr, H. Y. and Purcell, E. M. (1954). *Phys. Rev.* **94**, 630.
38. Meiboom, S. and Gill, D. (1959). *Rev. Sci. Instr.* **29**, 688.
39. Frahm, J., Merboldt, H-D., Hanicke, W., and Haase, A. (1985). *J. Magn. Reson.* **64**, 81.
40. Hinshaw, W. S. (1976). *J. Appl. Phys.* **47**, 3709.
41. Mansfield, P. and Morris P. G. (1982). *NMR imaging in biomedicine*, Academic Press, New York.
42. Becker, E. D., Ferretti, J. A. and Farrar, J. C. (1969). *J. Am. Chem. Soc.* **91**, 7784.
43. Shoup, R. R., Becker, E. D. and Farrar, T. C. (1972). *J. Magn. Reson.* **8**, 298.
44. Jeener, J. (1971). *Ampère International Summer School*, Basko Polje, Yugoslavia.
45. Aue, W. P., Bartholdi, E. and Ernst, R. R. (1976). *J. Chem. Phys.* **64**, 2229.
46. Bodenhausen, G. (1981). *Progr. NMR. Spectr.* **14**. 137.
47. Weitekamp, D. P. (1983). *Adv. Magn. Reson.* **11**, 111.
48. Piantini, U., Sorensen, O. W. and Ernst, R. R. (1982). *J. Am. Chem. Soc.* **104**, 6800.
49. Shaka, A. J. and Freeman, R. (1983). *J. Magn. Reson.* **51**, 169.
50. Rance, M., Sorensen, O. W., Bodenhausen, G., Wagner, G., Ernst, R. R. and Wütrich, K. (1983). *Biochem, Biophys. Res, Commun.* **117**, 479.
51. Hall, L. D. and Norwood, T. J. (1985). *J. Magn. Reson.* **67**, 382.
52. Sanctuary, B. C. (1983). *Mol. Phys.* **48**, 1155.
53. Ostroff E. D. and Waugh, J. S. (1966). *Phys. Rev. Lett.* **16**, 1097.
54. Mansfield, P. and Ware, D. (1966). *Phys. Lett.* **162**, 209.
55. Haeberlen, U. and Waugh, J. S. (1968). *Phys. Rev.* **175**, 453.
56. Waugh, J. S., Huber, L. M. and Haeberlen. U. (1968). *Phys. Rev. Lett.* **20**, 180.
57. Mansfield, P. (1971). *J. Phys. C.* **4**, 1444.
58. Mansfield, P., Orchard, M. J., Stalker, D. C. and Richards, K. H. B. (1973). *Phys. Rev.* **B7**, 90.
59. Rhim, W. K., Elleman, D. D. and Vaughan, R. W. (1973). *J. Chem. Phys.* **59**, 777.
60. Andrew, E. R., Bradbury, A. and Eades, R. G. (1958). *Nature* **182**, 1659.
61. Andrew, E. R., Bradbury, A. and Eades, R. G. (1959). *Nature* **183**, 1802.
62. Andrew, E. R. (1971). *Progr. NMR Spectr.* **8**, 1.

3
THE INFLUENCE OF MAGNETIC FIELD GRADIENTS

3.1 Spin density and k-space

In conventional NMR spectroscopy the spectrum of nuclear precession frequencies gives information about the chemical environment of the spins and in order to ensure that each chemically identical nucleus has a common behaviour, it is important to arrange that the polarizing magnetic field, B_0, should vary as little as possible across the sample. The removal of field inhomogeneities by careful adjustment of the currents in the magnet shim coils represents a challenge both to the magnet manufacturer and the NMR spectroscopist alike.

In this book we are concerned with magnetic field profiles which have been equally carefully tailored but which, by contrast, have been purposely designed to vary linearly across the sample space. This means that the Larmor frequencies of the spins will show a similar spatial dependence. The linearly varying field, known as a field gradient, is applied independently of the much larger polarizing field by means of specially shaped coils. Because the gradient causes additional fields much smaller than the polarizing field magnitude, B_0, the Larmor frequency is affected only by any components parallel to $\mathbf{B}_0$. Orthogonal components only have the effect of slightly tilting the net field direction. Hence we define the local Larmor frequency as

$$\omega(\mathbf{r}) = \gamma B_0 + \gamma \mathbf{G} \cdot \mathbf{r} \tag{3.1}$$

where $\mathbf{G}$ is defined in the usual manner as the **grad** of the pulsed gradient field component parallel to $\mathbf{B}_0$. This simple linear relation between the Larmor frequency and the nuclear spin coordinates, $\mathbf{r}$, lies at the heart of the imaging principle. In this chapter we shall examine how an image of the spin density can be reconstructed from the NMR signal. Since the original imaging papers in 1973,[1,2] various imaging methods have been proposed. These are discussed in a number of reviews[3-8] and we shall here be concerned only with those methods of sufficient sensitivity that they represent viable approaches to microscopy.

3.1.1 *Conjugate spaces in reconstruction*

Consider the nuclear spins at position $\mathbf{r}$ in the sample, occupying a small element of volume dV. If the local spin density is $\rho(\mathbf{r})$ then there will be

$\rho(\mathbf{r})\mathrm{d}V$ spins in this element. Following eqn (2.42) the NMR signal from this element may be written as

$$\mathrm{d}S(\mathbf{G}, t) \propto \rho(\mathbf{r})\mathrm{d}V \exp[\mathrm{i}\omega(\mathbf{r})t]. \tag{3.2}$$

For simplicity we shall neglect the constant of proportionality in eqn (3.2) and write

$$\mathrm{d}S(\mathbf{G}, t) = \rho(\mathbf{r})\mathrm{d}V \exp[\mathrm{i}(\gamma B_0 + \gamma \mathbf{G}\cdot\mathbf{r})t]. \tag{3.3}$$

Since $\mathrm{d}S(\mathbf{G}, t)$ is real at the origin of time, eqn (3.3) assumes that the detector phase has been 'correctly set'. This is never a problem in practice since the adjustment may always be made at a later stage in the treatment of the data by the process of 'phasing' outlined in Section 2.3.3. Such a process ensures that when we come to compute $\rho(\mathbf{r})$ we will find it real rather than complex.

Of course, eqn (3.3) makes no allowance for the decay of the signal due to transverse relaxation. This is fine provided that our gradient is sufficiently large. Put another way, we can say that the dephasing of the transverse magnetization due to the spread in $\gamma\mathbf{G}\cdot\mathbf{r}$ is very much more rapid than that due to T_2. Chapter 4 deals with the effect of relaxation more precisely but for the moment it is convenient to make this simple approximation.

Finally, before embarking on our analysis of the effects of magnetic field gradients, we note that in phase-sensitive detection the radio-frequency signal is 'mixed' with a reference oscillation and the result is a signal at the difference frequency. This process is known as heterodyne mixing. By choosing the reference frequency to be γB_0, the so-called 'on-resonance' condition, the signal finally obtained oscillates at $\gamma\mathbf{G}\cdot\mathbf{r}$. This will be in the audio-frequency range and therefore easily amenable to direct analogue-to-digital conversion. In the present context, this detection process means that we may neglect the γB_0 term in eqn (3.3) and so write the integrated signal amplitude as

$$S(t) = \iiint \rho(\mathbf{r}) \exp[\mathrm{i}\gamma\mathbf{G}\cdot\mathbf{r}t]\,\mathrm{d}\mathbf{r} \tag{3.4}$$

where the symbol $\mathrm{d}\mathbf{r}$ is once again used to represent volume integration. This sum of oscillating terms has the form of a Fourier transformation. To make this more obvious, Mansfield[2,9-11] introduced the concept of a reciprocal space vector, $\mathbf{k}$, given by

$$\mathbf{k} = (2\pi)^{-1}\gamma\mathbf{G}t. \tag{3.5}$$

The $\mathbf{k}$-vector has a magnitude expressed in units of reciprocal space, m^{-1},and it is clear that $\mathbf{k}$-space may be traversed by moving either in time or in gradient magnitude. However, the direction of this traverse is determined by the direction of the gradient $\mathbf{G}$.

In the formalism of **k**-space, and using the concept of the Fourier transform and its inverse,

$$S(\mathbf{k}) = \iiint \rho(\mathbf{r}) \exp[\mathrm{i}2\pi\mathbf{k}\cdot\mathbf{r}]\, \mathrm{d}\mathbf{r} \tag{3.6a}$$

$$\rho(\mathbf{r}) = \iiint S(\mathbf{k}) \exp[-\mathrm{i}2\pi\mathbf{k}\cdot\mathbf{r}]\, \mathrm{d}\mathbf{k}. \tag{3.6b}$$

Eqn (3.6) states that the signal, $S(\mathbf{k})$, and the spin density, $\rho(\mathbf{r})$, are mutually conjugate. It is the fundamental relationship of NMR imaging. To use X-ray terminology, $S(\mathbf{k})$ represents the reciprocal lattice of $\rho(\mathbf{r})$. Of course eqns (3.6a) and (3.6b) represent ideality. In a real experiment the experimentally observed signal, $S(\mathbf{k})$, will not be a perfect representation of the Fourier transform of $\rho(\mathbf{r})$.

In practice, the sampling of **k**-space takes place as we sample the NMR signal at successive time intervals. Because of this, $S(\mathbf{k})$ is measured in the time domain while the Fourier transform, which yields $\rho(\mathbf{r})$, is therefore in the frequency domain. Consequently $\rho(\mathbf{r})$ is some sort of 'three-dimensional spectrum' of $S(\mathbf{k})$. In this simple sense we can say that there is a correspondence between real space and frequency, and between reciprocal space and time. Actually, as is clear from eqn (3.5), both **G** and t are involved in determining **k**-space.

Fig. 3.1 shows a one-dimensional visualization of the relationship expressed in eqn (3.6a). For convenience we use Cartesian coordinates and set **G** in the z-direction. Thus

$$S(k_z) = \int_{-\infty}^{\infty}\int_{-\infty}^{\infty}\left[\int_{-\infty}^{\infty} \rho(x,y,z) \exp[\mathrm{i}2\pi k_z z]\, \mathrm{d}z\right] \mathrm{d}x\, \mathrm{d}y \tag{3.7}$$

and the spectrum of this signal is

$$\mathfrak{F}\{S(k_z)\} = \int_{-\infty}^{\infty}\int_{-\infty}^{\infty} \rho(x,y,z)\, \mathrm{d}x\, \mathrm{d}y \tag{3.8}$$

where use has been made of the linearity of the Fourier transform and integral operations.

Eqn (3.8) states that the spectral data may, therefore, be regarded as the projection of the spin density on to the axis defined by the gradient direction. For this reason it is referred to as the 'projection profile'. The similarity with the photon intensity profile in X-ray tomography is obvious. However, it is important to note that, unlike X-ray tomography, the NMR data is acquired, in the first instance, in the conjugate domain.

In eqn (3.2) it was assumed that the NMR signal was simply proportional

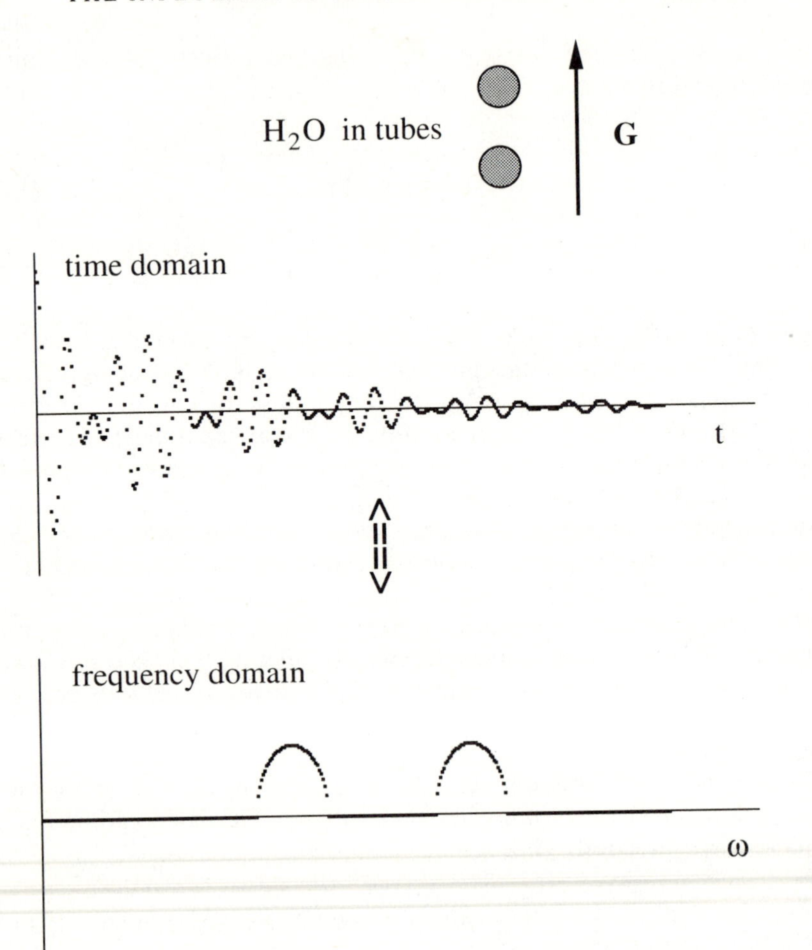

Fig. 3.1 Fourier relationship between time domain signal and the one-dimensional spin density when the FID is acquired in the presence of a magnetic field gradient. The frequency domain spectrum is a profile corresponding to $\rho(\mathbf{r})$ projected normal to the direction of the gradient, **G**.

to the spin density, $\rho(\mathbf{r})$, where the constant of proportionality had no spatial dependence. In Chapter 2 we saw that there are many physical parameters which can affect the NMR signal phase and amplitude due to spin interactions. Examples include spin relaxation, dipolar and scalar coupling interactions, and spin translation in the presence of magnetic field gradients. In every case such modulation of the signal requires a finite time evolution for effect so that, in principle, by acquiring the signal sufficiently rapidly after the excitation pulse, these effects can be overwhelmed and $\rho(\mathbf{r})$ is imaged. In many situations, however, we shall wish to deliberately measure the spatial dependence of these interactions by inserting some suitable evolution time in the imaging experiment. The details need not concern us here except to observe that spatially dependent modulation of the proportionality in eqn (3.2) may be incorporated by including a 'contrast factor' $C(\mathbf{r})$. This does not in any way change the arguments which lead to eqn (3.6b) except that what is imaged will be $C(\mathbf{r})\rho(\mathbf{r})$, the contrasted spin density, rather than the simple spin density, $\rho(\mathbf{r})$. Clearly, by normalizing images obtained with and without the contrast effect, a map of $C(\mathbf{r})$ can be obtained.

3.1.2 *Efficiency*

The procedure represented by eqn (3.8) is often referred to as one-dimensional imaging and leads to the projected distribution of spins rather than the exact point-by-point density, $\rho(x, y, z)$ (or alternatively the exact point-by-point contrast $C(x, y, z)$). To achieve such a three-dimensional representation we can take one of two paths. The first, known as 'line-scanning imaging' involves selectively exciting a thin strip of spins along the z-direction.[12-17] For the moment we shall not be concerned with the means by which such excitation might take place. If this strip is moved stepwise, a series of projection profiles could be obtained which would, eventually, cover the entire three-dimensional spin distribution, but such a line-scanning method is very wasteful of time since one only ever acquires a signal from part of the sample during each part of the measurement. Efficient signal acquisition is fundamental to achieving high-resolution images.

In the second, more efficient, approach the signal is acquired simultaneously from all spins in the region of interest and must be mathematically processed in such a way as to unravel the spin density. Generally the region of interest will be a plane of the sample. This is because we will probably process and display our image in two dimensions. For example, if we want 256^2 of 16-bit words, that involves 128-kBytes. By contrast 256^3 corresponds to 32-MBytes! The processing time for such a three-dimensional array can be formidable. None the less, if we wish ultimately to obtain a full three-dimensional density distribution, complete sample excitation and signal acquisition has several advantages in microscopy; not the least of which is the simplicity of the NMR apparatus (if not the computer!) required to perform such an operation. Such methods will be discussed in Section 3.4. For the moment we will focus our attention on the more common two-dimensional approach. The key to this is the excitation of a single, predetermined, layer of spins, a process known as selective excitation.[18-25]

3.2 Selective excitation

3.2.1 *Soft and hard pulses*

Selective excitation involves applying an r.f. pulse which affects only a specific region of the NMR frequency spectrum. By this means only nuclei of a certain chemical shift may be disturbed or, when the spectral properties of the spins are dominated by the spread of Larmor frequencies in the presence of a magnetic field gradient, the selective r.f. pulse may be used to excite only those spins within some specified layer of the sample. We shall refer to these respective uses as chemical selection and slice selection.

Efficient and precise selective excitation is a vital component of most NMR imaging techniques. The principle underlying the excitation of spins in a specified region of the spectrum is as follows. The bandwidth of frequencies contained in an excitation pulse is inversely proportional to the pulse duration. For example, if the 90° pulse has a duration, T, of order 1 ms, then only those spins with a resonant frequency within approximately a 1 kHz bandwidth of the radio-frequency will be stimulated in an appreciable manner. In normal NMR spectroscopy the pulse duration is made sufficiently short that the associated bandwidth covers the chemical shifts of all spins of a given nuclear species. As discussed in Section 2.4.8, the bandwidth of the pulse is simply related to the r.f. amplitude. A non-selective (broadband) 90° pulse will have a very much larger magnitude, B_1, than a selective (narrowband) 90° r.f. pulse since the turn angle is determined by the product $\gamma B_1 T$.

The crucial factor in determining the effect on other spins is the relative size of the r.f. field precession frequency, γB_1, and the resonance offset, $\Delta\omega_0$. Suppose that an r.f. pulse is applied with a reduced bandwidth, but with a magnitude and duration so as to cause 'on-resonant' spins to precess through 90° into the transverse plane. Spins with Larmor frequencies far from the r.f. excitation frequency ($\Delta\omega_0 \gg \gamma B_1$) will experience an effective field in the rotating frame considerably tilted towards the longitudinal axis as shown in Fig. 3.2. Unlike the on-resonant spins they will therefore be tipped through an angle less than 90°. Intense broadband excitation pulses are termed 'hard' pulses while the weak, narrowband pulse is termed 'soft'.

The simplest form of soft pulse is obtained by simply reducing the amplitude and extending the duration of the usual rectangular time domain profile. The corresponding frequency spectrum of this pulse is given by the Fourier transform, namely the sinc function shown in Fig. 3.3(a). Clearly, the weak rectangular pulse suffers from having side lobes, so that, while the majority of the excitation is close to the central frequency, extensive excitation due to the lobes occurs over a wide bandwidth. One solution

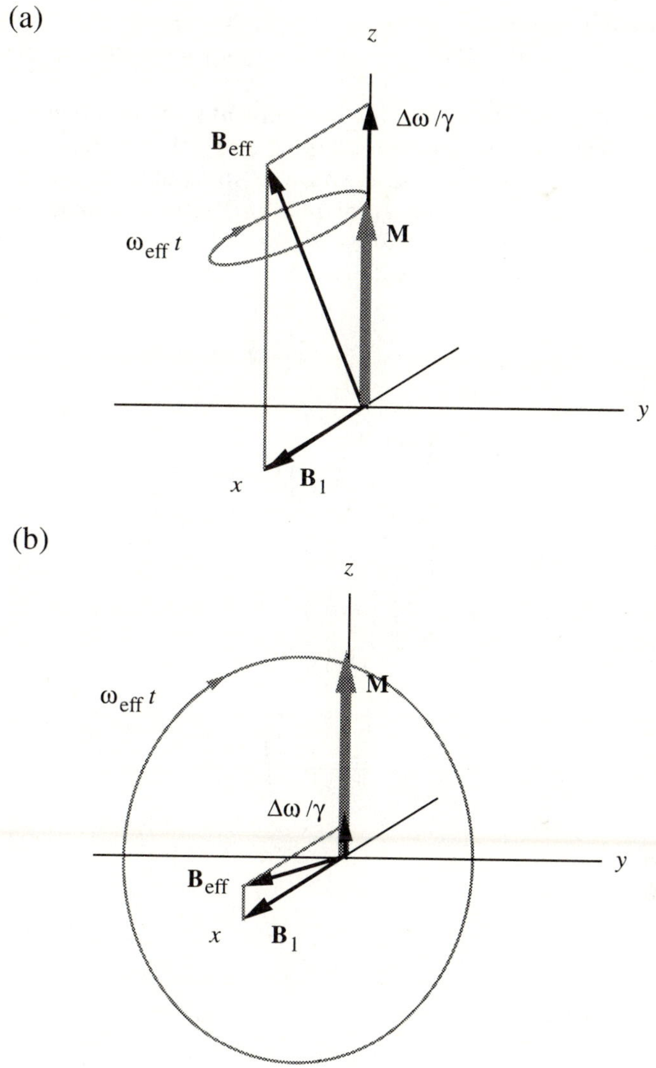

Fig. 3.2 Soft pulse effect showing different magnetization precession in the rotating frame for spins which are (a) off-resonant and (b) on-resonant. The tilting of $\mathbf{B}_{\mathrm{eff}}$ towards the longitudinal axis for off-resonant spins where B_1 is weak, means that these spins are only slightly disturbed from equilibrium. By contrast a hard pulse, with much larger B_1, would have very similar $\mathbf{B}_{\mathrm{eff}}$ orientation for both sets of spins.

is to employ a composite train of rectangular pulses, for example, the Redfield '241', sequence.[26,27] This is an attractive alternative for spectrometers which do not have r.f. modulation capability, but in imaging systems such modulation facilities are always available and lobe suppression is more easily achieved in a single pulse by 'softening' the edges of the r.f. pulse, for example, by the use of Gaussian shaping in the time domain, as shown in Fig. 3.3(b).

Gaussian pulses are very effective in chemical shift selective excitations, discussed in Section 3.5.3 and in Chapter 5. In this section we shall be principally concerned with the application of soft pulses in the presence of a magnetic field gradient. If this gradient is sufficient to spread the Larmor spectrum to a width greater than the excitation bandwidth, then spins in different regions of the sample will be excited differently. Ideally, we would

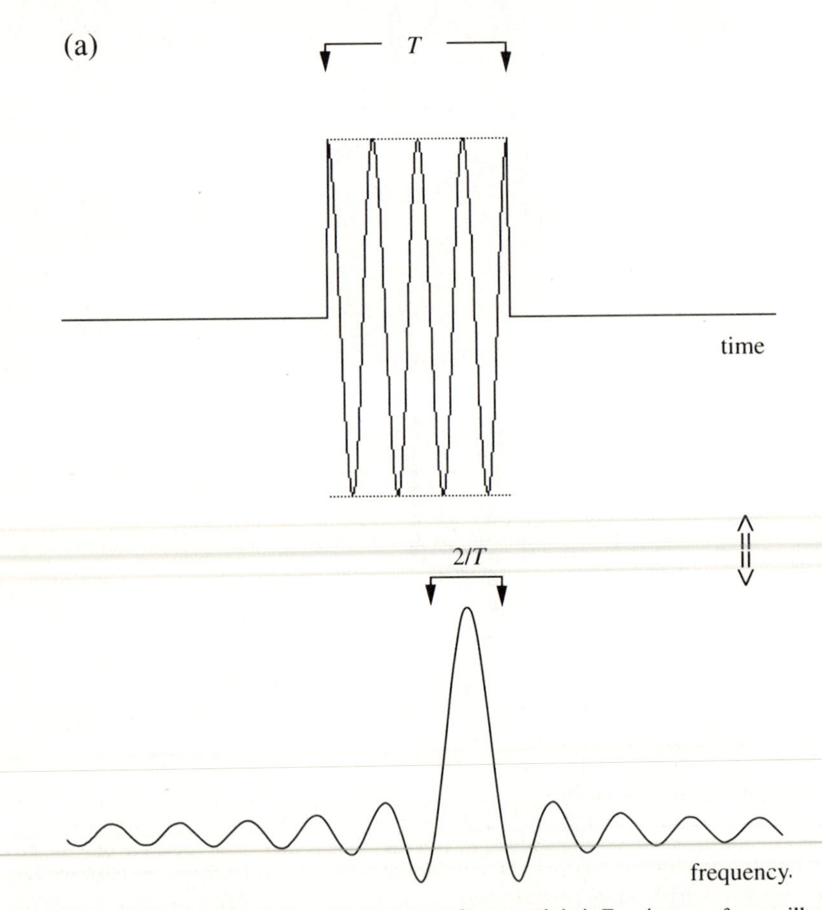

Fig. 3.3 Amplitude-modulated oscillatory wave-forms and their Fourier transforms, illustrating the effect of r.f. pulse shaping. In each case the carrier frequency is deliberately made small so that the effect of the modulation envelope (dotted line) can be seen. Below each time domain wave-form is the associated frequency spectrum: (a) rectangular pulse, (b) Gaussian pulse (width defined by e^{-1} attenuation points) and (c) double lobe truncated sinc pulse. Note the oscillatory modulation in the ideally rectangular frequency response which is caused by the sinc truncation.

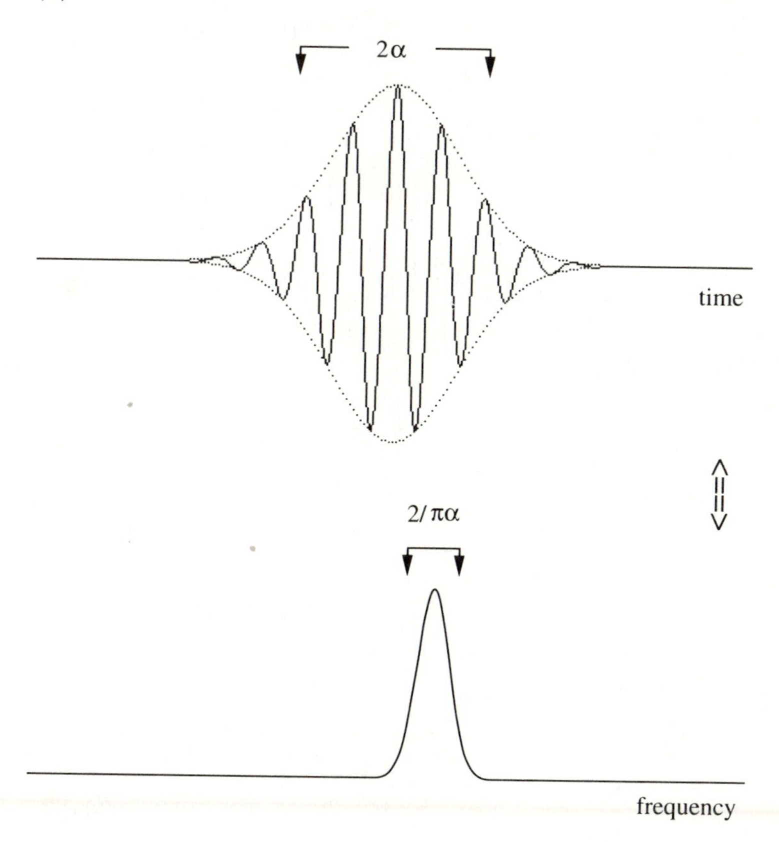

Fig. 3.3(b)

like to precisely stimulate a 'rectangular' slice for which the Gaussian pulse profile is not ideal. Fig. 3.3(c) suggests a means by which this might be performed using sinc modulation.

3.2.2 *Evolution of the magnetization during selective excitation*

Rectangular excitation requires 90° precessions for spins within a well-defined range normal to the slice plane and zero precession for spins outside. This implies that the frequency response of the r.f. pulse should have a rectangular profile and so, in the time domain, should be modulated in sinc ($\sin(at)/(at)$) form, the Fourier transform of the hat function. However, such a model assumes that the nuclear spins behave as a linear system whereas a linear response can only be true for small excitation angles.

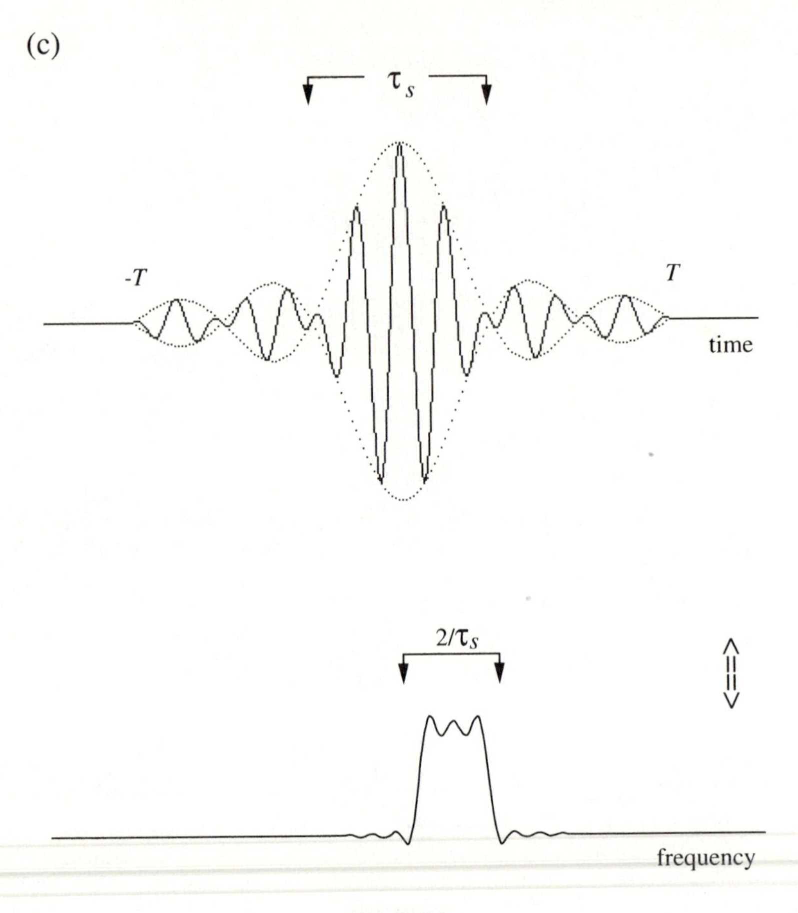

Fig. 3.3(c)

Despite this, the linear assumption is very helpful and we will adopt it for the moment. Fig. 3.4 illustrates this approach. A very nice mathematical formulation of this idea has been given by Bailes and Bryant[8] and is reproduced here.

As noted in Chapter 2, for simplicity we use the same symbols to represent independent coordinates in the laboratory and rotating frames. Magnetization directions, (M_x, M_y, M_z), refer to the rotating frame, while gradient directions, (G_x, G_y, G_z), refer to the laboratory frame and always result in an addition to the polarizing field. Thus the gradient whatever its orientation, always contributes to the z-axis field in the rotating frame. We will assume for convenience that the selection gradient happens to be G_z which means that the z-component of the field in the rotating frame will vary with

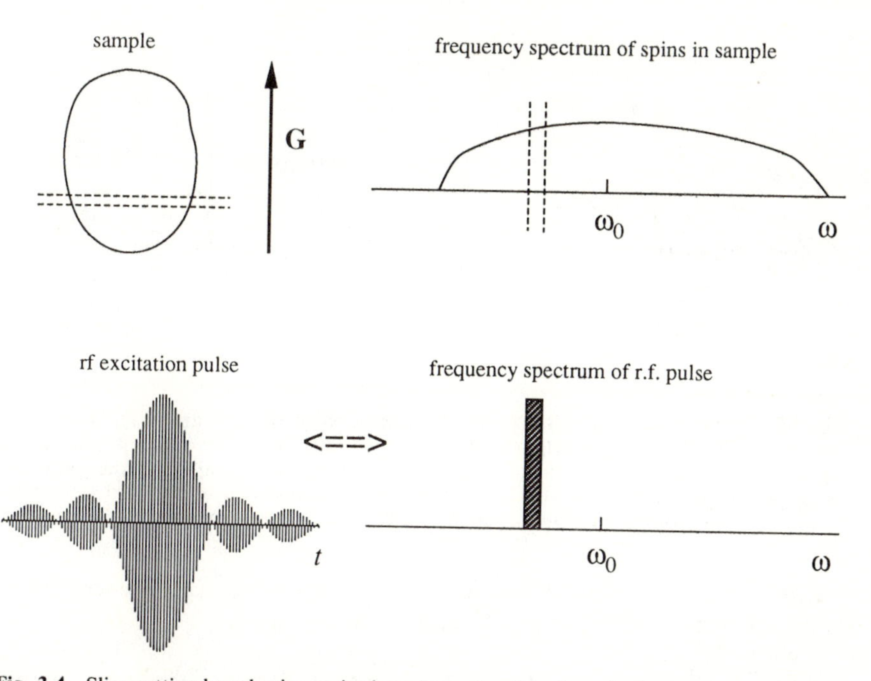

Fig. 3.4 Slice cutting by selective excitation. Assuming that the spins have a linear response, the desired r.f. pulse frequency spectrum should correspond to the desired region to be excited in the Larmor spectrum due to the slice selection gradient, **G**. ω_0 corresponds to the resonant frequency when $G=0$.

position along the laboratory z-axis. The r.f. field defines the x-direction of the rotating frame. Since we are modulating the r.f., B_x will vary with time. This situation is already complicated so we will neglect relaxation to keep things as simple as possible. The Bloch equations of eqn (2.40) now give

$$\frac{\mathrm{d}M_x}{\mathrm{d}t} = \gamma M_y G_z z \tag{3.9a}$$

$$\frac{\mathrm{d}M_y}{\mathrm{d}t} = \gamma (M_z B_1(t) - M_x G_z z) \tag{3.9b}$$

$$\frac{\mathrm{d}M_z}{\mathrm{d}t} = -\gamma M_y B_1(t). \tag{3.9c}$$

The Bailes and Bryant trick involves considering this situation from another frame of reference which rotates about the z-axis at an angular frequency $\gamma G_z z$. The relative orientations of the frames are shown in Fig. 3.5. Of course, each slice normal to z will have a different reference frame (x', y', z')

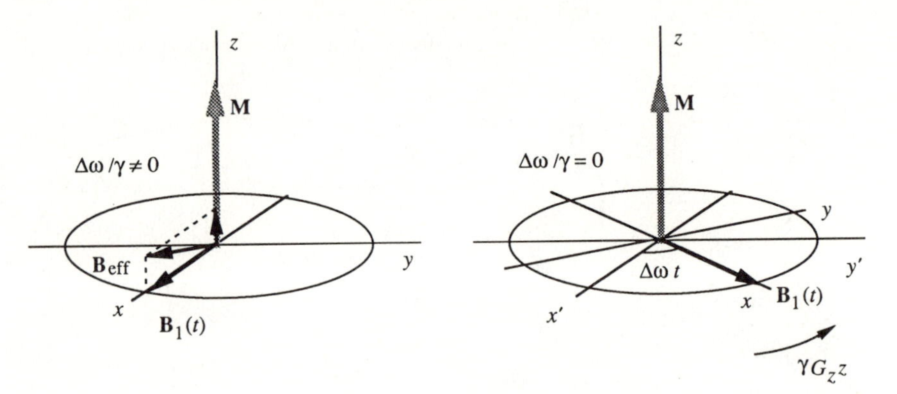

Fig. 3.5 Rotating frames of reference for off-resonant spins corresponding to the isochromat, $\Delta\omega = \gamma G_z z$. The frame on the left is the usual rotating frame (x, y, z) in which the direction of $\mathbf{B}_1$ is stationary but the off-resonant longitudinal field component, $\Delta\omega/\gamma$, causes precession about an inclined effective field. On the right, the magnetization is described in a frame (x', y', z') in which the spins are on-resonant but the B_1 vector is rotating at frequency $\Delta\omega$.

but for the moment we will just focus on a single value of z. The new frame is useful because it makes eqn (3.9) more symmetrical.

$$\frac{\mathrm{d}M_{x'}}{\mathrm{d}t} = -\gamma M_{z'} B_{y'} = -\gamma M_{z'} B_1(t) \sin[\gamma G_z z(t + T)] \tag{3.10a}$$

$$\frac{\mathrm{d}M_{y'}}{\mathrm{d}t} = \gamma M_{z'} B_{x'} = \gamma M_{z'} B_1(t) \cos[\gamma G_z z(t + T)] \tag{3.10b}$$

$$\frac{\mathrm{d}M_{z'}}{\mathrm{d}t} = \gamma (M_{x'} B_{y'} - M_{y'} B_{x'}) \tag{3.10c}$$

where the two reference frames coincide at the start of the pulse when $t = -T$. We will let the pulse end when $t = T$.

The linearity assumption corresponds to saying that the z-component of magnetization changes very little so that $\mathrm{d}M_{z'}/\mathrm{d}t$ is zero and $M_{z'}$ takes the constant, equilibrium, value of M_0. This leads to

$$\frac{\mathrm{d}M_{x'}}{\mathrm{d}t} = -\gamma M_0 B_1(t) \sin[\gamma G_z z(t + T)] \tag{3.11a}$$

$$\frac{\mathrm{d}M_{y'}}{\mathrm{d}t} = \gamma M_0 B_1(t) \cos[\gamma G_z z(t + T)] \tag{3.11b}$$

$$\frac{\mathrm{d}M_{z'}}{\mathrm{d}t} = 0. \tag{3.11c}$$

Eqns (3.11) are easily solved, especially if we treat $M_{x'}$ and $M_{y'}$ as real and imaginary parts of a complex number, M'_+. They become

$$\frac{\mathrm{d}M'_+}{\mathrm{d}t} = \mathrm{i}\gamma M_0 B_1(t)\exp[\mathrm{i}\gamma G_z z(t+T)] \tag{3.12}$$

Integrating and returning to the rotating frame ($M_+(T) = M'_+(T)\exp(-\mathrm{i}\gamma G_z z 2T)$) gives

$$M_+ = \mathrm{i}\gamma M_0 \exp[-\mathrm{i}\gamma G_z zT]\int_{-T}^{T} B_1(t)\exp[\mathrm{i}\gamma G_z zt]\,\mathrm{d}t. \tag{3.13}$$

The integral is simply the Fourier transform, or spectrum, of the r.f. pulse of finite duration (from $-T$ to T). The frequency scale for this spectrum is $\gamma G_z z$, so that the integral represents the amplitude of the r.f. corresponding to the Larmor frequency of the spin plane labelled by z.

Eqn (3.13) is very helpful. It states that M_+, which gives us the FID signal, is proportional to the amplitude of the r.f. spectrum at z. If we want to excite a rectangular slice then we will need a rectangular spectrum. Second, it tells us that the magnetization for a spin plane normal to z has a net phase shift, $\gamma G_z zT$, which is a nuisance since the phase shift will vary across a slice of finite thickness. It is, however, easy to remove, since all we have to do is apply an opposite sign z gradient of magnitude $-G_z$ for a time T. Alternatively we can invert all the spin phases with a 180° r.f. pulse and then apply the identical sign gradient for this same time T. This spin echo technique means that gradient sign reversal is not necessary in the gradient switching apparatus.[21,22,24,28] The significance of these alternative refocusing schemes is discussed in the next section.

An elegant alternative explanation of the selective excitation process and the residual phase shift has been given by Mansfield.[29] His description starts by considering the selective excitation as comprising a number of equally spaced, discrete-frequency, pulsed oscillators, as shown in Fig. 3.6. If they are all initially in phase then the corresponding time domain modulation is a sinc function with maximum value at the time origin. The complete modulation wave-form is therefore a 'split-sinc' function with duration $2T$ such that $\Delta\omega 2T = 2\pi$. The spin phase diagram after switch-off consists of a lobe with net y-magnetization but no x-magnetization. Such an excitation has no residual phase shift but the depicted r.f. modulation is difficult to perform.

The usual sinc modulation, shown in Fig. 3.6, has initially zero amplitude so that the oscillators must start in antiphase. The effect is to place half the y-magnetization along the y-axis and half along the $-y$-axis, leaving zero net magnetization at time $2T$. Refocusing results by reversal of the selection gradient for time T, or the application of a 180° r.f. pulse and a second gradient of similar sign. The term 'spin echo' is commonly used to refer to sequences which use r.f. pulses to invert spin phase although the expression 'r.f. echo' might seem a more appropriate way for these to be distinguished from gradient echo methods.

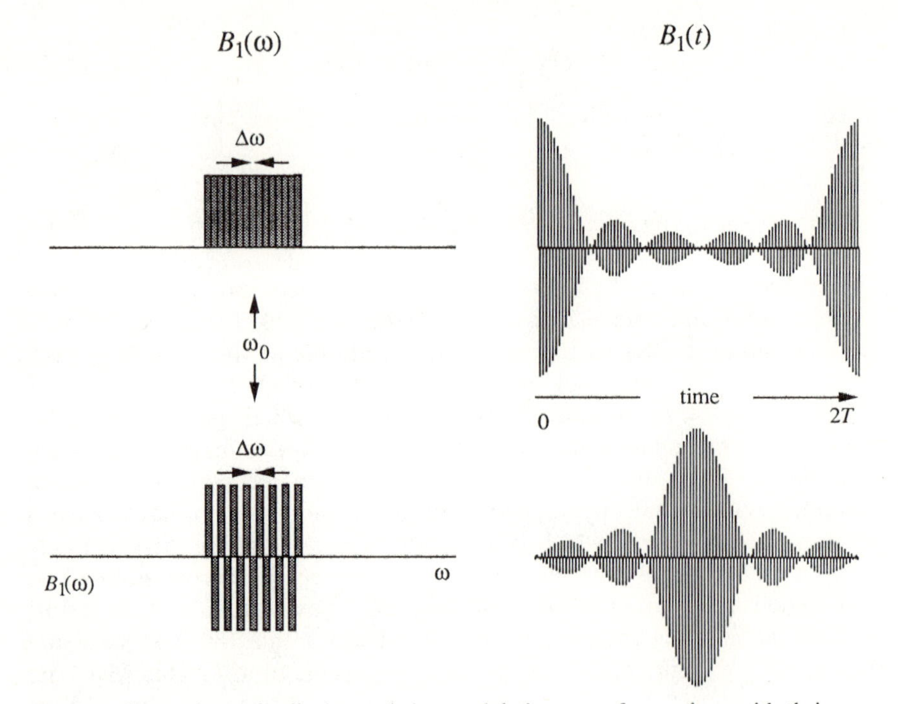

Fig. 3.6 Comparison of split-sinc and sinc modulation wave-forms along with their corresponding frequency domain isochromats. The sinc wave-form is easier to implement but requires that the alternating inverted-phase isochromats have their magnetizations restored from $-y$ to y in the rotating frame. This can be performed using a reversed gradient for a period T. All isochromats along y correspond to frequency offsets which are even multiples of $\Delta\omega$, so that after time T these have precessed an integral number of cycles. By contrast, isochromats along $-y$ are those whose frequency offets are odd multiples of $\Delta\omega$ and thus precess a half-integral number of cycles, returning to y at the end of the rephasing gradient.

Of course, all the theory from eqn (3.11) onwards applies only for small tip angles and certainly not for 90° pulses. A variety of mathematical tricks is available to solve the more exact problem, including the use of perturbation theory.[23] Probably the most direct approach is to find a numerical solution to the Bloch equations of eqn (3.10). Fig. 3.7(a) shows the result of such a calculation in which M_x, M_y, and M_z are plotted as a function of position (i.e., offset frequency) for tip angles of 30° and 90°. Remarkably, the result is not too dissimilar to the linear prediction, hence justifying this rather simple but illuminating approach.

A couple of features of Fig. 3.7(a) are worth noting. The small oscillation in the magnetization along the spatial axis is caused by the need to truncate the sinc r.f. pulse. In this case just two side-lobes are used. The 'non-linearity' of the spin response shows up in the 'out-of-phase' magnetization, M_x. According to eqn (3.13), if $B_1(t)$ is symmetric about $t = 0$ then the

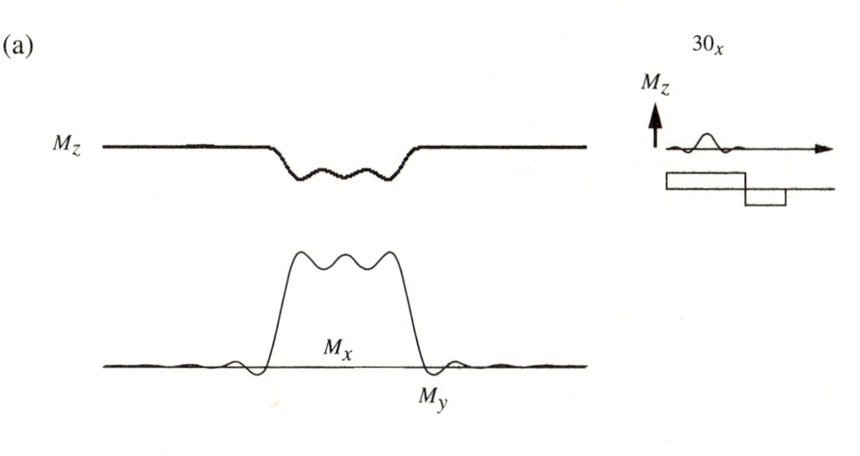

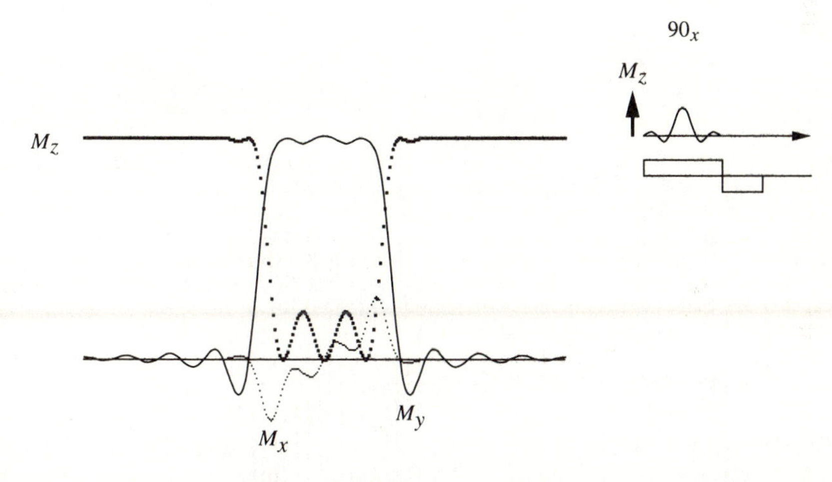

Fig. 3.7 (a) Magnetization profiles along the offset frequency axis for 30_x and 90_x double side-lobe sinc pulses, calculated using the Bloch equations where the initial values of M_x, M_y, and M_z are, respectively, 0, 0, and M_0. In each case the first-order phase shift is refocused by allowing a period, T, with reversed gradient. M_x, M_y, and M_z are shown as small dots, solid curve, and large dots, respectively. (b) As for (a) but for a 180_x pulse with refocusing period where the initial values of M_x, M_y, and M_z are, respectively, 0, 0, and M_0. Precisely the same M_z profile is achieved if the refocusing (negative gradient) period is omitted. The lower part of the diagram shows a symmetric 180_x pulse without a refocussing period where the initial values of M_x, M_y, and M_z are respectively, 0, M_0, and 0. As a result of this latter pulse any extra-slice transverse magnetization is dephased while the magnetization within the slice is reversed as desired.

(b)

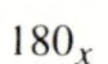

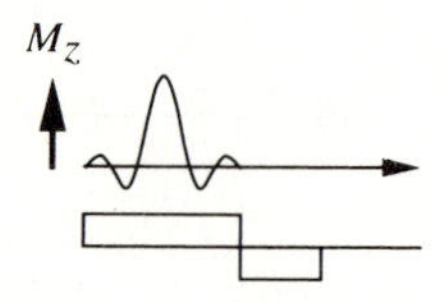

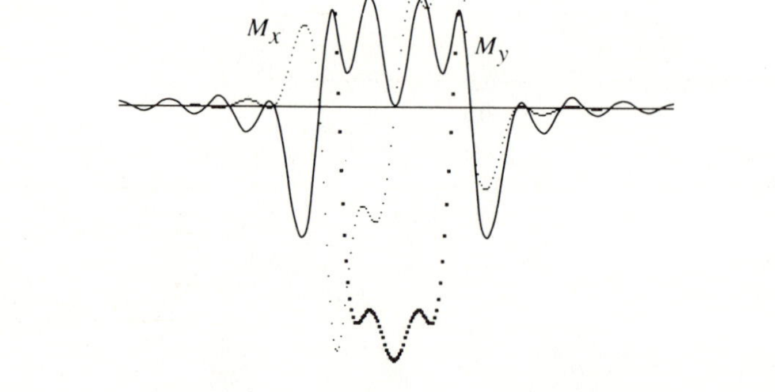

180_x

M_y

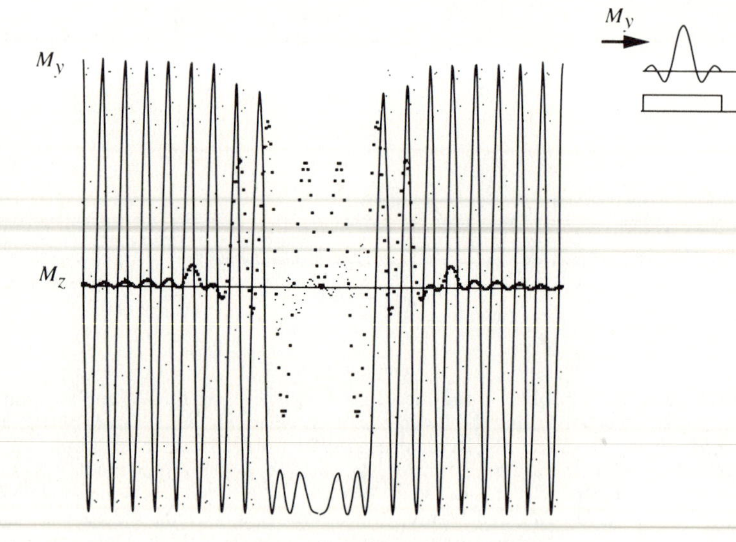

Fig. 3.7(b)

spectrum of B_1 should be real and M_+ should be purely in-phase. One of the consequences of non-linear behaviour is the presence of such phase distortion. For small turn angles the predictions of linear response theory are quite accurate so that the out-of-phase M_x component is insignificant in the case of the 30° pulse whereas this component is clearly apparent in the excitation caused by the 90° pulse. None the less, the 90° pulse M_y distribution is remarkably close to ideally rectangular, so that the truncated sinc modulation suffices quite well in practice.

A solution to the problem of out-of-phase magnetization generated by sinc-modulated 90° pulses has been suggested by Silver *et al.*[30] Their solution is to slightly time-shift the sinc wave-form. The slight out-of-phase distortion which results from sinc modulation can therefore be corrected by introducing a small delay (about 10% of the sinc pulse FWHM) into the start of data accumulation.

One very nice experiment enables us to observe directly the distribution of transverse magnetization excited by a given r.f. pulse. If, instead of removing the rephasing gradient, we sample the time domain echo signal in its presence, this echo has the time dependence which exactly corresponds to the magnetization profile, an effect first demonstrated by Hoult.[23] The method is particularly effective in revealing the effect of non-linearity distortion as the magnetization tip angles are increased.

Fig. 3.7(b) shows the effect of 180° selective r.f. pulses for two different starting magnetizations. The upper part of the diagram corresponds to excitation from thermal equilibrium with the initial values of M_x, M_y, and M_z being respectively 0, 0, and M_0. Such a pulse should, ideally, result in M_z inversion in a rectangular slice with no resultant transverse magnetization. Note that, while the pulse produces selective inversion of M_z, there is significant transverse magnetization generated within the slice, albeit with close-to-zero integral. In fact the negative gradient refocusing period is unnecessary for M_z inversion and the M_z profile is identical in the case that this evolution period is omitted. The lower part of Fig. 3.7(b) shows the effect of a 'refocusing' 180° sinc-modulated r.f. pulse, applicable in slice-selective spin echo applications, where the starting M_x, M_y, and M_z magnetization components are, respectively, 0, M_0, and 0. Note the self-refocusing action of this latter pulse, there being no subsequent negative gradient pulse. Because the y-magnetization is required to traverse a trajectory which is time symmetric, the 180° pulse must also be symmetric in time. In consequence there is no need for an additional gradient echo (or spin echo). While the pulse shown is 180_x, the behaviour of a 180_y pulse is equivalent, the M_y magnetization being inverted for 180_x and the M_x magnetization being inverted for 180_y. Outside the slice the transverse magnetization exhibits the phase twist characteristic of gradient-induced dephasing. While it appears large, its net contribution arises from the integral across the sample and is therefore zero. This dephasing of extra-slice magnetization will only occur when this magnetization is initially placed in the transverse plane, perhaps by means of a preceding hard pulse. When the 180° refocusing pulse is used in conjunction with an initial slice-selective 90° excitation, the extra-slice magnetization remains along the z-axis and is unperturbed by the action of the refocusing pulse.

While the description of selective 90° pulses using a linear response theory is similar to the exact solution, for 180° selective pulses the effect of non-linearity distortions is far more severe and some major phase distortions result[31,32] as is apparent in Fig. 3.7(b). However, for many pulse trains requiring 180° pulses the slice selection can be performed with the initial 90° pulse excitation and any subsequent inversion of the magnetization which is required can be performed non-selectively. Of course, such non-selective pulses always result in some residual transverse magnetization due to spins outside the slice because r.f. inhomogeneity leads to turn angles other than 180°. Judicious use of field gradient pulses can always cancel such effects. For example, because these out-of-slice spins have not experienced the phase shift associated with the initial selective excitation, they can be made to suffer dephasing rather than rephasing effects when a second gradient pulse is applied after the non-selective 180° pulse. This approach can ensure that only the slice spin magnetization will contribute to the final signal.

In multislicing experiments where extra-slice magnetization needs to be preserved for later use, selective pulses must be used exclusively. As apparent in Fig. 3.7(b), there are several problems which arise when sinc wave-forms are used in selective 180° excitation (M_z inversion) or in selective 180° refocusing (M_y or M_x inversion). The inversion of magnetization within the slice is inefficient, leading to reduced signal-to-noise ratios, significant out-of-phase magnetization, and the 'tailing off' of the excitation near the edges of the desired slice plane.[31] One solution to the latter problem involves increasing the bandwidth of the 180° pulse by narrowing its time profile. This has the effect of disturbing the magnetization on either side of the slice, a particular nuisance where multislice experiments are to be performed. For clean multislicing, 180° pulses can be avoided entirely by the use of gradient echoes, or by employing stimulated echoes[33] rather than Hahn echoes. For accurate selective 180° inversion the hyperbolic secant modulation proposed by Silver, Joseph, and Hoult[30] provides an effective alternative. This latter pulse requires both amplitude and phase modulation such that the complex pulse wave-form, $B_1(t)$, is given by

$$B_{1x}(t) + \mathrm{i}B_{1y}(t) = B_1[\operatorname{sech}(\beta t)]^{1+\mathrm{i}\mu} \tag{3.14}$$

where an excellent 180° pulse results if μ is set to 5, the frequency width of the selected region being given by

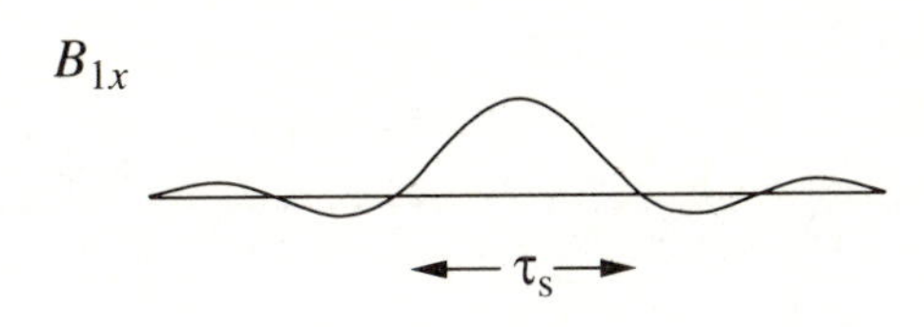

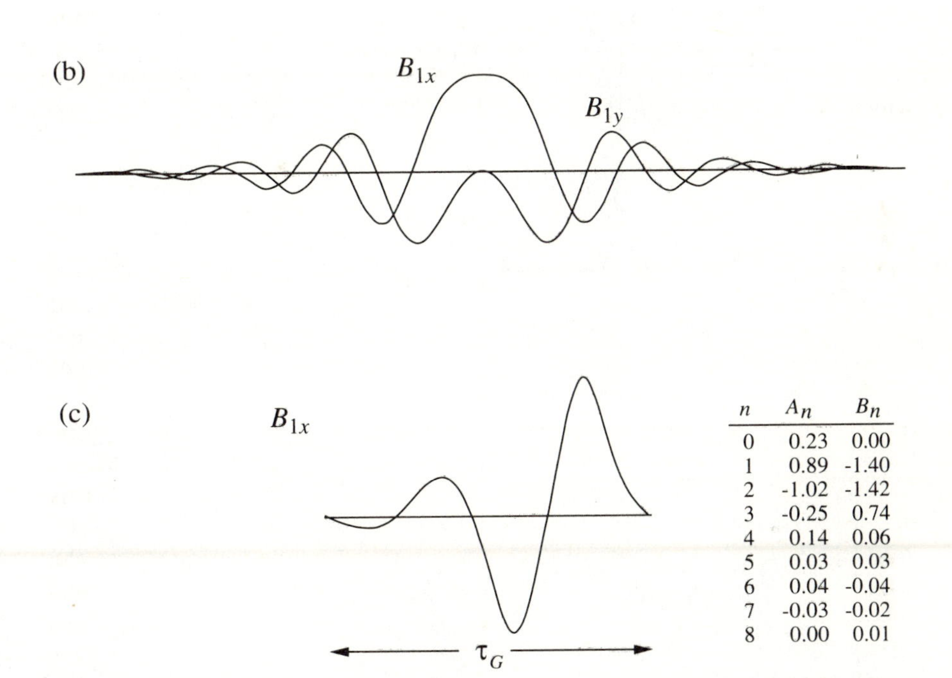

n	A_n	B_n
0	0.23	0.00
1	0.89	-1.40
2	-1.02	-1.42
3	-0.25	0.74
4	0.14	0.06
5	0.03	0.03
6	0.04	-0.04
7	-0.03	-0.02
8	0.00	0.01

Fig. 3.8 r.f. modulation wave-forms for three pulse types, each of which have identical bandwidths: (a) 90_x sinc wave-form with bandwidth $\Delta f = 2/\tau_S$, (b) hyperbolic secant 180_x inversion pulse with bandwidth $\mu\beta/\pi$ where $\mu = 5.0$, and (c) Geen *et al.* self-refocusing 90_x pulse with bandwidth $5/\tau_G$. The A_n and B_n Fourier coefficients are given in the accompanying table.

$$\Delta f = \frac{\beta\mu}{\pi}. \tag{3.15}$$

The in-phase and quadrature components of $B_1(t)$ are shown in Fig. 3.8 along with a corresponding sinc pulse with the same frequency width. Fig. 3.9(a) shows theoretical magnetization components, M_x, M_y, and M_z using a Bloch equation calculation, starting with initial $M_z = M_0$ polarization. Fig. 3.9(b) shows the experimental M_z polarization which results when

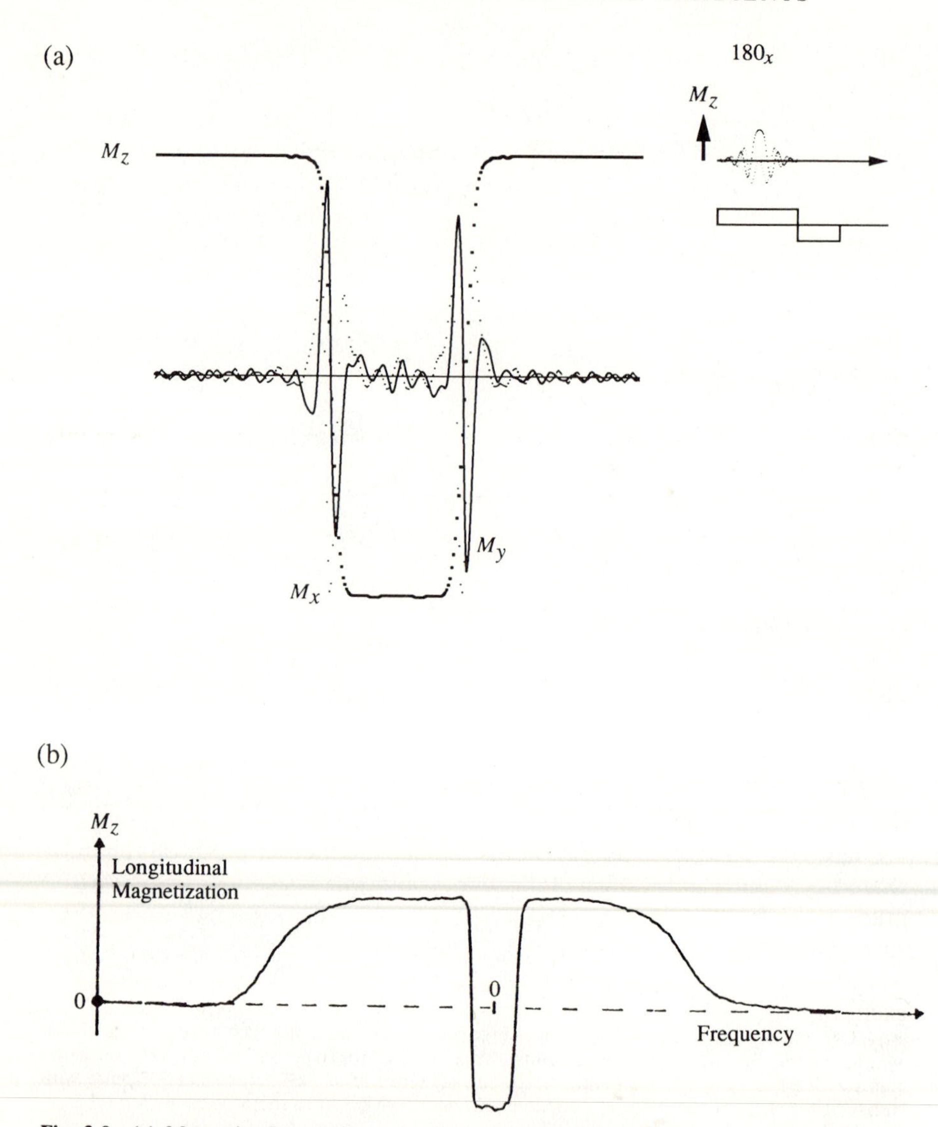

Fig. 3.9 (a) Magnetization profiles along the offset frequency axis for a 180_x hyperbolic secant pulse, calculated using the Bloch equations where the initial values of M_x, M_y, and M_z are, respectively, 0, 0, and M_0. The first-order phase shift is refocused by allowing a period, T, with reversed gradient. (b) Experimentally determined selective population inversion created by a complex hyperbolic secant pulse, (from Silver *et al.*[30])

the hyperbolic secant pulse is applied to a long cylindrical sample in the presence of a magnetic field gradient along the cylinder axis and then the magnetization is subsequently 'interrogated' in the present of the gradient by applying a hard 90° pulse. It is clear that clean inversion has occurred within a very well-defined slice.

One remarkable feature of hyperbolic secant pulses is the independence of turn angle on the amplitude, B_1, in eqn (3.14). Provided $\gamma B_1 > 2\beta$ (i.e., $0.4\Delta\omega$) the spin isochromats within the slice region rotate by 180° despite the fact that their trajectories in reaching this rotation may depend upon B_1.[30] This very useful feature makes these pulses very attractive, despite the extra difficulty involved in requiring the pulse modulation equipment to have phase as well as amplitude modulation available. It should be noted that while the hyperbolic secant pulse provides highly effective magnetization inversion, starting from a state of z-polarization, it does not possess universal rotation properties and cannot, for example, be used as a transverse magnetization inversion pulse in a slice-selective spin echo experiment.

3.2.3 *Alternative refocusing methods*

Fig. 3.10 shows four examples of selective excitations in which the 90° pulse rephasing gradient is applied in different ways. In the first two sequences the spins are refocused using a 'gradient echo' by reversing the sign of the slice selection gradient, in the first case without a subsequent 180° pulse, and in the second with such a pulse. Notice that the selective 180° pulse is self-refocusing and requires no additional rephasing gradient pulse. In the third example rephasing following selective excitation is achieved by using a gradient of the same sign applied after the hard 180° pulse has inverted the slice spin phase. This final gradient pulse also has the effect of destroying transverse magnetization arising from spins outside the slice which have been excited by the hard pulse. The final example in Fig. 3.10 is equivalent to the third except that a soft 180° pulse is used to form the spin echo.

The refocusing schemes illustrated in Fig. 3.10, while ostensibly similar in their outcomes, have in fact very different side effects. Because of the finite time over which the refocused magnetization vector is formed, other terms in the Hamiltonian have an opportunity to induce phase evolution. Such terms might include chemical shifts or Zeeman broadening due to magnetic field imperfections or susceptibility variations. Generally, the time to selectively excite is sufficiently small that such effects may be neglected. In longer echo sequences these additional dephasing effects may be very important, as discussed in Section 3.5. Gradient reversal, shown in Fig. 3.10(a), causes refocusing only of those phase shifts which were associated with the original gradient applied during the r.f. excitation. By contrast, the 180° r.f. pulse

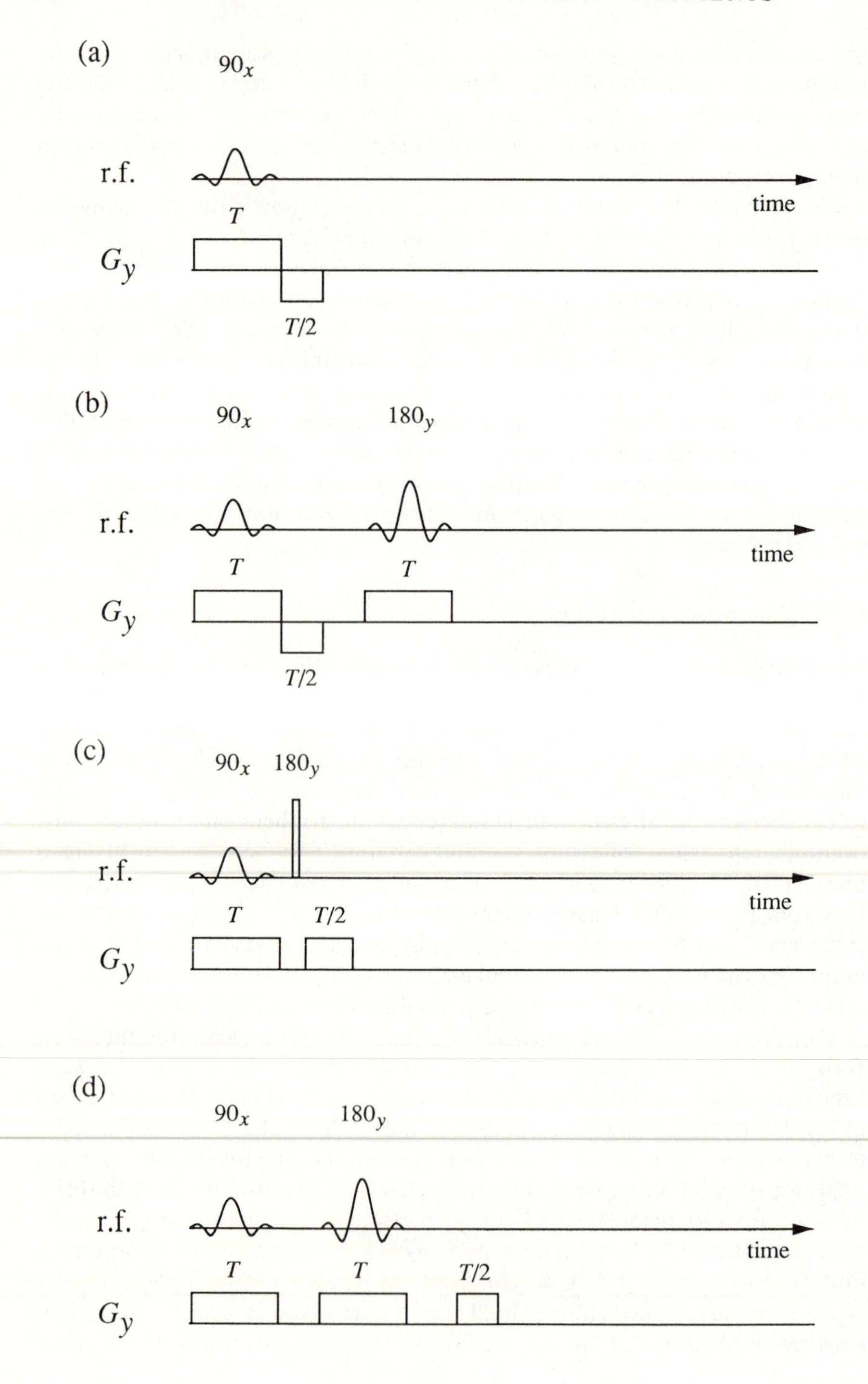

(a)
90x
r.f.
time
T
Gy
T/2
(b)
90x
180y
r.f.
time
T
T
Gy
T/2
(c)
90x
180y
r.f.
time
T
T/2
Gy
(d)
90x
180y
r.f.
time
T
T
T/2
Gy

methods, shown in Fig. 3.10(c) and (d), are capable of refocusing all shifts which are 'Zeeman-like' in origin. Note, however, that the final method suffers from the disadvantage that a greater time elapses before selection gradient phase restoration occurs, thus making the signal more susceptible to decay due to molecular self-diffusion. This diffusive-attenuation effect is discussed in detail in Section 3.6.2. Example (c) has the advantage that it provides Zeeman refocussing as well as minimizing evolution time, thus avoiding the diffusive delay.

3.2.4 *Self-refocused pulses*

The frequency-dependent phase shift which occurs when a spin system is excited using a sinc-modulated r.f. pulse in the presence of a magnetic field gradient is a feature of all time-symmetric amplitude-modulated r.f. pulses. Fig. 3.11(a) shows the magnetization profiles which result when a 90_x sinc wave-form is applied without subsequent refocusing. The strong transverse magnetization phase twist across the slice is clearly apparent. This effect may be contrasted directly with the profiles shown in Fig. 3.7(a) in which a gradient echo is employed. While phase twist removal using spin or gradient echoes may seem straightforward, these approaches do, however, suffer from some disadvantages. The first-order phase shifts are inherent to the frequency spectrum irrespective of the gradient applied. This means, for example, that chemical shift spreads will also suffer the phase twist and in this case a reversed gradient pulse will not cause phase restoration. While a spin echo can suffice for any Zeeman features in the spectrum, it cannot refocus phase twists associated with the spectral effects of homonuclear scalar coupling. Furthermore, as indicated in the previous sections, the application of hard 180° pulses excites extra-slice transverse magnetization, while selective 180° r.f. pulses cause slice distortion and are costly in evolution time, leading to transverse relaxation and loss of signal amplitude.

An alternative approach to selective excitation has been proposed by Geen, Wimperis, and Freeman[34] in which the r.f. pulse modulation wave-form is tailored to remove the phase twist, resulting in a 'self-refocused' excitation. These authors adjusted sixteen Fourier coefficients A_n and B_n of an arbitrary modulation, $\Sigma_n A_n \cos(n2\pi t/\tau_G) + B_n \sin(n2\pi t/\tau_G)$, until the excitation, as computed using the Bloch equations, matched a target profile which was chosen to have $M_y = M_0$ within a bandwidth $4/\tau_G$, and

---

Fig. 3.10 Alternative refocusing schemes following selective excitation: (a) gradient echo, (b) gradient echo with additional 180° refocussing of Zeeman broadening terms (note that the soft 180° pulse is self-refocusing provided that the preceding magnetization is in the transverse plane, (c) spin echo refocusing using a hard 180° r.f. pulse, and (d) spin echo refocusing using a soft 180° r.f. pulse.

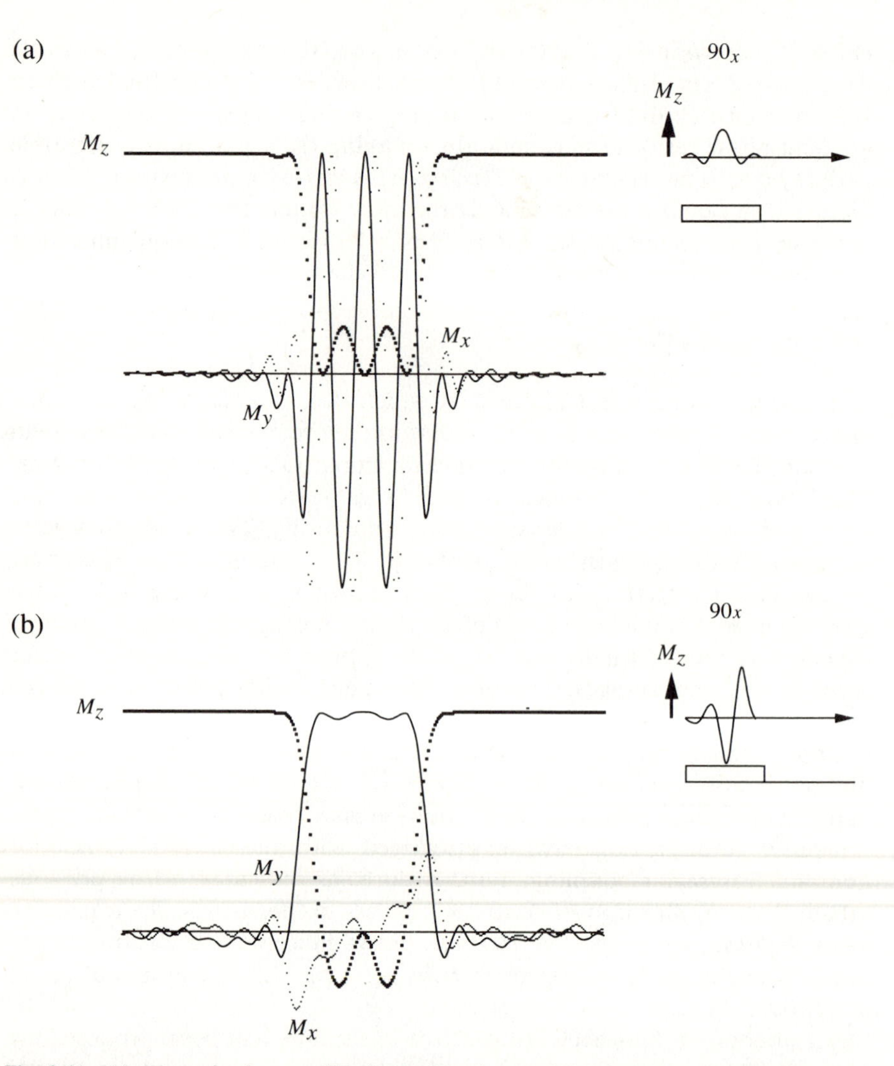

Fig. 3.11 (a) Magnetization profile for a 90_x sinc-modulated pulse without refocusing gradient, calculated using the Bloch equations where the initial values of M_x, M_y, and M_z are, respectively, 0, 0, and M_0. (b) Comparative magnetization profiles along the offset frequency axis for a 90_x self-refocusing pulse of Geen *et al.*[34] Note that there is no need for refocusing by allowing a period of evolution under a reversed gradient.

$M_x = M_y = 0$ outside a bandwidth $6/\tau_G$, where τ_G is the pulse duration. (For convenience we can describe the excitation bandwidth of the pulse as $5/\tau_G$.) Optimization of the A_n and B_n parameters requires special care to avoid local minima. The global optimization method used by Geen *et al.* is known as 'simulated annealing' in which steps towards minima are always permitted but steps away from minima are also occasionally permitted, with a Boltzmann-like probability in which the optimization function is treated as an energy surface and the temperature at which the overall energy minimization occurs is slowly adjusted downward.

The 90_x pulse of Geen *et al.* is shown in Fig. 3.8(c), along with the table of relevant Fourier coefficients. Fig. 3.11(b) shows the magnetization profiles resulting from application of this pulse. These profiles exhibit smaller residual transverse magnetization artefacts than the refocused 90_x sinc pulse profiles shown in Fig. 3.7 and it is clear that self-refocused pulses show considerable promise. One point which must be borne in mind concerns practical implementation. Whereas gradient or spin echo refocused sinc pulses are tolerant of some amplitude distortion, self-refocusing requires precise and absolute B_1 amplitude modulation. This can be achieved by avoiding non-linear elements in the r.f. transmitter circuits, or by using variable-width hard pulses in a DANTE sequence (discussed in Section 3.2.6).

3.2.5 *Spatially selective destruction and localized spectroscopy*

The selective excitation procedures discussed above are aimed at disturbing the equilibrium magnetization of a well-defined slice. This approach is ideal for imaging in the plane but in localized spectroscopy experiments the slice excitation approach is less appropriate. Here, the aim of a measurement is to obtain a high-resolution NMR spectrum from a selected volume element in the sample and the procedure employed is based upon the successive application of pulse trains which leave orthogonal planes of spins undisturbed, so that the only remaining z-magnetization exists in the intersection voxel. This voxel magnetization may then be disturbed by a hard pulse and the subsequent FID sampled in the absence of a magnetic field gradient. The essence of this approach is selective destruction of the z-magnetization in the extra-slice spins.

Two effective schemes for extra-slice magnetization destruction developed by Doddrell and co-workers are SPACE and DIGGER whose pulse trains are illustrated in Fig. 3.12. The SPACE scheme[35] relies on the selective excitation of a transverse magnetization coherence followed by a hard 90° pulse in the presence of a spoiler gradient. Application of the 90° pulse at the moment when the slice magnetization refocuses, stores this magnetization along the z-axis for later recall. At the same time all other extra-slice spins are excited by this pulse and have their magnetization destroyed by the spoiler gradient. The process is then repeated along the orthogonal gradient axes so that at the end of the triple axis sequence the stored magnetization arises from the orthogonal plane intersection voxel.

The DIGGER sequence[36,37] uses only soft pulses to destroy the extra-slice magnetization. These pulses, named sine–sinc, have a modulation

(a)

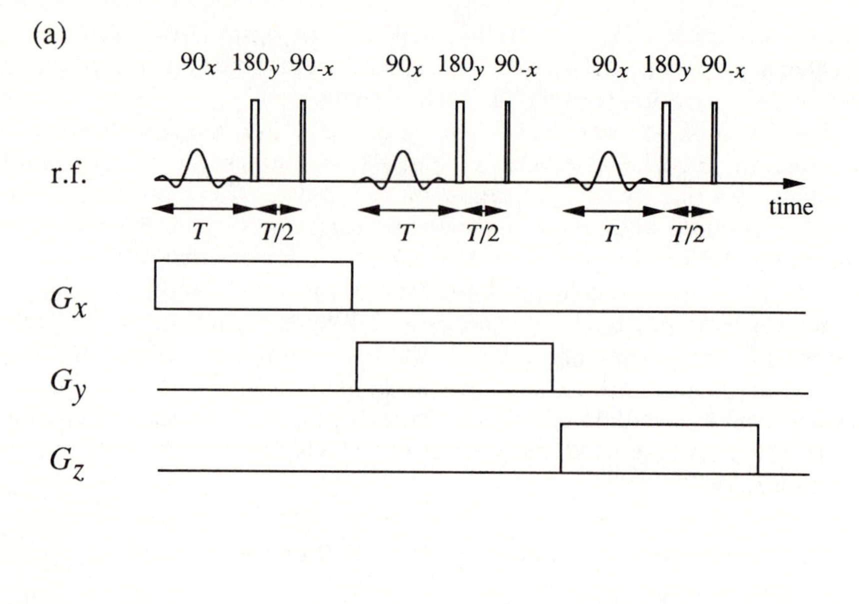

(b)

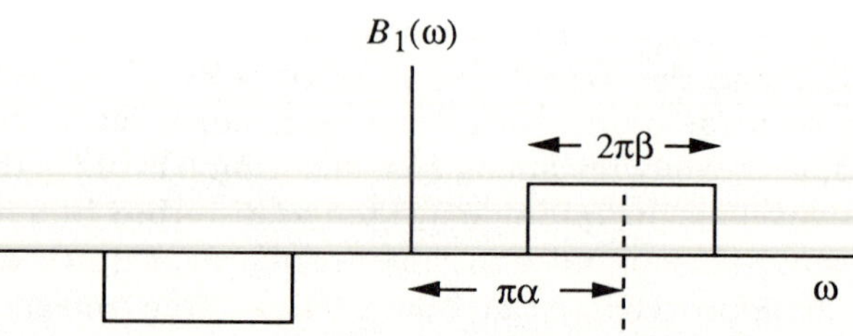

Fig. 3.12 (a) SPACE pulse sequence in which selected slice magnetization is stored along the z-axis while extra-slice magnetization is destroyed. Successive application with gradients along three orthogonal axes results in the storage for later recall of magnetization from the intersecting voxel. (b) Sine–sinc pulse frequency response. (c) Calculated magnetization profile resulting from the application of a double sine–sinc pulse in a DIGGER sequence (from Doddrell *et al.*[36])

envelope $B_1(t)$ given by the the Fourier transform of the double hat function shown in Fig. 3.12(b), namely,

$$B_1(t) = \frac{-\mathrm{i}\sin(\pi\alpha t)\sin(\pi\beta t)}{\pi t}. \tag{3.16}$$

The z-magnetization shown in Fig. 3.12(c) was produced by a double sine–sinc sequence with α and β set to 3.5 and 3.0, respectively. The delay

(c)

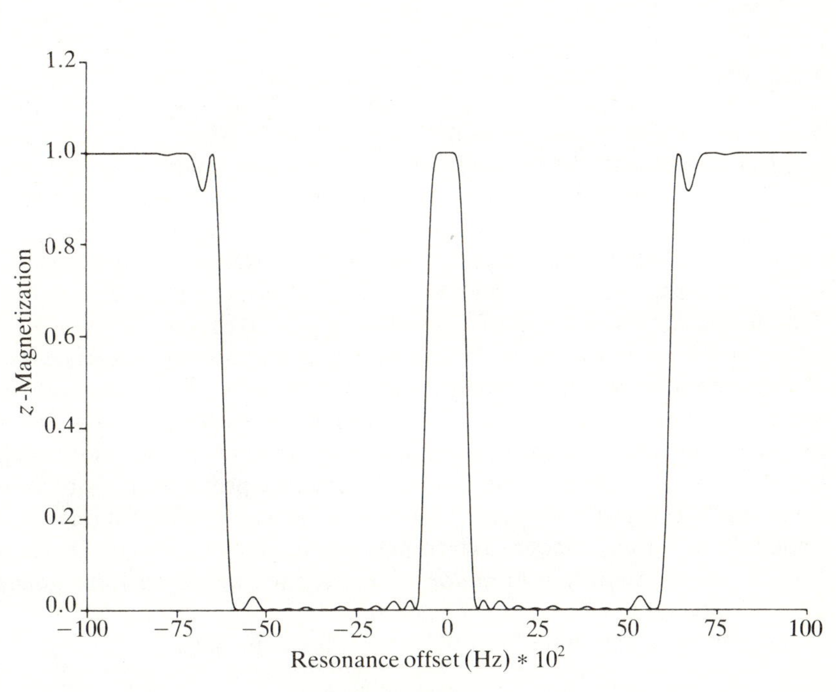

Fig. 3.12(c)

between the two sine–sinc pulses allows for destruction of the extra-slice transverse magnetization by the slice selection gradient, any remaining extra-slice z-magnetization being tipped into the transverse plane by the second selective pulse and then destroyed by the second gradient dephasing period. The method suffers from the effects of residual z-magnetization distant from the slice and outside the bandwidth of the sine–sinc pulse, but this can be removed by appropriate addition subtraction phase cycling.[37,38]

Methods which rely on phase cycling for the correction of artefacts are inconvenient in applications where on-line processing of a single transient is required. One obvious example concerns the process of magnet shimming in order to optimize homogeneity in the selected volume of interest. A highly effective method of single transient volume-selective excitation is the VOSY (Volume Selective Spectroscopy) method of Kimmich and Hoepfel[39] which relies on the three r.f. pulses of a stimulated echo to separately define othogonal slice planes which intersect in the chosen voxel. Because of the multiplicity of echoes generated by the three-pulse sequence a considerable variety of spatial selectivity is possible simply by altering the way in which the gradients are applied. Furthermore, the asymmetric manner in which the three slice selection gradients are applied causes unwanted coherences to be automatically spoiled.

3.2.6 *The DANTE sequence*

Soft pulses suffer from a major practical difficulty. Most spectrometers are highly non-linear in the transmission of r.f. to the sample coil. Generally the final high-power transmitter amplifier operates in a Class C mode and the duplexor which enables the same coil to be used both as a receiver and transmitter contains non-linear elements (for example, crossed diodes) which act as short circuits to high-level r.f. and open circuits to low-level r.f. This means that the pulse-shaping process, which takes place before the final power amplification stage, will often bear only a remote resemblence to the B_l field modulation finally experienced by the nuclei. As an alternative to using low-level, long-duration, r.f. fields to provide selectivity, it is possible to achieve the same result with a multiple pulse train of shorter, more intense pulses. This is the basis of the DANTE sequence proposed by Morris and Freeman.[40] The sequence name (Delays Alternating with Nutations for Tailored Excitation) alludes to the repetitive circular journeys by Dante and Virgil in Dante Alighieri's *Purgatorio*, akin to the trajectories undergone by off-resonant spins.

A simple explanation is obtained by using linear response theory. Fig. 3.13

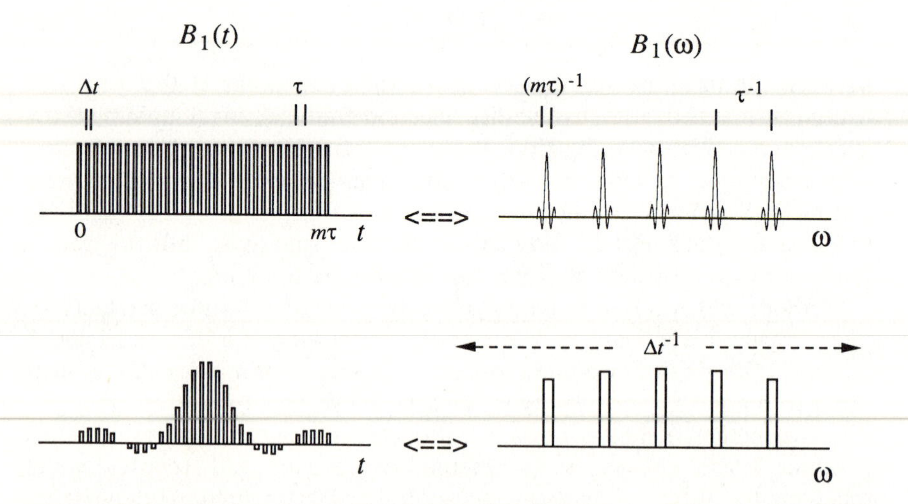

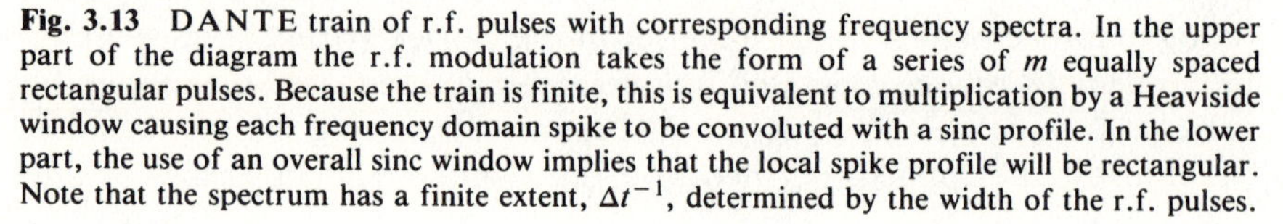

Fig. 3.13 DANTE train of r.f. pulses with corresponding frequency spectra. In the upper part of the diagram the r.f. modulation takes the form of a series of m equally spaced rectangular pulses. Because the train is finite, this is equivalent to multiplication by a Heaviside window causing each frequency domain spike to be convoluted with a sinc profile. In the lower part, the use of an overall sinc window implies that the local spike profile will be rectangular. Note that the spectrum has a finite extent, Δt^{-1}, determined by the width of the r.f. pulses.

shows a series of m r.f. pulses of duration Δt and spacing τ for which a spin isochromat on-resonance experiences a total flip angle $m\gamma B_1 \Delta t$. The frequency response of this train is shown alongside and consists of a comb of sidebands spaced by τ^{-1} with each spike in the comb having the form of a sinc function of width $(m\tau)^{-1}$, the total comb width being of order $(\Delta t)^{-1}$. For a specific total turn angle, such as $\pi/2$, B_1 can be made arbitrarily large by making Δt sufficiently short, within the constraints set by the transmitter bandwidth. By choosing τ sufficiently short that the off-resonant sidebands are outside the spectral range of the spins, the selective excitation will be confined to the central on-resonant comb element with an excitation bandwidth given by $(m\tau)^{-1}$. Because of the sharp discrete nature of the spin precession steps, it is possible to use a value of B_1 which is larger by a factor of $\tau/\Delta t$ than that used in continuous r.f. excitation, thus avoiding the non-linear transmission problem which is so severe at low levels of B_1. The additional transmitted power is, of course, dissipated in the outer sidebands and, for a specific total turn angle θ, represents no more, in total, than would be applied to the coil in a hard θ pulse at the same B_1 value.

It is not necessary to restrict the DANTE sequence to equal amplitude (or phase) r.f. pulses. Indeed, where an ideal rectangular frequency spread is desired in the selective excitation, the train can be sinc modulated as shown in the lower part of Fig. 3.13. DANTE trains are particularly useful in selective signal excitation or suppression in NMR spectroscopy. In NMR microscopy they are helpful in assisting slice selection in the solid state, and in providing a grid marker in the detection of fluid motion by NMR imaging, applications which are referred to in Chapters 5 and 8.

3.3 Reconstruction in two dimensions

3.3.1 *Coordinate definition*

If the equilibrium magnetization in a three-dimensional sample is disturbed by selective excitation, the reconstruction may be performed in the two dimensions of the slice plane. The resulting image is an average of contributions from across the layer. The volume element (or voxel) will then have its depth dimension determined by the slice thickness.

As explained earlier, reconstruction requires that we sample **k**-space and perform a Fourier transformation. This sampling involves the application of gradients in the plane of the slice. By choosing an appropriate direction for the slice selection gradient the orientation of this plane may be chosen at will. Such freedom is one of the great advantages of MRI over X-ray CT where slices are necessarily normal to the detector plane axis.

When dealing with human samples the names axial, coronal, and saggittal are used to describe the slice orientation. In NMR microscopy, samples are usually inserted down a specific axis which we will call the 'access axis' and this will depend on the geometry of the polarizing magnet. In an electromagnet the access axis is generally labeled 'y' and is normal to the polarizing field. In most superconducting solenoids the access axis coincides with the solenoidal axis and is therefore z. In wide-bore systems, however, it is possible to use a transverse axis for sample entry, once the NMR probe has been removed. In NMR imaging, therefore, there are two 'transverse planes': one is that normal to the polarizing field, the plane in which our transverse magnetization is described in the rotating frame; the other is that normal to the slice selection gradient, the so-called slice or excitation plane. We have previously used the coordinates (x, y) to represent the direction transverse to the main field axis. For convenience we will also describe the slice plane by (x, y) which means that we shall routinely represent the slice selection gradient by G_z. In some superconducting magnet geometries the main field direction and the slice normal will happen to coincide. That is not important. The orientation of the magnetization and excitation planes is entirely independent and is determined in practice by the choice of gradient orientation.

Having looked at the problem of exciting a plane of spins we now turn our attention to image reconstruction within the plane. **k**-space sampling is performed using a finite number of sample points, typically a 256^2 array. The way in which these points are distributed is known as a raster. Two obvious rasters are those based on Cartesian and polar coordinates and in the history of NMR imaging the polar raster came first.[41-45] In terms of Fourier theory, however, the simplest to understand is the Cartesian raster due to Kumar, Welti, and Ernst[46,47] and known as Fourier Imaging or FI. FI encodes phase and amplitude information in the signal by the use of specifically designed evolution periods, an approach which forms the basis of two-dimensional spectroscopy methods.[48-51] In the following description of the imaging method we retain the continuous coordinate description. Digitization leads to some important effects which we will discuss in the next chapter.

3.3.2 *Two-dimensional Fourier imaging (FI)*

When the FID is sampled in the presence of a gradient, signal points are obtained along a single line in **k**-space. In FI this line is oriented along one of the Cartesian axes and the associated gradient is known as the 'read' gradient. We will ascribe the x-coordinate to this direction. The intercept of this line along the orthogonal axis can be changed by imposing the G_y gradient for a fixed period before sampling begins. The G_y gradient is then named the 'phase' gradient since it imparts a phase modulation to the signal, dependent on the position of volume elements along the y-axis.

Following eqn (3.6), the signal is therefore

$$S(k_x, k_y) = \int_{-a/2}^{a/2} \left[\int_{-\infty}^{\infty} \int_{-\infty}^{\infty} \rho(x, y, z) \exp\{i2\pi(k_x x + k_y y)\} \, dx \, dy \right] dz \tag{3.17}$$

where a is the slice thickness. For convenience we will ignore the outer integral which merely represents the process of averaging across the slice, and write

$$S(k_x, k_y) = \int_{-\infty}^{\infty} \int_{-\infty}^{\infty} \rho(x, y) \exp[i2\pi(k_x x + k_y y)] dx \, dy. \tag{3.18}$$

It is clear that $S(k_x, k_y)$ is the two-dimensional Fourier transform of the spin area density function $\rho(x, y)$ (i.e. the volume density function $\rho(x, y, z)$ averaged normal to the slice). Reconstruction of $\rho(x, y)$ from $S(k_x, k_y)$ simply requires that we calculate the inverse Fourier transform

$$\rho(x, y) = \int_{-\infty}^{\infty} \int_{-\infty}^{\infty} S(k_x, k_y) \exp[-i2\pi(k_x x + k_y y)] dk_x \, dk_y. \tag{3.19}$$

This process is carried out using a computer, subsequent to obtaining the two-dimensional signal, $S(k_x, k_y)$. It is the acquisition of this signal which is at the heart of the FI imaging method.

k_x and k_y are given by $(2\pi)^{-1}\gamma G_x t_x$ and $(2\pi)^{-1}\gamma G_y t_y$. It may seem a little odd to associate time coordinates with a spatial direction but this labelling simply emphasizes that t_x and t_y are different periods of time and involve evolutions under the influence of different gradients. When $k_x = 0$ and $k_y \neq 0$ the spins evolve in **k**-space along the positive y-axis as time advances. This is the phase period. Subsequently, when $k_x \neq 0$ and $k_y = 0$, the signal is mapped along the positive x-axis in **k**-space at an intercept along the y-axis set by the previous evolution. This is the read period. The intercept may be chosen at will by using a fixed phase gradient and adjusting the time of the phase evolution[46,47] or, as is more commonly the case, using a fixed phase period and adjusting the magnitude of the phase gradient. This latter approach is commonly termed 'spin warp imaging'[52,53] but we shall retain the label FI. An example of the two-dimensional Cartesian Fourier transform process is shown for a circular object in Fig. 4.4.

The process of sampling **k**-space along the x-axis under a fixed read gradient is repeated for successively increased values of k_y intercept so that the first quadrant of **k**-space is sampled on a raster with each FID along k_x

being stored as a row in a matrix. Reversing the y-gradient allows the fourth quadrant to be mapped. Because the image density $\rho(x, y)$ is ideally a real function, the signal $S(k_x, k_y)$ is subject to the symmetry $S(-k_x, -k_y) = S^*(k_x, k_y)$ where S^* is the complex conjugate of S. This reality of $\rho(x, y)$ depends on the the use of the correct zero reference phase, in the phase-sensitive detection process. While this is a straightforward process in practice, one must be aware of the assumptions inherent in two-quadrant sampling. In Chapter 8 we will meet an example where such 'phasing of the spectrum' is intrinsically impossible so that full four-quadrant sampling is essential.

A complete pulse sequence and associated raster for FI reconstruction is shown in Fig. 3.14. Notice that the slice, phase, and read gradients are arbitrarily assigned to z-, y-, and x-axes, respectively, a convention which is repeated in later pulse sequences. Of course, in any given experiment these assignments may be changed at will.

Because sampling is delayed until the beginning of the read gradient, some T_2 relaxation will occur following the excitation r.f. pulse. Typically the delay between excitation and reading will be of order 2 to 10 ms in a microscopic imaging system where rapid gradient switching is possible because of the small size of the coils. In a sample where T_2 is on a similar time-scale there will be some inevitable loss of sensitivity. Under such conditions the variable t_y method will create artefacts in the image. These are avoided by varying the magnitude of the phase gradient, G_y, rather than the phase evolution period, t_y. Furthermore, the phase shifts due to residual magnet inhomogeneity which would vary with increasing t_y, are fixed in the spin warp method, thus avoiding any inhomogeneity distortion along the phase axis following reconstruction. For these reasons the spin warp variant is always preferred.

3.3.3 *Two-dimensional projection reconstruction (PR)*

If the FID is acquired while the imaging gradient $\mathbf{G}$ is applied in some arbitrary direction, a series of signal points are obtained along a radial line in $\mathbf{k}$-space. By varying the orientation of $\mathbf{G}$, $\mathbf{k}$-space may be sampled on a polar raster as shown in Fig. 3.15. In practice this is done by applying G_x and G_y simultaneously and adjusting their relative magnitudes so that the variable polar angle and constant magnitude are, respectively, given by

$$\phi = \tan^{-1}(G_y/G_x) \qquad \text{and} \qquad |\mathbf{G}| = (G_x^2 + G_y^2)^{1/2}. \tag{3.20}$$

This polar raster is particularly simple to implement in practice since it could be performed with a fixed magnitude gradient while the sample is successively rotated!

Having obtained the signal on a polar raster, two options are available.

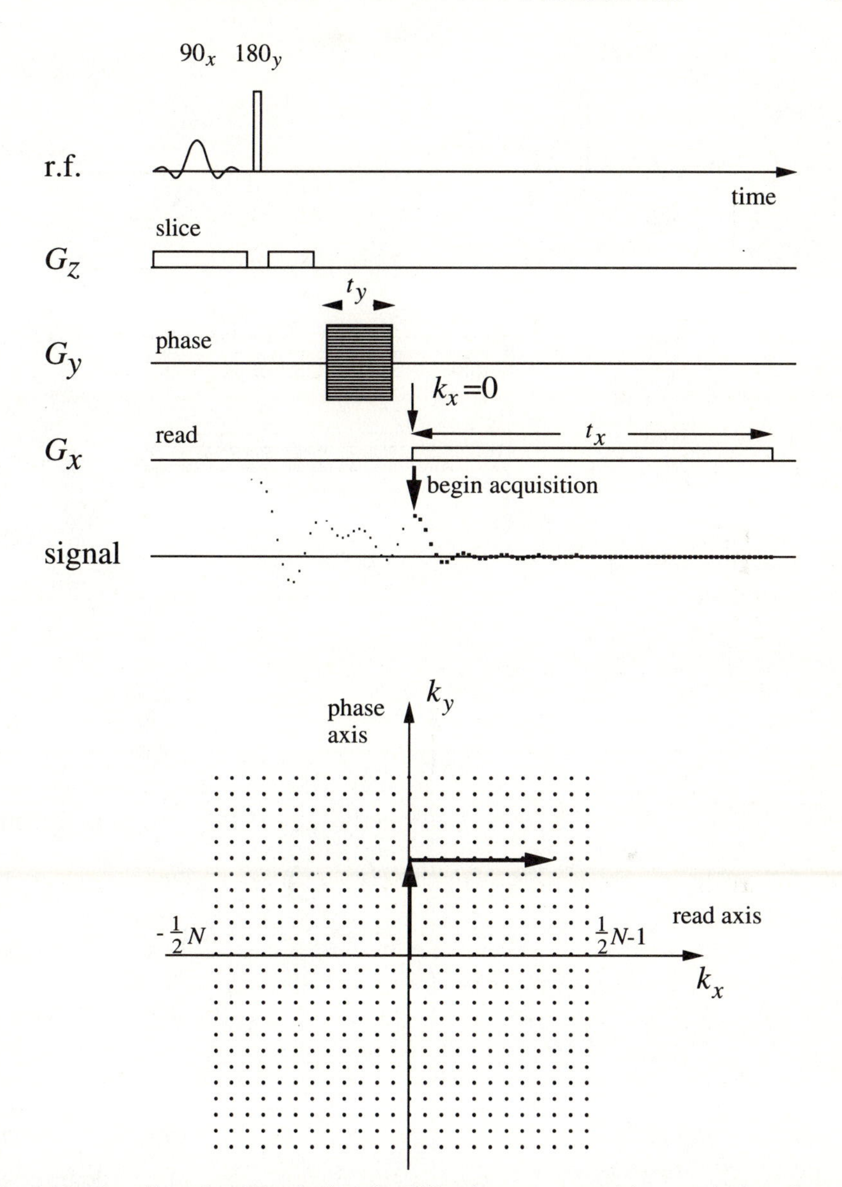

Fig. 3.14 Two-dimensional Fourier imaging (F I) pulse sequence and data acquisition scheme. The y- and x-gradients are responsible for phase and frequency encoding, respectively, the latter being termed a read gradient since the signal is acquired in its presence. Because the phase and read gradients are applied sequentially, the phase and read periods, t_y and t_x, are associated with evolutions through orthogonal directions in **k**-space. During the read period N data points of the FID are acquired and the first $N/2$ points placed in one quadrant of **k**-space. Alternatively, an N-points Fourier transform may be computed from the complete FID and the resultant spectrum placed across two quadrants. Note that the pulse sequence is schematic regarding the relative time durations. The F I sequence can be compressed by applying the phase and slice refocusing gradients simultaneously.

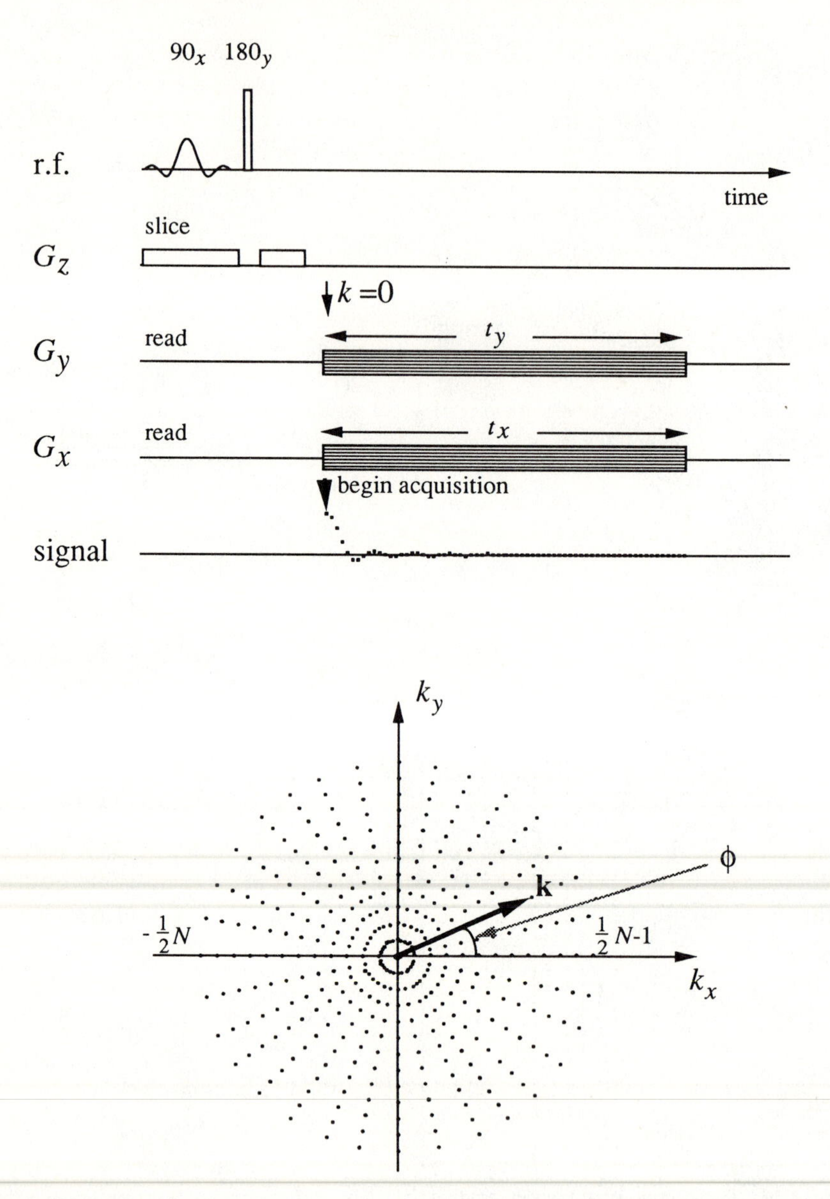

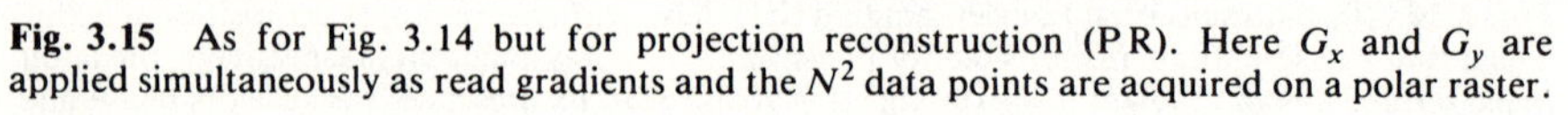

Fig. 3.15 As for Fig. 3.14 but for projection reconstruction (PR). Here G_x and G_y are applied simultaneously as read gradients and the N^2 data points are acquired on a polar raster.

Both will involve an interpolation process if the image is to be displayed using computer graphics, since monitor screens are inherently Cartesian. The first process involves an initial interpolation on to a Cartesian raster and subsequent Fourier transformation with respect to k_x and k_y. The interpolation error leads to significant artefacts in image space and the entire process is slow and clumsy. The alternative, the Projection Reconstruction (PR) process follows the imaging algorithm of X-ray CT, outlined in Chapter 1. In such an approach the interpolation is performed in real space where a simple nearest-neighbour shift may be employed without leading to significant artefacts. The method has advantages and disadvantages with respect to FI, some of which will be discussed in Chapter 4. One obvious advantage is the real-time nature of the reconstruction. NMR microscopy involves imaging times of several tens or hundreds of minutes so problems which are not revealed, as in FI, until the end of the total acquisition can be extremely time wasting. By contrast, in PR, the experimenter is able to view the current state of reconstruction at all stages as the raster is rotated. For an experienced observer, the incompletely reconstructed image is a useful indication as to the progress of image acquisition and can act as an early warning in the event of problems in the measurement.

In order to see how projection reconstruction applies to NMR imaging we rewrite eqn (3.18) as

$$S(k,\phi) = \int_{-\infty}^{\infty}\int_{-\infty}^{\infty} \rho(x,y)\exp[\mathrm{i}2\pi\mathbf{k}\cdot\mathbf{r}]\,\mathrm{d}x\,\mathrm{d}y \tag{3.21}$$

where $\mathbf{k} = (2\pi)^{-1}\gamma\mathbf{G}t$ while $\mathbf{G}$ and ϕ are as defined in eqn (3.20). Having acquired $S(k,\phi)$ in polar coordinates we should write the inverse Fourier transform in polar form as

$$\rho(x,y) = \int_0^{\pi}\left\{\int_{-\infty}^{\infty} S(k,\phi)\exp[-\mathrm{i}2\pi\mathbf{k}\cdot\mathbf{r}]\,|k|\,\mathrm{d}k\right\}\mathrm{d}\phi. \tag{3.22}$$

Eqn (3.22) is identical to eqn (1.33) and is the standard PR reconstruction relation. The inner integral is the filtered profile $\rho_\phi(r)$ given by

$$\rho_\phi(r) = \int_{-\infty}^{\infty} S(k,\phi)\exp[-\mathrm{i}2\pi kr]\,|k|\,\mathrm{d}k \tag{3.23}$$

where r is the component of radial displacement along the gradient direction. $\rho_\phi(r)$ belongs to all points (x,y) satisfying

$$r = (x\cos\phi + y\sin\phi). \tag{3.24}$$

In practice, therefore, the gradient is successively swept in angular increments over two quadrants. At each position the filtered profile, $\rho_\phi(r)$, is calculated. For each position (x, y) in the image matrix the projection, r, along **G** is calculated and the closest available value of $\rho_\phi(r)$ assigned. This is the back-projection and interpolation process. An example of back-projection for a circular object is shown in Fig. 4.3.

Although the PR and FI approaches are very different in practice, comparison of the respective rasters shown in Fig. 3.14(b) and 3.15(b) shows that, fundamentally, they differ only in the nature of the coordinate systems employed. In X-ray tomography, where the data is acquired as a real space projection, the connection between FI and PR is somewhat obscure. By contrast, in the NMR variant of this method, where the data is acquired in conjugate space, the connection between FI and PR is explicit and the title 'projection reconstruction' serves not so much to emphasize the way in which the data is acquired, as the way in which it is finally interpolated on to a Cartesian grid.

Notice that both FI and PR use read gradients. These cause the spectrum of the NMR signal to be widely spread both above and below the transmitter frequency, which means that in the heterodyne detection frame, both positive and negative frequencies will be present. These can only be distinguished if both in-phase and quadrature signals are acquired. Such dual phase detection is an essential prerequisite if read gradients are to be employed.

3.3.4 *Frequency and time domain relationships*

Before leaving the topic of image reconstruction it is worth re-emphasizing that **k**-space and image space correspond in practice to the time and frequency domains in NMR spectroscopy. The various integrals relating to imaging can therefore be recast in the familiar language of NMR. In this case explicit time and frequency (t, f) coordinates are used rather than (k, r). The connection between these pairs is via the proportionality constant $(2\pi)^{-1}\gamma G$. For example, the spin density projection is the direct Fourier transform

$$P_\phi(f) = \int_{-\infty}^{\infty} S(t, \phi) \exp[-i2\pi f t] \, dt. \tag{3.25}$$

It is interesting to pursue this relationship a little further in terms of the spectral description used in NMR. $S(t, \phi)$ is the signal acquired during the application of the gradient **G** at direction ϕ. Suppose that the quadrature detection phase is 'correctly set' as described in Section 3.1. Since $\rho(x, y)$ is real, $P_\phi(f)$ must be real and absorptive. Curiously, the integral of eqn (3.25) is taken over 'negative time', a process which can only be carried out in practice if the signal is acquired in both the leading and trailing periods of a spin echo. Where the time domain data is acquired only for $t > 0$, both real (R) and imaginary (I) components of $P_\phi(f)$ are generated such that

$$P_\phi^{\mathrm{R}}(f) = \tfrac{1}{2} P_\phi(f). \tag{3.26}$$

The factor of $\frac{1}{2}$ reflects the loss of the signal from the negative time region. $P_\phi^{\mathrm{R}}(f)$ and $P_\phi^{\mathrm{I}}(f)$ are closely related through the Kramers–Kronig relation

$$P_\phi^{\mathrm{I}}(f) = -(1/\pi)\,\mathcal{P} \int_{-\infty}^{\infty} \frac{P_\phi^{\mathrm{R}}(f') - P_\phi^{\mathrm{R}}(\infty)}{f' - f}\, \mathrm{d}f'. \tag{3.27}$$

In PR it is the filtered transform $\rho_\phi^{\mathrm{R}}(f)$ which is back-projected and summed over orientations ϕ to produce the final image. It is clear from eqns (1.34) and (3.27) that for data in the range $t > 0$

$$\rho_\phi^{\mathrm{R}}(f) = -(1/2\pi)\, \mathrm{d}/\mathrm{d}f\, P_\phi^{\mathrm{R}}(f). \tag{3.28}$$

Hence the filtered transform is related to the spin density projection via the derivative of the Kramers–Kronig integral

$$\rho_\phi^{\mathrm{R}}(f) = (1/2\pi)\,\frac{\mathrm{d}}{\mathrm{d}f}\left[(1/\pi)\,\mathcal{P} \int_{-\infty}^{\infty} \frac{P_\phi^{\mathrm{R}}(f') - P_\phi^{\mathrm{R}}(\infty)}{f' - f}\, \mathrm{d}f'\right]. \tag{3.29}$$

In practice we do not use eqns (3.28) or (3.29) in reconstructing the image. These equations are, however, useful in enabling us to calculate theoretical filtered profiles for objects where $P_\phi(f)$ is known a priori. They also relate the filtered profile to the familiar dispersion spectrum in nuclear magnetic resonance. As indicated in Section 2.6.4, it is indeed possible to acquire the signal for $t < 0$ in such a manner that the dispersion signal is absent. This is an important feature since it enables us to calculate the modulus image, so avoiding many phase artefact problems in the image reconstruction.

3.4 Alternative reconstruction methods

3.4.1 *Three-dimensional imaging*

Two-dimensional imaging suffers from one severe disadvantage, namely the finite time required for the selective excitation process. During the time delay which elapses during the extended r.f. pulse and the subsequent gradient or spin echo essential to phase restoration, the magnetization is subject to transverse spin relaxation. In NMR microscopy, where T_2 values are often quite short, this can present a particular problem. For example, in the imaging of solids where T_2 values may be as short as a few tens of microseconds, the usual selective excitation methods completely fail. Solutions to this problem are discussed in Chapter 5. In elastomers or in biological systems, where bound water is being imaged, T_2 values may be a few milliseconds. In practice it is difficult to reduce the selective excitation period to much less than one millisecond so that in these systems two-dimensional imaging methods may result in significant T_2 contrast.

An alternative solution to this problem is to employ three-dimensional imaging in which the entire sample is excited by a 'hard' pulse of sufficiently short duration that no transverse relaxation occurs. Three-dimensional imaging may be performed using either PR or FI signal encoding as shown in Fig. 3.16 although it is clear that PR offers the significant advantage that no phase gradient delay is needed before the signal acquisition commences. Furthermore, the use of a 90° hard pulse avoids the problem of soft pulse non-linearities and is therefore inherently more efficient.

The mathematics of three-dimensional projection reconstruction are essentially the same as for two dimensions except that now a spherical polar rather than cylindrical polar coordinate system is used. The equivalent to eqn (3.22) is

$$\rho(x, y, z) = \int_0^{\pi} \int_0^{\pi} \left\{ \int_{-\infty}^{\infty} S(k, \theta, \phi) \exp[-\mathrm{i}2\pi \mathbf{k} \cdot \mathbf{r}]\, k^2\, \mathrm{d}k \right\} \sin\theta\, \mathrm{d}\theta\, \mathrm{d}\phi. \quad (3.30)$$

Note that the volume element in spherical polars is now $k^2 \mathrm{d}k \sin\theta\, \mathrm{d}\theta\, \mathrm{d}\phi$ so that the inner integral representing the filtered profile, $\rho_{\theta\phi}(r)$, contains the ramp k^2 and is defined by

$$\rho_{\theta\phi}(r) = \int_{-\infty}^{\infty} S(k, \theta, \phi) \exp[-\mathrm{i}2\pi k r]\, k^2\, \mathrm{d}k. \quad (3.31)$$

where r is the component of radial displacement along the gradient direction. $\rho_{\theta\phi}(r)$ belongs to all points (x, y, z) satisfying

$$r = (x \sin\theta \cos\phi + y \sin\theta \sin\phi + z \cos\theta). \quad (3.32)$$

In other respects the acquisition and reconstruction process is akin to that used in two-dimensional PR except that now the gradient direction is successively reoriented in both polar and azimuthal directions, covering at least four octants in three-dimensional space. However, to ensure isotropic resolution, the gradient orientations must be uniformly distributed over the surface of a sphere. Since the elemental area on the surface of a sphere is proportional to $\sin\theta\, \mathrm{d}\theta \mathrm{d}\phi$ this implies that the polar angle θ may be stepped uniformly but the azimuth, ϕ, should be incremented as $1/\sin\theta$.

The three-dimensional projection reconstruction approach to NMR microscopy has been advocated and demonstrated by Lauterbur and

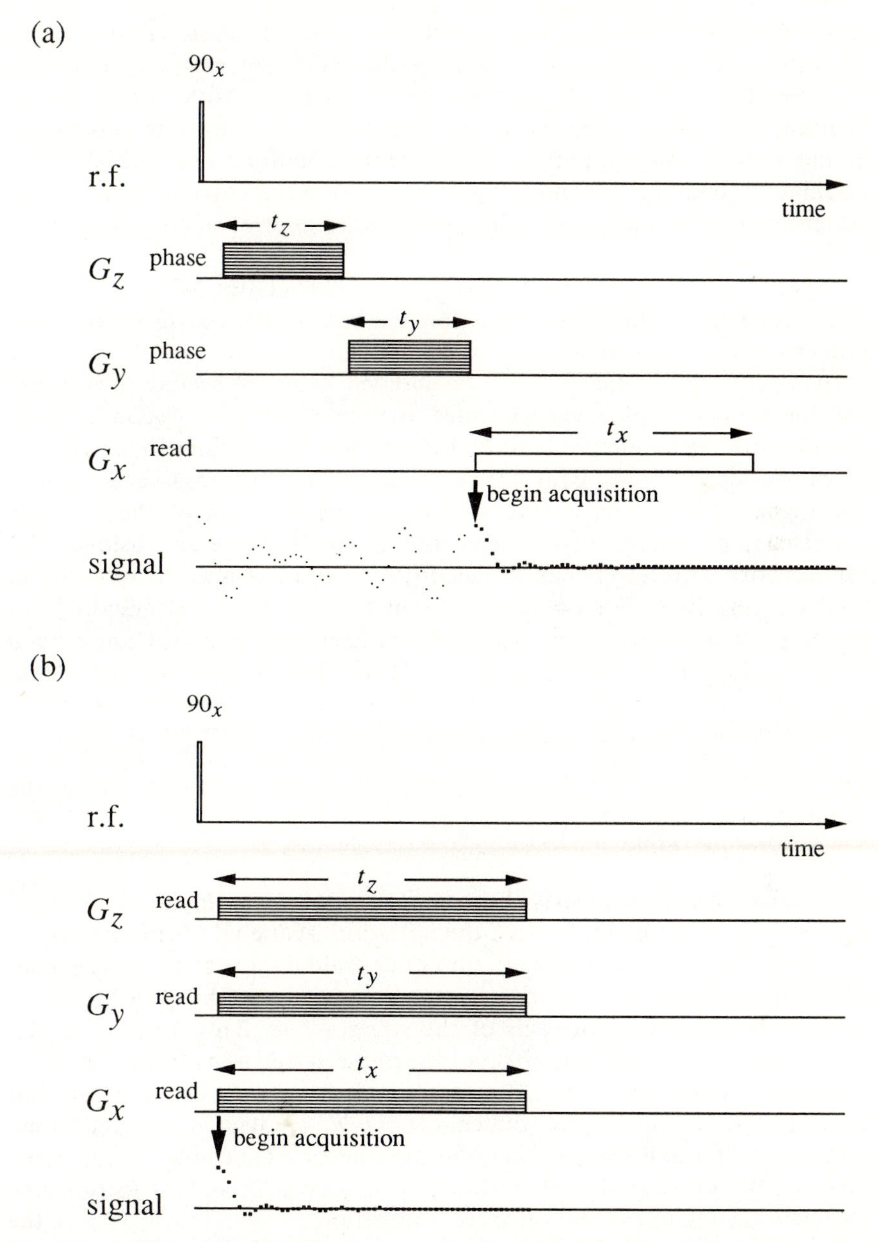

Fig. 3.16 (a) FI and (b) PR pulse sequences for three-dimensional reconstruction. No slice selection is required so that a hard 90_x pulse is used. Although the G_z and G_y phase encoding gradients in FI are shown as being applied sequentially, they can in principle be applied simultaneously. Clearly PR enjoys the advantage of no phase gradient delay.

co-workers[54–58]. Lauterbur points out that three-dimensional projection reconstruction represents hardware 'gentleness'.[59] For example, it avoids the need for complicated r.f. pulse shaping and permits imaging using constant rather than pulsed gradients, thus reducing hardware complexity at the expense of computing sophistication, 'shifting the burden from machine engineering to computing'.[60] In an era when computing is becoming cheaper while hardware is becoming more expensive, this philosophy should be taken seriously.

There are, however, two fundamental difficulties with the three-dimensional projection reconstruction method. First, the method works properly only if the sample is entirely enclosed within the r.f. coil. If frequency foldback effects are to be avoided, selective excitation is essential for samples which are extended in one dimension. Second, three-dimensional PR requires at least N^2 independent acquisitions to produce an N^3-dimensional image array. While intrinsically as sensitive as planar imaging using N signal averaging acquisitions for each of the N phase gradient steps, unlike two-dimensional methods, three-dimensional PR cannot work with fewer than N^2 acquisitions and this may prove costly in total imaging time. For 64^3 arrays the method is not too demanding but when $N = 256$ the time taken to acquire an image may be several hours. Such long imaging times are common in NMR microscopy so that three-dimensional PR is likely to be most useful in high-resolution imaging where the resolution is strongly limited by sensitivity factors.

3.4.2 *Rotating frame imaging*

An alternative method of spatial encoding using gradients in the r.f. field has been suggested by Hoult.[61] Here the variation of the r.f. field amplitude, B_1, across the sample causes an amplitude modulation of the subsequent FID which relates to the differing turn angles experienced by spins according to their position along the axis of the r.f. gradient. This form of spatial encoding is analogous to the effect of the phase gradient in Fourier imaging. Varying the duration of the r.f. pulse is equivalent to varying the duration or magnitude of the phase gradient. If the B_1 gradient is directed along some particular axis (say, z) then the subsequent FID can be 'read' in the presence of orthogonal, conventional B_0 gradients (x and y) in order to complete the reconstruction in three dimensions.

One useful feature of rotating frame phase encoding is that the B_0 gradients need not be switched. While this may be advantageous in medical applications it is not so important in microscopy. Furthermore, the rotating frame method suffers from the difficulties associated with producing large linear B_1 gradients. One of the most profitable medical applications of rotating frame methods has been in the area of spatially resolved chemical shift spectroscopy.[62-65]

3.5 The use of echoes in imaging experiments

3.5.1 *The echo sampling scheme*

The FI pulse sequence and **k**-space map illustrated in Fig. 3.14 correspond to positive time sampling. Following the convention used here that G_x is the read gradient and G_y the phase gradient, the sampling origin in each acquisition corresponds to $k_x = 0$. In Section 2.6.4 and again in Section 3.3.4 we saw that there is a factor of $\frac{1}{2}$ penalty in spectrum amplitude which results when only positive time intervals are sampled. The process of acquiring data for positive and negative k values requires the use of an echo. Full echo sampling corresponds to shifting the beginning of sampling to $k_x = -(1/2\pi)\gamma G_x(\frac{1}{2}N)T$ and acquiring N read points up to $k_x = (1/2\pi)\gamma G_x(\frac{1}{2}N - 1)T$, where T is the sampling interval. This process is illustrated in Fig. 3.17(a). In two-dimensional imaging the use of full echoes is equivalent to a full four-quadrant mapping of **k**-space. Fig. 3.14 shows the alternative approach in which N data points are acquired from $k_x = 0$ to $k_x = (1/2\pi)\gamma G_x(N - 1)T$.

Of course, what concerns us in practice is our desire to reconstruct the real spin density, $\rho(\mathbf{r})$, while at the same time optimizing the signal-to-noise ratio. The discussion in Section 2.6.4 is equally applicable when the signal is sampled in the presence of a magnetic field gradient. The full echo sampling scheme of Fig. 3.17(a) therefore, confers two advantages over that shown in Fig. 3.14. First, there is a factor of 2 signal-to-noise gain. Second, there is no dispersion spectrum. This avoids the need to calculate the real absorption image by 'phasing the spectrum' since it is possible to calculate $\rho(\mathbf{r})$ by taking the modulus of the real and imaginary parts of the resulting transforms. This can be quite helpful where phase artefacts are present in the image due to sample movement or field inhomogeneities. Provided that the whole of **k**-space is sampled, the phase shifts result in a multiplication factor of $\exp(i\phi)$ in the image pixels, an effect which is removed by obtaining the image modulus. Two-quadrant sampling cannot be converted to modulus data because of the distortion caused by the dispersion image. Where one wishes to avoid modulus calculation, for example in inversion recovery experiments where the image sign is important, it is possible, in principle, to use automatic phase correction algorithms[66] although their use still depends on the acquisition of the full echo signal.

There is, however, a difficulty with the full echo sampling scheme where T_2 relaxation is significant. First, the delay before $t_x = 0$ leads to a decay of the signal at the origin of **k**-space. It is this amplitude which will determine

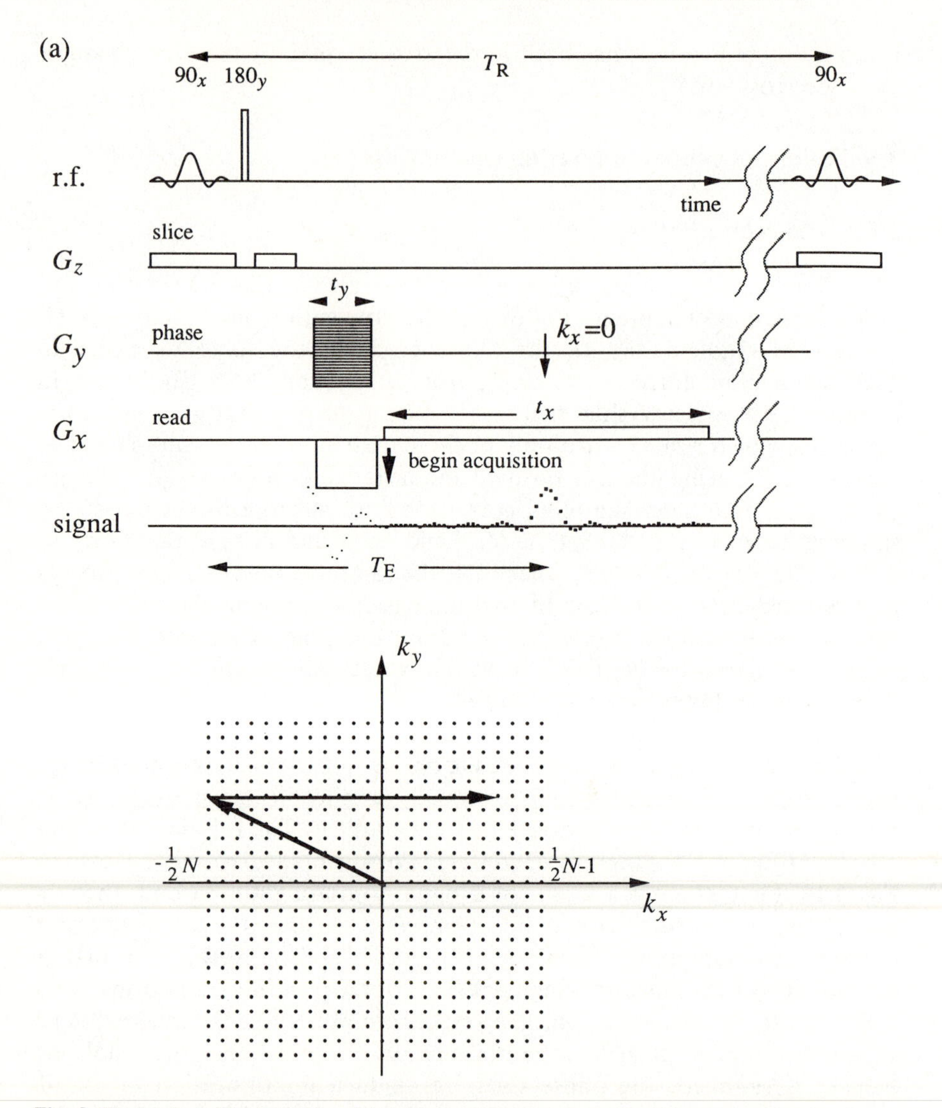

Fig. 3.17 (a) As in Fig. 3.14 but with full echo sampling. The use of a negative precursor read gradient causes a delayed 'gradient echo' to be formed, thus shifting the origin of **k**-space well away from the gradient pulse leading edge. Note that the gradient echo is the second of two echoes, the first being the spin echo formed by the 180_y r.f. pulse following dephasing during slice selection. The gradient echo formation and cycle repetition times, T_E, and T_R, are defined. (b) As in (a) but using spin echoes for both the slice selection and precursor read rephasing. In the **k**-space map the grey arrow represents the phase inversion caused by the 180_y pulse. (c) Echo advancing by means of a reduced area precursor read gradient pulse. This has the effect that a signal at the origin of **k**-space suffers less from T_2 relaxation than in normal full echo sampling, but, apart from the loss of some early noisy signal, the 'complete' echo is acquired, albeit with a time shift which induces first-order phase twisting. In each of the sequences the spin echo used in slice-selection can be conveniently replaced by a gradient echo as shown in Fig. 3.10(a).

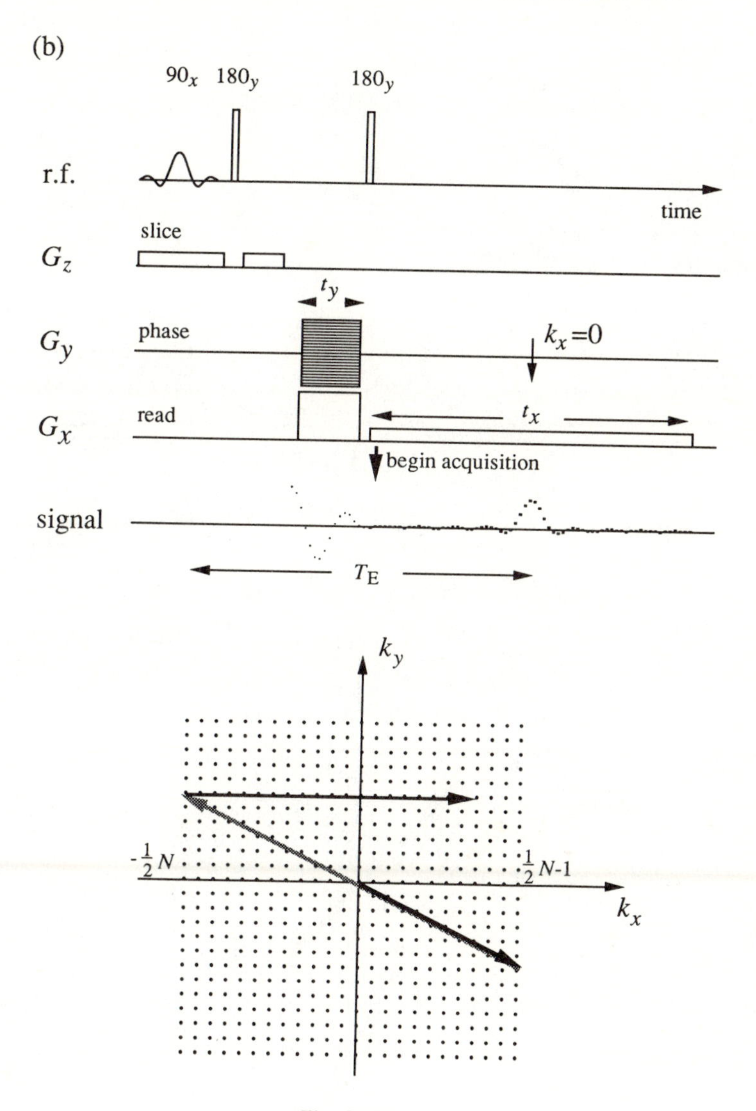

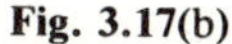

Fig. 3.17(b)

the image signal-to-noise ratio. T_2 decay over the period before the **k**-space origin leads to reduced signal-to-noise and hence reduced resolution. Second, since the signal will be modulated by a continuously decaying function, which is clearly not symmetric about $t_x = 0$, this will lead to convolution of the image by a rather unusual point spread function. Finally, any spatial inhomogeneity in T_2 will lead to the imposition of T_2 contrast effects in the image, an effect which is useful when comparison with a true spin

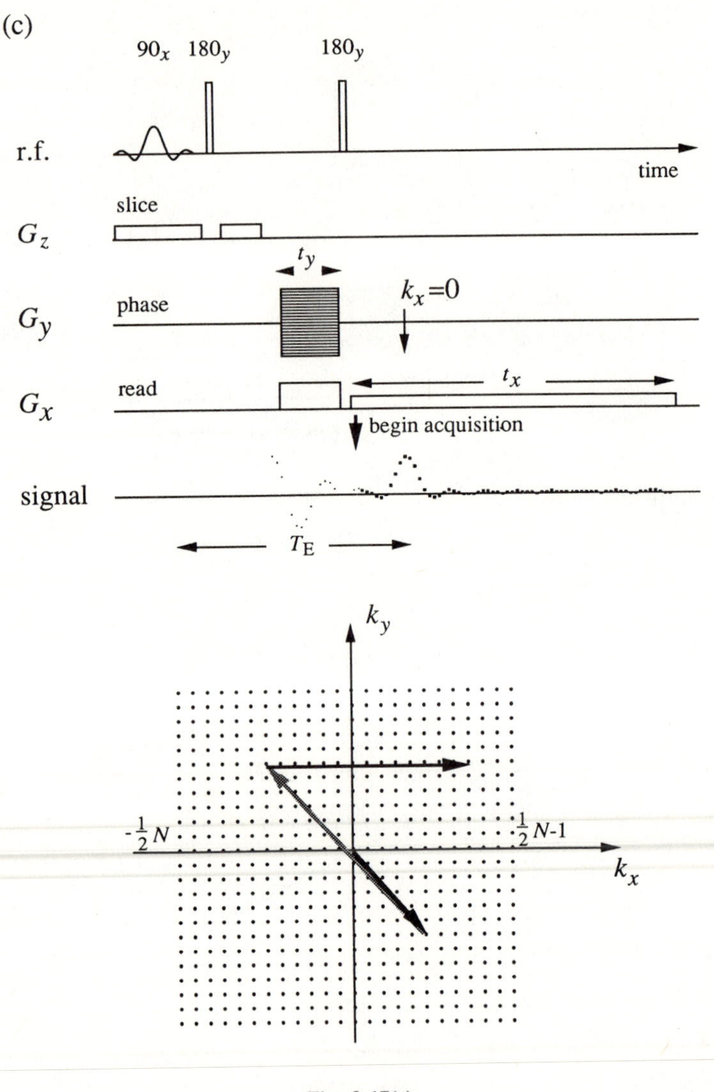

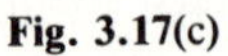
Fig. 3.17(c)

density image is available, but when taken alone can lead to difficulties in image interpretation.

The pulse sequence shown in Fig. 3.17(a) uses gradient reversal to induce negative phase shifts. Fig. 3.17(b) shows an alternative scheme using another 180_y hard r.f. pulse so that Figs. 3.17(a) and 3.17(b) illustrate, respectively the gradient echo and the spin echo. In the r.f. phase inversion method the precursor read gradient pulses are applied before the 180° pulse, causing k

evolution. The 180° pulse then inverts the phase shifts associated with the precursor gradient pulse.

The full echo and FID sampling schemes of Figs 3.17 and 3.14 suffer from the respective disadvantages of T_2 distortion on the one hand, and, on the other, a signal-to-noise loss coupled with the need to compute the real image. An alternative compromise is shown in Fig. 3.17(c). Here sampling begins before $k_x = 0$ but sufficiently close to $k_x = 0$ that relaxation effects are insignificant, and sufficiently far from $k_x = 0$ that the signal is still within the noise level and the resulting dispersion signal is negligible. Of course, the effect of a time shift in the origin of sampling is to introduce a 'phase twist' across the image, according to the shift theorem outlined in Section 1.2. This can be removed from the final image by computing the image modulus.

Even if the sampling origin is placed at $k_x = 0$ there are other important reasons why echoes should be used to form the signal. In the pulse sequence of Fig. 3.14 the origin of **k**-space sampling always occurs at the gradient leading edge. In all coils the gradient rise time is necessarily finite, and in some cases the leading edge will be accompanied by switching transients. For this reason it is preferable to avoid sampling the initial point while the gradient is switching and the pulse schemes of Figs. 3.17(a) and 3.17(b) avoid this by employing an echo. Furthermore where a spin echo is formed using a hard 180° r.f. pulse, this always produces a substantial transverse magnetization from spins outside the slice for which precise magnetization inversion has not occurred, either for reasons of r.f. inhomogeneity, or because the spins arise in a region of the sample outside the r.f. coil centre and thus are subject to a turn angle smaller than 180°. It is clear therefore that the use of an echo sampling scheme with a precursor G_x pulse confers an additional advantage. The compensating G_x evolution, before the origin of sampling, will destroy the unwanted magnetization caused by the non-selective r.f. pulse, but refocus that created initially by the selective pulse. Only this slice magnetization has the required phase shift label which will lead to a perfectly rephased signal at $k_x = 0$. For these reasons half-echo sampling, with acquisition beginning at $k_x = 0$ is better performed using the pulse sequence of Fig. 3.17(c) than with that shown in 3.14. Gradient pulses which destroy unwanted magnetization are sometimes referred to as 'spoiler' or 'crusher' pulses.

Finally, it is important to re-emphasize the difference in outcomes of the alternative approaches to echo formation embodied in the gradient echo and the spin echo. Gradient reversal will refocus only those precessions caused by the identical gradient coil in the 'previous history' of the sequence. Attenuation or phase shift effects due to susceptibility or polarizing field variations will persist. So will the evolution due to the chemical shift. This independence can often be an advantage, for example, where contrast relating to these effects is deliberately sought. In other cases it will be a severe disadvantage and r.f. refocusing will be preferred. One particular advantage of gradient reversal methods in medical imaging concerns the need to avoid excess r.f. power dissipation in the patient, an effect associated with multiple echo formation using hard r.f. pulses. This problem will concern us less in NMR microscopy, so this approach to imaging offers a greater degree of flexibility.

Another way of describing the difference between gradient echo and spin echo refocusing is to say that the gradient echo suffers from T_2^* relaxation in which both homogeneous and inhomogeneous broadening plays a role, whereas the spin echo is attenuated only by T_2. The degree of relaxation in each case will be given by the attenuation factor $\exp(-T_E/T_2^*)$ or $\exp(-T_E/T_2)$, respectively, where T_E is the time elapsed between the creation of transverse magnetization in the selection excitation process and the moment in the acquisition corresponding to the centre of **k**-space, the echo centre.

From the moment that the spins are disturbed from equilibrium by the initial excitation pulse, the longitudinal magnetization begins to recover by T_1 relaxation. The time over which this recovery occurs is determined by the cycle time, T_R the delay between successive signal averaging repetitions. T_E and T_R are indicated in Fig. 3.17(a). In subsequent depictions of pulse sequences, T_R will not be explicitly shown but is taken to be the time between the start of one sequence and the start of the next.

3.5.2 *Multislicing and STEAM*

In multislicing experiments it is possible to usefully employ the waiting time interval T_R which is needed to allow T_1 recovery. The method involves sequential imaging of adjacent or interleaved slices in a cycle time sufficient for the first slice in the sequence to recover. This facility relies on the fact that selective excitation ideally disturbs only one slice in the sample so that nuclear spins in other potential slices will remain in thermal equilibrium, available for immediate excitation following the first acquisition. None of the imaging sequences shown in Fig. 3.17 is entirely appropriate for this purpose. Gradient echoes suffer from T_2^* effects while the r.f. echo shown employs a non-selective 180° pulse which will clearly disturb all spins in the sample. It is possible in principle to use selective 180° pulses but these suffer from non-linearity artefacts as discussed in Section 3.2.3 and the usual solution of bandwidth spreading causes disturbance in adjacent slices. An elegant solution to this problem involves the use of stimulated echoes,[33] dubbed STEAM for STimulated Echo Acquisition Mode. These involve three r.f. pulses (see Section 2.6.5), any one of which may be made selective as shown in Fig. 3.18.

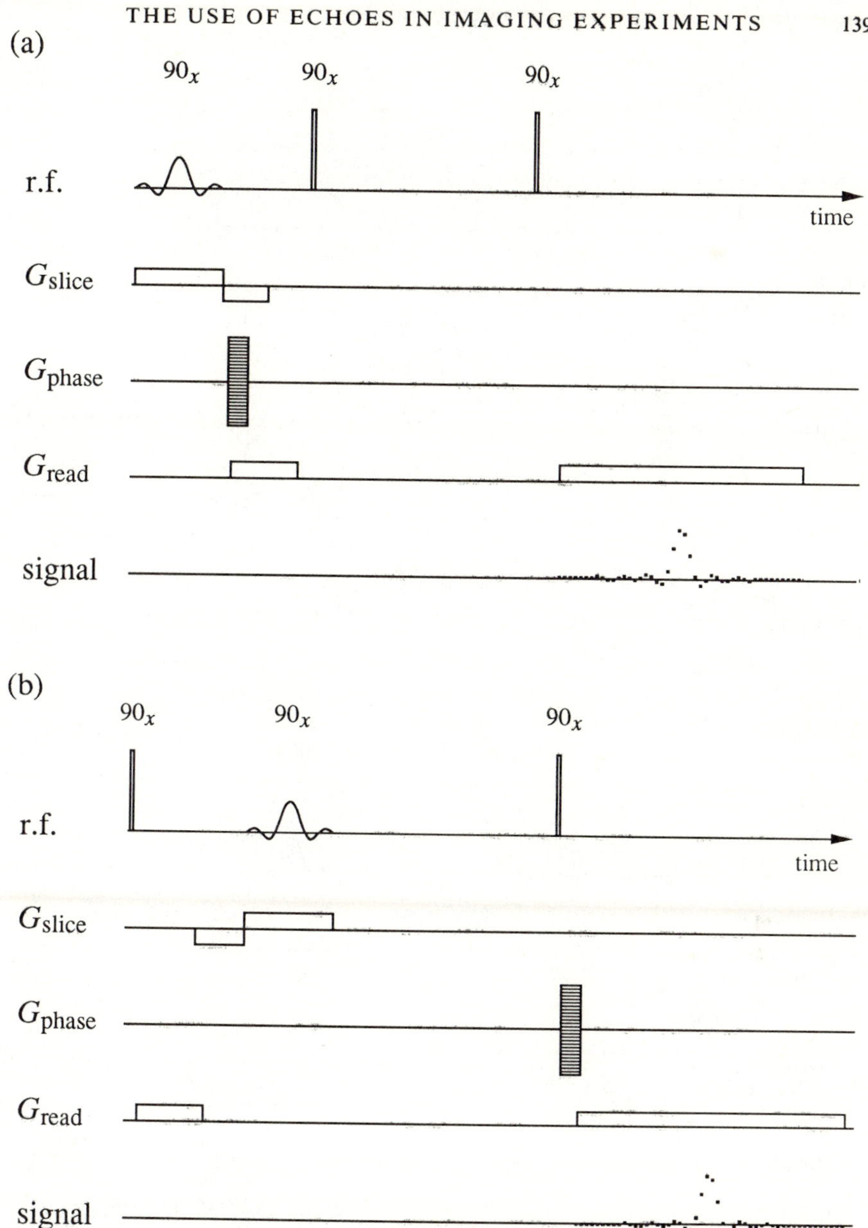

Fig. 3.18 STEAM imaging sequences adapted from Frahm *et al.*[33] In (a), (b), and (c) the slice-selective pulse is placed in the first, second, and third positions, respectively. The refocusing gradient is placed before the soft slice-selective 90_x pulse in (b) and (c) with the sign of the gradient dependent on the sign of phase spreading in the respective periods. (d) Shows a sequence in which all three r.f. pulses are slice selective. Note that any pair are self-compensating, a point which is easily understood by reference to the Mansfield alternating phase isochromats. In this and subsequent diagrams gradients are labelled as slice, phase, and read, with the axis assignment being arbitrary.

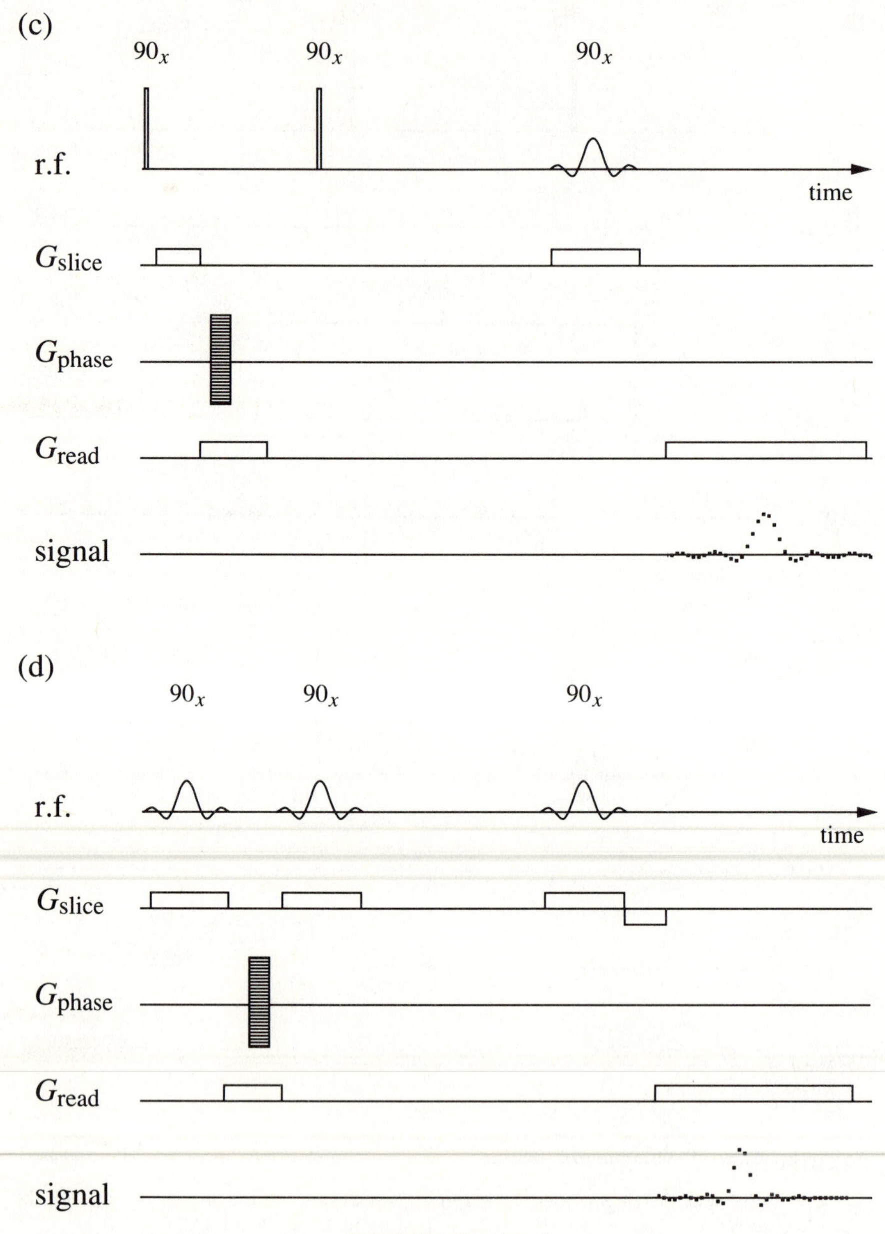

Fig. 3.18 *Continued*

The sequence for multislicing is shown in Fig. 3.19 and relies on the fact that the first two pulses 'store' the phase-encoded magnetization for the entire sample along the z-axis where it is subject to T_1 relaxation only, slices of which can be successively recalled using selective 90° pulses with the r.f. frequency being stepped before each sinc pulse and restored to normal for data read-out. The price paid in using stimulated echoes is the loss of a factor of 2 in signal-to-noise ratio. On the other hand, the sequence offers the advantage that r.f. refocusing is available without the need for selective or non-selective 180° pulses. The sequence also offers the possibility of obtaining a 'transmission' (i.e., non-slice-selective) image using the normal spin echo which appears between the first two pulses. This facility to obtain two echoes from the one initial excitation can also be used to reconstruct images at two different gradient strengths, thus providing differing degrees of magnification. An example of this 'zoom' imaging method is shown in Fig. 3.19(b).

In Fig. 3.18 it is clear that the existence of three r.f. pulses in the basic excitation sequence offers great flexibility. Pulses may be chosen to be spatially selective or chemical shift selective. For example, selective voxel excitation is easily performed in a single shot using a triple spatially selective pulse sequence as shown in Fig. 3.18(d), where the slice gradient is alternated along the three axes.

3.5.3 *Chemical shift selective imaging using stimulated echoes: CHESS*

The STEAM sequence provides an ideal solution to the problem of producing spatially and chemically selective excitation. Because three pulses are available, one may be used for slice selection while another can be made frequency selective, resulting in an echo from one particular region of chemical shift in the NMR spectrum.[67] Fig. 3.20 shows how a chemical shift selective (CHESS) pulse can be used for the initial excitation while the remaining two pulses consist of one spatially selective pulse and one non-selective pulse. In Fig. 3.20(b) it is clear that the multislicing facility inherent in STEAM is still available and it is therefore possible to use the method to obtain chemical shift selective multislice images in a very efficient manner.

Fig. 3.20(c) shows a sequence in which a single slice is excited to give a normal non-chemically selective image in the spin echo while a sequence of successively frequency-shifted CHESS pulses can yield images of different chosen lines in the NMR spectrum. Finally, Fig. 3.20(d) shows a double CHESS excitation of two different spectral lines (e.g., water and fat) so that in the subsequent multislicing recall, two stimulated echoes appear temporally separated for independent acquisition. In this example successively frequency-shifted SLICE pulses (this time transmitted in the presence

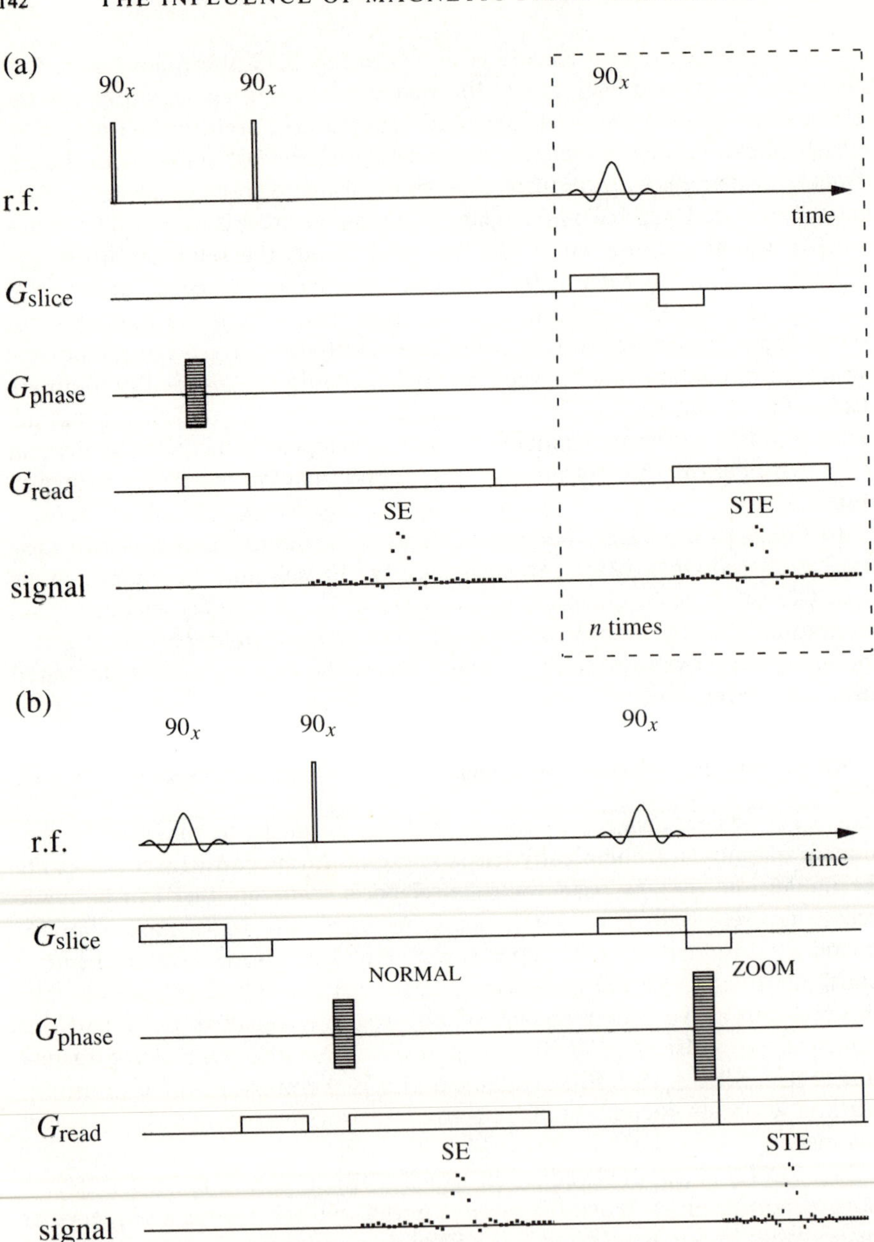

Fig. 3.19 (a) Multislice imaging using stimulated echoes, adapted from Frahm *et al.*[33] The first two hard 90_x pulses store phase-encoded magnetization from the entire sample along the z-axis. The final part of the sequence performs the slice selection and is stepped n times with the transmitter frequency being successively shifted to recall magnetization from adjacent slices. Note that the sequence allows the formation of a spin echo (SE) from the entire sample as well as the formation of a stimulated echo (STE) from the chosen slice. The SE signal may

of the slice gradient) allow the successive recall of signals from differing slices. The two separated echoes enable separate water and fat images to be obtained for each slice. One complication in this method is the different T_2 relaxation times experienced by the two spectral lines but this is a small price to pay for the capacity to obtain efficient multislice images from two independent components of the proton NMR spectrum.

3.5.4 *Echo summation*

Signal averaging provides a means of enhancing the signal-to-noise ratio in NMR by co-adding the FID signals in successive experiments, a method discussed in Section 2.6. Before repeating a pulse train and thereby gaining a new FID, it is necessary to wait for a sufficient period of time, of order T_1, for the thermal equilibrium nuclear magnetization to be substantially restored, or to use reduced turn angles where incomplete spin–lattice relaxation occurs. The discussion so far has neglected the possibility of employing residual transverse magnetization coherence.

In cases where the total sampling time is shorter than T_2, some transverse magnetization 'coherence' remains to be used again. The gradients used in the **k**-space evolution will have dephased this magnetization but the NMR signal can be restored by means of a phase-inverting 180° pulse followed by another gradient evolution, as illustrated in Fig. 3.2.1. Of course, some T_2 relaxation will have occurred so that the signal magnitude will be somewhat attenuated but, so long a substantial signal remains, the restoration of the transverse magnetization can be successively repeated in a CPMG train and the sensitivity enhanced by the summation of successive echoes. Once thermal equilibrium is restored the process is then repeated in the usual manner.

Note that two alternative signal-averaging options exist. Either the echo signals can be added before image reconstruction or separate images can be obtained from each echo for later addition. This latter option offers greater flexibility in the choice of analysis. Where T_2 contrast information is required instead of co-adding the images obtained from successive echoes the set can be analysed to yield a T_2 map and an initial magnetization map.

In order to understand the signal-to-noise gain, consider the pulse sequence of Fig. 3.21 where there are N points per full echo acquisition and the sampling interval is T. Suppose that the initial signal begins on the FID at the centre of the echo formed soon after the selective excitation. The

---

be used to reconstruct a 'transmission' image from all spins in pixels projected normal to the slice axis. (b) The use of both SE and STE signals to obtain images at two differing magnifications, an example of 'zoom' imaging (adapted from Frahm *et al.*[33])

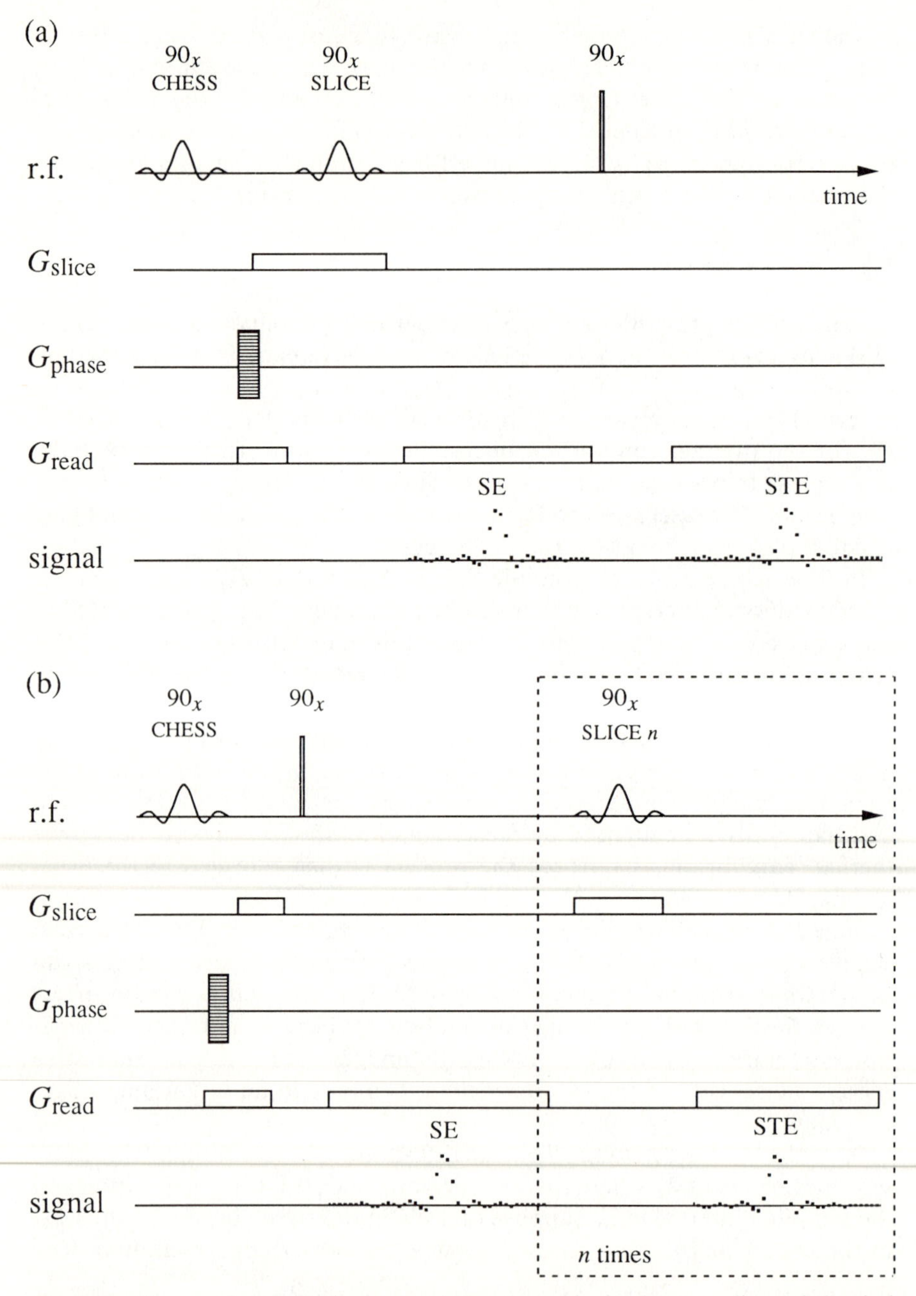

Fig. 3.20 The use of stimulated echo acquisition to enable simultaneous slice and chemical shift selection after Haase and Frahm.[67] (a) Chemical shift selective first pulse and slice-selective second pulse leading to spin echo (SE) and stimulated echo (STE) chemical and slice-selective imaging. (b) Scheme for multislicing with chemical shift selectivity. (c) Slice selection on the first pulse and chemical shift selection on the third pulse leading to an SE image without

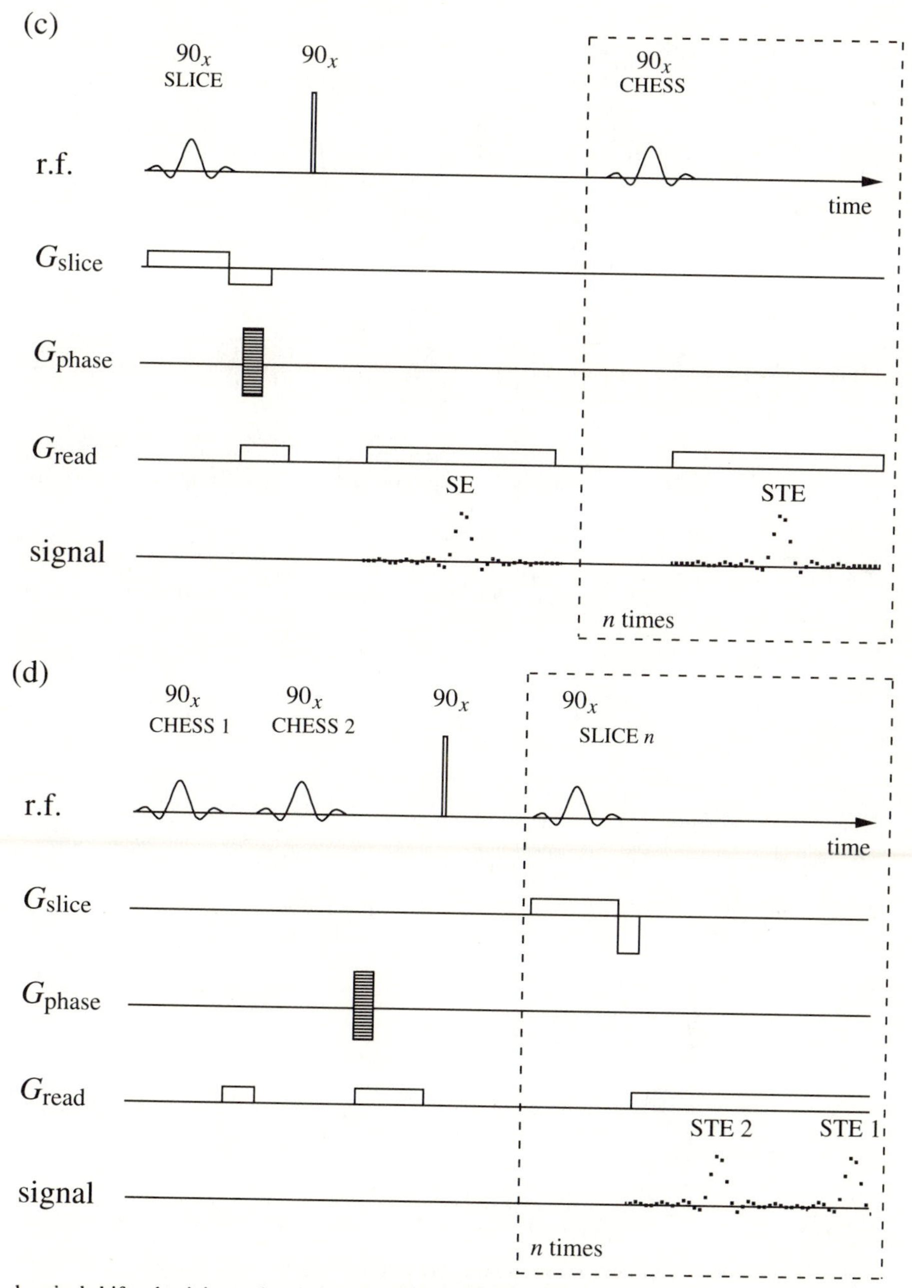

chemical shift selectivity and a chemically selective STE. By shifting the transmitter frequency successive lines in the spectrum may be recalled by the last pulse, a procedure which corresponds to chemical shift multislicing. (d) Multislice double chemical shift selection. The two STE images correspond respectively to the chemical shifts excited by CHESS 2 and CHESS 1.

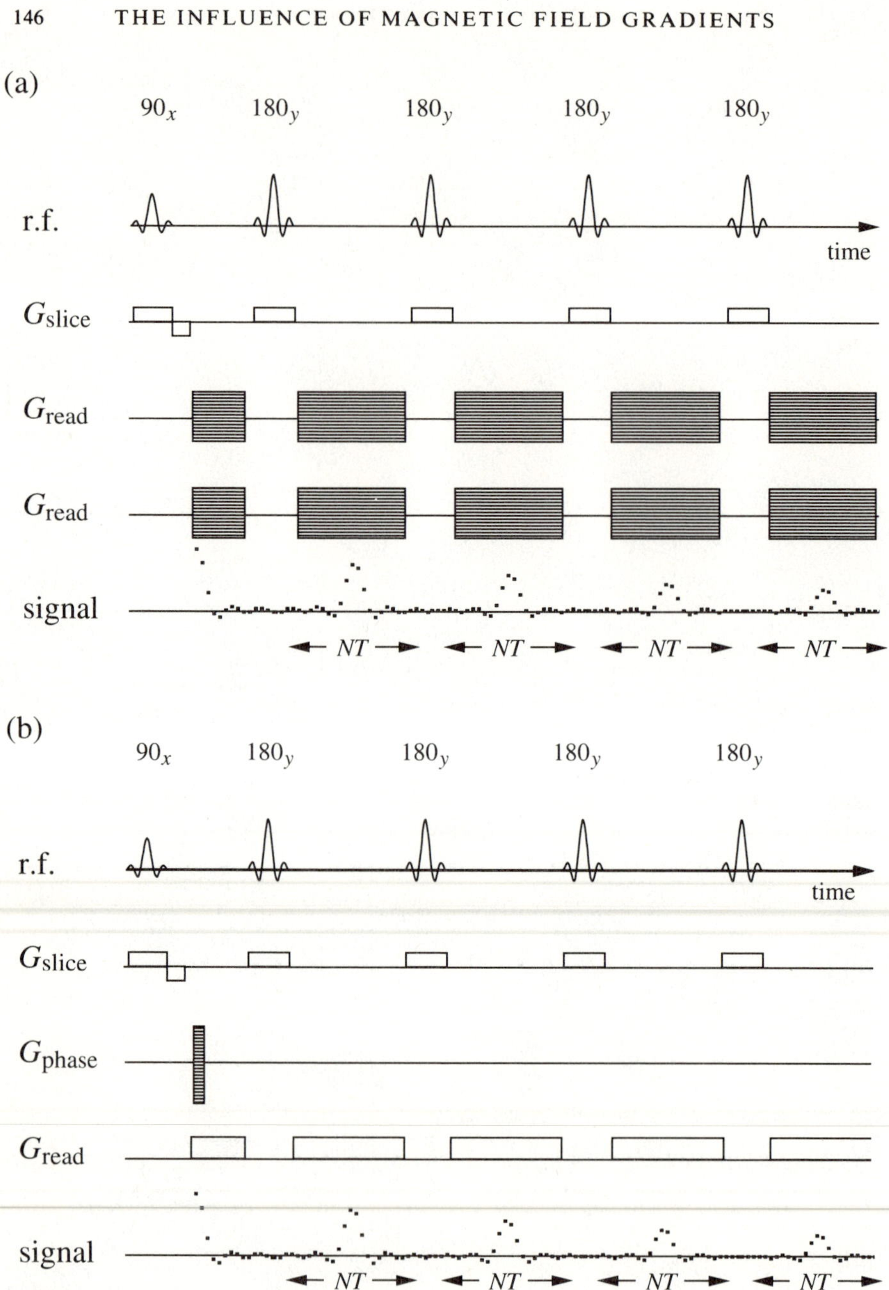

Fig. 3.21 Multiple echo data acquisition in (a) PR and (b) FI pulse schemes. Echoes are separated by NT where T is the sampling interval.

subsequent echoes can be regarded as comprising two similar 'FID signals' each of $N/2$ sampling points. In considering the role of T_2 relaxation we will neglect relaxation during one echo but consider the effect on the magnitude of the signal at each successive origin in **k**-space. We will find that, for optimum sensitivity, we may arbitrarily decrease the duration of each echo period provided that the overall attenuation after all echoes is fixed. In consequence, our assumption of little attenuation during an echo turns out to be very reasonable. The sequence shown in Fig. 3.21 uses soft 180° r.f. pulses in order to minimize any interference which might be caused by excitation of magnetization from outside the slice of interest. In the FI scheme these pulses cause successive acquisitions to have opposite sign phase-encoding, thus resulting in images which are successively reflected in the phase direction. This effect can be subsequently corrected by reflecting images prior to co-addition.

Compared with a single FID acquisition, the total signal-to-noise ratio for m echoes is enhanced by

$$(2m+1)^{-\frac{1}{2}}\left[1+\sum_{n=1}^{m} 2\exp(-nNT/T_2)\right] \tag{3.33}$$

where the denominator factor reflects the co-addition of the noise power. Obviously the advantage to be gained from co-addition decreases as the signal decays relative to the noise amplitude. Alternatively, full echo sampling can be used, starting with the first complete echo, in which case the total signal-to-noise ratio for m echoes is enhanced by

$$m^{-\frac{1}{2}}\sum_{n=1}^{m}\exp(-nNT/T_2). \tag{3.34}$$

The signal averaging enhancement can be maximized with respect to m to yield an optimum value which, for both modes of acquisition, and for $NT/T_2 \ll 1$, is given by[68]

$$m = 1.26\ T_2/NT. \tag{3.35}$$

This corresponds to an attenuation of the last (m^{th}) echo by $\exp(-1.26)$. Of course, in practice only integer values of m are possible. Multiple echo summation not only confers a signal-to-noise enhancement in NMR microscopy but also offers a means of avoiding certain image artefacts. These important benefits will be discussed further in Chapter 4.

3.6 Rapid sampling of k-space

The **k**-space sampling schemes shown in Figs. 3.14 and 3.15 both involve the acquisition of data along a single line. Because the image cannot be reconstructed until the **k**-space raster is completed, a multiplicity of FID signals is essential. The **k**-space rasters discussed above employ successive readings along a single dimension and a waiting period of order T_1 before the stepping of the second dimension. As a consequence, the minimum time for an entire raster sweep on an N^2 array is of order NT_1, typically in the region of 5 to 10 minutes. In NMR microscopy this time can be expected to be longer because of the need to enhance the signal-to-noise ratios by extensive signal averaging. In medical imaging, where voxel sizes are considerably larger and the signal is abundant, it is often feasible to trade off image signal-to-noise in return for a gain in imaging speed. This can be particularly advantageous when there is a need to image moving tissue such as the stomach or heart. To be useful, however, the raster sampling time needs to be reduced by several orders of magnitude.

3.6.1 *Low-angle excitation: FLASH imaging*

In Chapter 2 we saw that it is possible to optimize the turn angle for a given repetition rate ratio, T_R/T_1, and where the repetition time is made quite small and successive signals are averaged, a modest signal-to-noise gain is possible. Haase *et al.*[69] have shown, in an alternative to signal averaging, that another advantage is possible where small turn angles are used for selective excitation in imaging. The signal resulting from a turn angle θ is proportional to $\sin\theta$, while the z-magnetization which remains after this pulse is proportional to $\cos\theta$. In consequence, it is possible to use a rapid succession of small-angle excitations to sweep out the entire **k**-space evolution in a short period of time. This method has been termed Fast Low-Angle SHot imaging or FLASH. Of course, the repetitive 'reading' of the residual z-magnetization causes depletion but this may be compensated to a degree by spin–lattice relaxation. In practice quite small values of T_R/T_1 are employed so that the signal will not be described by the steady-state amplitude given in eqn (2.81) but rather by a simple transient expression. After n r.f. pulses of turn angle θ the FID amplitude will be given by $M_0 \cos^n\theta \sin\theta$. Typically θ is around 5°. In reducing the turn angle the total image acquisition time of FLASH is reduced at the expense of the image signal magnitude. Thus it is possible to trade off imaging speed and signal-to-noise ratios in a controllable manner.

Such rapid repetition experiments can result in interference from echo coherences in the transverse magnetization. In NMR imaging it might appear that this problem is alleviated by the effect of the read gradient dephasing the transverse magnetization before onset of the following low-angle excitation pulse. However, where full echo acquisition is used, this dephasing gradient will be matched by an equivalent refocusing period in the next cycle. On the other hand, phase gradient incrementing in each cycle prevents phase shift matching and so helps to suppress echo formation. To ensure complete echo suppression a deliberate spoiling pulse can be applied at the end of the read periods, a modification known as spoiled FLASH.[70]

A high-gradient, rapid-acquisition application of this method, snapshot FLASH,[71, 72] enables an entire **k**-space map to be obtained in a time of order 100 ms. As with Echo planar imaging, discussed in the next section, snapshot FLASH requires fast (<1 ms) gradient pulse switching, a feature which is difficult to achieve in large-scale medical imaging applications, but quite straightforward in micro-imaging. In the sequence reported in reference 71 the repetition time was 3 ms, allowing an image formed from 64 phase-encoding steps to be acquired in 192 ms. One consequence of such rapid repetition is that the data acquisition period is substantially reduced, in this case to 1.6 ms for 128 data points, a sampling rate corresponding to a bandwidth of 80 kHz. The cost of such high acquisition bandwidths is an increased noise component, an effect which leads to degraded resolution. One nice feature of FLASH however is the facility to operate over a range of acquisition intervals. Where slow motions are to be observed the value of T_R can be accordingly increased so as to take advantage of a reduced acquisition bandwidth.

Because T_1 relaxation effects are insignificant in rapid FLASH, the snapshot image will have little intrinsic T_1-weighting. Where such weighting is desired the FLASH acquisition can be preceeded by an inversion-recovery sequence in order to impose T_1-contrast. A Hahn echo precursor can be used where T_2-contrast is desired.

FLASH necessarily uses gradient echo refocusing since the imposition of phase-inverting r.f. pulses causes unwanted disturbance of the stored z-magnetization. As a consequence, FLASH is subject to T_2^* relaxation although this effect may be small where the acquisition time is very short, as in the snapshot version. The restriction to gradient echoes can be a disadvantage where it is necessary to elucidate the role of susceptibility inhomogeneity by comparing the gradient echo and r.f. echo images under identical conditions of T_1 and T_2 contrast.

In NMR microscopy, where voxel sizes are small and signal-to-noise is not willingly sacrificed, the use of such rapid **k**-space acquisition is less attractive and the reduction of the image amplitude by the factor $\sin\theta$ will generally be a severe disadvantage. However, it is possible to co-add successive **k**-space maps, thus providing an opportunity to gain the modest signal-to-noise improvement which results from signal averaging. For a moving sample this addition would only be possible for periodic motion where appropriate synchronous acquisition was used. Rapid acquisition is also relevant to microscopy when we consider the possibility of signal averaging in the

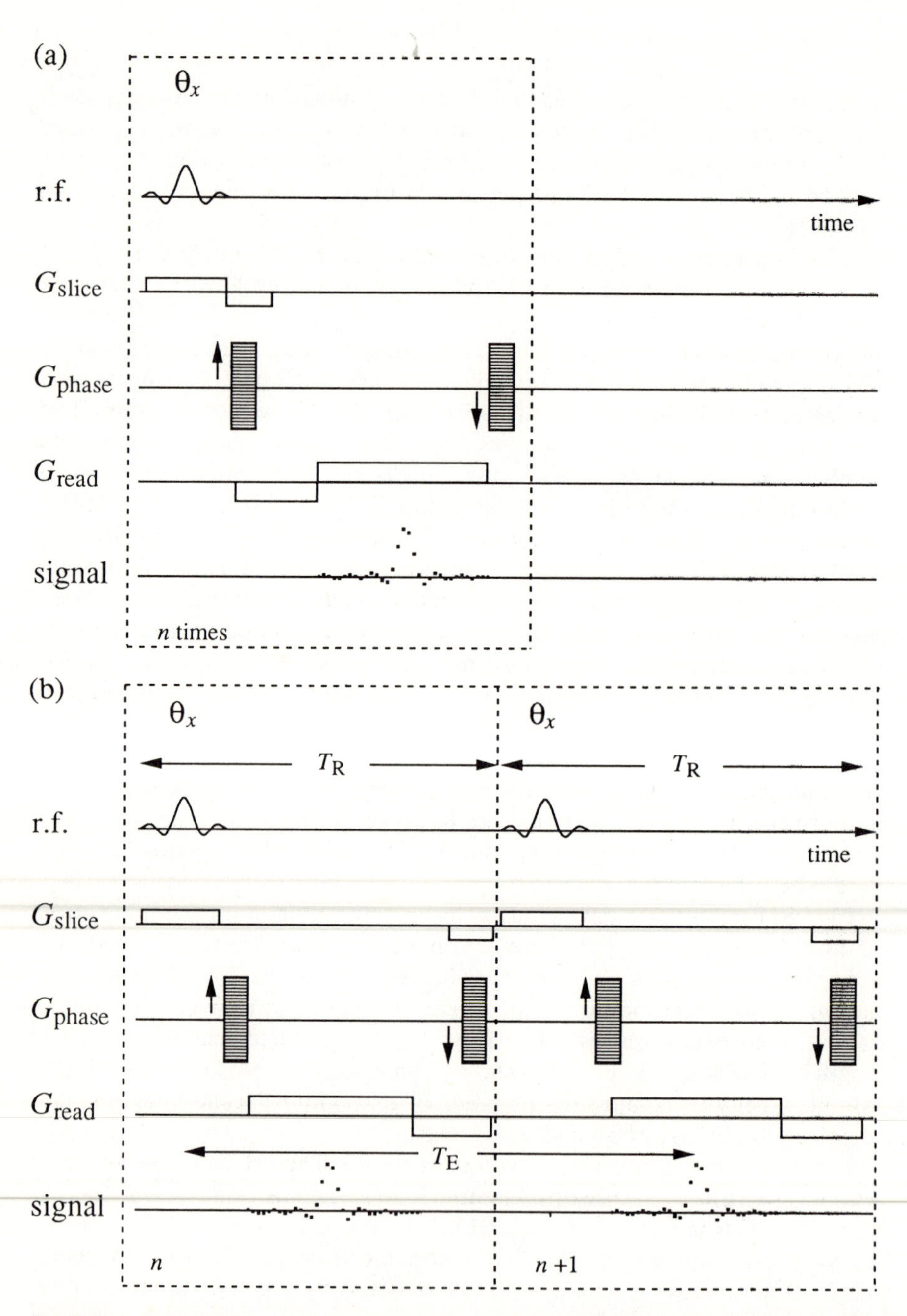

Fig. 3.22 (a) FAST and (b) CE-FAST pulse sequences due to Glyngell.[73] In the FAST SSFP imaging scheme the signal arises from the FID of the current cycle, whereas in CE-FAST the echo from the preceding cycle is acquired. (c) The FADE SSFP sequence of Redpath and Jones[74] in which the FAST and CE-FAST signals are simultaneously acquired as echo 1 and echo 2.

imaging of stationary samples. This matter will be discussed in some detail in chapter 4.

3.6.2 *Steady-state free precession: FAST, CE-FAST, FISP, and FADE*

Rather than suppressing the transverse coherence as is customary in FLASH, it is possible to carefully preserve it so as to generate a steady-state free precession (SSFP) condition. As pointed out in Chapter 2, coherences will occur for specific frequency offsets $2\pi n/\tau$, where τ is the repetition time, T_R. Provided that the spins are 'on-resonance' in the absence of the gradient, SSFP conditions could be achieved were it not for the phase gradient incrementing in successive cycles. In 1988 M. Glyngell[73] suggested the use of a rewinding phase gradient in each cycle as shown in Fig. 3.22(a) and (b). Two variants of SSFP sampling are shown in which the rephasing of the slice selection gradient phase shift is placed in the current and preceding cycles, respectively. In the first (known as FAST) the slice selection rephasing acts only on the FID generated by the r.f. pulse which begins the current cycle. The SSFP echo is therefore suppressed. In the second (known as CE-FAST or Contrast Enhanced FAST) the rephasing pulse is placed at the end of the preceding cycle so that the SSFP echo will be focused but the

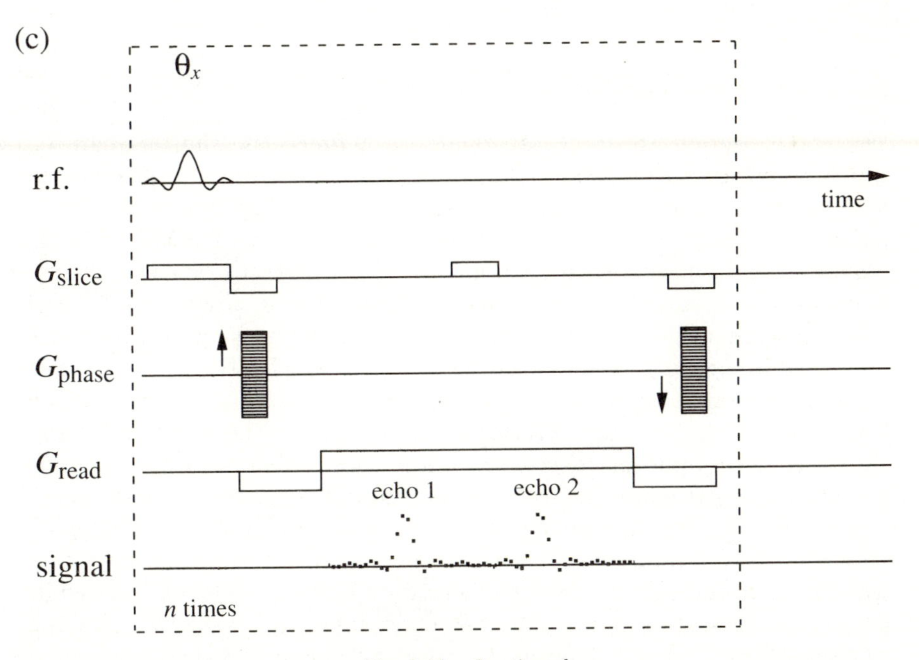

Fig. 3.22 *Continued*

current SSFP FID will be suppressed. FAST imaging returns a rapidly acquired image in a manner similar to FLASH but with the signal amplitude containing a degree of T_2 weighting (eqn. (2.87)). On the other hand, the imaging of the echo in CE-FAST results in an additional $\exp(-2T_R/T_2)$ contrast so that comparison of images obtained by the two methods can be very useful.

In conventional non-imaging SSFP, the FID and echo signals appear immediately after and immediately before the r.f. pulse, respectively, the echo being in essence a 'time-reversed FID' which rises up to the pulse. The use of the read gradient inversion pulse has the effect of shifting these signals in time, as is apparent in Figs. 3.22(a) and (b), so that use of the terms 'FID' and 'echo' to describe SSFP imaging signals may appear misleading. In practice it is possible to acquire both the echo and FID signals in superposition by employing a read gradient whose time integral is zero in each cycle, a method suggested independently by Oppelt *et al.* and known as FISP.[75] However it is advantageous to separate the echo and FID signals into two gradient echoes (sometimes labelled echo 1 and echo 2) by extending the central portion of the read gradient so that both may be obtained simultaneously and independently. This modification, termed FADE for Fast Acquisition Double Echo, by Redpath and Jones[74] combines the advantages of FAST and CE-FAST in a single experiment. The FADE pulse sequence is shown in Fig. 3.22(c).

3.6.3 *Echo planar imaging*

The magnetization coherence available on a time-scale shorter than T_2, which was discussed in Section 3.5, affords an opportunity to produce an entire **k**-space raster in a single-shot sequence following one selective r.f. pulse, a possibility which has enabled the production of NMR imaging movies! This remarkable application of spin echoes to NMR imaging, known as echo planar imaging, is due to Mansfield and co-workers.[76-79] and two typical echo planar pulse schemes with associated **k**-space rasters are illustrated in Fig. 3.23. The first of these, known as FLEET, for Fast Low-angle Excitation Echo planar Technique,[80] involves a zig-zag sweep though **k**-space. Of course the correct raster consists of parallel lines so that the zig-zag sweep needs appropriate compensation. This is obtained by performing a double zig-zag in which the second acquisition is carried out with the raster moving into the opposite quadrant by reversing the sign of the starting G_y gradient. This second sweep is performed in direct succession without waiting for restoration of the z-magnetization by relaxation and this can be achieved because the first r.f. pulse has a turn angle, α, less than 90°, leaving a component of z-magnetization ($M_0 \cos\alpha$) available for use by the second

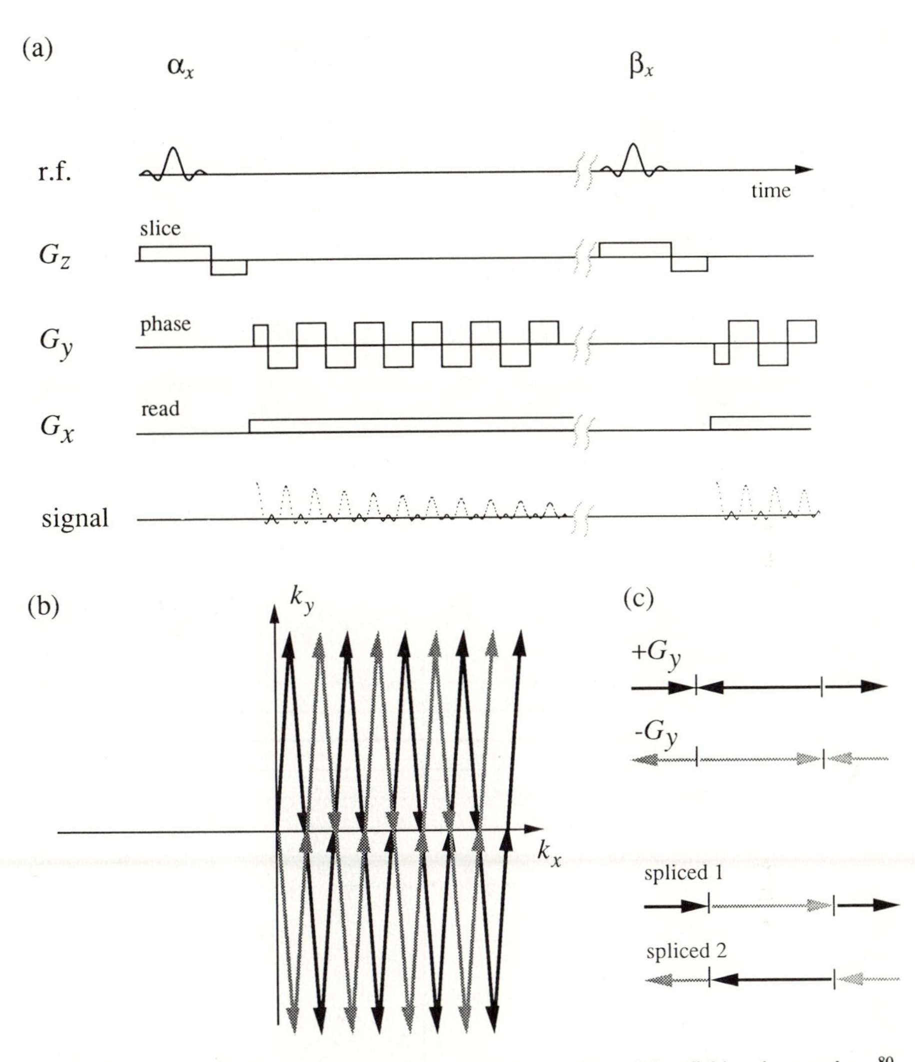

Fig. 3.23 (a) FLEET echo planar imaging pulse sequence due to Mansfield and co-workers.[80] (b) **k**-space FLEET raster. (c) Data splicing.

r.f. pulse of turn angle β. In order to ensure that the signal amplitudes are the same in the first and second sweeps, the pulse turn angles must obey

$$\tan \alpha = \sin \beta \tag{3.36}$$

the usual solution being to set $\alpha = 45°$ and $\beta = 90°$. The curious shape of the raster shown in Fig. 3.23(b) is compensated by splicing the data[29] as shown in Fig. 3.23(c). This creates two sets of data, one with the arrows

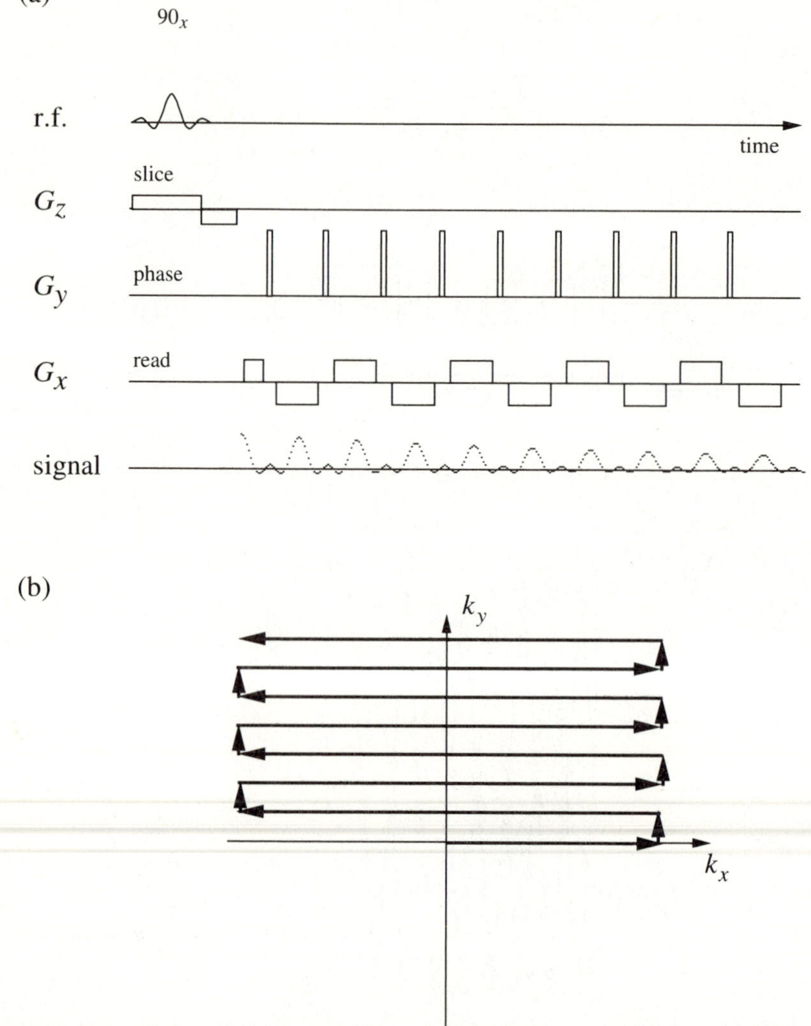

Fig. 3.24 (a) BEST echo planar imaging pulse sequence due to Mansfield and co-workers.[78] (b) BEST **k**-space raster. (c) MBEST echo planar imaging pulse sequence.[81] (d) MBEST **k**-space raster.

pointing left to right and vice versa. The two sets of data with oppositely directed arrows are then one-dimensionally Fourier transformed along the k_y dimension. Because the **k**-space evolution is opposite in each case, the resulting transformed arrays are left/right mirror images. These may be co-added following reflection of one of the arrays with respect to the k_x

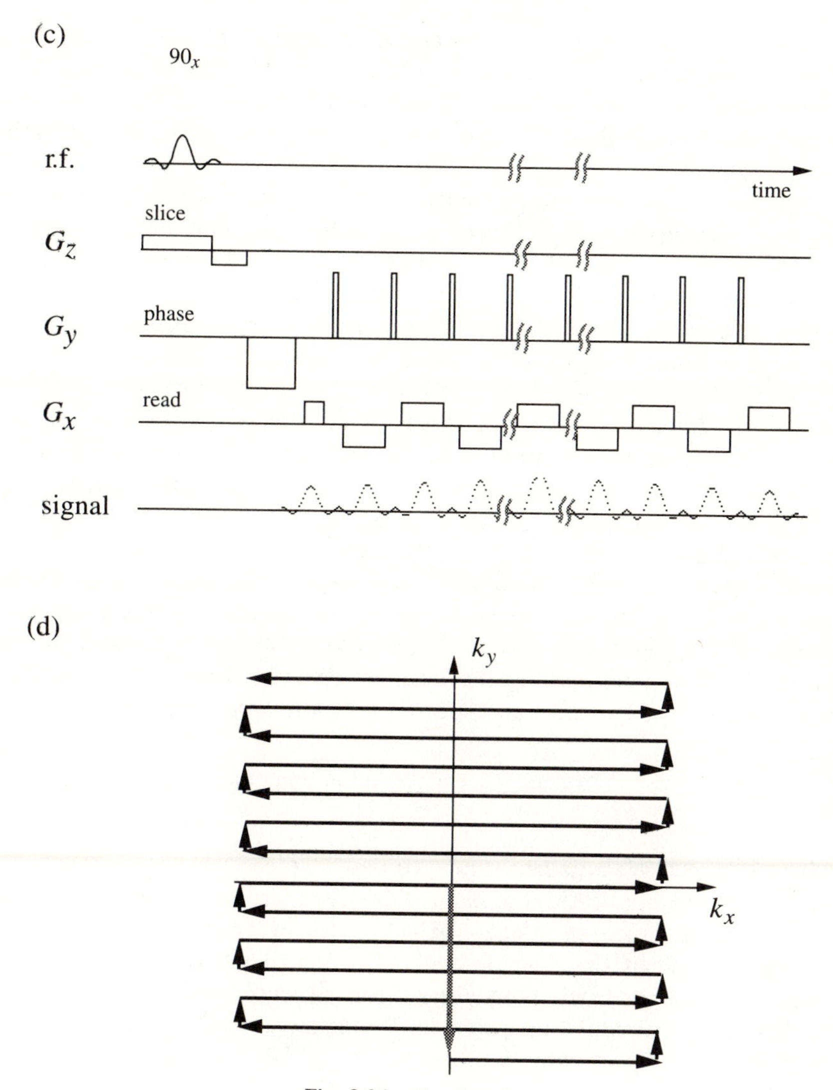

Fig. 3.24 *Continued*

axis. This co-addition process helps correct for artefacts which arise from the non-rectangular character of the **k**-space sampling.

The second echo planar acquisition scheme, shown in Fig. 3.24, is more rapid than FLEET. Known as BEST for Blipped Echo planar Single-pulse Technique,[78] the need for a double sweep of **k**-space is avoided by ensuring that the raster is rectangular as shown in Fig. 3.24(a). This is achieved by avoiding the steady k_x evolution associated with a constant G_x gradient by applying G_x as a series of short pulses, or 'blips', at the end of each k_y

evolution period. Of course, the alternate reversal of the data-sampling direction, apparent in Fig. 3.24(b), means that alternate data lines in the array must be reversed before Fourier transformation. The total image acquisition time for BEST is typically 32 ms (with, of course, twice this time being required for FLEET).

The BEST sequence samples only two quadrants of **k**-space thus enforcing the computation of the real image with the resulting problems, as discussed in Section 3.5.1. A modified version known as MBEST[81] (Modulus BEST) uses full **k**-space sampling by applying a precursor G_x phase gradient as shown in Figs. 3.24(c) and (d). This sequence doubles the acquisition time to around 65 ms but is relatively free from phase artefacts. As with all such echo-sampling schemes, the displacement of the **k**-space origin confers significant T_2 contrast on the image. In the gradient echo MBEST scheme shown in Fig. 3.24 the contrast arises inherently from T_2^*. Of course the MBEST raster can also be implemented using spin echoes in which case the image is T_2-weighted.

One feature of these rasters is the need for rapid gradient switching, especially in the BEST and MBEST sequences. The finite rise times associated with the G_y reversals causes the gradient wave-form to be somewhat trapezoidal, an effect which can be compensated for by using non-linear signal sampling along the k_y axis[82] ensuring that k_y is advanced in equal increments by unequally spacing the time according to

$$\Delta k_y = \frac{\gamma}{2\pi} \int_{t_m}^{t_n} G_y(t) \, \mathrm{d}t = \text{constant} \tag{3.37}$$

where t_m and t_n are adjacent sample times.[81] Indeed eqn (3.37) can be used to enable the application of sinusoidal read gradients which are much simpler to implement in practice. A key factor in the success of echo planar imaging has been the development of screened gradient coils[83-85] which enable rapid current switching without consequential eddy current gradients arising in the surrounding probe and magnet assembly. These screening methods are of particular importance in NMR microscopy and will be discussed in Chapter 9.

It should be emphasized that the price paid for rapid **k**-space sampling is the reduced image signal-to-noise ratio which results from the large acquisition bandwidths which are required. In this sense echo planar imaging (EPI) suffers in the same sense as FLASH although it should be noted that FLASH is inherently less efficient because of the need to repeat the selective excitation process and to use low angle excitation. Another difference between the methods concerns the timescale over which the raster may be applied. EPI is restricted to a time of order T_2^* or less whereas FLASH may take advantage of longer times, provided the sample motion is sufficiently small. Clearly the FLASH and Echo Planar Imaging methods are complementary with EPI being advantageous where an extremely rapid raster is required.

Echo planar NMR movies are out of the question in the microscopic realm because of the sensitivity problem. However, where sample movement is either cyclic or steady, motion can be studied using successive image averaging. The use of EPI and other motion contrast schemes in NMR microscopy is dealt with in Chapter 8.

3.7 Translational motion of the spins

3.7.1 *The influence of diffusion in the presence of field gradients*

The formation of an echo in the presence of a magnetic field gradient is predicated on the assumption that nuclei experience the same local Larmor frequency during the successive dephasing and rephasing parts of the cycle. Since there is a correspondence between the nuclear position and the local magnetic field, this assumption is equivalent to requiring that the nuclei do not move in translation along the gradient direction. In fact, since we are dealing with signals arising from nuclei in liquid state molecules, some translational motion is inevitable because of molecular self-diffusion. This movement leads to random fluctuations in Larmor frequency and hence to a distribution of residual phase shifts at the echo centre. The influence of self-diffusion on spin echo amplitudes was apparent to Hahn in his original 1950 paper.[86] Indeed the multiple echo scheme of Carr and Purcell[87] was proposed as a means of minimizing such effects.

There are two major reasons why phase shifts and echo attenuation arising from molecular translation are important to NMR microscopy. Self-diffusion causes a fundamental limitation to resolution which, in many cases, may dominate the intrinsic sensitivity limit. On the positive side, however, the effect of these modulations on the echo amplitude can provide a contrast mechanism which makes possible the imaging of the molecular displacement spectrum.

In this section we shall be concerned with molecular self-diffusion alone; more complex motion will be treated in Chapter 6. For the moment we shall be concerned with understanding what happens to the NMR signal when the spins diffuse in the presence of a magnetic field gradient. The most convenient way to understand molecular self-diffusion is to picture it as a succession of discrete hops with motion resolved in one dimension, the direction of the field gradient. The independence of motion in other directions means that we treat dimensions separately, obtaining total displacements where necessary by adding quadratically according to the Pythagoras theorem.

Let the mean time between steps be τ_s and the r.m.s. displacement in one dimension be ξ. A molecule has equal probability of jumping to the left or right so that the distance travelled after n jumps at time $t = n\tau_s$ is given by

$$Z(n\tau_s) = \sum_{i=1}^{n} \xi\, a_i \tag{3.38}$$

where a_i is a random number equal to $\pm$ 1. Z represents the z-axis displacement of molecules from their respective origins. Thus

$$\overline{Z^2(n\tau_s)} = \sum_{i=1}^{n}\sum_{j=1}^{n} \xi^2\, \overline{a_i a_j} \tag{3.39}$$

where horizontal bar represents an ensemble average. Because a_i is randomly ± 1, $\overline{a_i a_j} = 0$ unless $i = j$ so that all cross-terms cancel. Thus

$$\overline{Z^2(n\tau_s)} = \sum_{i=1}^{n} \xi^2\, \overline{a_i^2} = \xi^2 \sum_{i=1}^{n} 1 = n\xi^2. \tag{3.40}$$

Defining the self-diffusion coefficient as

$$D = \xi^2/2\tau_s \tag{3.41}$$

we obtain

$$\overline{Z^2(t)} = 2Dt \tag{3.42}$$

and, for three dimensions,

$$\overline{R^2(t)} = 6Dt. \tag{3.43}$$

Now we calculate the influence of this motion on the coherence of transverse magnetization in the case where a magnetic field gradient is present. Our approach follows the method originally used by Carr and Purcell in their classic paper on spin echoes.[87] The only motion which will concern us is that along the field gradient axis which we will label z. For convenience we consider the influence of diffusion on the transverse magnetization of spins originating at $z = 0$. Then the local Larmor frequency in each case is

$$\omega(n\tau_s) = \gamma B_0 + \gamma G \sum_{i=1}^{n} \xi\, a_i \tag{3.44}$$

So that the cumulative phase angle after time $\tau = n\tau_s$ is

$$\phi(t) = \gamma B_0\, n\tau_s + \sum_{m=1}^{n} \gamma G \tau_s \sum_{i=1}^{m} \xi\, a_i. \tag{3.45}$$

The first term in eqn 3.45 represents the constant average Larmor precession and is of no particular interest here. We shall be concerned only with the second, phase deviation, term which varies randomly across the ensemble and so causes dephasing. We may write it

$$\Delta\phi(t) = \gamma G \tau_s \xi \sum_{i=1}^{n} (n + 1 - i)\, a_i. \tag{3.46}$$

The proof of this relationship is best demonstrated pictorially as in Fig. 3.25 where each row in the triangle indicates the successive, random, value of a_i up to time $m\tau_s$ and the horizontal sum of these elements gives the phase shift which occurs in the interval τ_s while the spin is at the current location. The total phase shift accumulating at time t is given by the area of the triangle, multiplied by $\gamma G \tau_s \xi$.

What we wish to calculate is $\overline{\exp(i\Delta\phi)}$, the coefficient by which the ensemble-averaged transverse magnetization will be phase modulated as a result of the diffusional motion in the presence of the gradient. Of course, $\Delta\phi(t)$ varies randomly over the ensemble. We will assume that its distribution, $P(\Delta\phi)$, is Gaussian which is a good assumption when dealing with the sum of very many randomly varying quantities, a result known as the 'central limit theorem'.[88] Thus

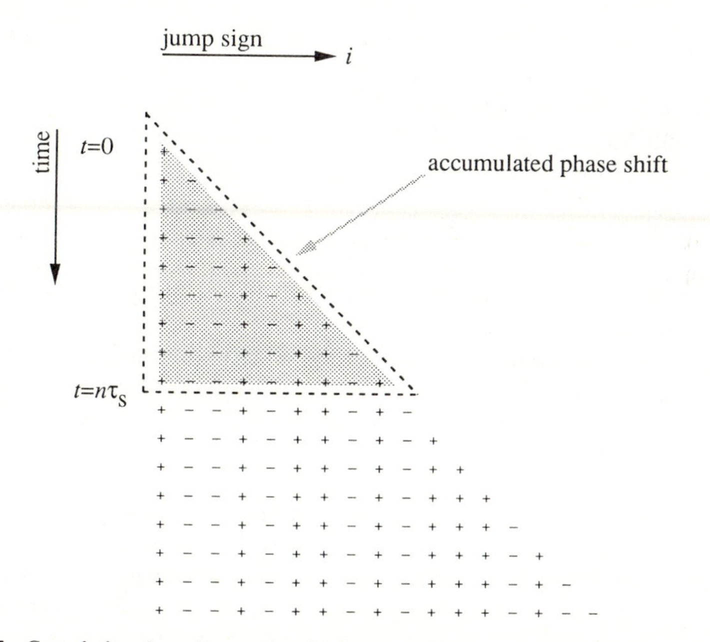

Fig. 3.25 Cumulative phase diagram in which the sign at the end of each row indicates whether the jumps in the current τ_s interval are positive or negative with respect to the gradient direction. The precessional phase shift which results during that interval is determined by the net position in the gradient and is therefore calculated by summing jumps along the row. The cumulative shift at time t is given by the sum within the shaded triangle.

$$\overline{\exp(\mathrm{i}\Delta\phi)} = \int_{-\infty}^{\infty} P(\Delta\phi)\exp(\mathrm{i}\Delta\phi)\,\mathrm{d}(\Delta\phi). \tag{3.47}$$

Since $P(\Delta\phi)$ is given by the normalized Gaussian function, $(2\pi\,\overline{\Delta\phi^2})^{-\frac{1}{2}}$ $\exp(-\Delta\phi^2/2\,\overline{\Delta\phi^2})$, eqn 3.47 yields

$$\overline{\exp(\mathrm{i}\Delta\phi)} = \exp(-\overline{\Delta\phi^2}/2). \tag{3.48}$$

This coefficient represents a simple attenuation of the magnetization and hence the signal at time t. In order to evaluate $\overline{\Delta\phi^2}$ we square eqn 3.46 and take an ensemble average. Once again, cross-terms disappear. Thus

$$\begin{aligned}\overline{\Delta\phi^2} &= \gamma^2 G^2 \tau_s^2 \xi^2 \sum_{i=1}^{n} (n+1-i)^2 \\ &= \gamma^2 G^2 \tau_s^2 \xi^2 \sum_{j=1}^{n} j^2 \\ &= \frac{1}{3}\gamma^2 G^2 \tau_s^2 \xi^2 n^3 \end{aligned} \tag{3.49}$$

where the last sum is evaluated assuming that n is large. (Note that this is the standard mean-squared phase shift for a triangular section of steps as shown in Fig. 3.25, where n is the numbers of steps in the orthogonal sides.) Substituting eqn (3.41) we find that the signal attenuation due to diffusion is expressed by the coefficient

$$\overline{\exp(\mathrm{i}\Delta\phi)} = \exp(-\tfrac{1}{3}\gamma^2 G^2 D t^3). \tag{3.50}$$

This t^3 dependence is characteristic of self-diffusion in the presence of a steady gradient.

Suppose we now consider the dephasing effect which remains at the time of formation of an echo at time $2t$ when a steady gradient is present. The phase step diagram for this is shown in Fig. 3.26. Here the 180° pulse has reversed all phase shifts which existed before time t, where t is the separation of the first and second r.f. pulse.

Clearly the second evolution period contains a section which completely cancels the net phase shift which occurred before the 180° pulse. The residual phase shift at the time of the echo is given by the sum of the shaded triangular region. This region now involves two uncorrelated triangular segments, each with orthogonal sides of n steps. Using our previous observation concerning $\overline{\Delta\phi^2}$ in such triangular sections, we can see that the value of $\overline{\Delta\phi^2}$ which applies in the case of the echo is exactly double that which applied in

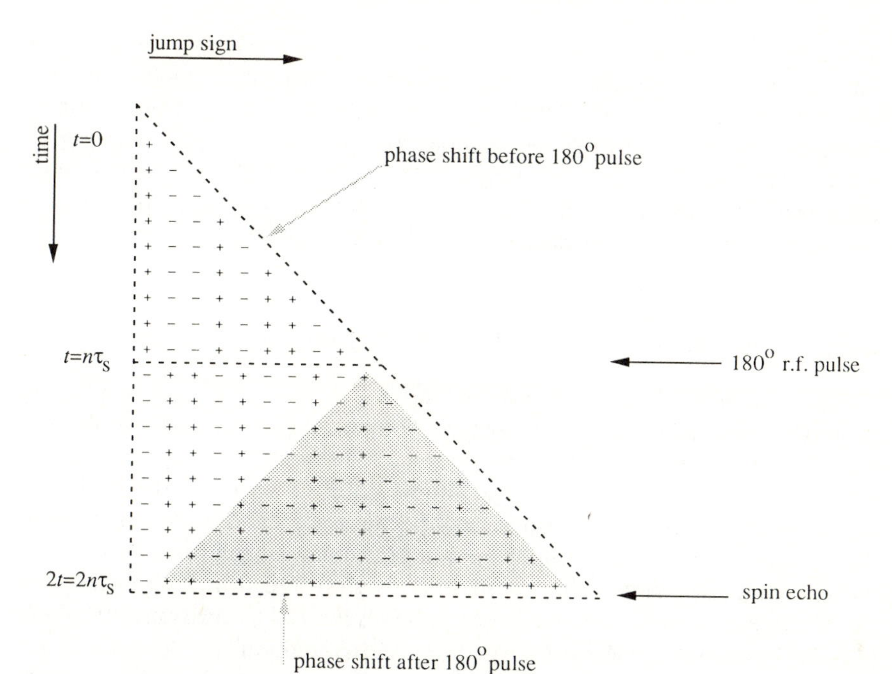

Fig. 3.26 Cumulative phase diagram for spin echo formation in a steady magnetic field gradient. The net phase shift is the sum of that occurring before and after the 180° r.f. pulse and corresponds to the shaded region.

Fig. 3.25 (and twice that which occurs during the period t between the first two r.f. pulses.) Consequently, for the echo at $2t$

$$\begin{aligned}\overline{\exp(\mathrm{i}\Delta\phi)} &= \exp(-\tfrac{2}{3}\gamma^2 G^2 D t^3) \\ &= \exp(-\tfrac{1}{12}\gamma^2 G^2 D (2t)^3) \end{aligned} \tag{3.51}$$

where the second form expresses this attenuation in terms of the time $2t$ from the first r.f. pulse to the echo centre. This constant-gradient spin echo method has been used to obtain self-diffusion coefficients in simple liquids with an accuracy of order 0.1%.[89]

Eqn (3.51) is easily generalized to a multiple Carr–Purcell echo sequence. For m echoes there are m such triangular phase step regions in the diagram equivalent to Fig. 3.26. Consequently the attenuation coefficient is

$$\begin{aligned}\overline{\exp(\mathrm{i}\Delta\phi)} &= \exp\left(-\frac{2m}{3}\gamma^2 G^2 D t^3\right) \\ &= \exp(-\tfrac{2}{3}\gamma^2 G^2 D t^2 (mt)). \end{aligned} \tag{3.52}$$

Now the dependence of the attenuation on the total time mt is linear rather than cubic as in the case of the simple spin echo. This means that the attenuation effect at some total time, mt, may be arbitrarily reduced by decreasing the interval t between echoes and increasing m, the number of echoes. This was the method suggested by Carr and Purcell to eliminate additional relaxation of the transverse magnetization due to diffusive motion occurring in the presence of field inhomogeneities.

3.7.2 *The pulsed magnetic field gradient: diffusion and velocity*

The attenuation of the spin echo which arises from diffusive dephasing under the influence of a steady gradient may be used to measure molecular self-diffusion. This method has been effective in providing very precise self-diffusion coefficients in a variety of simple liquids. However, the method is inconvenient where the motion is slow because the steady gradient spreads the Larmor spectrum at all times, and in particular during the period of r.f. pulse transmission and during the period of signal detection. This means that the maximum gradient which can be applied is limited by the transmitter and receiver bandwidths. McCall, Douglass, and Anderson[90] suggested in 1963 that the gradient might be applied in the form of rectangular pulses inserted, respectively, in the dephasing and rephasing parts of the echo sequence, but gated off during r.f. pulse transmission and signal detection. The pulsed gradient spin echo (PGSE) sequence, first demonstrated by Stejskal and Tanner in 1965,[91] is shown in Fig. 3.27(a) and the equivalent phase step diagram in Fig. 3.28. The use of such a phase step approach in analysing the PGSE experiment was first proposed by James and Macdonald.[92] Here we denote the pulse duration as δ and the separation as Δ. The gradient magnitude is labelled g and should be distinguished from the upper case symbol used to represent our imaging gradients.

Following the previous argument it is apparent that the phase shifts associated with the unshaded areas cancel and the net phase shift is obtained by summing two uncorrelated triangular regions, each with mean square phase shift $\frac{1}{3}\gamma^2 g^2 \tau_s^2 \xi^2 n^3$, along with one uncorrelated rectangular region with mean square phase shift $\gamma^2 g^2 \tau_s^2 \xi^2 n^2 (p - n)$. The net mean square phase shift is therefore

$$\begin{aligned}\overline{\Delta\phi^2} &= \gamma^2 g^2 \tau_s^2 \xi^2 n^2 (p - n + \tfrac{2}{3}n) \\ &= 2\gamma^2 g^2 \delta^2 D(\Delta - \delta/3).\end{aligned} \tag{3.53}$$

Using eqn 3.48 we obtain the well-known Stejskal–Tanner equation[91] for the attenuation of the echo amplitude.

$$S(g)/S(0) = \exp[-\gamma^2 g^2 \delta^2 D(\Delta - \delta/3)]. \tag{3.54}$$

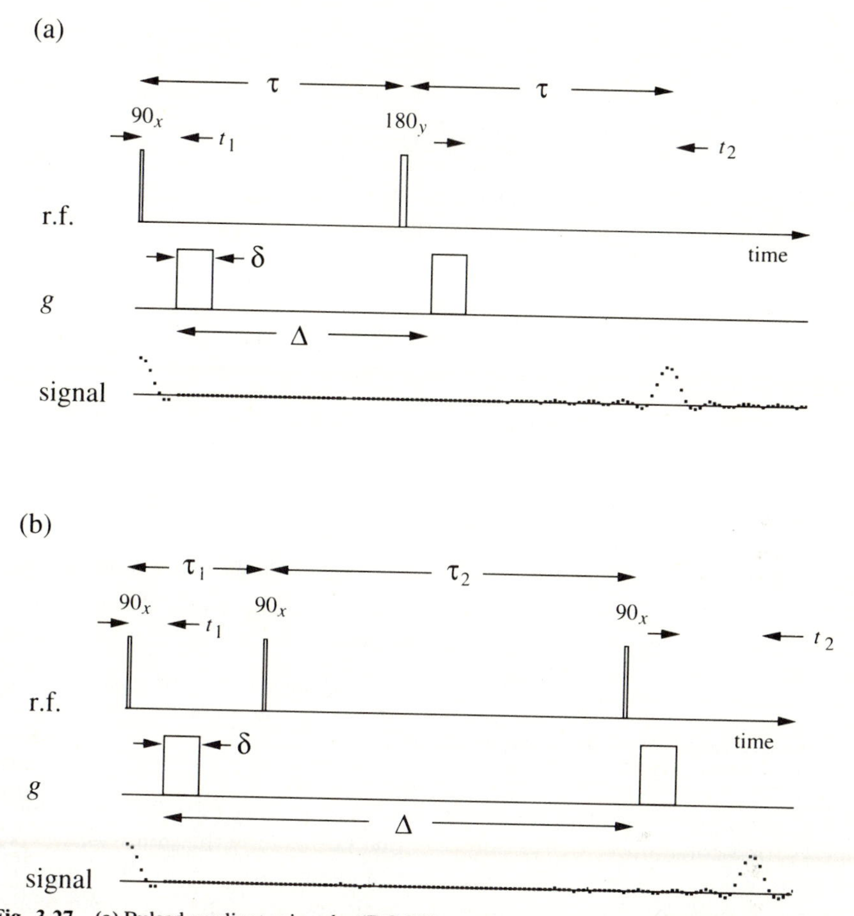

Fig. 3.27 (a) Pulsed gradient spin echo (PGSE) sequence with gradient amplitude, g, gradient pulse duration, δ, and gradient pulse spacing, Δ. τ is the time between the 90_x and 180_y r.f. pulses and corresponds to half the spin echo formation time, T_E. (b) Stimulated echo version of PGSE sequence.

The ratio of the echo amplitude at gradient g to that at zero gradient, will henceforth be labelled $E(g)$.

Note that the exponent in the Stejskal–Tanner equation is proportional to the mean-squared displacement of the molecules over an effective time-scale $(\Delta - \delta/3)$. The attenuation of the echo arises because of the incoherent nature of the phase shifts across the nuclear ensemble. Suppose, by contrast, that there is an additional coherent motion due to each molecule being additionally displaced by an identical amount. This, of course, corresponds to the superposition of a uniform velocity $\mathbf{v}$, so that the extra displacement moved per incremental step is $\mathbf{v}\tau_s$. The effect is easy to calculate. Consider

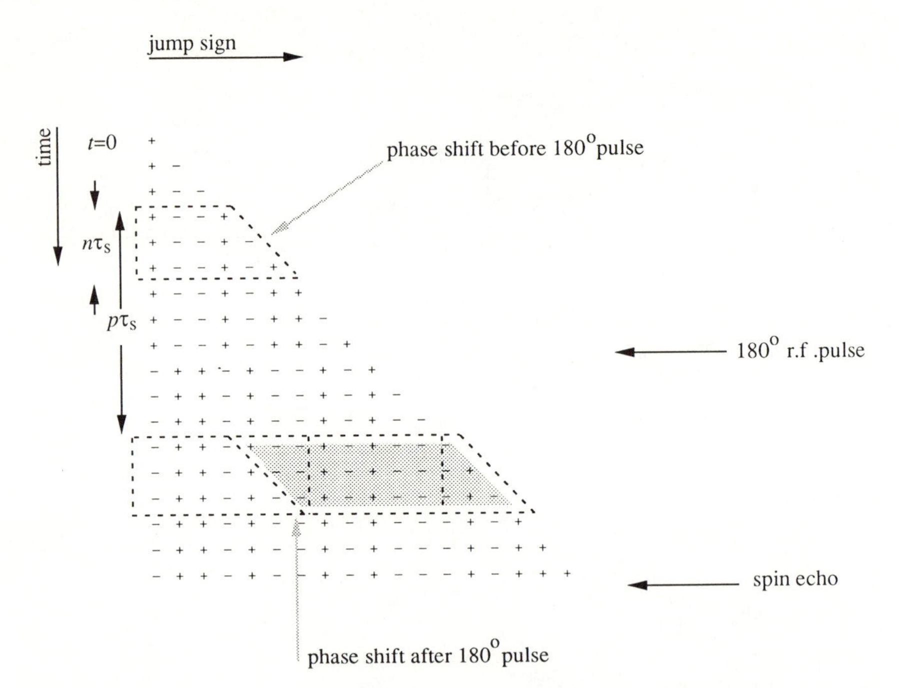

Fig. 3.28 Cumulative phase diagram for spin echo formation in the PGSE experiment. The net phase shift is the sum of that occurring before and after the 180° r.f. pulse and corresponds to the shaded region. Note that $\delta = n\tau_s$ and $\Delta = p\tau_s$.

the PGSE phase step diagram of Fig. 3.28. The shape of the diagram which allows for a constant velocity component will be exactly the same. However, all steps will now have an additional, identical magnitude contribution, with negative sign before the 180° pulse and positive sign after. The mean phase shift corresponds to the sum in the shaded area. This is $\gamma\tau_s \mathbf{g}\cdot(\mathbf{v}\tau_s)pn$ or $\gamma\delta\mathbf{g}\cdot\mathbf{v}\Delta$. Consequently a phase shift of $\exp(i\gamma\delta\mathbf{g}\cdot\mathbf{v}\Delta)$ is common to all spins in the ensemble, and may be factorized out of the signal, leaving the diffusive contribution exactly as before. The combined effect of diffusion and flow is therefore a phase shift due to flow and an attenuation due to diffusion given by

$$E(\mathbf{g}) = \exp[i\gamma\delta\mathbf{g}\cdot\mathbf{v}\Delta - \gamma^2\delta^2 g^2 D(\Delta - \delta/3)]. \tag{3.55}$$

For completeness, we note that where a steady gradient $\mathbf{G}_0$ is also present the exact form of the signal is given by[91,93]

$$E(\mathbf{g}) = \exp(i\gamma\delta\mathbf{g}\cdot\mathbf{v}\Delta + i\gamma\tau^2\mathbf{G}_0\cdot\mathbf{v} - \gamma^2 D[g^2\delta^2(\Delta - \delta/3) - \tfrac{2}{3}G_0^2\tau^3 - \mathbf{g}\cdot\mathbf{G}_0\delta\{t_1^2 + t_2^2 + \delta(t_1 + t_2) + \tfrac{2}{3}\delta^2 - 2\tau^2\}]) \tag{3.56}$$

with t_1 being the delay between the 90_x pulse and the start of the first

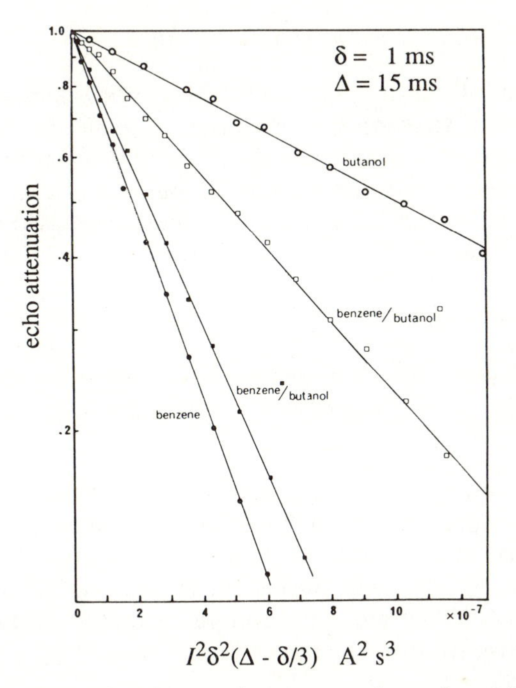

Fig. 3.29 Echo attenuation data for pure benzene, pure butanol, and for the benzene and butanol components in an equimolar mixture. The self-diffusion coefficients are, respectively, $(2.23 \pm 0.02) \times 10^{-9}$, $(0.43 \pm 0.01) \times 10^{-9}$, $(1.83 \pm 0.02) \times 10^{-9}$, and $(0.9 \pm 0.01) \times 10^{-9}$ m^2s^{-1}. (From Callaghan *et al.*[97])

gradient pulse and t_2 being the delay between the end of the second gradient pulse and the centre of the echo. This result is useful in dealing with an apparatus where the residual gradients are poorly 'shimmed' but in well-constructed systems the additional terms involving G_0 in eqn 3.54 can be neglected.

Eqn 3.54 provides a precise description of the influence of self-diffusion in the PGSE experiment and is the basis of a considerable literature pertaining to this technique.[94-96] An example of self-diffusion measurement is shown in Fig 3.29 in which spectrally separated molecular species in the same liquid mixture have their respective echo attenuations plotted as a function of squared gradient amplitude. The spectral separation is possible because the echo signal is sampled in the absence of the magnetic field gradient. The PGSE method may be used to precisely measure diffusion down to 10^{-14} m^2s^{-1}, some five orders of magnitude slower than that exhibited in Fig. 3.29.

3.7.3 *Reducing the influence of transverse relaxation*

The measurement of self-diffusion by PGSE NMR is limited by the loss of phase coherence due to transverse relaxation. One nice method of avoiding this problem is to store the spatially encoded magnetization along the longitudinal axis by means of a second 90° pulse as shown in Fig. 3.27(b). The magnetization is recalled at a later time and rephased in a stimulated echo. Over the storage period the spins are subject to T_1 relaxation which is generally slower than T_2. The expression equivalent to eqn 3.56 for the case of the stimulated echo is[98]

$$E(\mathbf{g}) = \exp(\mathrm{i}\gamma\delta\mathbf{g}\cdot\mathbf{v}\Delta + \mathrm{i}\gamma\tau_1(\tau_1 + \tau_2)\mathbf{G}_0\cdot\mathbf{v} - \gamma^2 D[g^2\delta^2(\Delta - \delta/3) - \tfrac{2}{3}G_0^2\tau_1^2(\tau_2 - \tfrac{1}{3}\tau_1) - \mathbf{g}\cdot\mathbf{G}_0\delta\{t_1^2 + t_2^2 + \delta(t_1 + t_2) + \tfrac{2}{3}\delta^2 - 2\tau_1\tau_2\}]). \tag{3.57}$$

For small background gradients this expression is entirely equivalent to that of the spin echo expression, the only penalty paid being the loss of a factor of 2 in the signal intensity due to projection of only half the transverse magnetization along the z-axis during the storage period. Comparison of eqns 3.56 and 3.57 illustrates the reduced influence of $\mathbf{G}_0$ in the latter expression because of the replacement of τ^3 and τ^2 by $\tau_1^2(\tau_2 - \frac{1}{3}\tau_1)$ and $\tau_1\tau_2$, respectively.

Another approach[99] to the reduction of residual gradient influence is to employ a CPMG echo train but with a single pair of gradient pulses inserted in appropriately spaced locations between the 180° r.f. pulses. This sequence has been analysed in detail by Williams *et al.*[100] While the CPMG train method does not suffer from the factor of 2 signal-to-noise ratio loss of the stimulated echo it does suffer T_2 relaxation. An alternative method[101] of retaining the full signal amplitude while avoiding transverse relaxation is to employ spin-locking pulses over the waiting period between the two gradient pulses in a normal spin echo. Under a sustained spin-lock pulse the magnetization decays as $T_{1\rho}$ which, for sufficiently powerful r.f. fields, can be made much longer than T_2.

In the context of NMR microscopy, PGSE NMR is significant in two respects. First, it represents a curious form of imaging in its own right, giving images in a spatial domain corresponding to the distance moved by molecules over the time-scale Δ. Second, it provides a remarkable contrast mechanism in ordinary NMR microscopy, in which the molecular motions in each pixel of the image may be examined. This process can provide maps of the velocity and diffusion distribution in an inhomogeneous sample. Both of these ideas will be explored in detail in later chapters.

The stimulated echo method has the additional advantage that the magnetization is not only protected from T_2 relaxation during the 'storage' period but is also subject to reduced precessional dephasing caused bv residual gradients.

3.7.4 *Generalized treatment of diffusion and flow: the Bloch–Torrey equations*

The effect of molecular self-diffusion and velocity can be accounted for in an alternative description due to Torrey[102] in which an additional term is introduced in the Bloch equations (eqn 2.40) so that self-diffusion and flow is represented as a 'transport of magnetization'. Thus, for example, the equation for the x-component contains additional terms viz,

$$\frac{\mathrm{d}M_x}{\mathrm{d}t} = \gamma M_y(B_0 - \omega/\gamma) - M_x/T_2 + \nabla\cdot\mathbf{D}\cdot\nabla M_x - \nabla\cdot\mathbf{v}M_x. \tag{3.58}$$

We are interested in the evolution of the magnetization in the rotating frame and in the absence of the r.f. field. We shall also be concerned here only with isotropic diffusion and spatially independent velocities. With these simplifications the evolution of the complex magnetization vector M_+ in the rotating frame of reference may be written

$$\frac{\partial M_+}{\partial t} = -\,\mathrm{i}\gamma\mathbf{r}\cdot\mathbf{g}M_+ - M_+/T_2 + D\nabla^2 M_+ - \nabla\cdot\mathbf{v}M_+. \tag{3.59}$$

It is clear that M_+ is a function of both $\mathbf{r}$ and t and eqn (3.59) is solved by making the substitution[91]

$$M_+(\mathbf{r},t) = A(t)\exp\left[-\mathrm{i}\gamma\mathbf{r}\cdot\int_0^t \mathbf{g}(t')\,\mathrm{d}t'\right]\exp(-t/T_2). \tag{3.60}$$

The condition for a gradient echo at time t is simply $\int_0^t \mathbf{g}(t')\,\mathrm{d}t' = 0$ so that at the echo centre $M_+(\mathbf{r},t)$ reduces to $A(t)\exp(-t/T_2)$, independent of the local coordinates. Thus $A(t)\exp(-t/T_2)$ is the echo amplitude. Of course, our treatment here neglects the role of the r.f. pulses so that it would appear that we are unable to handle the case of the spin echo. In fact the same arguments apply for spin echoes if the appropriate phase inversions[77] are taken into account in computing $\int_0^t \mathbf{g}(t')\,\mathrm{d}t'$. To allow for this we employ an 'effective gradient', $\mathbf{g}^*$. This quantity is defined precisely in Chapter 6 (Section 6.4.1) but for the moment we shall simply regard $\mathbf{g}^*$ in the case of the spin echo as being the 'gradient echo equivalent'.

Substitution in eqn 3.59 leads therefore to

$$\frac{\partial A(t)}{\partial t} = -\,D\gamma^2\left(\int_0^t \mathbf{g}^*(t')\,\mathrm{d}t'\right)^2 A(t) \tag{3.61}$$

with solution[90]

$$A(t) = \exp\left[-D\gamma^2 \int_0^t \left(\int_0^{t'} \mathbf{g}^*(t'')\,\mathrm{d}t''\right)^2 \mathrm{d}t'\right] \exp\left[\mathrm{i}\gamma\mathbf{v}\cdot \int_0^t \left(\int_0^{t'} \mathbf{g}^*(t'')\,\mathrm{d}t''\right) \mathrm{d}t'\right]. \tag{3.62}$$

This is a useful result for the particular case of self-diffusion and flow and it is generally applicable for any echo sequence as defined by the function $\mathbf{g}(t)$. Consider, for example, the pulsed gradient spin echo experiment in which the gradient is applied as in Fig. 3.27(a). The gradient echo equivalent, $\mathbf{g}^*$, is shown in Fig. 3.30.

The relevant integrals are

$$\int_0^{2\tau} \mathbf{g}^*(t')\,\mathrm{d}t' = 0 \tag{3.63}$$

$$\begin{aligned}\int_0^t \left(\int_0^{t'} \mathbf{g}^*(t'')\,\mathrm{d}t''\right) \mathrm{d}t' &= -\int_d^{d+\delta} g(t'-d)\,\mathrm{d}t' - \int_{d+\delta}^{d+\Delta} g\delta\,\mathrm{d}t' \\ &+ \int_{d+\Delta}^{d+\Delta+\delta} [-g\delta + g(t'-d-\Delta)]\,\mathrm{d}t' \\ &+ \int_{d+\Delta+\delta}^{2\tau} [-g\delta + g\delta]\,\mathrm{d}t' \\ &= -g\delta\,\Delta \end{aligned} \tag{3.64}$$

$$\int_0^t \left(\int_0^{t'} \mathbf{g}^*(t'')\,\mathrm{d}t''\right)^2 \mathrm{d}t' = -\int_d^{d+\delta} g^2(t'-d)^2\,\mathrm{d}t' - \int_{d+\delta}^{d+\Delta} g^2\delta^2\,\mathrm{d}t'$$

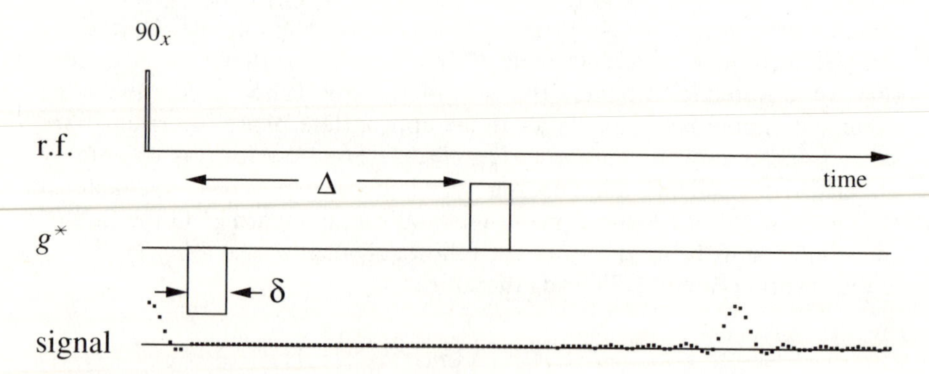

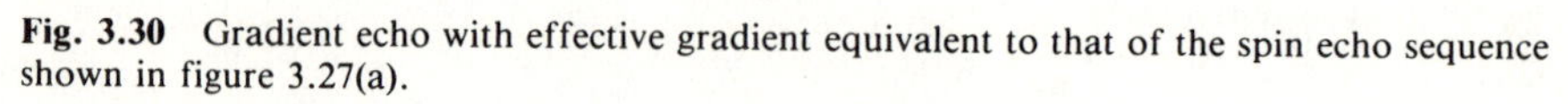

Fig. 3.30 Gradient echo with effective gradient equivalent to that of the spin echo sequence shown in figure 3.27(a).

$$+\int_{d+\Delta}^{d+\Delta+\delta}[g^2\delta^2-2g^2(t'-d-\Delta)+g^2(t'-d-\Delta)^2]\,\mathrm{d}t'$$

$$=g^2\delta^2(\Delta-\delta/3). \tag{3.65}$$

Eqn (3.63) is simply the condition for the formation of an echo. Eqns (3.64) and (3.65) taken together with eqn (3.62) reproduce the earlier expression for the velocity-induced echo phase shift and diffusion-induced echo attenuation given in eqn (3.55).

Of course, diffusion and flow represent an important, but none the less specific, type of molecular motion. For more general motions it is not possible to obtain a closed-form expression for the echo amplitude under all possible effective gradients, $\mathbf{g}^*(t)$ in the manner of eqn (3.62). However, for the particular example of the PGSE experiment in the limit as $\delta \to 0$, general analytic expressions are available. The measurement of generalized motion is the subject of Chapter 6.

3.8 References

1. Lauterbur, P. C. (1973). *Nature* **242**, 190.
2. Mansfield, P. and Grannell, P. K. (1973). *J. Phys. C* **6**, L422.
3. Mansfield, P. and Morris, P. G. (1982). *NMR imaging in biomedicine*, Academic Press, New York.
4. Morris, P. G. (1986). *NMR imaging in biology and medicine*, Oxford University Press.
5. Ernst, R. R., Bodenhausen, G. and Wokaun, A. (1987). *Principles of nuclear magnetic resonance in one and two dimensions*, Clarendon Press, Oxford.
6. Bottomley, P. A. (1982). *Rev. Sci. Instr.* **53**, 1319.
7. Ernst, R. R. (1987). *Q. Rev. Biophysics* **19**, 183.
8. Bailes, D. R. and Bryant, D. J. (1984). *Contemp. Phys.* **25**, 441.
9. Mansfield, P. and Grannell, P. K. (1975). *Phys. Rev.* **12**, 3618.
10. Mansfield, P. (1976). *Contemp. Phys.* **17**, 553.
11. Mansfield, P. (1988). *J. Phys. E.* **21**, 18.
12. Mansfield, P., Maudsley, A. A. and Baines, T. (1976). *J. Phys. E.* **9**, 271.
13. Hinshaw, W. S., Bottomley, P. A. and Holland, G. N. (1977). *Nature* **270**, 722.
14. Hinshaw, W. S. Andrew, E. R., Bottomley, P. A., Holland, G. N., Moore, W. S. and Worthington, B. S. (1978). *Br. J. Radiol.* **51**, 273.
15. Holland, G. N., Bottomley, P. A. and Hinshaw, W. S. (1977). *J. Magn. Reson.* **28**, 133.
16. Brooker, H. R. and Hinshaw, W. S. (1978). *J. Magn. Reson.* **30**, 129.
17. Hinshaw, W. S. (1983). *Proc. IEEE,* **71**, 338.
18. Tomlinson, B. L. and Hill, H. D. W. (1973). *J. Chem. Phys.* **59**, 1775.
19. Garroway, A. N., Grannell, P. K. and Mansfield, P. (1974). *J. Phys C* **7**, L457.
20. Mansfield, P. and Maudsley, A. A. (1977). *J. Magn. Reson.* **27**, 101.
21. Hoult, D. I. (1977). *J. Magn. Reson.* **26**, 165.

22. Hutchison, J. M. S. Sutherland, R. J. and Mallard, J. R. (1978). *J. Phys. E* **11**, 217.
23. Hoult, D. I. (1979). *J. Magn. Reson.* **35**, 69.
24. Mansfield, P., Maudsley, A. A., Morris, P. G. and Pykett, I. L. (1979). *J. Magn. Reson.* **33**, 261.
25. Hutchison, J. M. S, Edelstein, W. A. and Johnson, G. (1980). *J. Phys. E* **13**, 947.
26. Redfield, A. G., Kunz, S. D. and Ralph, E. K. (1975). *J. Magn. Reson.* **19**, 114.
27. Redfield, A. G. (1978). *Methods Enzymol.* **49**, 253.
28. Sutherland, R. J. and Hutchison, J. M. S. (1978). *J. Phys. E* **11**, 79.
29. Mansfield, P. (1988). *Physics of NMR Spectroscopy in biology and medicine*, p. 345, Soc. Italiana di Fisica, Bologna.
30. Silver, M. S., Joseph, R. I. and Hoult, D. I. (1984). *J. Magn. Reson.* **59**, 347.
31. Frahm, J. and Hanicke, W. (1984). *J. Magn. Reson.* **60**, 320.
32. Locher, P. R. (1980). *Phil. Trans. Roy. Soc. Lond.* **B289**, 537.
33. Frahm, J., Merboldt, K. D. Hanicke, W. and Haase, A. (1985). *J. Magn. Reson.* **64**, 81.
34. Geen, H., Wimperis, S. and Freeman, R. (1989). *J. Magn. Reson.* **85**, 620.
35. Doddrell, D. M., Brooks, W. M., Bulsing, J. M., Field, J., Irving, M. G. and Baddeley, H. (1986). *J. Magn. Reson.* **68**, 367.
36. Doddrell, D. M., Bulsing, J. M., Galloway, G. J., Brooks, W. M., Field, J., Irving, M. G. and Baddeley, H. (1986). *J. Magn. Reson.* **70**, 319.
37. Galloway, G. J., Brooks, W. M., Bulsing, J. M., Brereton, I. M., Field, J., Irving, M. G., Baddeley, H. and Doddrell, D. M. (1987). *J. Magn. Reson.* **73**, 360.
38. Brooks, W. M., Brereton, I. M. and Doddrell, D. M. (1987). *Magn. Reson. Med.* **5**, 191.
39. Kimmich, R. and Hoepfel, D. (1987). *J. Magn. Reson.* **72**, 379.
40. Morris, G. A and Freeman, R. (1978). *J. Magn. Reson.* **29**, 433.
41. Lauterbur, P. C. (1973). *Proc. First Int. Conf. on Stable Isoptopes in Biol. and Medicine.*
42. Lauterbur, P. C. (1974). *Pure Appl. Chem.* **40**, 149.
43. Lauterbur, P. C., Kramer, D. M., House, W. V. and Chen, C. N. (1975). *J. Am. Chem. Soc.* **97**, 6866.
44. Lauterbur, P. C. (1977). In NMR in biology, (ed. R. A. Dwek, I. D. Campbell, R. E. Richards and R. J. P. Williams), p. 323, Academic Press, London.
45. Lauterbur, P. C. (1979). *IEEE Trans. Nucl. Sci.* **NS26**, 2808.
46. Kumar, A., Welti, D. and Ernst, R. R. (1975). *Naturwiss.* **62**, 34.
47. Kumar, A., Welti, D. and Ernst, R. R. (1975). *J. Magn. Reson.* **18**, 69.
48. Jeener, J. *Ampère International Summer School*, (1971). Basko Polje, Yugoslavia.
49. Aue, W. P., Bartholdi, E. and Ernst, R. R. (1976). *J. Chem. Phys.* **64**, 2229.
50. Bodenhausen, G. (1981). *Progr. NMR. Spectr.* **14**, 137.
51. Bax, A. (1984). *Two-dimensional nuclear magnetic resonance in liquids*, Delft University Press, Dordrecht.
52. Edelstein, W. A., Hutchison, J. MS., Johnson, G. and Redpath, T. W. (1980). *Phys. Med. Biol.* **25**, 751.
53. Johnson, G., Hutchison, J. M. S., Redpath, T. W. and Eastwood, L. M. (1983). *J. Magn. Reson.* **54**, 374.

54. Lai, C. M. and Lauterbur, P. C. (1980). *J. Phys. E.* **13**, 747.
55. Lai, C. M. and Lauterbur, P. C. (1981). *Phys. Med. Biol* **26**, 851.
56. Lai, C. M. and Lauterbur, P. C. (1981). *J. Phys. E* **14**, 874.
57. Lauterbur, P. C. (1981). *J. Comput. Assisted Tomography* **5**, 285.
58. Lauterbur, P. C. (1986). In *NMR in biology and medicine* (ed. S. Chien and C. Ho) Raven Press, New York.
59. Lauterbur, P. C. (1989). *International Society For Magnetic Resonance Meeting*, Morzine.
60. Lauterbur, P. C. (1989). *British Radio Spectroscopy Group Meeting*, Cambridge, Abstracts, p 8.
61. Hoult, D. I. (1979). *J. Magn. Reson.* **33**, 183.
62. Cox, S. and Styles, P. (1980). *J. Magn. Reson.* **40**, 209.
63. Haase, A., Malloy, C. and Radda, G. K. (1983). *J. Magn. Reson.* **55**, 164.
64. Styles, P., Blackledge, M. J., Moonen, C. T. W. and Radda, G. K. (1987). *Ann. N. Y. Acad. Sci.* **508**, 349.
65. Styles, P., Cadoux-Hudson, T. and Hogan, G. *Magn. Reson. in Med. and Biol.* In press.
66. Liu, J. and Koenig, J. L. (1990). *J. Magn. Reson.* **86**, 593.
67. Haase, A. and Frahm, J. (1985). *J. Magn. Reson.* **64**, 94.
68. Callaghan, P. T. and Eccles, C. D. (1987). *J. Magn. Reson.* **71**, 426.
69. Haase, A., Frahm, J., Matthaei, D., Hanicke, W. and Merboldt, K. D. (1986). *J. Magn. Reson.* **67**, 258.
70. Frahm, J., Hanicke, W. and Merboldt, K. D. (1987). *J. Magn. Reson.* **72**, 307.
71. Haase, A. (1990). *Magn. Reson. Med.* **13**, 77.
72. Haase, A., Matthaei, D., Bartkowski, R., Duhmke, E. and Leibfritz, D. (1989). *J. Comput. Assisted Tomography* **13**, 1036.
73. Glyngell, M. (1988). *Magn. Reson. Imaging* **6**, 415.
74. Redpath, T. W. and Jones, R. A. (1988). *Magn. Reson. Med.* **6**, 224.
75. Oppelt, A., Graumann, R., Baruss, H., Fischer, H., Hartl, W. and Shajor, W. (1986). *Electromedica* **54**, 15.
76. Mansfield, P. (1977). *J. Phys. C.* **10**, L55.
77. Mansfield, P. and Pykett, I. L. (1978). *J. Magn. Reson.* **29**, 355.
78. Doyle, M., Chapman, B., Turner, R., Ordidge, R. J., Cawley, M., Coxon, R., Glover, P., Coupland, R. E., Morris, G. K., Worthington, B. S. and Mansfield, P. (1986). *Lancet* **2**, 682.
79. Ordidge, R. J., Howseman, A., Coxon, R., Turner, R., Chapman, B., Glover, P., Stehling, M. and Mansfield, P. (1989). *Mag. Reson. Med.* **10**, 227.
80. Chapman, B., Turner, R., Ordidge, R. J., Doyle, M., Cawley, M., Coxon, R., Glover, P. and Mansfield, P. (1987). *Magn. Reson. Med.* **5**, 246.
81. Houseman, A. M., Stehling, M. K., Chapman, B., Coxon, R., Turner, R., Ordidge, R. J., Cawley, M. G., Glover, P., Mansfield, P. and Coupland, R. E. (1988). *Br. J. Radiol,* **61**, 822.
82. Ordidge, R. J. and Mansfield, P. (1985). *NMR Methods, U.S. Patent No.*, 4 509 015.
83. Mansfield, P. and Chapman, B. (1986). *J. Magn. Reson.* **66**, 573.
84. Mansfield, P. and Chapman, B. (1986). *J. Phys E.* **19**, 540.
85. Mansfield, P. and Chapman, B. (1987). *J. Magn. Reson.* **72**, 211.
86. Hahn, E. L. (1950). *Phys. Rev.* **80**, 580.
87. Carr, H. Y. and Purcell, E. M. (1954). *Phys. Rev.* **94**, 630.

88. Reif, F. *Fundamentals of statistical and thermal physics*, (1965). McGraw Hill, New York.
89. Harris, K. R., Mills, R., Back, P. J. and Webster, D. S. (1978). *J. Magn. Reson.* **29**, 473.
90. McCall, D. W., Douglass, D. C. and Anderson, E. W. (1963). *Ber. Bunsenges. Physik. Chem.* **67**, 336.
91. Stejskal, E. O. and Tanner, J. E. (1965). *J. Chem. Phys.* **42**, 288.
92. James, T. L. and McDonald, G. G. (1973). *J. Magn. Reson.* **11**, 58.
93. Stejskal, E. O. (1965). *J. Chem. Phys.* **43**, 3597.
94. Kärger, J., Pfeifer, H. and Heink, W. (1988). *Advances in Magn. Reson.* **12**, 1.
95. Callaghan, P. T. (1984). *Austr. J. Phys.* **37**, 359.
96. Stilbs, P. (1987). *Prog. Nucl. Magn. Reson. Spectr.* **19**, 1.
97. Callaghan, P. T., Trotter, C. M. and Jolley, K. W. (1980). *J. Magn. Reson.* **37**, 247.
98. Tanner, J. E. (1970). *J. Chem. Phys.* **52**, 2523.
99. Packer, K. J., Rees, C. and Tomlinson, D. J. (1970). *Mol. Phys.* **18**, 421.
100. Williams, W. D., Seymour, E. F. W. and Cotts, R. M. (1978). *J. Magn. Reson.* **31**, 271.
101. Germanus, A., Pfeifer, H., Heink, W. and Kärger, J. (1983). *Ann. Phys. (Leipzig)* **40**, 161.
102. Torrey, H. C. (1956). *Phys. Rev.* **104**, 563.

4
HIGH-RESOLUTION k-SPACE IMAGING

4.1 Sensitivity, motion, and resolution

The oscillating e.m.f. induced in a receiver coil by precessing magnetization is superposed on the noise e.m.f. arising from random thermal motion of electrons in the receiver coil wire and the comparative size of the signal and noise e.m.f.s is known as the signal-to-noise ratio (or just plain signal-to-noise). As we have seen in Chapter 2, this ratio may be improved by co-adding signals from successive experiments, a technique known as signal averaging. However, the repetition rate is limited by the time required for spin relaxation. Since the signal-to-noise ratio improves only as the square root of the number of co-added experiments, we find, for example, that if T_1 is ~ 0.5 s, then an improvement over a single experiment by a factor of 10 takes of order 1 minute, an improvement of 100 takes of order 1 hour while an improvement of 1000 takes of order one week. In practice, improvements of between one and two orders of magnitude are as much as can be realistically achieved by such means.

The signal-to-noise ratio provides a fundamental limitation to resolution in magnetic resonance microscopy since the signal available from each volume element decreases as the voxel size is reduced.[1,2] Without reference to specific imaging techniques, Mansfield and Morris have shown by a simple argument that the acquisition time for a sample of volume $(\Delta x)^3$ is given by

$$t = \left(\frac{\rho_0}{\sigma_f}\right)^2 a^2 \left(\frac{T_1}{T_2}\right) \frac{2.8 \times 10^{-15}}{f^{7/2}} \left(\frac{1}{\Delta x}\right)^6 \tag{4.1}$$

where (ρ_0/σ_f) is the desired proton NMR signal-to-noise ratio, a is the r.f. coil radius, and f the spectrometer frequency in MHz. This implied dependence of the imaging co-addition time on the sixth power of the resolution demonstrates that the sensitivity limit acts as a sort of 'brick wall' which stands in the way of further improvements in resolving power by the method of signal averaging. In this chapter we examine in detail just where this wall is positioned and how that position depends on the various NMR parameters. In particular, it will be clear that the resolution limit is determined by the intrinsic spectral linewidth. This is narrowest for rapidly tumbling molecules so that the best resolution is achieved in the liquid state. The scaling of resolution with sensitivity is well understood and we will find that the linewidth/sensitivity limit in an optimally designed experiment using proton NMR is a few micrometres.

Suppose that we are imaging the spin density associated with protons in water molecules. The acquisition of the nuclear signal in **k**-space lasts typically for a few milliseconds. A free water molecule at room temperature will diffuse a few microns over this time so that the question arises as to whether such motion will affect the resolution which can be achieved by NMR microscopy. From the viewpoint of the final image it might appear that the movement of an individual water molecule is unimportant provided the overall distribution of water remains constant. In fact, such motion, occurring in the presence of the imaging field gradient, causes a loss of phase coherence in the transverse magnetization and a consequent broadening of the spectral line. Clearly nature is a little unkind in this regard. The intrinsic relaxation linewidth is smallest for molecules in the liquid state yet the translational motion associated with this state creates new problems.

The resolution loss[3,4] associated with these effects is subtle. In Section 4.5 we examine how the diffusive limit is related to the optimal relaxation limit and how the achievable resolution is determined by the choice of pulse sequence.

4.2 The signal-to-noise ratio in NMR

The thermal equilibrium magnetization for spins of gyromagnetic ratio γ and spin quantum number I in a longitudinal field B_0 is determined by the relative population in the substates $|m\rangle$. These populations (the diagonal elements of the density matrix) are given by the Boltzmann factors

$$\overline{|a_m|^2} = \frac{\exp(-\hbar\gamma m B_0/k_B T_s)}{\sum_{m=-I}^{I}\exp(-\hbar\gamma m B_0/k_B T_s)} \tag{4.2}$$

where T_s is the sample temperature. In the high-temperature approximation[5,6]

$$\overline{|a_m|^2} = (1 - \hbar\gamma m B_0/k_B T_s)/(2I+1) \tag{4.3}$$

and, following eqn (2.26) the magnetization is

$$\begin{aligned} M_0 &= N_S \hbar\gamma \mathrm{Tr}(\rho I_z) \\ &= N_S \sum_{m=-I}^{I} \overline{|a_m|^2}\hbar\gamma m \\ &= N_S \gamma^2\hbar^2 I(I+1)B_0/3k_B T_s \end{aligned} \tag{4.4}$$

where N_S is the number of spins per unit volume.

Using the complex notation described in Section 2.3.3, the transverse magnetization is given by $M_+ = M_0 e^{i\omega_0 t}$. In order to calculate the e.m.f. induced in the receiver coil following a 90° pulse we apply Lenz's law which states that this e.m.f. is proportional to the time rate of change of field within the coil. Consequently the signal amplitude will be proportional to $M_0\omega_0$ or $\gamma^3 I(I+1)B_0^2$. Clearly the best result is achieved for large magnetic fields and large gyromagnetic ratios and this explains the high sensitivity of the proton as compared with other nuclei. The proton has the further advantage of 100% isotopic abundance. The relative sensitivities[7] of ^{1}H and other nuclei relevant to NMR microscopy are shown in Table 4.1.

An instructive method of calculating the induced e.m.f. has been used by Hoult and Richards.[8] They employed the principle of reciprocity to show that the signal induced by a magnetic dipole $\mathbf{M}\,\mathrm{d}V_\mathrm{s}$ placed in a volume element $\mathrm{d}V_\mathrm{s}$ at a point within the receiver coil is simply related to the magnetic field $(\mathbf{B}_1/i)$ which unit current flowing in the coil produces at that point, namely

$$S = -(\partial/\partial t)\int_{\text{sample}} (\mathbf{B}_1/i)\cdot\mathbf{M}\,\mathrm{d}V_\mathrm{s}. \tag{4.5}$$

This equation predicts an in-phase signal component

$$S = \omega_0 K (B_1/i)_{xy} M_0 V_\mathrm{s} \cos\omega_0 t \tag{4.6}$$

where $(B_1/i)_{xy}$ is the transverse component of $(\mathbf{B}_1/i)$ at the centre of the receiver coil and K represents the result of integrating over the field.

The noise e.m.f. against which the signal must compete is calculated as follows. The thermal noise power per unit frequency bandwidth, Δf, is determined by the coil resistance R and the coil temperature T_c. The time domain r.m.s. noise e.m.f. is then[9]

$$\sigma_t = (4k_\mathrm{B}T_\mathrm{c}\Delta f R)^{\frac{1}{2}}. \tag{4.7}$$

Δf will depend on the chosen bandpass filter which should be set exactly equal to the sampling bandwidth as discussed in Section 1.2.4. The assignment of detection circuit resistance to that of the r.f. coil neglects the role played by r.f.-field energy losses in the sample. Of these the most significant loss is that arising from magnetically-induced eddy currents. In practice these dominate the resistance only for large samples. For example, assuming a conductivity equivalent to 100 mM saline solution, at 500 MHz proton frequencies the coil resistance is predominant for samples smaller than 10 mm diameter whilst at 100 MHz it dominates for samples smaller than 20 mm diameter. This means that whereas inductive losses must be accounted for in developing signal-to-noise expressions in large-scale medical imaging, in NMR microscopy they may normally be neglected. Further discussion of this point may be found in Chapter 9.

At NMR frequencies of several MHz, the current in the receiver coil is confined to a skin depth[10] thickness on the coil surface so that the effective coil resistance is

Table 4.1. *NMR properties of some stable nuclei.*

Property	^1H	^2H	^7Li	^{13}C	^{14}N	^{15}N	^{17}O	^{19}F	^{23}Na	^{31}P
spin	$\frac{1}{2}$	1	$\frac{3}{2}$	$\frac{1}{2}$	1	$\frac{1}{2}$	$\frac{5}{2}$	$\frac{1}{2}$	$\frac{3}{2}$	$\frac{1}{2}$
$\gamma (10^8 s^{-1} T^{-1})$	2.675	0.410	1.037	0.673	0.193	0.272	0.365	2.517	0.708	1.081
NMR Frequency at 9.4 T (MHz)	400	61.4	155.1	100.6	28.9	40.5	54.2	376.3	105.8	161.9
Chemical shift range (p.p.m)	20[a]	20[a]	4	300	600	600	1500	1200	80	700
Quadrupole moment (electron cm^2)	-	0.0027	−0.03	-	0.010	-	−0.026	-	0.110	-
Natural abundance	99.985	0.015	92.58	1.108	99.63	0.365	0.037	100	100	100
Relative sensitivity[b] for equal numbers of spins	1.00	9.65×10^{-3}	0.293	1.59×10^{-2}	1.01×10^{-3}	1.04×10^{-3}	2.91×10^{-2}	0.833	9.27×10^{-2}	6.59×10^{-2}
Sensistivity × natural abundance	1.00	1.45×10^{-6}	0.271	1.76×10^{-4}	1.01×10^{-3}	3.85×10^{-6}	1.08×10^{-5}	0.833	9.27×10^{-3}	6.59×10^{-2}

[a] Note that most hydrogen chemical shifts fall within 10 p.p.m.

[b] Sensitivity is calculated using the factor $\gamma^3 I(I+1)$. Strictly speaking the dependence should be $\gamma^{11/4} I(I+1)$

$$R = (l/p)\,(\mu_r\mu_0\omega_0\rho\,(T_c)/2)^{\frac{1}{2}} \tag{4.8}$$

where l is the conductor length, p its circumference, $\mu_r\mu_0$ its permeability, and $\rho(T_c)$ its resistivity at temperature T_c. This simple picture is modified by a factor σ which allows for the reduction in skin depth arising from proximity effects between conductors.[8] For a typical single-layer solenoidal receiver coil, σ is about 5.

Clearly the choice of receiver coil geometry is of great significance in determining the final signal-to-noise ratio. Hoult and Richards have considered two winding geometries as shown in Fig. 4.1 in which they optimize dimensions for both homogeneity and proximity effects. The standard coil configuration has $2a \approx L$ and, for the saddle coil, an angular width of 120°. The distance between centres of each turn should be about three times the wire radius r. They find that the signal-to-noise performance of the solenoid is about three times better than the saddle coil. Because of the use of the reciprocity theorem in the derivation above, it is simple to find a relationship between the signal-to-noise ratio and the r.f. field amplitude resulting from a transmitter pulse of given power. It may be shown that they are simply proportional which means that the 90° pulse length when a saddle coil is used is typically three times longer than that obtainable with a solenoid.

We will follow through the signal-to-noise calculation for a sample enclosed by an ideal solenoid for which the (B_1/i) field is approximately uniform and $K \approx 1$. For such a coil with N_t turns

$$\begin{aligned}(B_1/i)_{xy} &= \mu_0 N_t/(4a^2 + L^2)^{\frac{1}{2}} \\ &= \mu_0 N_t/\sqrt{2}L. \end{aligned} \tag{4.9}$$

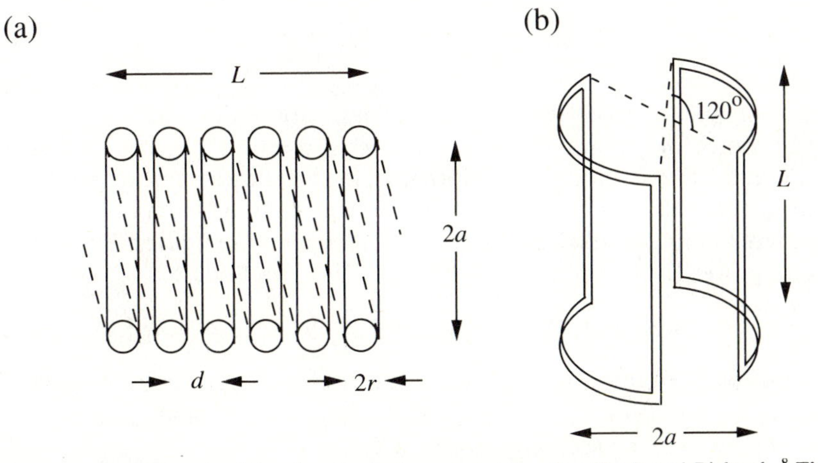

Fig. 4.1 Solenoidal and saddle r.f. coil winding geometries after Hoult and Richards.[8] The optimum wire spacing is approximately equal to the wire radius.

Combining eqns (4.6) to (4.9) an expression for the (peak signal)/(r.m.s. noise) ratio is obtained

$$\frac{S_0}{\sigma_t} = \frac{V_s(B_1/i)_{xy}N_S\gamma\hbar^2 I(I+1)\omega_0^{7/4}}{5.05\,k_B T_s}\left(\frac{p}{k_B T_c \sigma l \Delta f}\right)^{\frac{1}{2}}\left(\frac{1}{\mu_r\mu_0\rho}\right)^{\frac{1}{4}}. \tag{4.10}$$

It is interesting to note that, strictly speaking, the signal-to-noise ratio depends on $\omega_0^{7/4}$ and not ω_0^2 as implied earlier. For different nuclei at the same polarizing field the dependence will therefore be $\gamma^{11/4}I(I+1)$.

Eqn (4.10) can be simplified if the 'standard coil' configuration is used. If $2a = L$ then we have the following approximate relationships derived by Hoult and Richards,

$$l = 6.3\,N_t a$$

$$p = 4.2\,a/N_t \tag{4.11}$$

and, combining these results with eqn (4.10), one obtains

$$\frac{S_0}{\sigma_t} = \frac{\mu_0 V_s(\chi_0/\gamma)\omega_0^{7/4}}{5.8a}\left(\frac{1}{k_B T_c \sigma \Delta f F}\right)^{\frac{1}{2}}\left(\frac{1}{\mu_0\rho}\right)^{\frac{1}{4}} \tag{4.12}$$

where the susceptibility, $\chi_0 = M_0/B_0$, is defined by eqn (4.4). As an exercise we obtain a value for the signal-to-noise ratio in the case of water protons at room temperature in a solenoidal copper coil, assuming that the coil and sample temperatures are the same. This figure represents a best case for room temperature sensitivity and can be used as a reference from which to scale other examples. Substituting the appropriate parameters gives

$$\frac{S_0}{\sigma_t} = 2.72 \times 10^{-3}\,V_s f_0^{7/4} a^{-1} \Delta f^{-\frac{1}{2}} \sigma^{-\frac{1}{2}}. \tag{4.13}$$

For example, at 60 MHz the 'single accumulation' time domain signal-to-noise ratio for 1 mm^3 of water protons in a coil of 2.5 mm radius will be ≈ 200 if the detection bandwidth is 20 kHz and we assume $\sigma = 5$. Taking into account the signal-to-noise enhancement resulting from co-adding N_{acc} accumulations and allowing for the spectrometer noise figure F by which the noise power from the coil must be multiplied, we find

$$S_0/\sigma_t = 2.72 \times 10^{-3} a^{-1} F^{-\frac{1}{2}} \sigma^{-\frac{1}{2}}\, V_s f_0^{7/4} \Delta f^{-\frac{1}{2}}\, N_{acc}^{1/2}. \tag{4.14}$$

The result for other nuclei (at the same frequency f_0) can be obtained by scaling by the factor $N_S\gamma I(I+1)$ while the result for saddle coil geometry can be obtained by scaling down by a factor of 3.

4.3 Frequency domain: discrete transformation in two dimensions

The signal-to-noise ratio given in eqn (4.14) applies to the time domain in which the signal is initially acquired. The next step in the analysis is to investigate the processing of both the signal and the noise during image reconstruction. The effects of two-dimensional FI and PR reconstruction in this regard are slightly different and it is instructive to compare them. First, however, we need to express these transformations in digital form. In doing so we retain the explicit time and frequency conjugate spaces. We will presume a raster consisting of N time domain points sampled at interval T for which the conjugate Fourier transformations are given by eqns (1.14) and (1.15).

The FI raster uses a Cartesian space (l, m) as shown in Fig. 4.2(a) while the raster for PR uses a polar space $(l, m\pi/N)$ and is shown in Fig. 4.2(b). Ideally the *read* data in both methods is acquired from $m = -\frac{1}{2}N$ to $\frac{1}{2}N - 1$. This can be achieved using a full spin echo. However, as explained in Chapter 3, full echo sampling has the effect of introducing an asymmetric T_2 relaxation envelope with a finite decay at the time 'origin'. For the moment we will attempt to compare intrinsic sensitivities independent of such asymmetry or decay. This would be possible by using the FID or half echo sampling, illustrated in Figs. 3.14 and 3.15. Of course, in signal-to-noise terms these are poorer than full echo acquisition by a factor of 2 and introduce a dispersive component in the image. An alternative approach in which the signal-to-noise loss is only $\sqrt{2}$ is half echo sampling using $\frac{1}{2}N$ data points followed by either zero-filling or symmetrization, as discussed in Section 2.6.4. For simplicity we will take as our reference the full echo sampling scheme but assume the simpler relaxation envelope of the half echo. While such a reference represents an ideal we shall find that it is a straightforward matter to compare this with the results achieved with specific pulse sequences. Furthermore, as shown in Section 4.5.3, this optimum sensitivity is indeed achievable in practice.

We are now in a position to compare the intrinsic sensitivities of FI and PR. Changing eqn (3.16) to digital form we obtain for FI,

$$\rho(n_x/NT, n_y/NT) = \sum_{l=-\frac{1}{2}N}^{\frac{1}{2}N-1} \sum_{m=-\frac{1}{2}N}^{\frac{1}{2}N-1} \mathrm{S}(lT, mT) \exp[-\mathrm{i}2\pi l n_x/N - \mathrm{i}2\pi m n_y/N]$$
$$n_x, n_y = -\tfrac{1}{2}N, \ldots, \tfrac{1}{2}N - 1 \qquad (4.15)$$

while for PR we write

$$\rho(n_x/NT, n_y/NT) = \sum_{l=0}^{N_p-1} (\pi/N_p) \sum_{m=-\frac{1}{2}N}^{\frac{1}{2}N-1} mS(mT, l\pi/N_p) \exp[-\mathrm{i}2\pi l n_\mathrm{R}/N]$$
$$n_x \cos\phi + n_y \sin\phi = n_\mathrm{R}. \qquad (4.16)$$

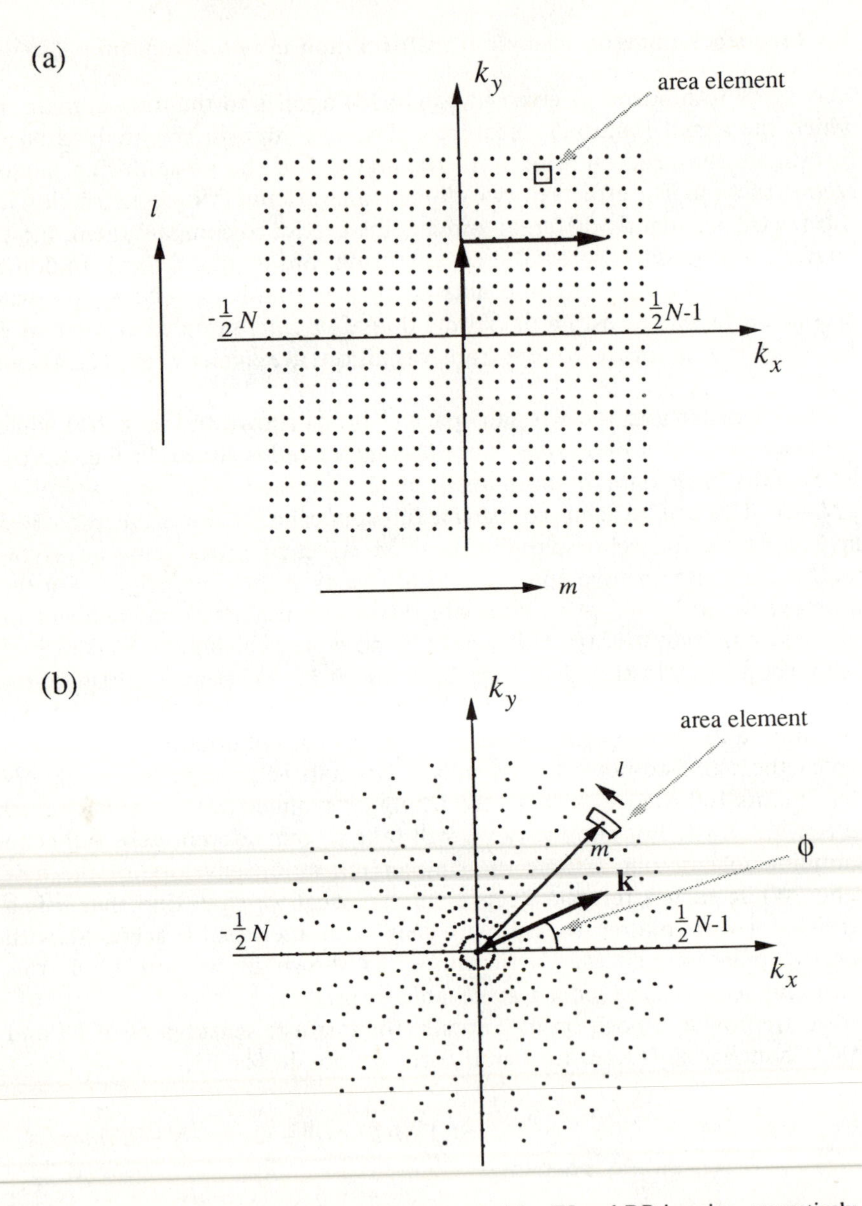

Fig. 4.2 (a) Cartesian raster and (b) polar raster used for FI and PR imaging, respectively. Note that the digital area associated with a given sampling point (m, l) is $m(\pi/N)$ compared with unity in FI.

A detailed comparison of signal-to-noise ratios in the two reconstruction methods requires some suitable spin density profile to represent the image. The simplest choice is a uniform cylindrical object. This has spin density $\rho(f_x,f_y)$ equal to the constant value ρ_0 inside the frequency radius f_s. Of course, f_s is related to the actual dimensions by $f_s = (2\pi)^{-1}\gamma G R_0$, where R_0 is the object radius and G the applied gradient. We seek the relationships between the time and frequency domain signals shown in Fig. 4.3.

Consider the projected spin density $P_\phi(f)$ along some axis at azimuthal angle ϕ. At each point in frequency space this density is proportional to the length of a chord through the object and so

$$P_\phi(f) = \begin{cases} S_0'(1 - f^2/f_s^2)^{\frac{1}{2}} & -f_s \leq f \leq f_s \\ 0 & \text{elsewhere} \end{cases} \tag{4.17}$$

and

$$S_0' = 2 f_s \rho_0. \tag{4.18}$$

To help find the filtered transform we use the Kramers–Kronig relationship[6] in a derivation which is detailed in reference 3. It is shown that

$$\begin{aligned} \mathrm{Re}(\rho_\phi^{\mathrm{R}}) &= (2\pi)^{-1}\mathrm{d}/\mathrm{d}f[(2\pi)^{-1}f_s^{-1}S_0'f\pi] \\ &= (4\pi)^{-1}f_s^{-1}S_0' \end{aligned} \tag{4.19}$$

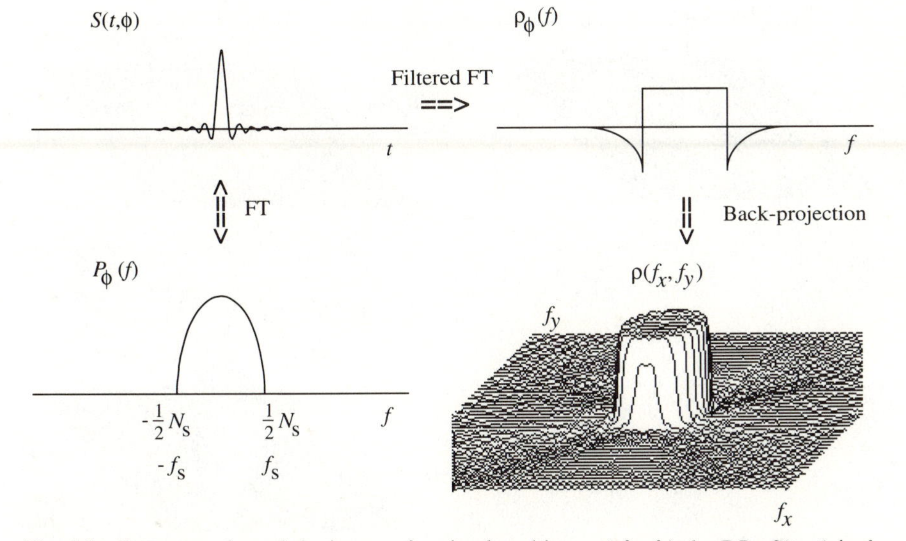

Fig. 4.3 Reconstruction of the image of a circular object, $\rho(f_x, f_y)$, by PR. $S(t, \phi)$ is the initial (time domain) data acquired under the influence of a gradient applied at an angle ϕ to the laboratory x-axis, as defined by eqn (3.20). $P_\phi(f)$ is the one-dimensional Fourier transform and corresponds to the projected spin density along the gradient axis. $\rho_\phi(f)$ is the filtered transform which yields the image on back-projection and summation over all ϕ. Note that the noise in the image is introduced by the process of interpolating from the initial polar raster to the final Cartesian grid.

where ρ_ϕ^{R} is given by

$$\rho_\phi^{\mathrm{R}}(r) = \int_0^\infty S(k, \phi) \exp[-\mathrm{i}2\pi kr]\, k\, \mathrm{d}k. \tag{4.20}$$

For a signal obeying real image symmetry this is precisely half the filtered profile, ρ_ϕ, defined in eqn (3.23). For 'complete' data acquisition therefore the filtered transform is given by

$$\rho_\phi = (2\pi)^{-1} f_{\mathrm{s}}^{-1} S_0'. \tag{4.21}$$

In calculating and comparing the final image signal-to-noise ratios in FI and PR it is helpful to begin with a common time domain reference. In both techniques S_0 is the signal amplitude at the **k**-space origin while σ_{t} is the time domain r.m.s. noise.

In FI we compute the Fourier transformation in two steps: $S(t_x, t_y) \Rightarrow S'(t_x, f_y) \Rightarrow S''(f_x, f_y)$ as shown in Fig. 4.4. $S''(f_x, f_y)$ is, of course, ρ_0 within f_{s} and the diameter of the object corresponds to N_{s} data points, where

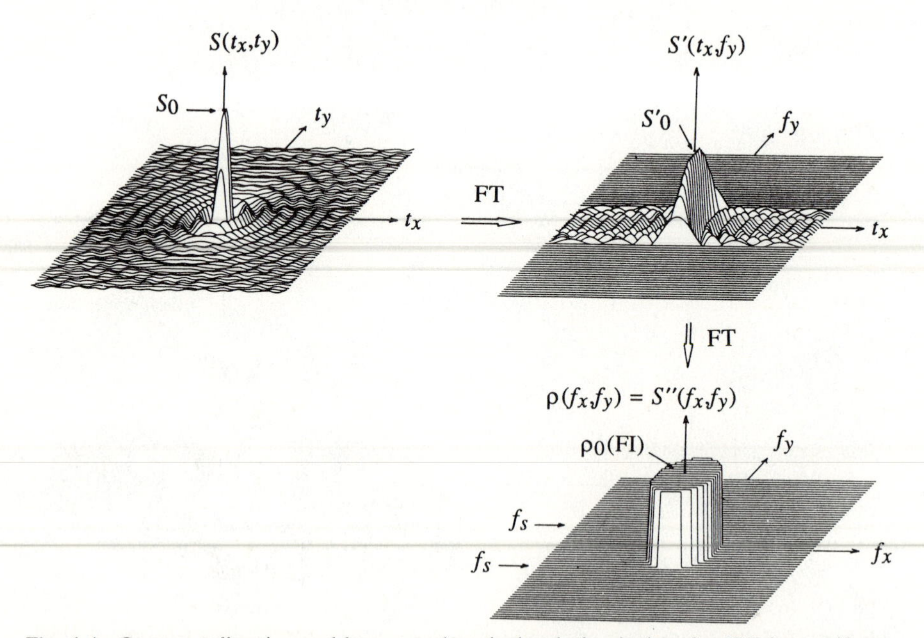

Fig. 4.4 Corresponding time and frequency domain data in fourier imaging (FI) for a uniform circular object. The time domain $S(t_x, t_y)$ has amplitude S_0 at the origin. $S'(t_x, f_y)$ and $S''(f_x, f_y)$ are spectra arising from successive one- and two-dimensional Fourier transformations. (From Callaghan and Eccles[3])

$$N_s = (2\pi)^{-1}\gamma|G|2R_0NT. \tag{4.22}$$

From eqn (1.15) the inverse transformation of $S'(t_x, f_y)$ gives

$$S(0,0) = N^{-1} \sum_{n_y=-\frac{1}{2}N_s}^{\frac{1}{2}N_s-1} S'(0, n_y/NT). \tag{4.23}$$

The data $S'(0, n_y/NT)$ corresponds to the spin density projection spectrum. For a uniform circular object this is given by eqn (4.18) where $f_y = n_y(NT)^{-1}$ and $f_s = N_s(NT)^{-1}$. Evaluation of the sum yields

$$S_0' = (4/\pi)NN_s^{-1}S_0. \tag{4.24}$$

The final image in FI is related to S_0 by applying a second inverse transformation

$$S'(0,0) = N^{-1} \sum_{n_x=-\frac{1}{2}N_s}^{\frac{1}{2}N_s-1} \rho_0(\mathrm{FI}) \tag{4.25}$$

whence

$$\rho_0(\mathrm{FI}) = NN_s^{-1}S_0' = (4/\pi)N^2N_s^{-2}S_0. \tag{4.26}$$

The r.m.s. noise in the two-dimensional frequency domain may be similarly estimated using eqn (1.14) where we introduce frequency and time domain noise functions $\mathfrak{N}_f$ and $\mathfrak{N}_t$

$$\mathfrak{N}_f(n_x/NT, n_y/NT) = \sum_{l=-\frac{1}{2}N}^{\frac{1}{2}N-1} \sum_{m=-\frac{1}{2}N}^{\frac{1}{2}N-1} \mathfrak{N}_t(mT, lT)\exp[-\mathrm{i}2\pi mn_x/N - \mathrm{i}2\pi ln_y/N]. \tag{4.27}$$

Given that the noise is random, we obtain the average

$$\overline{|\mathfrak{N}_f|^2} = N^2\,\overline{|\mathfrak{N}_t|^2} \tag{4.28}$$

or

$$\sigma_f(\mathrm{FI}) = N\sigma_t \tag{4.29}$$

where $\sigma_f(\mathrm{FI})$ is the r.m.s. noise in the FI image space.

Combining eqns (4.26) and (4.29), the final image signal amplitude to r.m.s. noise ratio is

$$\rho_0(\mathrm{FI})/\sigma_f(\mathrm{FI}) = (4/\pi)NN_s^{-2}(S_0/\sigma_t) \tag{4.30}$$

where (S_0/σ_t) is the reference time domain signal amplitude to noise ratio which was obtained earlier as eqn (4.14).

A similar calculation can be performed for PR. The time and frequency domain relationships are depicted in Fig. 4.3. From eqn (4.21) we have, using $f_s = N_s/2N$ (in digital units) and $S_0' = (4/\pi)NN_s^{-1}S_0$,

$$\rho_\phi(n_R/NT) = (1/\pi)[(4/\pi)NN_s^{-2}S_0] \qquad \text{for } n_R < N_s. \tag{4.31}$$

Summing over N_p projections gives

$$\rho_0(\mathrm{PR}) = (N_p/\pi)[(4/\pi)N^2N_s^{-2}S_0]. \tag{4.32}$$

Direct integration of the noise power from $-\frac{1}{2}N$ to $\frac{1}{2}N - 1$ for the filtered transformation and subsequent summation over N_p projections gives for the r.m.s. noise in the PR image

$$\sigma_f(\mathrm{PR}) = (1/\sqrt{12})N_p^{\frac{1}{2}}N^{\frac{3}{2}}\sigma_t. \tag{4.33}$$

Hence

$$\rho_0(\mathrm{PR})/\sigma_f(\mathrm{PR}) = (\sqrt{12}/\pi)[(4/\pi)N_p^{\frac{1}{2}}N^{\frac{1}{2}}N_s^{-2}(S_0/\sigma_t)]. \tag{4.34}$$

For an equivalent data acquisition time in the two methods we require $N_p = N$. The signal-to-noise in PR is therefore superior to that in FI by the factor $\sqrt{12}/\pi$. This factor arises because PR treats the noise in a rather curious way. While the digital area, $f(m, l)$, associated with each point in **k**-space is unity in FI, it varies for PR as

$$f(m, l) = m(\pi/N). \tag{4.35}$$

These $f(m, l)$ values are used to weight the data in image reconstruction. While the area weighting factors give the correct result for the signal in each method, the noise is treated anomalously since it is the noise amplitude rather than the noise power which is weighted by the variable $f(m, l)$ in PR. The total noise power for FI is given by

$$\sum_{l=-\frac{1}{2}N}^{\frac{1}{2}N-1} \sum_{m=-\frac{1}{2}N}^{\frac{1}{2}N-1} f^2(m, l)\sigma_t^2 = N^2\sigma_t^2 \tag{4.36}$$

and for PR the result is

$$\sum_{l=0}^{N-1} 2 \sum_{m=0}^{\frac{1}{2}N-1} f^2(m, l)\sigma_t^2 = \sum_{l=0}^{N-1} 2 \sum_{m=0}^{\frac{1}{2}N-1} (m\pi/N)^2\sigma_t^2 = (\pi^2/12)N^2\sigma_t^2. \tag{4.37}$$

It is apparent that the r.m.s. noise amplitude in PR is diminished by $\pi/\sqrt{12}$. While this factor is apparently subtle it points to an important difference between the methods which arises because of the different rasters employed. This difference leads to more dramatic effects where smoothing filters are employed in order to reduce the influence of noise power at the outside of the raster where the signal is weakest.

4.4 Influence of smoothing filters

It is customary in one-dimensional NMR to multiply the time domain data by an appropriately shaped function known as an apodization filter.[11,12] This filter has the effect of convoluting the NMR spectrum with the transform of the apodization function. The procedure is carried out in order to reduce the noise amplitude in the frequency domain, a process sometimes called smoothing. A wide variety of filters is possible[13-16], and each filter has its own special virtues. Here we shall examine one of the most commonly used examples of a smoothing filter, an exponential decay in the time domain as shown in Fig. 4.5. Such filtering always leads to a broadening in the frequency domain because of the convolution process. For example, an exponential filter, $F(m)$, of decay time τ will cause the frequency domain data to be broadened by the FWHM $(1/\pi\tau)$ of the equivalent Lorentzian. In digital applications where $\tau = pT$, the filter $\exp(-mT/pT) = \exp(-m/p)$ would be used, giving a point spread function broadening, $1/(\pi pT)$. Since the pixel spacing in the frequency domain is $1/NT$, this corresponds to $N/(\pi p)$ pixels.

In imaging there are two distinct broadening regimes, $N/(\pi p) \lesssim 1$ and $N/(\pi p) \gtrsim 1$. For $N/(\pi p) < 1$, the broadening introduced is less than the separation of a pair of pixels. In this regime we may increase the broadening, with consequent improvement in signal-to-noise, and without sacrifice in resolution. Eventually, however, we enter the $N/(\pi p) \gtrsim 1$ regime where the broadening causes the resolution to deteriorate and may therefore outweigh the benefits of noise power reduction.

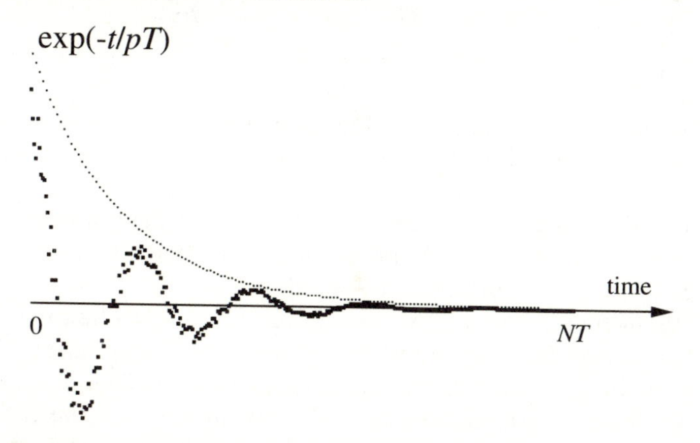

Fig. 4.5 Effect of apodization filter, $\exp(-t/pT)$ in the time domain on a simulated cosine oscillation with noise.

A filter which optimizes the sensitivity of a one-dimensional NMR experiment has been shown by Ernst[11] to be one which matches the envelope decay of the FID. For decay due to transverse relaxation (in the BPP limit) this optimal filter is therefore an exponential with time constant T_2 and it is sensible therefore to choose this filter as a reference. It also turns out to be optimal for imaging purposes. Apodization filtering is very important in NMR microscopy since it provides an effective means of limiting the noise. We need, however, to develop the concept for two-dimensional computation. Here the filter is $F(m, l)$ and an obvious extension of the one-dimensional exponential filter to the case of FI would suggest

$$F(m, l) = \exp\left\{-\frac{|m| + |l|}{p}\right\} \tag{4.38}$$

where p, as before, characterizes the severity of the filter. $F(m, l)$ is designed so as to have little effect at low values of **k**-space where the 'most important' signal data occurs, but to significantly attenuate the signal and noise at high values. Since the r.m.s. noise level is independent of spatial frequency, the application of this filter will therefore result in a reduction in the noise level across the image. The function represented by eqn (4.38) is attractive because it is easily applied to the data, each horizontal scan of **k**-space being multiplied by $\exp(-|m|/p)$ with the function $\exp(-|l|/p)$ being applied in the transverse direction. This multiplication causes a convolution of $\mathfrak{F}\{F(m, l)\}$ with the image. However, $F(m,l)$ and its transform do not have azimuthal symmetry and its use may lead to distortions in the image. In particular, the transverse resolution will now depend on the angle ϕ. Ideally, we seek a circularly symmetric point spread function, an example being

$$F(m, l) = \exp\{-(m^2 + l^2)^{\frac{1}{2}}/p\}. \tag{4.39}$$

This function and its transform, the apodization point spread function, are illustrated in Fig. 4.6.

For PR the exponential decay filter is naturally applied in the radial **k**-space raster exactly as in the one-dimensional case. The back-projection summation over all azimuthal angles ensures that the PR filter has the desired circular symmetry. It is, in fact, entirely equivalent to that suggested for FI in eqn (4.39). Notice that its profile in the image domain is not, of course, the Lorentzian applicable to one-dimensional transformation.

It is obvious that in the case of the circular object calculation of Section 4.3, application of a smoothing filter will not significantly alter the image amplitude within the circular boundary $|n_R| < \frac{1}{2}N_s$. This is because the image is uniform within this region and convolution with the filter point spread function will have no effect. The filter's influence on the signal-to-noise ratio can therefore be ascertained by considering its effect on the noise

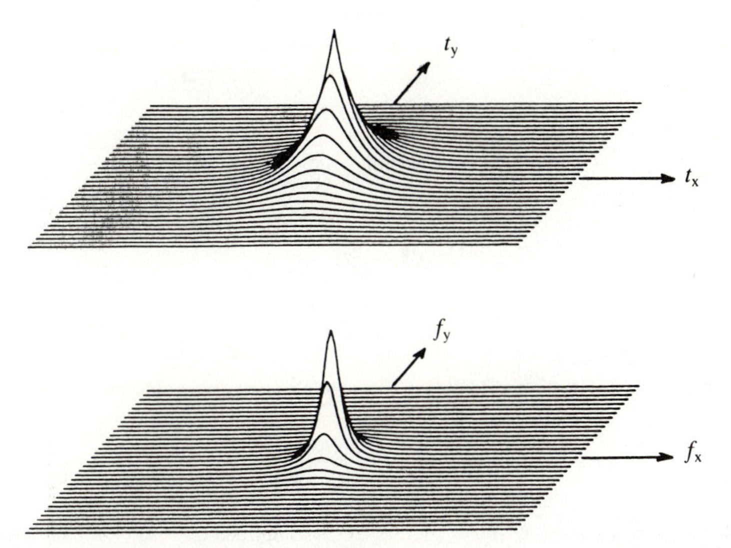

Fig. 4.6 Corresponding time and frequency domain filters used in FI (or PR) image reconstruction. (From Callaghan and Eccles[3]).

alone. Eqn (4.27) is modified to include the effect of apodization so that for FI the noise power is

$$\overline{|\mathfrak{N}_f(n_x/NT, n_y/NT)|^2} = 4 \sum_{m=0}^{\frac{1}{2}N-1} \sum_{l=0}^{\frac{1}{2}N-1} \exp\{-2(m^2+l^2)^{\frac{1}{2}}/p\} \overline{|\mathfrak{N}_t(mT, lT)|^2} \tag{4.40}$$

or

$$\sigma_f(\mathrm{FI}) = N\sigma_t S_{\mathrm{FI}}(N/p)^{-1} \tag{4.41}$$

where

$$S_{\mathrm{FI}}(N/p) = \left[\tfrac{1}{2}\pi\left\{\frac{p^2}{N^2} - \mathrm{e}^{-N/p}\left(\frac{p^2}{N^2} + \frac{p}{N}\right)\right\}\right]^{-\frac{1}{2}}. \tag{4.42}$$

Comparison with eqn (4.29) shows that the filter suppresses the noise by the smoothing factor $S_{\mathrm{FI}}(N/p)$ shown in eqn (4.42). In fact this analytic expression for $S_{\mathrm{FI}}(N/p)$ was obtained using polar coordinates. While correct for $p \ll N$, when $p \gg N$ it underestimates the noise power by $\pi/4$, the ratio of the areas of a circle and a square. Fig. 4.7 shows the exact dependence of $S_{\mathrm{FI}}(N/p)$ on N/p obtained by numerical summation in Cartesian coordinates.

For PR, application of the simple one-dimensional filter results in the noise power.

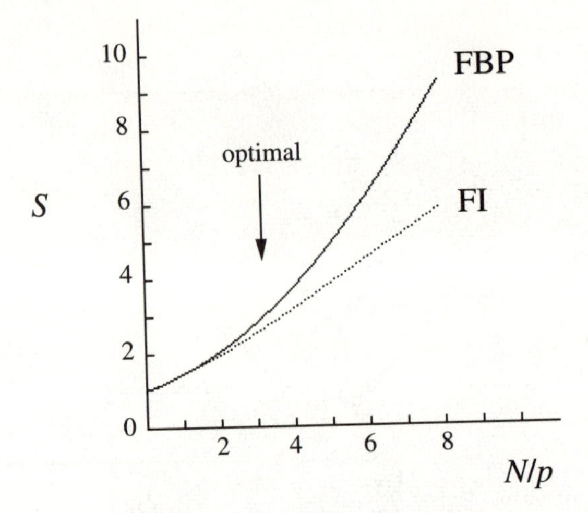

Fig. 4.7 Smoothing factors $S_{\mathrm{FI}}(N/p)$ and $S_{\mathrm{PR}}(N/p)$ as defined by eqns (4.42) and (4.45). At optimal apodization broadening $p=N/\pi$. Note the dependence of $S_{\mathrm{FI}}(N/p)$ on (N/p) and $S_{\mathrm{PR}}(N/p)$ on $(N/p)^{3/2}$ as N/p becomes large.

$$\overline{|\mathfrak{N}_f|^2} = 2N_p \sum_{m=0}^{\frac{1}{2}N-1} m^2 \exp(-2m/p) \tag{4.43}$$

whence

$$\sigma_f(\mathrm{PR}) = \frac{1}{\sqrt{12}} N_p^{\frac{1}{2}} N^{\frac{3}{2}} \sigma_t S_{\mathrm{PR}}(N/p)^{-1} \tag{4.44}$$

where

$$S_{\mathrm{PR}}(N/p) = \left[6\left(\frac{p}{N}\right)^3 - \mathrm{e}^{-N/p}\left(6\frac{p^3}{N^3} + 6\frac{p^2}{N^2} + 3\frac{p}{N}\right)\right]^{-\frac{1}{2}}. \tag{4.45}$$

The dependence of the smoothing noise suppression factor $S_{\mathrm{PR}}(N/p)$ on N/p is also shown in Fig. 4.7.

In summary the following expressions apply to the circular object.

Filtered signal-to-noise: FI

$$\frac{\rho_0(\mathrm{FI})}{\sigma_f(\mathrm{FI})} = \left[\left(\frac{4}{\pi}\right) N N_s^{-2} \left(\frac{S_0}{\sigma_t}\right)\right] S_{\mathrm{FI}}(N/p) \tag{4.46}$$

Filtered signal-to-noise: PR

$$\frac{\rho_0(\mathrm{PR})}{\sigma_f(\mathrm{PR})} = \left(\frac{\sqrt{12}}{\pi}\right)\left[\left(\frac{4}{\pi}\right) N_p^{\frac{1}{2}} N^{\frac{1}{2}} N_s^{-2} \left(\frac{S_0}{\sigma_t}\right)\right] S_{\mathrm{PR}}(N/p) \tag{4.47}$$

It is apparent in Fig. 4.7 that as p is shortened the PR signal-to-noise ratio improves asymptotically as $(N/p)^{3/2}$ whereas for FI it improves as (N/p). In this asymptotic limit the signal-to-noise ratio is superior to that in FI by the factor $(1/\sqrt{\pi})(N/p)^{\frac{1}{2}}$. This is the regime where the filter dominates the broadening and so determines the resolution. In this asymptotic regime where the resolution is proportional to the point spread function width, $N/(\pi p)$, we have the following.

Filtered signal-to-noise: FI *with* $p \ll N$

$$\frac{\rho_0(\mathrm{FI})}{\sigma_f(\mathrm{FI})} = \left[\left(\frac{4}{\pi}\right) N N_s^{-2} \left(\frac{S_0}{\sigma_t}\right)\right] \left(\frac{\pi}{2} \frac{p^2}{N^2}\right)^{-\frac{1}{2}} \tag{4.48}$$

Filtered signal-to-noise: PR *with* $p \ll N$

$$\frac{\rho_0(\mathrm{PR})}{\sigma_f(\mathrm{PR})} = \left(\frac{\sqrt{12}}{\pi}\right) \left[\left(\frac{4}{\pi}\right) N_p^{\frac{1}{2}} N^{\frac{1}{2}} N_s^{-2} \left(\frac{S_0}{\sigma_t}\right)\right] \left(\frac{6p^3}{N^3}\right)^{-\frac{1}{2}}. \tag{4.49}$$

Given $N_s \sim N$ it is clear that the image signal-to-noise is independent of the number of data points, N, in the asymptotic smoothing regime. This is not surprising since, for $p \ll N$, additional sampling points have effectively zero signal and noise so that the process of increasing N merely constitutes zero-filling.

If all other parameters remain constant, the relationship between FI and PR signal-to-noise expressions are different in the asymptotic regime. It is clear that, for FI, the image spectral resolution, $N/(\pi p)$, and image signal-to-noise, $\rho_0(\mathrm{FI})/\sigma_f(\mathrm{FI})$, vary in inverse proportion. In PR, however, a greater enhancement in signal-to-noise is possible given the same sacrifice in resolution.

4.5 T_2-limited resolution

4.5.1 *T_2-optimal bandwidth*

Any discussion of resolution must begin with an effective criterion. In practice we will seek an optimum broadening which allows the resolution of next-nearest neighbours. This is the best screen resolution which can be achieved in practice since we require some modulation of intensity between three adjacent pixels to identify that the outer two represent separated features in the image. The Rayleigh criterion[17] states that two pixels will be just resolved when the contrast ratio, as defined in Fig. 4.8 is 0.81. The value of N/p which produces this contrast ratio may be found by convoluting the Fourier transform of $F(m, l)$ with two delta functions spaced two pixels apart. Rayleigh broadening occurs when $N/p = 6.4$.

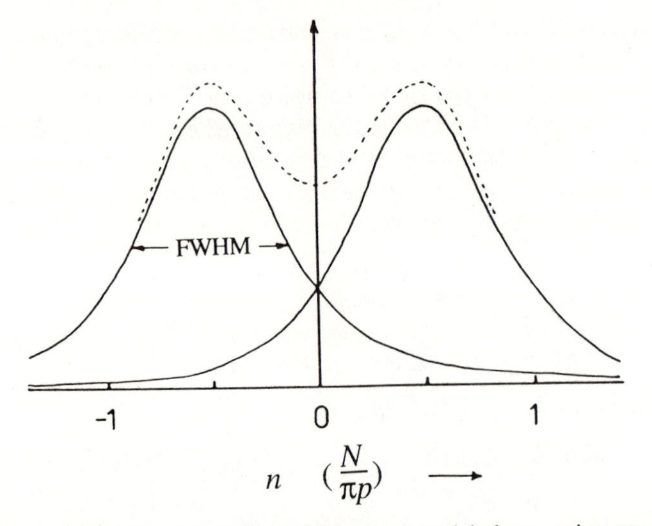

Fig. 4.8 One-dimensional sections through the exponential decay point spread function depicted in the lower part of Fig. 4.6. The two curves shown have centres separated by two pixels and have widths determined by the Rayleigh resolution criterion which requires that $N/p = 6.4$. (From Callaghan and Eccles[3]).

Of course the T_2 relaxation contributes a fundamental linewidth which also influences the resolution. This influence is insignificant if the bandwidth is sufficiently large that the T_2 broadening is less than one pixel. For narrower bandwidths it is necessary to establish which apodization filter, when combined with T_2 relaxation, will give a broadening that just satisfies the Rayleigh criterion. Such a filter is easy to write down if we remember that the relaxation occurs only along the read axis in **k**-space and so produces a decay $\exp(-m/p_2)$, where $p_2 = T_2/T$. The filter required is $\exp[2m/p_2 - 2(m^2 + l^2)^{1/2}/p]$, where $p = N/2\pi$. Following the method of eqn (4.40) we obtain,

$$S_{\mathrm{FI}}(\Delta f) = N\left\{4 \sum_{m=0}^{\frac{1}{2}N-1} \sum_{l=0}^{\frac{1}{2}N-1} \exp\left[2m/p_2 - 12.8(m^2 + l^2)^{\frac{1}{2}}/N\right]\right\}^{-\frac{1}{2}}. \tag{4.50}$$

Clearly $S_{\mathrm{FI}}(\Delta f)$ is dependent on the bandwidth since $p_2 = T_2\Delta f$.

Having determined the correct filter for next-nearest neighbour resolution, the next step is to adjust the bandwidth for optimal resolution in the T_2 linewidth limit. This is done by first substituting the relevant time domain signal-to-noise expression into eqn (4.46). The sample volume is taken to be that for a circular object of radius R_0 and slice thickness Δz. Incorporation of eqn (4.22) implies that R_0 is eliminated. We may then express the gradient magnitude $|G|$ in terms of the desired frequency domain signal-to-noise, and, writing the resolution Δx in the optimal limit as $(2/NT)(2\pi/\gamma G)$, we obtain

$$\Delta x = 38.3\, a^{\frac{1}{2}} F^{\frac{1}{4}} \sigma^{\frac{1}{4}} \left(\frac{\rho_0}{\sigma_f}\right)^{\frac{1}{2}} \Delta z^{-\frac{1}{2}} f^{-\frac{7}{8}} N_{\mathrm{acc}}^{-\frac{1}{4}} N^{-\frac{1}{2}} \Delta f^{\frac{1}{4}} S_{\mathrm{FI}}(\Delta f)^{-\frac{1}{2}}. \quad (4.51)$$

This result can be optimized by minimizing $\Delta f^{\frac{1}{4}} S_{\mathrm{FI}}(\Delta f)^{-\frac{1}{2}}$ to find the optimal bandwidth. The minimum occurs for a value of Δf where the separation of points in the frequency domain is approximately the relaxation linewidth, $1/\pi T_2$. Consider the case relating to the k_x (read) direction where both T_2 and filter broadening contribute to the linewidth. It is clear that at the optimal bandwidth (under the Rayleigh resolution criterion) the filter broadening and the relaxation broadening each correspond to approximately one pixel separation, $1/NT$, and the resolution in frequency units is $2/NT$. This is in fact the matched filter criterion. In SI units the spatial resolution is given by the following expressions.

Optimal broadening (at optimal bandwidth $1/T = N/\pi T_2$).

$$\Delta x(\mathrm{FI}) = 16 a^{\frac{1}{2}} F^{\frac{1}{4}} \sigma^{\frac{1}{4}} \left(\frac{\rho_0}{\sigma_f}\right)^{\frac{1}{2}} \Delta z^{-\frac{1}{2}} f^{-\frac{7}{8}} N_{\mathrm{acc}}^{-\frac{1}{4}} N^{-\frac{1}{4}} T_2^{-\frac{1}{4}} \quad (4.52)$$

$$\Delta x(\mathrm{PR}) = 16 a^{\frac{1}{2}} F^{\frac{1}{4}} \sigma^{\frac{1}{4}} \left(\frac{\rho_0}{\sigma_f}\right)^{\frac{1}{2}} \Delta z^{-\frac{1}{2}} f^{-\frac{7}{8}} N_{\mathrm{acc}}^{-\frac{1}{4}} N_p^{-\frac{1}{4}} T_2^{-\frac{1}{4}}. \quad (4.53)$$

The corresponding expressions for asymptotic broadening are as follows.

Asymptotic broadening

$$\Delta x(\mathrm{FI}) = 6.1 a^{\frac{1}{2}} F^{\frac{1}{4}} \sigma^{\frac{1}{4}} \left(\frac{\rho_0}{\sigma_f}\right)^{\frac{1}{2}} \Delta z^{-\frac{1}{2}} f^{-\frac{7}{8}} N_{\mathrm{acc}}^{-\frac{1}{4}} \Delta f^{\frac{1}{4}} N^{\frac{1}{2}} p^{-1} S_{\mathrm{FI}}(N/p)^{-\frac{1}{2}}$$

$$= 6.8 a^{\frac{1}{2}} F^{\frac{1}{4}} \sigma^{\frac{1}{4}} \left(\frac{\rho_0}{\sigma_f}\right)^{\frac{1}{2}} \Delta z^{-\frac{1}{2}} f^{-\frac{7}{8}} N_{\mathrm{acc}}^{-\frac{1}{4}} \Delta f^{\frac{1}{4}} p^{-\frac{1}{2}} \quad (4.54)$$

$$\Delta x(\mathrm{PR}) = 5.8 a^{\frac{1}{2}} F^{\frac{1}{4}} \sigma^{\frac{1}{4}} \left(\frac{\rho_0}{\sigma_f}\right)^{\frac{1}{2}} \Delta z^{-\frac{1}{2}} f^{-\frac{7}{8}} N_{\mathrm{acc}}^{-\frac{1}{2}} \Delta f^{\frac{1}{4}} N_p^{-\frac{1}{4}} N^{\frac{3}{4}} p^{-1} S_{\mathrm{PR}}(N/p)^{-\frac{1}{2}}$$

$$= 9.1 a^{\frac{1}{2}} F^{\frac{1}{4}} \sigma^{\frac{1}{4}} \left(\frac{\rho_0}{\sigma_f}\right)^{\frac{1}{2}} \Delta z^{-\frac{1}{2}} f^{-\frac{7}{8}} N_{\mathrm{acc}}^{-\frac{1}{4}} \Delta f^{\frac{1}{4}} N_p^{-\frac{1}{4}} p^{-\frac{1}{4}}. \quad (4.55)$$

The optimal bandwidth expressions are useful in helping predict the best possible resolution in NMR microscopy. To do this we continue our 'idealized' approach. An optimized detection system would have $F = 1$ and the smallest possible coil, namely $a = N\Delta x/4$, where Δx corresponds to

Table 4.2. *Optimal transverse and voxel resolution for an image signal-to-noise ratio of 20 and total accumulation number ($N_{acc}N$) of 1000. Intrinsic spectral broadening due to T_2 relaxation alone is assumed.*

		Resolution (μm) [and gradient (G cm^{-1})]			
		at 60 MHz		at 400 MHz	
T_2(ms)	optimal bandwidth (Hz) ($N = 256$)	100 μm slice	cubic voxel	100 μm slice	cubic voxel
500	163	16 [.19]	29 [.10]	3.0 [1.0]	9.6 [.31]
100	815	23 [.65]	38 [.39]	4.7 [3.2]	13 [1.1]
50	1630	27 [1.1]	42 [.71]	5.2 [5.7]	14 [2.1]
10	8150	42 [3.5]	56 [2.7]	7.6 [20]	18 [8.3]

two pixels. σ is typically 5 for ideal solenoidal geometry. In the expression for optimal resolution, only variation of N_{acc} and N (or N_p) offers the possibility of enhancing the resolution for a given slice of nuclear spins in a given detector. In both cases the improvement arises from an increased number of acquisitions and in consequence an increased experimental time-scale. The resolution is, not surprisingly, fixed by the available experimental time and the spin–lattice relaxation time. If we presume a minimum practical coil size of 2 mm, a total time of 1500 s, $T_1, T_2 = 0.5$ s with $T_R = 3T_1$, a signal-to-noise ratio of 20, $N = N_p = 256$, and a slice thickness of 100 μm, then eqns (4.52) and (4.53) predict an optimal transverse resolution of 16 and 3 μm for protons at 60 and 400 MHz, respectively. In fact, T_2, may be much shorter, of order 10 ms, in microscopic imaging. Table 4.2 summarizes the transverse and voxel resolutions pertaining to a range of experimental conditions.

Putting aside for the moment any consideration of rapid acquisition methods it is clear that the number of accumulations, N_{acc}, which are available in any given imaging time T_I, will be inversely proportional to the pulse repetition time T_R and hence to the spin–lattice relaxation time T_1. In consequence, N_{acc} is of order T_I/T_1 and inspection of eqns (4.52) and (4.53) shows that the transverse resolution for a fixed time experiment varies as $(T_1/T_2)^{-\frac{1}{4}}$. We can tolerate T_2 short provided that T_1 is similarly reduced. Of course, the worst case obtains when T_1 is long and T_2 is short. This is precisely the relaxation behaviour exhibited by nuclei in solid state environments.

The transverse resolution given by eqns (4.52) and (4.53) correspond to an optimal experiment where the resolution limit is determined by T_2 and the bandwidth has been adjusted to the value $(1/\pi T_2)$. The gradient magnitude has been eliminated from our resolution equations but by implication it has been adjusted to give the desired signal-to-noise ratio in the frequency (image) domain. The largest gradient value shown in Table 4.2 is $20\,\mathrm{G\,cm^{-1}}$. While such a gradient is quite large in comparison with those used for medical imaging a gradient of few tens of $\mathrm{G\,cm^{-1}}$ is by no means difficult to achieve in small-scale probe assemblies. On this basis, resolution in NMR microscopy is not field gradient limited!

It is helpful to retrace the steps in the argument in such a way that the interdependence of the applied gradient G and the bandwidth $(1/T)$ is more explicitly obvious. For the moment we shan't be concerned with the role of optimal filtering. This analysis starts by examining the dependence of the image signal-to-noise on G and T. The magnitude of the maximum **k**-space vector is proportional to GT and, in two-dimensional imaging, the image signal is spread over the spatial domain with an amplitude proportional to $G^{-2}(1/T)^2$. This product is obvious when one realizes that the spectral spread for an object in one frequency dimension is proportional to the gradient and inversely proportional to the bandwidth. If an analogue bandpass filter of bandwidth $(1/T)$ is applied then the signal-to-noise reduction consequent on decreasing the sampling bandwidth is partially compensated, leading to an overall image signal-to-noise proportional to $G^{-2}(1/T)^{3/2}$.

First we consider the regime where the resolution is limited by the spacing of a pair of pixels. When the intrinsic linewidth is less than the pixel spacing $(1/NT)$, the transverse distance corresponding to nearest neighbour pixels is given by $(1/NT)(2\pi/\gamma G)$. At fixed signal-to-noise the gradient and bandwidth must therefore be reduced to improve the resolution. This line of constant signal-to-noise is shown in Fig. 4.9 in which the pixel separation is plotted against the gradient on a long/log scale, so exhibiting a slope of $\frac{4}{3}$. In the lower part of the diagram we see the corresponding resolution with slope $\frac{1}{3}$.

By contrast, once the pixel separation is smaller than the linewidth, the smallest distinguishable element is linewidth-limited with a dimension given by $(1/\pi T_2)(2\pi/\gamma G)$. This width in frequency space is gradient independent but the spatial resolution deteriorates as G is decreased further as shown in Fig. 4.9. The minimum in Δx occurs at an optimal bandwidth $1/T = N\pi/T_2$. Similarly an optimum gradient is defined via the constant signal-to-noise product, $G^{-2}(1/T)^{3/2}$, an optimal bandwidth. The more detailed argument presented earlier takes account of optimal filtering and uses the *next*-nearest neighbour pixel separation in determining the resolution.

While the intrinsic linewidth enters the resolution expression as a parameter it is clear from the discussion above that the ultimate limit to resolution is determined by the intrinsic sensitivity of proton NMR detection, since this determines how large we may make our gradient before the signal-to-noise deteriorates unacceptably. The T_2 limit does, however, provide a useful reference and henceforth we refer to it as Δx_{opt}.

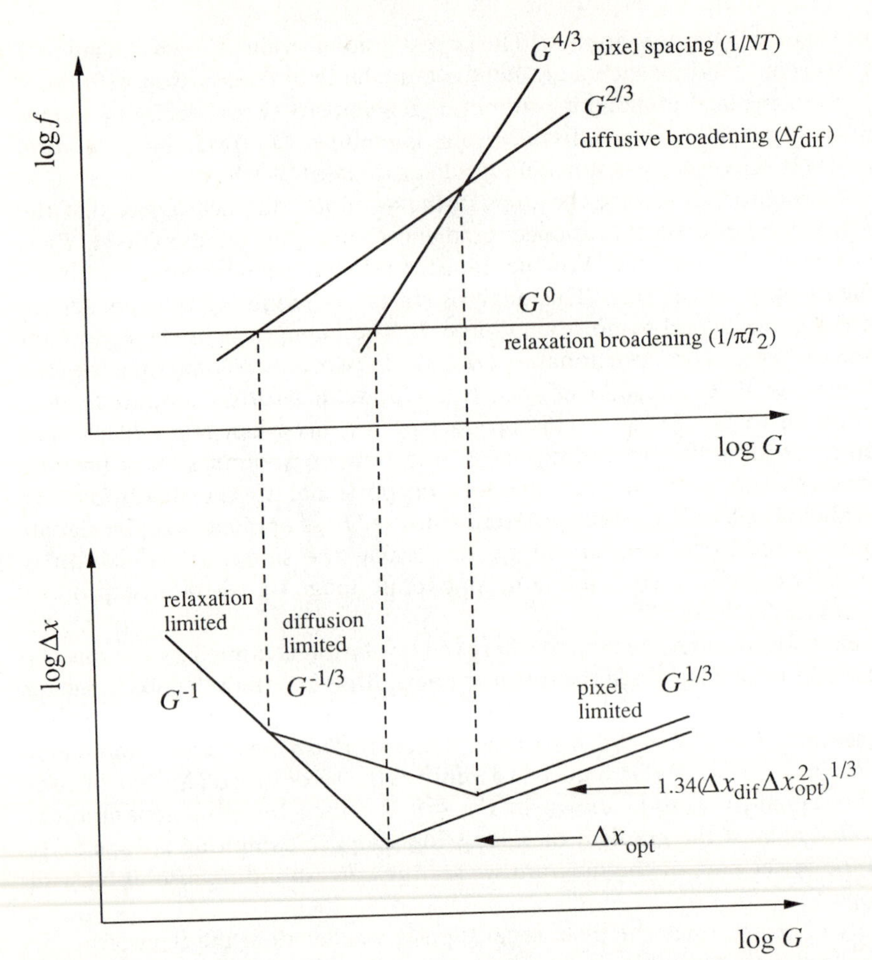

Fig. 4.9 Schematic log log plot of (a) frequency broadening and (b) spatial resolution, Δx, as a function of magnetic field gradient amplitude, G, at fixed signal-to-noise ratio. In (a) the relaxation broadening is fixed by T_2 while the pixel spacing, $1/NT$, at the chosen image signal-to-noise ratio depends on the system sensitivity. These values, along with the molecular self-diffusion coefficient, determine the cross-overs between various regimes. The arrows define optimal T_2- and diffusion-limited resolutions. (This latter limit is discussed in Section 4.6.1.) Point 1 is the cross-over from relaxation-limited to diffusion-limited resolution. (From Callaghan and Eccles[4]).

4.5.2 *Multiple echo summation*

The precise optimal resolution values shown in Table 4.2 are consistent with the simple expression given in eqn. (4.1). They refer, none the less, to a reference experiment which is, strictly speaking, somewhat idealized. There are, however, a number of practical acquisition schemes which are worthy contenders. These include the FLASH, DEFT, SSFP, and CPMG pulse sequences discussed in Chapter 3. In Section 4.5.3 we shall show that, of these, CPMG echo summation is the most attractive and, indeed, allows the limit set by Δx_{opt} to be most closely approached. As indicated in Chapter 3, echo signals may be directly added or alternatively a set of independent echo images may be reconstructed, the latter approach allowing greater flexibility in the subsequent analysis.

The CPMG method of sensitivity enhancement relies on the recycling of magnetization, an effect which is only possible if homonuclear couplings are not significant in the proton spin Hamiltonian, since these interactions will not refocus in a simple spin echo. In conventional NMR spectroscopy the use of this method of sensitivity enhancement is therefore restricted to dilute nuclei such as ^{13}C, where homonuclear couplings are unimportant, and is known by the acronym SEFT, for Spin Echo Fourier Transform spectroscopy.[18] Of course, the method will be applicable for proton NMR provided that *J*-couplings are unimportant, as in the case of water or fat molecules.

We will choose full echo acquisition, neglecting the initial FID which is insignificant when m is much larger than 1. The total signal is then weighted by $\Sigma_{n=1}^{m} \exp(-nNT/T_2)$ while the total signal-to-noise ratio, is proportional to $m^{-\frac{1}{2}} \Sigma_{n=1}^{m} \exp(-nNT/T_2)$ as given in eqn (3.34). Using the optimal echo number given by eqn (3.28), $m = 1.26(T_2/NT)$, this ratio becomes

$$m^{-\frac{1}{2}} \sum_{n=1}^{m} \exp\left(-\frac{nNT}{T_2}\right) = 0.72\left(\frac{NT}{1.26T_2}\right)^{\frac{1}{2}} \frac{\exp(-NT/T_2)}{1\text{-}\exp(-NT/T_2)}. \quad (4.56)$$

Table 4.3 shows the dependence of the overall signal-to-noise ratio on the number of co-added echoes. As m increases so does the acquisition bandwidth and, as a consequence, the resulting noise increase competes with the signal averaging effect as more echoes are co-added. The column on the far right shows this effect by multiplying the signal averaging enhancement of eqn (3.26) by the bandwidth noise factor. This factor is determined by the

Table 4.3. Optimal CPMG echo addition.

$\frac{NT}{T_2}$	practical optimum number of echoes	signal averaging enhancement	net signal-to-noise ratio *c.f.* full echo at optimal BW
1	1	0.37	0.21
0.4	3	0.82	0.29
0.2	6	1.30	0.33
0.1	13	2.68	0.34
0.04	31	3.14	0.35

ratio of acquisition bandwidth, $1/T$, to the T_2-optimal value, $N/\pi T_2$, i.e., $(NT/\pi T_2)^{\frac{1}{2}}$.

At first sight it would seem that the result of CPMG co-addition is an inferior signal-to-noise ratio in comparison with the optimal bandwidth case, and hence that the resolution will be necessarily poorer. However, it should be remembered that in this large-bandwidth regime, the linewidth will be determined by $1/\pi pT$ rather than $2/\pi T_2$. If we adjust the filter constant p so that the resolution limit is, as before, two pixels, then the apodization factor N/p is 2π, twice that used in the T_2-optimal case. This is a major advantage since, unlike the effect of relaxation, filter broadening is also associated with noise suppression.

Eqn (4.51) has the noise bandwidth effect already accounted for. Using eqns (4.54) and (4.55) we can incorporate the signal averaging enhancement (for $m \gg 1$ this is approximately $0.65(T_2/NT)^{\frac{1}{2}}$) and so obtain the following expressions for the resolution with FI and PR.

Asymptotic broadening with echo summation

$$\Delta x(\mathrm{FI}) = 7.5 a^{\frac{1}{2}} F^{\frac{1}{4}} \sigma^{\frac{1}{4}} \left(\frac{\rho_0}{\sigma_f}\right)^{\frac{1}{2}} \Delta z^{-\frac{1}{2}} f^{-\frac{7}{8}} N_{\mathrm{acc}}^{-\frac{1}{4}} N^{\frac{3}{4}} p^{-1} T_2^{-\frac{1}{4}} S_{\mathrm{FI}} (N/p)^{-\frac{1}{2}} \tag{4.57}$$

$$\Delta x(\mathrm{PR}) = 7.2 a^{\frac{1}{2}} F^{\frac{1}{4}} \sigma^{\frac{1}{4}} \left(\frac{\rho_0}{\sigma_f}\right)^{\frac{1}{2}} \Delta z^{-\frac{1}{2}} f^{-\frac{7}{8}} N_{\mathrm{acc}}^{-\frac{1}{4}} N_p^{-\frac{1}{4}} N p^{-1} T_2^{-\frac{1}{4}} S_{\mathrm{PR}} (N/p)^{-\frac{1}{2}} \tag{4.58}$$

Allowing $(N/p) = 2\pi$, eqns (4.47) and (4.58) give resolution limits of around $1.4\Delta x_{\mathrm{opt}}$ for FI and $1.1\Delta x_{\mathrm{opt}}$ for PR. It is apparent that echo summation is of slightly greater value when employed with the PR reconstruction method.

In summary, CPMG echo summation offers advantages only under high-bandwidth conditions where the sampling period is less than T_2 so that the optimum m exceeds unity. Provided N/p is set to the smallest value for which the filter limits the resolution, namely 2π, the transverse resolution is approximately Δx_{opt} for FI and PR. For neither image reconstruction technique does the method of echo summing significantly improve the resolution. Its attraction lies in the facility to achieve optimal resolution over a range of operating bandwidths. This is a very useful result. In the case of long relaxation times the optimal bandwidth may be a few hundred Hz or less. Unless the intrinsic field homogeneity corresponds to a linewidth of less than 1 Hz, sampling at such a bandwidth may cause severe image distortion. It is clearly advantageous to be able to achieve optimal resolution at a higher sampling rate. Finally, we note that the use of large acquisition bandwidths necessarily demands proportionately larger magnetic field gradients than those quoted in Table 4.2.

4.5.3 *Comparison of rapid acquisition methods*

The multiple echo CPMG method allows the acquisition bandwidth to be increased, while still retaining the optimal resolution associated with sampling the signal at a much slower rate over a time comparable with T_2. This is possible because the loss in sensitivity as the acquisition bandwidth is increased is compensated by the co-addition of echoes resulting from magnetization recycling. The concept can be generalized in order to compare other rapid acquisition methods in which such signal averaging is employed. These include FLASH, Partial Saturation (PS), DEFT, and SSFP. Three factors must therefore be accounted for in making signal-to-noise comparisons, namely, the mean signal amplitude $\overline{M}_y$, the sampling bandwidth $1/T$, and the repetition time T_R. The figure of merit of any method is therefore given by $\overline{M}_y(T/T_R)^{\frac{1}{2}}$. Each of the experiments to be considered in this section provides a practical approach to optimizing sensitivity.

We have seen in the CPMG example that an equivalent sensitivity factor of around 0.35 can indeed result in optimal resolution provided that the appropriate apodization filter is used. In all the examples compared in this section, high bandwidths, and hence asymptotic broadening conditions, apply. The results are therefore directly comparable with the CPMG analysis previously carried out. A comparison with the optimal resolution limit can be made by accounting for the equivalent sensitivity factors, f_s and relative apodization noise factors $S_{FI}(2\pi)/S_{FI}(\pi)$ or $S_{PR}(2\pi)/S_{PR}(\pi)$. Thus

$$\Delta x = \Delta x_{opt} f_s^{-\frac{1}{2}} \left(\frac{S(2\pi)}{S(\pi)}\right)^{-\frac{1}{2}}. \tag{4.59}$$

Eqn (4.59) gives $0.7\Delta x_{opt} f_s^{-\frac{1}{2}}$ for PR and $0.9\Delta x_{opt} f_s^{-\frac{1}{2}}$ for FI.

In order to compare potential resolution with differing techniques it is necessary to define the relevant experiment in each case and then to calculate the relative sensitivity factors, f_s. The r.f. and acquisition parameters relevant to the imaging pulse sequences are shown schematically in Fig. 4.10. Not shown in the rapid FLASH sequence in which minimal T_1 recovery occurs between acquisitions. This sequence can be effective in generating signal-to-noise enhancement when $T_2 \ll T_1$. Because of the noise power arguments detailed in Table 2.5, the FID acquisitions (PS(FID) and SSFP) are assigned a signal-to-noise ratio lower by a factor of $\sqrt{2}$. All the echo schemes, while acquiring the full signal, suffer however from a resultant attenuation by $\exp(-T_E/T_2)$. Clearly this will be minimal when the bandwidth is large and hence $T_E/T_2 \ll 1$.

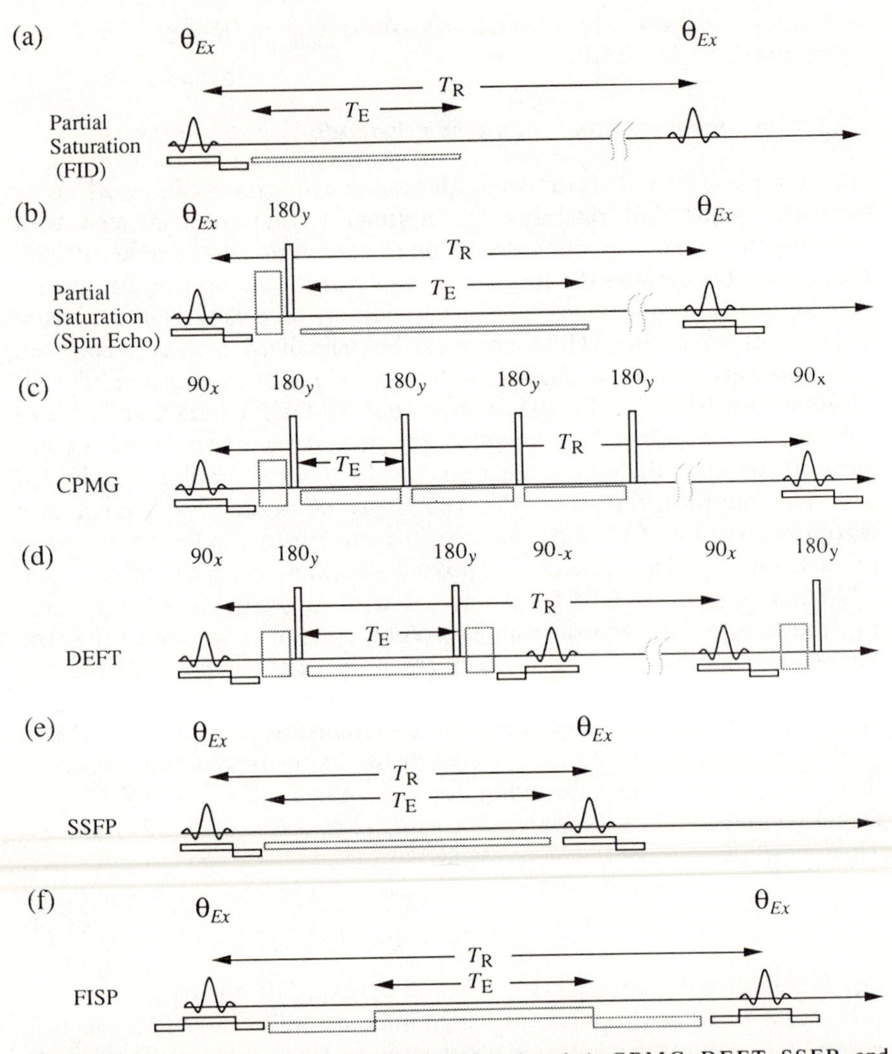

Fig. 4.10 r.f. pulse schemes for PS (FID), PS (spin echo), CPMG, DEFT, SSFP, and FISP imaging modes. The full echo or FID acquisition times are shown as T_E. In the echo methods this time corresponds to the echo delay and thus determines the signal attenuation due to transverse relaxation. In each case only read (dashed line) and slice (solid line) gradients are shown. For FISP the repetition time T_R is $2T_E$ compared with T_E in SSFP. Note that the 180° pulses are shown schematically as hard pulses for convenience. In practice soft pulses would be preferred.

In Fig. 4.10(a) and (b) the FLASH sequence is performed with a variable bandwidth and with multiple summation. Fig. 4.10(c) shows the full recovery CPMG (SEFT) sequence for which the mean value of M_y is found by averaging over the optimum number of echoes. Fig. 4.10(d) and (e) show DEFT and SSFP sequences, respectively. This imaging version of DEFT with spin echo sampling has been proposed by Maki *et al.*[19] while the mean value of M_y has been given by Shoup *et al.*[20] The SSFP imaging scheme employs FID sampling, whereas FISP has the SSFP echo and FID superposed.

PS(FID)

$$\overline{M_y} = \frac{1}{\sqrt{2}} M_0 \frac{1 - \exp(-T_R/T_1)}{1 - \cos\theta \exp(-T_R/T_1)} \sin\theta \qquad (4.60)$$

PS (spin echo)

$$\overline{M_y} = M_0 \frac{1 - 2\exp\{-(T_R - T_E/2)/T_1\} + \exp(-T_R/T_1)}{1 + \cos\theta \exp(-T_R/T_1)} \exp(-T_E/T_2) \sin\theta \qquad (4.61)$$

Full recovery CPMG(SEFT)

$$\overline{M_y} = \left(\frac{M_0}{m}\right) 0.72 \frac{\exp(-T_E/T_2)}{1 - \exp(-T_E/T_2)} \qquad (4.62)$$

where m is given by eqn (3.35).

DEFT

$$\overline{M_y} = M_0 \frac{1 - \exp\{-(T_R - 2T_E)/T_1\}}{1 - \exp(-2T_E/T_2)\exp(-(T_R - 2T_E)/T_1)} \exp(-T_E/T_2) \qquad (4.63)$$

SSFP

$$\overline{M_y} = \frac{1}{\sqrt{2}} M_0 \frac{1 - \exp(-T_E/T_1)}{[1 + \exp(-T_E/T_1)\exp(-T_E/T_2) - \cos\theta\{\exp(-T_E/T_1) + \exp(-T_E/T_2)\}]} \sin\theta \qquad (4.64)$$

FISP

$$\overline{M_y} = M_0 \frac{\{1 - \exp(-T_E/T_1)\}\exp(-T_E/2T_2)}{[1 + \exp(-T_E/T_1)\exp(-T_E/T_2) - \cos\theta\{\exp(-T_E/T_1) + \exp(-T_E/T_2)\}]} \sin\theta \qquad (4.65)$$

In order to compare relative sensitivities of these methods we must account for the role of the acquisition bandwidth and the co-addition of successive repetitions. In the reference full echo experiment, carried out at optimal bandwidth, it is convenient to imagine that the repetition time is set at $3T_1$. Even if T_2 is of order T_1, such a repetition time is consistent with a total sampling period of πT_2 and acquisition bandwidth, $N/\pi T_2$. Because the number of co-added signals will be proportional to $1/3T_1$, this particular idealized pulse sequence gives us a reference 'summation/bandwidth' sensitivity in terms of the ratio $(\pi T_2/3T_1 N)^{\frac{1}{2}}$. In the case of the CPMG sequence the sensitivity every repetition period of $3T_1$ is enhanced by $m^{\frac{1}{2}}$, the number of co-added echoes. The sampling time is T_E and the reception bandwidth is N/T_E. In the remaining cases where the bandwidth is larger than optimal because of rapid pulse repetition, we set the total sampling time, NT, to be the time period T_E, also giving a bandwidth, N/T_E. This corresponds to the smallest possible bandwidth and hence the optimum for each sequence. In PS and DEFT the 'summation/bandwidth' sensitivity is $(T_E/NT_R)^{\frac{1}{2}}$. In SSFP and FISP, where the repetition times are T_E and $2T_E$, respectively, this factor is simply $(1/\mathrm{N})^{\frac{1}{2}}$ and $(1/2\mathrm{N})^{\frac{1}{2}}$.

Using the idealized full echo experiment as a reference we obtain the following relative sensitivities. In the sequences where turn angles less than 90° are permitted θ has been set to the Ernst angle, θ_E. While this is strictly optimal only for the PS (FID) sequence, it represents a reasonable first approximation to optimal behaviour for the PS (spin echo) and SSFP sequences provided $T_1 \gg T_E$.

PS (FID)

$$f_s = \frac{1}{\sqrt{2}} \frac{1 - \exp(-T_R/T_1)}{1 - \cos\theta_E \exp(-T_R/T_1)} \sin\theta_E \left(\frac{3T_E T_1}{\pi T_2 T_R}\right)^{\frac{1}{2}} \tag{4.66}$$

PS(spin echo)

$$f_s = \frac{1 - 2\exp\{-(T_R - T_E/2)/T_1\} + \exp(-T_R/T_1)}{1 + \cos\theta_E \exp(-T_R/T_1)} \times \exp(-T_E/T_2)\sin\theta_E \left(\frac{3T_E T_1}{\pi T_2 T_R}\right)^{\frac{1}{2}} \tag{4.67}$$

Full recovery CPMG(SEFT)

$$f_s = 0.72 \frac{\exp(-T_E/T_2)}{1 - \exp(-T_E/T_2)} \left(\frac{T_E^2}{1.26\pi T_2^2}\right)^{\frac{1}{2}} \tag{4.68}$$

DEFT

$$f_s = \frac{1 - \exp\{-(T_R - 2T_E)/T_1\}}{1 - \exp(-2T_E/T_1)\exp\{-(T_R - 2T_E)/T_1\}} \exp(-T_E/T_2) \left(\frac{3T_E T_1}{\pi T_2 T_R}\right)^{\frac{1}{2}} \tag{4.69}$$

SSFP

$$f_s = \frac{1}{\sqrt{2}} \frac{1-\exp(-T_E/T_1)}{1+\exp(-T_E/T_1)\exp(-T_E/T_2)-\cos\theta_E\{\exp(-T_E/T_1)+\exp(-T_E/T_2)\}} \times \sin\theta_E \left(\frac{3T_1}{\pi T_2}\right)^{\frac{1}{2}} \tag{4.70}$$

FISP

$$f_s = \frac{\{1-\exp(-T_E/T_1)\}\exp(-T_E/2T_2)}{1+\exp(-T_E/T_1)\exp(-T_E/T_2)-\cos\theta_E\{\exp(-T_E/T_1)+\exp(-T_E/T_2)\}} \times \sin\theta_E \left(\frac{3T_1}{2\pi T_2}\right)^{\frac{1}{2}} \tag{4.71}$$

Fig. 4.11 shows the relative sensitivities and resulting resolution comparisons for each of the methods as both the echo and repetition times, T_E and T_R, are varied. All the echo schemes suffer for larger values of T_E/T_2 because of the echo attenuation factor. The most interesting sequences are those which are capable of achieving close-to-optimal resolution at large bandwidths (i.e. short T_E). Only DEFT and CPMG offer this facility although, in the case of DEFT, this advantage is confined to a narrow band of T_E values.

4.6 Diffusion-limited resolution

4.6.1 *Diffusion-optimal bandwidth*

The argument so far neglects diffusive effects. To clarify the expressions which follow, it is helpful to define another distance scale, the one-dimensional r.m.s. distance diffused in a time t by a molecule undergoing Brownian motion with a self-diffusion coefficient, D, given by $(2Dt)^{\frac{1}{2}}$. This distance we refer to as Δx_{dif}. t is the acquisition time and is πT_2 in the normal T_2-optimal experiment as described in Section 4.4. Note that both Δx_{opt} and Δx_{dif} are dependent on T_2.

Diffusion of spins in the presence of a magnetic field gradient results in magnetization dephasing, an effect akin to additional T_2 relaxation. As a consequence, diffusion can cause signal attenuation in spin echo imaging[21] and will always cause linewidth broadening in the read gradient irrespective of the chosen sampling scheme. Because the irreversible dephasing (see eqn (3.50)) depends on G^2t^3, whereas the desired spatial encoding phase shifts of spin echo imaging depend on the product Gt, the echo attenuation

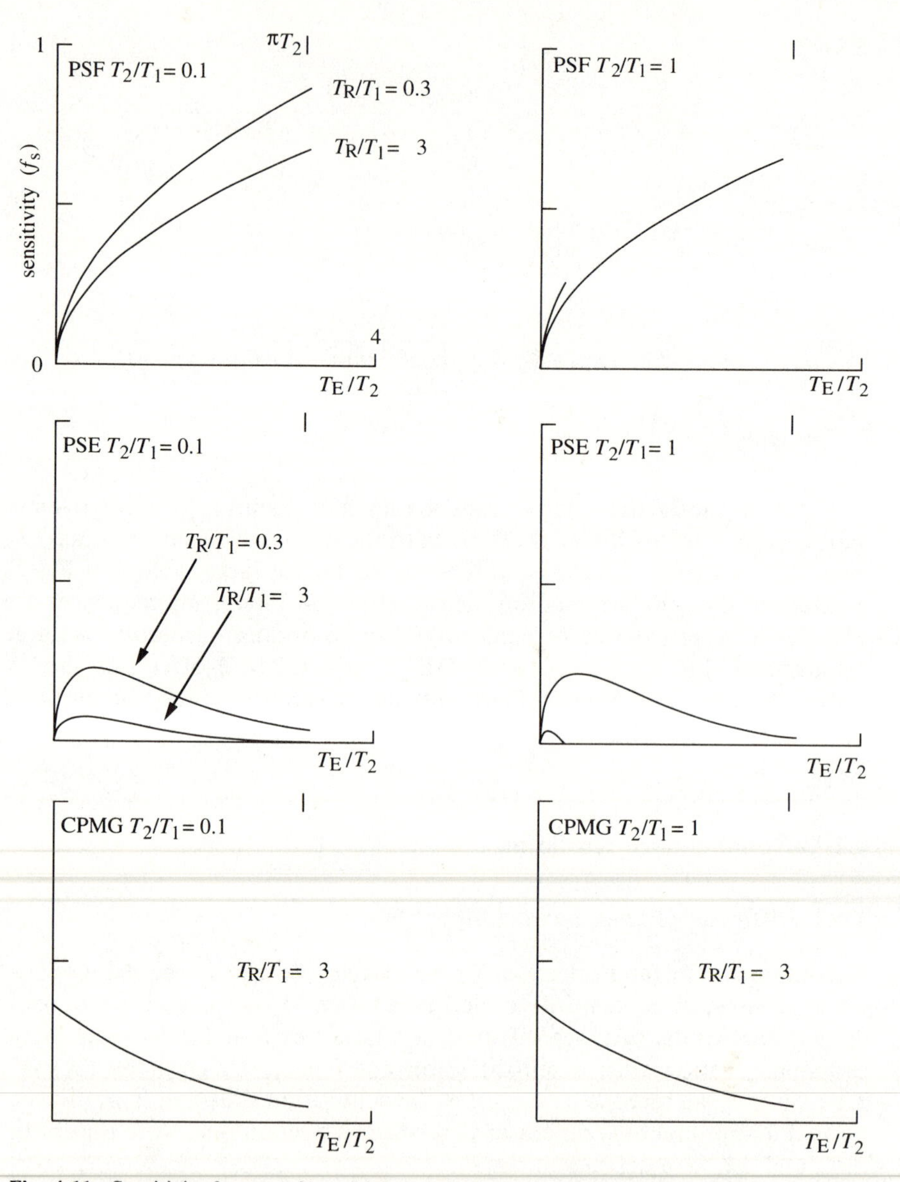

Fig. 4.11 Sensitivity factors (f_s) multiple acquisition methods (normalized against values for an idealized, full-echo-acquisition, reference experiment where T_R is set to $3T_1$). In (a) PSF and PSE refer to PS (FID) and PS (spin echo), respectively. For PS (FID), PS (spin echo), and DEFT, values are given for two repetition rates, $T_R/T_1 = 3$ and 0.3, while in the CPMG experiment T_R/T_1 is set to 3 and, in the steady-state free precession experiments, SSFP and FISP, T_R is, of course, determined by the echo time as T_E and $2T_E$, respectively. Note that the acquisition bandwidth is determined by T_E and that T_E cannot exceed T_R. In the PS (FID) experiment where no spin echo is employed, T_E may be interpreted as the acquisition time, NT. In each case the corresponding resolution may be calculated via eqn (4.59).

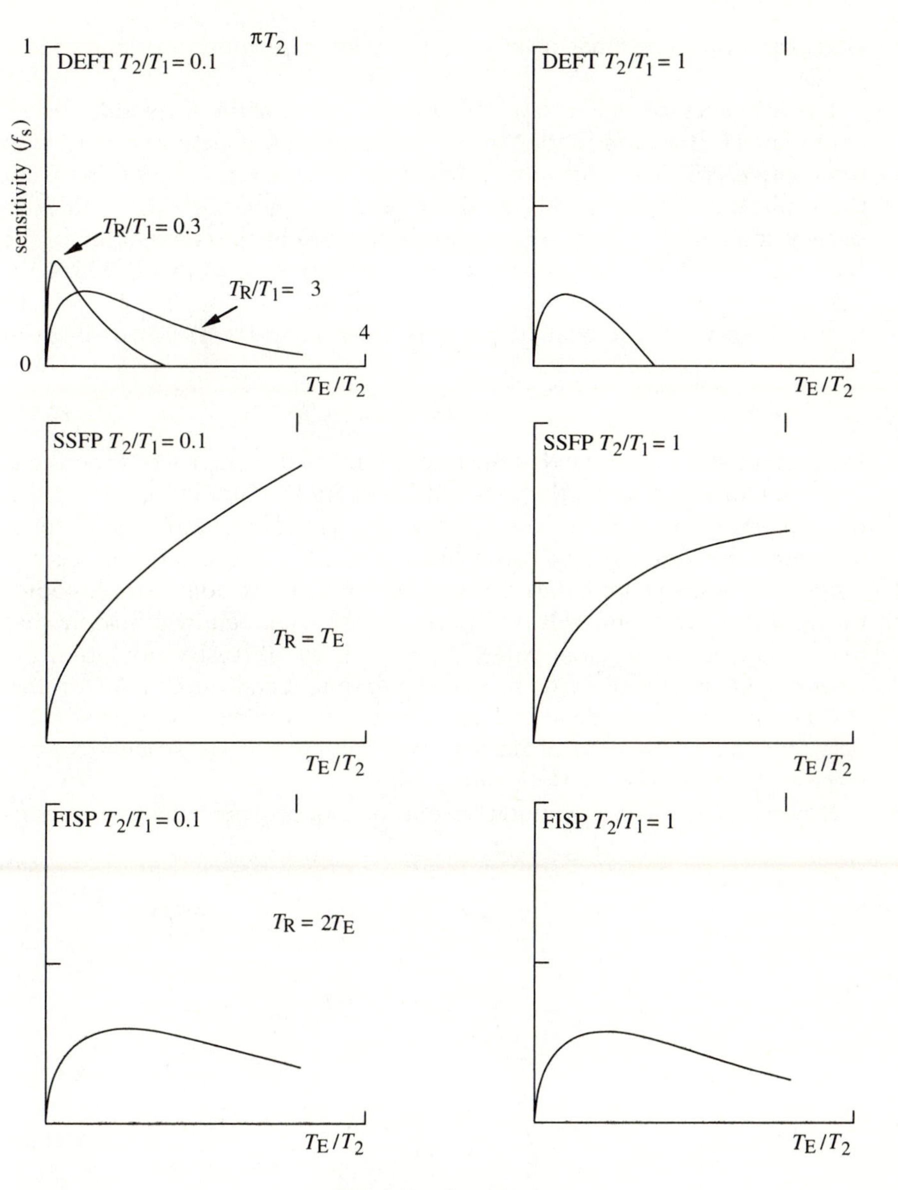

Fig. 4.11 *Continued*

effect can always be made arbitrarily small by employing short but intense gradient pulses.

The influence of molecular diffusion on the intrinsic linewidth can be appreciated by first considering the general imaging acquisition scheme, with total period NT. The additional attenuation time envelope, resulting from the application of the steady gradient, is $\exp(-\gamma^2 G^2 D t^3/3)$. In the frequency domain, therefore, the image is broadened by the convolution of the spectrum with this function whose cosine transform is shown in Fig. 4.12. The appropriate two-dimensional point spread function has a FWHM linewidth very close to that of the one-dimensional transform. It is given by[21]

$$\Delta f_{\mathrm{dif}} = 0.6(\gamma^2 G^2 D/3)^{1/3}. \tag{4.72}$$

In Section 4.4 it was shown that the resolution could be improved at constant signal-to-noise by altering the bandwidth and gradient until the pixel separation became comparable to the T_2 linewidth. However, unlike the intrinsic relaxation linewidth, the diffusive linewidth, Δf_{dif}, depends on the magnitude of the gradient so that as G increases this width steadily rises as shown in Fig. 4.9, intersecting with the pixel spacing at a bandwidth somewhat higher than the T_2-optimal value, $N/\pi T_2$. At the diffusive optimum, the pixel spacing is of order Δf_{dif} so that the optimal bandwidth, $1/T'$, in this case depends on the gradient and hence on the chosen image signal-to-noise. Selecting this optimal bandwidth is therefore much less straightforward when diffusive effects predominate.

However, the diffusion limited resolution can be calculated by compari-

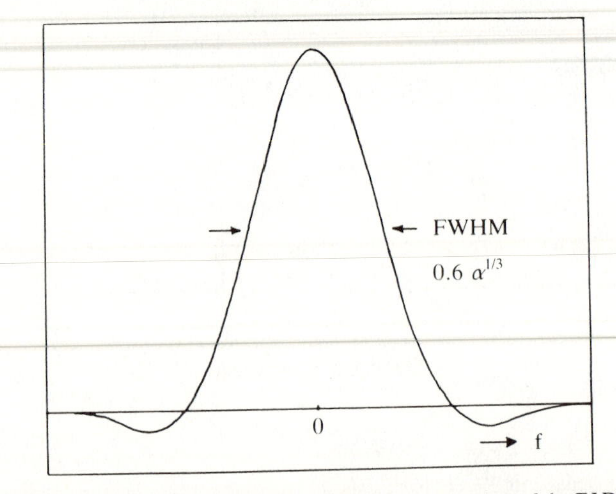

Fig. 4.12 Point spread function arising from diffusive attenuation of the FID. This lineshape, which is given by the real part of the Fourier transform of $\exp(-\alpha t^3)$, has a FWHM of $0.6\alpha^{1/3}$.

son with the T_2-optimal case. At the correct bandwidth, both the apodization filter broadening and the diffusive broadening correspond to the separation of a pixel pair, $1/NT'$, so that the resolution is the next-nearest neighbour spacing, $2/NT'$. The diffusive frequency spread can be regarded as arising from an effective T_2^* of $1/\pi\Delta f'_{\text{dif}}$ and it is tempting to use eqns (4.52) and (4.53) accordingly. Of course the gradient dependence of $\Delta f'_{\text{dif}}$ prevents us from adopting this simple-minded approach. Instead, we write down the resolution directly in terms of the bandwidth, $1/T'$, and the gradient, G', and find how these parameters scale, under constant signal-to-noise, in relation to the T_2 limit values for which we use unprimed parameters.

Thus

$$\begin{aligned}\Delta x &= (2/NT')\,(2\pi/\gamma G') \\ &= (2\Delta f'_{\text{dif}})\,(2\pi/\gamma G') \\ &= 1.20\,(\gamma^2 G^2 D/3)^{1/3}(2\pi/\gamma G)\,(G/G')^{\frac{1}{3}} \\ &= 1.78[\Delta x_{\text{dif}}^2\,\Delta x_{\text{opt}}]^{\frac{1}{3}}\,(G/G')^{\frac{1}{3}}.\end{aligned} \tag{4.73}$$

The gradient ratio is fixed by the constant signal-to-noise product, $G^{-2}(1/T)^{\frac{3}{2}}$, so that $(G/G') = (T/T')^{-\frac{3}{4}}$. Of course, the bandwidth ratio is just $\Delta f'_{\text{dif}}/(1/NT)$ and, after some manipulation, eqn (4.73) gives[4]

$$\Delta x = 1.34[\Delta x_{\text{dif}}\Delta x_{\text{opt}}^2]^{\frac{1}{3}}. \tag{4.74}$$

This is a curious relationship since it appears to say that the diffusive resolution limit depends on Δx_{opt}. While not a function of T_2 (since $\Delta x_{\text{dif}} \sim T_2^{1/2}$, whereas $\Delta x_{\text{opt}} \sim T_2^{-1/4}$), the resolution does depend on the desired signal-to-noise and other spectrometer characteristics evident in the expression for Δx_{opt}. This serves to emphasize the point that the diffusive resolution limit arises from consequential broadening and is not simply the distance travelled by the molecule over the experimental time-scale! By writing eqn (4.74) in this form we are able to make a ready comparison between diffusive and T_2 limits. By correct choice of optimal bandwidth it is clear that the diffusive resolution limit is geometrically weighted towards the T_2-optimal value so that the resolution loss due to diffusion is not as severe as that suggested by considering the distance moved by the molecules.

Fig. 4.13 compares the T_2 and diffusive limits in a $100\,\mu$m slice for a range of diffusion coefficients, the maximum corresponding to free water at room temperature. Note that the analysis above has assumed $\Delta f_{\text{dif}} > 1/\pi T_2$. For short T_2, small D values or low gradients (very high chosen signal-to-noise) the relaxation broadening will predominate and the resolution will be Δx_{opt}. Notice also that the imposition of the diffusion limit decreases the relative advantage of the higher spectrometer frequency because of the dependence of Δx on $\Delta x_{\text{opt}}^{2/3}$. In the example chosen the

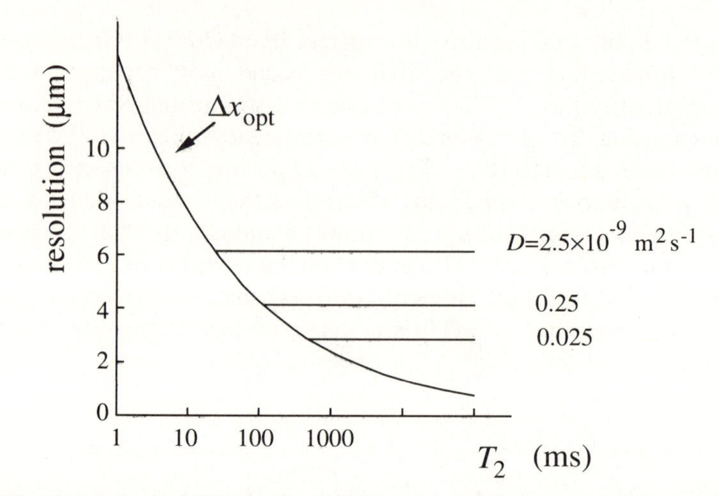

Fig. 4.13 Comparison of diffusive (eqn (4.74)) and T_2-optimal (eqn (4.52)) resolution limits for a 100 μm slice under the conditions which apply to Table 4.2 at 400 MHz. The horizontal lines correspond to diffusion coefficients ranging from that of free water to two orders of magnitude slower.

higher sensitivity resulting from using a Larmor frequency of 400 MHz rather than 60 MHz means that the T_2 optimal resolution is five times greater whereas in the diffusive limit the resolution is improved by only a factor of 3.

So far we have assumed that the diffusive broadening is present in two dimensions. This would indeed be the case for projection reconstruction and for the FI sampling scheme where the phase gradient is fixed and the phase duration is varied. However, in spin warp imaging the phase gradient can in principle be applied with arbitrarily large G and small t which, because of the dependence of dephasing on G^2t^3, means that broadening can be avoided along the phase axis. Of course, the same avoidance of phase axis broadening in FI also occurs in the T_2 limit although the advantages of such an asymmetric point spread function are somewhat dubious. However, the G^2t^3 relationship does suggest another remedy for the problem of diffusive broadening.

4.6.2 *Multiple echo summation*

The technique of reducing the appropriate time-scale in order to avoid dephasing due to diffusion was discussed in Chapter 2. There it was shown that the CPMG pulse sequence can be used to avoid diffusive spin echo attenuation and hence reveal the true T_2 relaxation. We also showed in Section 4.4.2, by means of echo summation using a CPMG sequence, that a resolution close to the T_2-optimal limit can be obtained regardless of bandwidth. Does such a sequence provide the means to avoid the limits to resolution imposed by molecular self-diffusion?

To understand this question we need to consider what happens as the sampling bandwidth is increased, making the sampling time per echo more rapid. One consequence of this increase in speed is that the echoes will be more closely spaced in time. In order to maintain resolution we must sample the same area of **k**-space so that G' and T' must vary inversely. The dependence of linewidth on $G'^2T'^3$ means that the broadening associated with diffusion can indeed be rendered insignificant in comparison with the apodization filter, provided T is sufficiently short. This means that we may calculate the resolution using the asymptotic broadening expression, eqns (4.54) and (4.55).

Despite appearances the diffusive problem has not been entirely eliminated. While the broadening has been avoided, the successive small attenuation of echoes reduces the overall signal-to-noise and so degrades the resolution. The effect is the same as if T_2 were artificially shortened. To see this we sum over m echoes exactly as before but now include the diffusive attenuation term. Note that the time t between the 180_y pulse and the echo centre in eqn (3.52) is $NT/2$. Following eqn (3.52) and (3.34) we find that the signal-to-noise enhancement is

$$\eta = m^{-\frac{1}{2}} \sum_{n=1}^{m} \exp[-n \frac{2}{3} \gamma^2 G^2 D (NT/2)^3] \exp[-nNT/T_2]$$

$$= m^{-\frac{1}{2}} \sum_{n=1}^{m} \exp[-nNT/T_2^*] \tag{4.75}$$

where

$$T_2^* = 1/T_2 + (1/12)\gamma^2 G^2 D(NT). \tag{4.76}$$

Notice that the resolution Δx varies as $T_2^{-\frac{1}{4}}$ in eqns (4.57) and (4.58). This leads, for the case of PR, to

$$\Delta x = 1.1(T_2^*/T_2)^{-\frac{1}{4}} \Delta x_{\text{opt}} \tag{4.77}$$

$$= 1.1[1 + (1/24\pi)\gamma^2 G^2 (NT)^2 (2D\pi T_2)]^{\frac{1}{4}} \Delta x_{\text{opt}}. \tag{4.78}$$

Using the relationship $\Delta x = (1/pT)(2\pi/\gamma G)$ and noting that the second term in the bracket predominates, we find

$$\Delta x = 1.8[\Delta x_{\text{dif}} \Delta x_{\text{opt}}^2]^{\frac{1}{3}}. \tag{4.79}$$

Remarkably, this result is very similar to eqn (4.74). The resolution is geometrically weighted towards the T_2 limit although the result is 30% poorer than can be achieved in a normal 'single acquisition' experiment where the gradient and bandwidth have been carefully adjusted until the diffusive broadening corresponds to a pair of pixels.

Even though the CPMG sequence does not avoid the resolution loss due to diffusion it is attractive in that it enables the same result to be achieved over a wide range of bandwidths. Hence it allows one to minimize distortion due to residual gradients by acquiring the image with large applied gradients. Furthermore it is far simpler in practice to adjust the experimental parameters for optimum conditions using echo summation than by bandwidth optimization, especially in the diffusive limit where the broadening is gradient dependent. The optimum number of echoes always corresponds to a final echo attenuation of exp (-1.26) regardless of conditions. This provides a simple practical criterion.

The echo summation results serve to emphasize yet again the role of intrinsic sensitivity in determining the resolution. Despite the appearance of Δx_{dif} in eqn (4.79), the influence of sensitivity predominates via Δx_{opt}.

4.7 Susceptibility-limited resolution

Atoms and molecules undergo magnetic polarization when placed in a magnetic field. In paramagnetic atoms, such as those in the lanthanide series, this magnetism arises from net atomic angular momentum and the induced dipolar fields may be very large. These are somewhat pathological cases in the case of NMR imaging and we shall not be concerned with paramagnetism except as it concerns dilute paramagnetic species which assist nuclear spin relaxation. Most biological and organic materials are diamagnetic and for these, while no net atomic angular momentum is present, the small disturbance to the electron orbits by the applied magnetic field results in a small induced local magnetic dipole. This effect is known as diamagnetic susceptibility and is the origin of the chemical shift effects discussed in Chapter 2.

The chemical shift represents the shielding effect at the nucleus due to the surrounding electron orbital which is perturbed in the presence of the polarizing field. There is also a macroscopic influence arising from the local magnetization associated with bulk susceptibility. It represents a 'through-space' interaction capable of influencing distant spins in differing molecular environments. The mean magnetic field in the sample, B_0^{s}, is offset from the free-space value B_0 according to

$$B_0^{\mathrm{s}} = (1 + \chi_m)B_0 \tag{4.80}$$

where χ_m is the diamagnetic susceptibility. In a real sample the value of χ_m is seldom constant since we are interested in imaging inhomogeneities! We might be tempted to write that the spatially dependent field, $B_0^{\mathrm{s}}(\mathbf{r})$, depends on susceptibility variations as

$$B_0^s(\mathbf{r}) = (1 + \chi_m(\mathbf{r}))B_0. \tag{4.81}$$

Such an equation is naive because it fails to take into account the boundaries between regions of differing susceptibility where Maxwell's equations must be satisfied. These boundaries have residual polarities which cause long-range field disturbances in the sample. The simplest example of this effect in classical electromagnetism is the well-known internal demagnetizing field[22] which arises from induced poles at the outer surface of a sample with uniform susceptibility placed in an external field. Because we wish to emphasize differences in Larmor frequency which arise from internal boundaries we shall not be concerned with this bulk demagnetizing effect but instead focus attention on superposed effects at small regions within the image. We will use the symbol χ_m^0 to describe the 'average' or background susceptibility of a sample and then $\Delta B_0(\mathbf{r})$ will be the local offset field from the average field $\chi_m^0 B_0$. The effect of $\Delta B_0(\mathbf{r})$ will be to introduce image distortions and the nature of these distortions will depend on the shape of the boundaries within the sample.

Complex local distortions have the effect of inducing artefacts in the image[23-25] and this leads to an inevitable resolution loss. In addition the diffusional motion of spins across regions of variable susceptibility results in fluctuating local fields, a phenomenon which causes additional transverse relaxation and consequential signal attenuation.[26-31] Furthermore these effects associated with susceptibility inhomogeneity will increase as the polarizing field, B_0, in increased. The additional T_2 relaxation arising from susceptibility variations are discussed in Section 4.7.3 while Section 4.7.1 deals with the problem of image distortion. In the earlier discussion of the role of sensitivity in determining resolution, the clear conclusion was that larger fields will give sharper images. The susceptibility problem casts some doubt on that conclusion, but given care and ingenuity we can restore the status quo. The tricks used to avoid susceptibility artefacts are discussed in Section 4.7.2. For the moment, however, we will explore their influence in the absence of these tricks. This is not just an academic exercise! Susceptibility artefacts are themselves a useful signature in the NMR micrograph because they tell us where the boundaries of sharp susceptibility difference are.

4.7.1 *Susceptibility artefacts*

The influence of the local field variations can be seen by adding the local precession frequency offset, $\gamma\Delta B_0(\mathbf{r})$, to eqn (3.1). In the heterodyne detection frame the time domain imaging signal becomes

$$S(t) = \iiint \rho(\mathbf{r}) \exp[i\gamma(\mathbf{G}\cdot\mathbf{r} + \Delta B_0(\mathbf{r}))t]\, d\mathbf{r}. \tag{4.82}$$

This result leads to a Fourier pair

$$S(\mathbf{k}) = \iiint \rho(\mathbf{r}) \exp[i\gamma \Delta B_0(\mathbf{r})t] \exp[i2\pi\mathbf{k}\cdot\mathbf{r}]\, d\mathbf{r} \tag{4.83}$$

$$\rho_I(\mathbf{r}) = \iiint S(\mathbf{k}) \exp[-i2\pi\mathbf{k}\cdot\mathbf{r}]\, d\mathbf{k} \tag{4.84}$$

Clearly $S(\mathbf{k})$ is not equal to the ideal reciprocal lattice, $S_0(\mathbf{k})$, and the reconstructed image, $\rho_I(\mathbf{r})$, is no longer a true representation of the original spin density, $\rho(\mathbf{r})$. The artefacts which are introduced will depend on the type of reconstruction employed. Rewriting eqns (3.15) and (3.18) we have, for two-dimensional Fourier imaging, using variable time, constant gradient phase encoding,

$$S(k_x, k_y) = \iint \rho(x, y) \exp[i2\pi k_x(x + \Delta B_0(x, y)/G_x)] \times \exp[i2\pi k_y(y + \Delta B_0(x, y)/G_y)]\, dx\, dy \tag{4.85}$$

and, for projection reconstruction,

$$S(k, \phi) = \iint \rho(x, y) \exp[i2\pi k(x \cos\phi + y \sin\phi + \Delta B_0(x, y)/G)]\, dx\, dy. \tag{4.86}$$

The simplest way to understand these equations is to think of an element of the object at (x_0, y_0) represented by the delta function $\delta(x - x_0, y - y_0)$. We can then obtain the contribution, $\delta\rho_I(x, y)$, of this element to the final image by obtaining the inverse (reconstruction) transformations of $S(k, \phi)$ and $S(k_x, k_y)$, respectively. Because $\delta\rho_I(x, y)$ gives the image plane representation of a single point in the original object it is equivalent to the point spread function. The case of FI is the easiest to understand and in the case of phase-encoding with a constant gradient, gives

$$\delta\rho_I(x, y) = \delta(x - [x_0 + \Delta B_0(x_0, y_0)/G_x], y - [y_0 + \Delta B_0(x_0, y_0)/G_y]). \tag{4.87}$$

This means that each element of the original object contributes to a single element in the final image displaced from the original position (x_0, y_0) by $(\Delta B_0(x_0, y_0)/G_x, \Delta B_0(x_0, y_0)/G_y)$. This simple effect is shown in Fig. 4.14(a). In the Spin Warp scheme the phase encoding uses a constant time period, t_y, and variable gradient, G_y. This means that element shifts occur only along the read direction. The influence of $\Delta B_0(x,y)$ under the phase gradient is to cause a local phase modulation $\exp(i\gamma\Delta B_0 t_y)$. This effect will generally be removed in calculating a modulus image but can lead to subtle attenuation effects where ΔB_0 strongly varies over one pixel.

For PR the situation is more complex. Here

$$\delta\rho_I(x, y) = \int_0^{\pi} \int_{-\infty}^{\infty} \exp[i2\pi k(x_0 \cos\phi + y_0 \sin\phi + \Delta B_0(x_0, y_0)/G)]$$

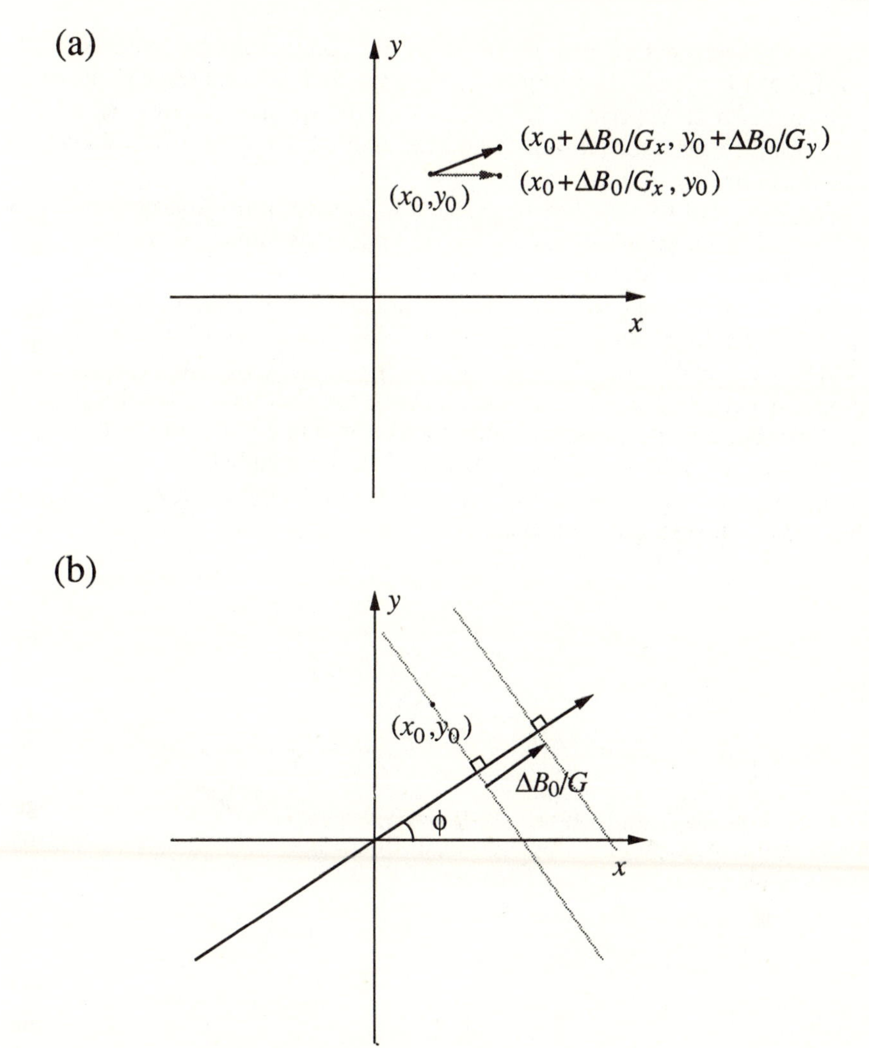

Fig. 4.14 Effect of magnetic field inhomogeneity on reconstruction in (a) FI and (b) PR. In (*a*) the horizontal dashed line corresponds to spin warp imaging.

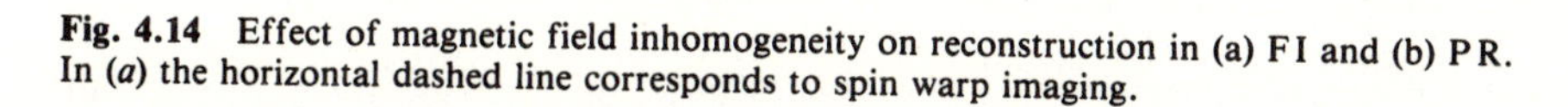

$$\times \exp[-i2\pi k(x\cos\phi + y\sin\phi)]|k| \, dk \, d\phi. \tag{4.88}$$

What this means is that each element at (x_0, y_0) contributes to a spectral component along the projection line oriented at ϕ but that this component is displaced along that line by $\Delta B_0(x_0, y_0)/G$ as illustrated in Fig. 4.14(b).

In the back-projection process this spectral component contributes to all positions (x, y) on the projection normal and the subsequent angular sum causes it to contribute to all parts of the image.[32] In this regard the susceptibility artefacts differ from the case of FI where the effects of distortion are more localized.[33]

The best way to visualize the effects of susceptibility inhomogeneity is by actual simulation on the basis of specific sample geometry.[34] Two important special cases are spherical and cylindrical regions of differing local susceptibility.[35] While the field profiles outside these objects are disturbed, the fields inside are uniform, a fact which forms the basis of high-resolution NMR using cylindrical NMR tubes with convex spherical bottoms! The model structures are shown in Fig. 4.15. In the first case a small sphere of uniform but differing susceptibility is embedded in a uniform environment. In the second a narrow but long uniform cylinder is embedded normal to the image plane. Depending on magnet configuration, the polarizing field may be either normal or transverse to the image slice. In both object

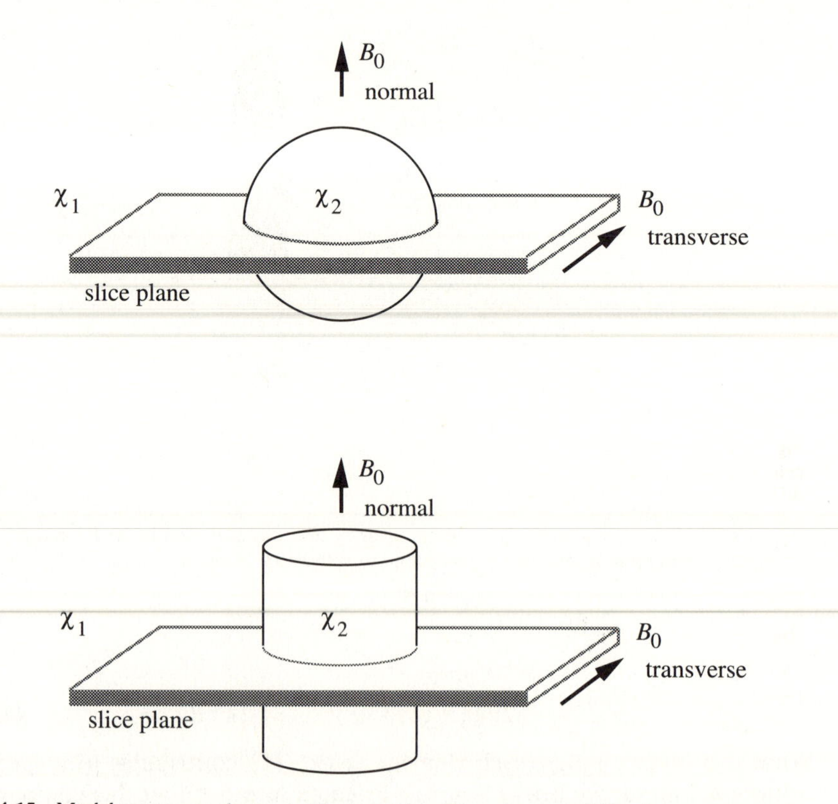

Fig. 4.15 Model structures chosen to illustrate the effect of susceptibility inhomogeneity. In each case the polarising field may be applied either transverse or normal to the slice plane.

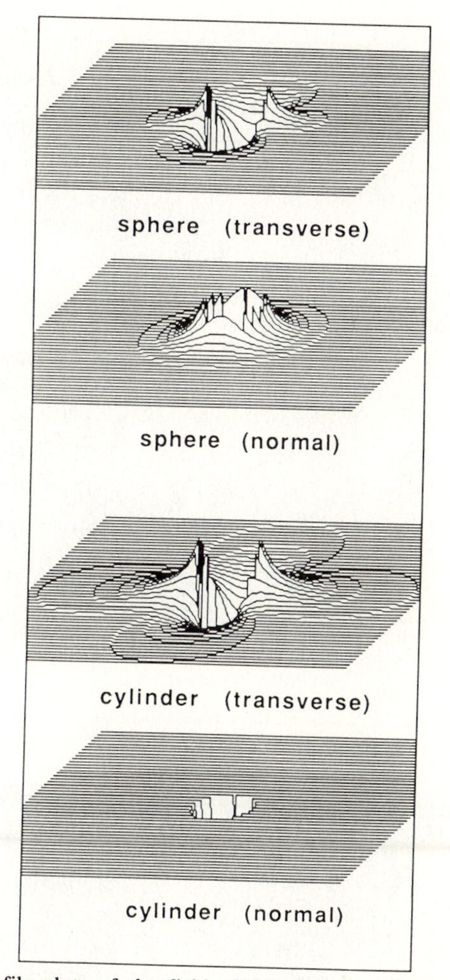

Fig. 4.16 Stacked profile plots of the field offsets, $\Delta B_0(\mathbf{r})$, arising from a sphere and a cylinder, shown in Fig. 4.15, embedded in a medium in which the diamagnetic susceptibility differs by 3 p.p.m. The image width is 20 kHz while the Larmor frequency is 200 MHz. (From Callaghan[34]).

geometries the central feature has constant susceptibility χ_{m2} while that of the surrounding medium is χ_{m1}. The field offset function, $\Delta B_0(\mathbf{r})$, is taken as the difference between the local field and the field in the sample distant from the disturbance caused by the embedded sphere or cylinder.

The embedded sphere and cylinder fields are shown in Fig. 4.16 as stacked profile plot images of $\Delta B_0(\mathbf{r})$. One interesting feature of susceptibility effects is that the absolute dimensions of the perturbing sphere or cylinder do not enter because the field distributions scale as r/a. The map of

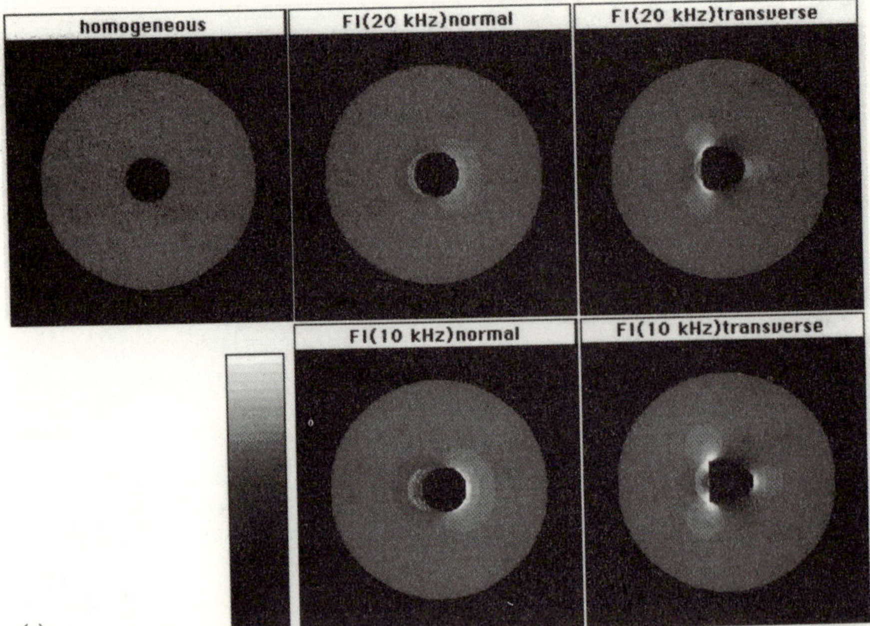
homogeneous
FI(20 kHz)normal
FI(20 kHz)transverse
FI(10 kHz)normal
FI(10 kHz)transverse

(a)

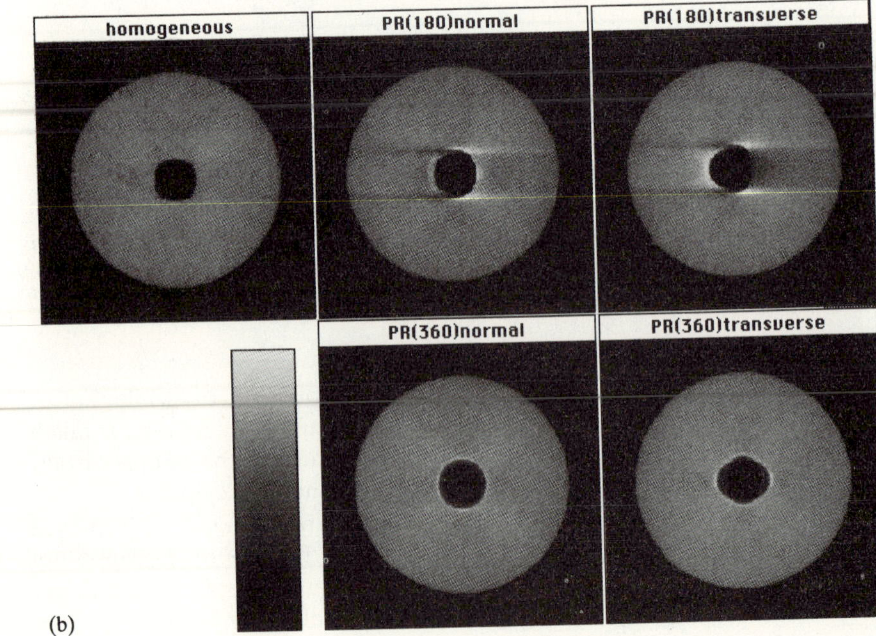
homogeneous
PR(180)normal
PR(180)transverse
PR(360)normal
PR(360)transverse

(b)

$\Delta B_0(\mathbf{r})$ is, of course, a map of the local Larmor frequency so that the significance of the disturbance will depend on the bandwidth across the NMR image. In the present case $(\chi_{m2} - \chi_{m1})$ is chosen as -3×10^{-6}, the Larmor frequency as 200 MHz, and the bandwidths as 20 kHz and 10 kHz. These bandwidths correspond to gradient strengths of 50 G cm^{-1} and 25 G cm^{-1} at an image width of 1 mm. Such a susceptibility inhomogeneity is not severe, representing, for example, the difference between water and some organic polymers. Between biological tissue and air the difference may be up to an order of magnitude larger!

Fig. 4.17(a) shows the initial object and the resulting FI reconstructions for the case of the embedded sphere in transverse and normal polarizing field geometries at 3 p.p.m. susceptibility difference. The sphere, which has been chosen to contribute less than its surroundings to the NMR signal, could represent a gel particle or glass bead surrounded by liquid, or a cell nucleus in the intracellular fluid. Each reconstruction uses the $\Delta B_0(\mathbf{r})$ profiles of Fig. 4.16 with acquisition bandwidths of both 20 kHz and 10 kHz at a Larmor frequency of 200 MHz. Spin Warp phase encoding is assumed for FI so that distortion along the vertical phase axis is minimal. Following the usual convention the read gradient is always positive so that the resulting image distortion is asymmetric along the read axis. The distortions introduced in the images are complicated and a simple visual interpretation is not available, an effect which leads to a loss of resolution in the image. This loss may be equated to the transverse distance over which the image intensity is disturbed by 50%. In the 20 kHz examples the size of the resolution loss introduced is of order 8 pixels, corresponding to 600 Hz, or 30 μm for a 1 mm field of view.

In Fig. 4.17(b) PR is used to obtain the corresponding images for a 3 p.p.m. susceptibility difference and 20 kHz bandwidth. PR can be carried out using either two or four quadrants of gradient vector orientation, two-quadrant gradient orientation being akin to the FI reconstruction examples.

Fig. 4.18 shows grey-scale images for an embedded cylinder with 3 p.p.m. susceptibility difference and 20 kHz bandwidth. The cylinder might represent a thin-walled tube of liquid surrounded by a different liquid or a long fibre bundle in a surrounding plant stem and here the field disturbances in

Fig. 4.17 (a) 200 MHz Grey-scale FI images of a slice through a sphere embedded in a surrounding medium. The image labelled 'homogeneous' was reconstructed using identical sphere and medium susceptibilities. The remaining images were reconstructed using field offsets arising from a 3 p.p.m. susceptibility difference, with the polarising field transverse and normal to the image plane, and with imaging bandwidths of 20 kHz and 10 kHz. (b) Grey-scale PR images of a slice through a sphere embedded in a surrounding medium. The image labelled 'homogeneous' was reconstructed using identical sphere and medium susceptibilities. The remaining images were reconstructed using field offsets arising from a 3 p.p.m. susceptibility difference. The upper set corresponds to 180° projection reconstruction and the lower to 360° PR.

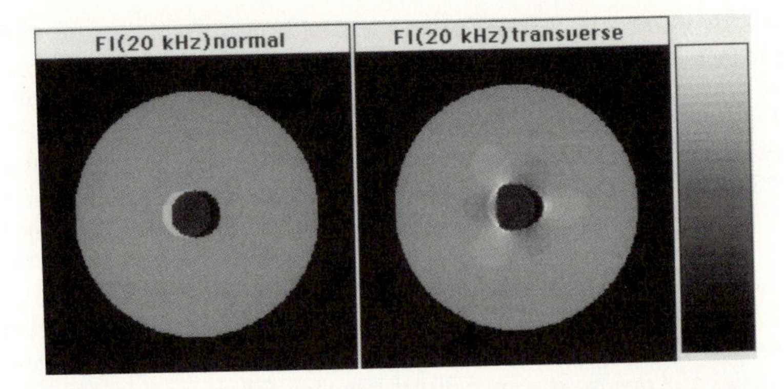

Fig. 4.18 Grey-scale F I images as for Fig. 4.17 (a) but for an embedded cylinder with 3 p.p.m. susceptibility difference and 20 kHz bandwidth. The artefact corresponds to a simple shift when the polarising field is applied along the cylindrical symmetry axis.

the cylindrical case are of somewhat longer range, although in these examples the size of the 50% disturbance is again of order 8 pixels, corresponding to 30 μm for a 1 mm field of view.

The gradients and bandwidths chosen to illustrate these effects are quite realistic. In fact, if the much smaller T_2-optimal bandwidths, recommended in Table 4.2, were chosen, the susceptibility artefacts would be far worse. The problem is illustrated by the fact that the spatial displacement which results from susceptibility inhomogeneity scales as $\chi_m(x, y)B_0/G$. Therefore distortion will increase as the polarizing field rises unless G is proportionally adjusted in compensation. This increasing value of G requires the use of larger acquisition bandwidths.

4.7.2 *Distortionless imaging*

The simplest way to avoid susceptibility distortion is to use larger imaging gradients. At 400 MHz Larmor frequency reduction of the distortion from a 1 p.p.m. susceptibility difference to the two-pixel level associated with intrinsic resolution would require a gradient and bandwidth of 150 G cm^{-1} and 60 kHz, respectively, for a 1 mm field of view. This is quite a large gradient for most micro-imaging systems. But the more serious problem is the need for a bandwidth required far above the optimal T_2 limit. Since the transverse resolution deteriorates as $G^{\frac{1}{3}}$ above the optimal bandwidth, the loss being contemplated is about an order of magnitude when T_2 is long.

In Section 4.6.2 it was demonstrated that optimal resolution was possible using high bandwidths by the method of echo summation. The obvious advantage accrued is that the effects of susceptibility and polarizing field inhomogeneity may be overwhelmed, simply by choosing a sufficiently large gradient. An example of such an effect is shown in Fig. 4.19 in which images of a plant stem cross-section obtained with different read and phase gradient amplitudes are compared. The image acquired with larger gradients is noticeably sharper.

In imaging systems where large gradients are unavailable more elegant methods are appropriate. A possible scheme for refocusing the phase shifts due to constant local field offsets has been proposed by Miller and Garroway.[36] In their method the gradient is applied as a sinusoidal oscillation with r.f. pulses situated at gradient nulls. A pulsed alternative is shown in Fig. 4.20 The spacing of the r.f. pulses shown is just the sampling interval T associated with the optimal bandwidth $1/T$. Because only one digitization sample is obtained per T interval, the usual bandpass filtering can be retained. In properly designed small gradient coils the speed of gradient switching would only become a problem at bandwidths in excess of optimal, for example at 20 kHz where $T = 50\,\mu$s.

The pulse scheme of Fig. 4.20 can, in principle, be used to provide distortionless sampling at optimal bandwidths by refocusing field offsets due to both susceptibility and polarizing field inhomogeneity. It can also be shown that the diffusive resolution limits associated with this scheme are no worse than those discussed in Section 4.6. One practical point arises when designing pulse sequences which avoid susceptibility artefacts. Both echo summation and gradient reversal trains imply the need for rapid gradient switching and hence the use of small shielded coils.

4.7.3. *Diffusive attenuation*

The translational diffusion of molecules in the presence of a spatially dependent magnetic field induces phase shifts in the spin transverse magnetization. Because of this effect susceptibility inhomogeneity in a sample will cause image attenuation if the molecules containing those spins are free to diffuse.[26] The effect of susceptibility inhomogeneity is examined in detail in Section 7.6.4. Two parameters are of importance in determining the size of this effect. These are the mean square Larmor frequency fluctuation, $\overline{\Delta\omega_0^2}$, due to local magnetic field variations, and the average fluctuation correlation time, τ_c, corresponding to the time taken for a spin to diffuse over the distance associated with the local field variation. Given these parameters, two regimes, fast motion and slow motion, may be defined according to $\overline{\Delta\omega_0^2}\tau_c^2 \ll 1$ or $\overline{\Delta\omega_0^2}\tau_c^2 \gg 1$. In practice, given the diffusion coefficient of free water and a typical susceptibility-induced frequency spread ($\gamma\Delta\chi_m B_0$), the distance scale which divides these regimes is around $2\,\mu$m. For susceptibility variations leading to very slowly varying fields (for example, those arising from bounding surfaces with radii of curvature much larger than $2\,\mu$m) the broadening is largely inhomogeneous and will therefore be refocused in a spin echo although the echo amplitude at 2τ will

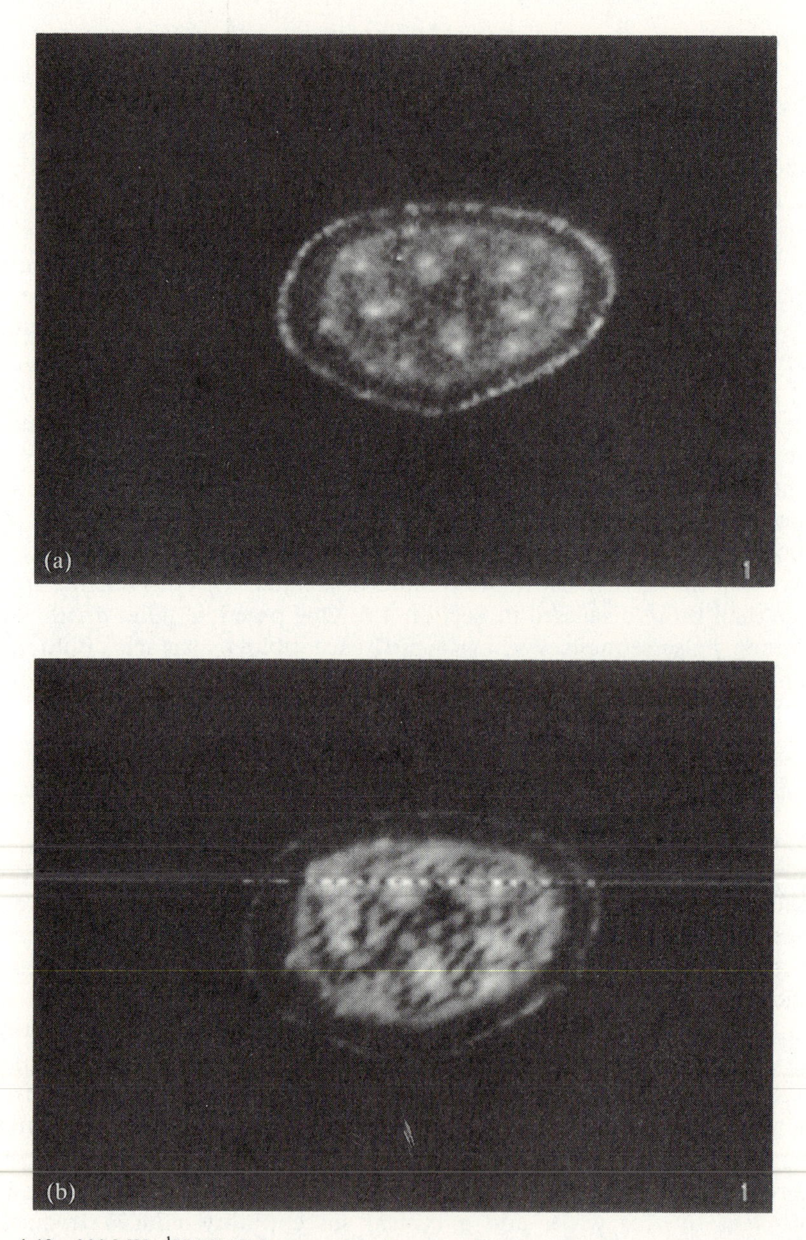

Fig. 4.19 300 MHz ^{1}H NMR images of a plant stem 800 μm in diameter. The slice thickness in each case is 400 μm while the pixel size is $(8\ \mu\text{m})^2$. The image above was obtained using read and (maximum) phase gradients of 120 G cm^{-1} with an acquisition bandwidth of 100 kHz. The lower image was obtained with gradients and bandwidth four times smaller. (The streak through the low gradient image is a zero frequency artefact).

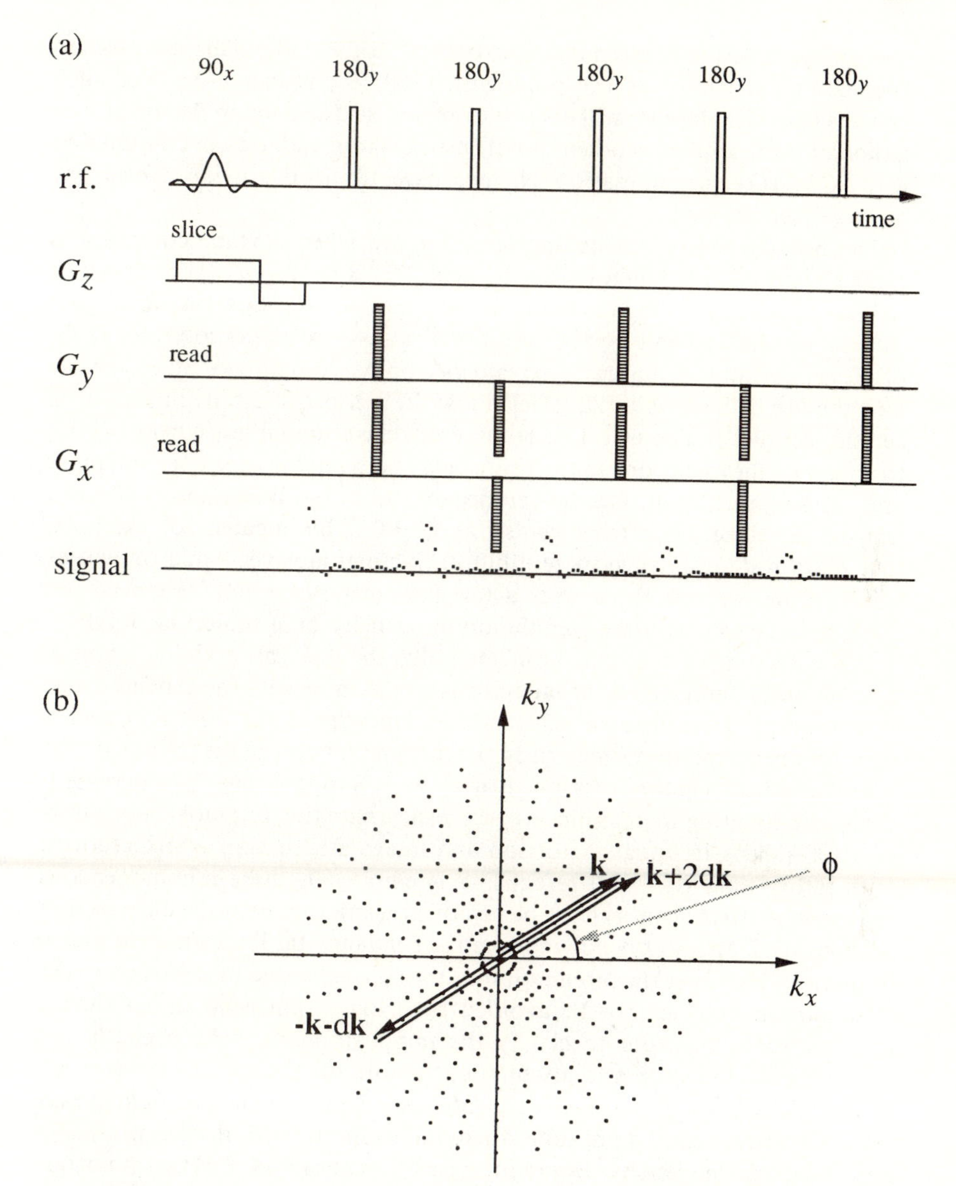

Fig. 4.20 PR pulse sequence involving gradient reversal in a CPMG pulse train. While progressive **k**-space evolution occurs all other Zeeman line-broadening is refocused.

also suffer a residual attenuation $\exp\left(-\frac{2}{3}\,\overline{\Delta\omega_0^2}\,\tau_c^{-1}\tau^3\right)$. This attenuation becomes more severe as the polarizing field, B_0, is increased. For such systems spin echo images will also be distorted as discussed in Section 4.7.1. although if the applied gradient is sufficiently large and used in conjunction with a CPMG echo summation scheme, susceptibility distortion effects may be largely eliminated.

The homogeneous broadening which arises when surface curvature is smaller than 2 μm is much harder to avoid. This is precisely the regime of cellular structures in living systems. In Section 7.6.4 we show that local field fluctuations lead to an effective additional transverse relaxation rate, $1/T_2$, of $\overline{\Delta\omega_0^2}\,\tau_c$. Again the quadratic dependence on $\overline{\Delta\omega_0^2}$ (and hence on B_0) shows how the use of high polarizing fields may in fact prove disadvantageous in certain samples.[37] The effect of transverse relaxation enhancement is both to attenuate the image due to the finite delay between the slice selection pulse and the **k**-space origin, and to deteriorate the T_2 resolution limit. Clearly, transverse relaxation effects vanish as $\tau_c \to 0$. This means, for example, that structural (and hence susceptibility) fluctuations over a micron will be troublesome whereas those over a few nanometers are not. We need not concern ourselves with the sample inhomogeneity on a molecular level!

It is interesting to note that, coincidentally, the size scale dividing inhomogeneous and homogeneous broadening is comparable with the existing limits to resolution. That distance scale, which depends via the diffusion coefficient on the correlation time, τ_c, is also dependent on the magnitude of the polarizing fields via the criterion, $\overline{\Delta\omega_0^2}\,\tau_c^2 \gg 1$. Suppose that B_0 is increased. While the resulting increase in $\Delta\omega_0$ will exacerbate the inhomogeneous linewidth, resulting in image distortion and image attenuation where gradient echo rather than r.f. echo refocusing is employed, these inhomogeneous broadening effects can at least be removed using the methods discussed in Section 4.7.2 and so do not represent a fundamental limitation on image resolution. However the attenuation of the spin echo due to diffusion in the slow motion limit cannot be avoided by using a spin echo rather than a gradient echo. In order to clarify the implications of field strength and distance scale the spin echo attenuation factors for the various regimes are shown in Table 4.4. For a given echo interval τ a higher magnetic field always results in lower signal amplitude where molecular diffusion takes place in a local field inhomogeneity. But reducing τ by means of a CPMG train offers the prospect of improvement, another manifestation of the subtle advantages which are offered by this method of acquisition.

One effective method of distinguishing whether homogeneous (T_2) or inhomogeneous (T_2^*) broadening contributes to attenuation is to compare images obtained using gradient echo and spin echo methods. Such an example is shown in Fig. 4.21.

The reduction in inhomogeneous broadening which results from rapid

Table 4.4. *The diffusive attenuation of the spin echo in various motional regimes associated with susceptibility inhomogeneity.*

→ increasing B_0

		fast motion (homogeneous broadening) $\overline{\Delta\omega_0^2}\tau_c^2 \ll 1$	slow motion (inhomogeneous broadening) $\overline{\Delta\omega_0^2}\tau_c^2 \gg 1$
↓	$\tau < \tau_c$	1	$\exp(-\frac{2}{3}\overline{\Delta\omega_0^2}\tau_c^{-1}\tau^3)$
increasing spin echo time	$\tau > \tau_c$	$\exp(-\overline{\Delta\omega_0^2}\tau_c\tau)$	0

motion over the domains of field variation is an example of fast-exchange averaging.[38] This effect is well known in NMR spectroscopy where a molecular species is in exchange between two states with associated spectral lines separated by angular frequency $\delta\omega$. Fast exchange is said to occur when the exchange time τ obeys the condition $\tau\delta\omega < 1$. One interesting point which arises in discussing the effect of diffusion is the degree to which the structural features of interest in an image may be 'damped' by such a process. The previous analysis suggests such effects are insignificant for regions separated by more than a few microns, at least when observing molecules diffusing no faster than free water. Clearly the greatest problem created by exchange effects will occur at the highest resolution when the diffusive 'exchange time' between structural features is a minimum, and the lowest bandwidths, where the frequency separation of spectral features, $\delta\omega$, is also minimized.

The significance of diffusive exchange averaging is best illustrated using a couple of examples. At 400 MHz with a voxel resolution of around 10 μm (diffusion time for free water $\approx$ 0.02 s) and an optimal bandwidth of around 200 Hz (two-pixel separation $\approx$ 2 Hz) the product $\tau\delta\omega$ is just less than unity so that exchange effects occur on a scale of order two pixels or less and are therefore difficult to observe. Of course, the pixel (and frequency) separation of the structural features could be spread, at constant signal-to-noise, by increasing the slice thickness. 3 μm resolution is possible if the slice is 100 μm. Then, however, $\tau\delta\omega$ increases and the exchange is no longer fast. Of course, these estimates are based on the 'best' possible conditions for observing exchange. When slower, and more realistic, water diffusion coefficients are taken into account and higher than optimal bandwidths employed, such effects are unlikely. Given the sensitivity limitations to resolution such fast exchange averaging does not present a problem.

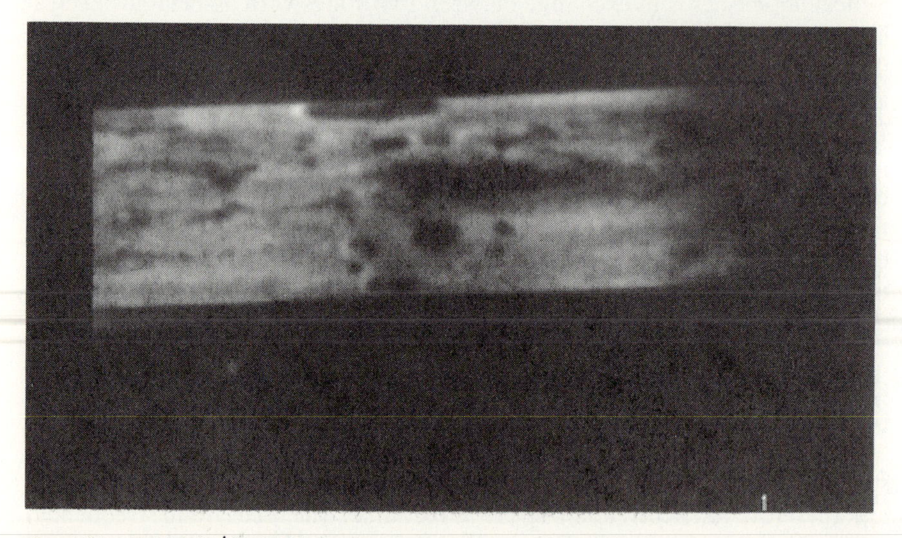

Fig. 4.21 300 MHz ^{1}H NMR images from a 34 μm thick slice of spring onion tissue with $(8.7\ \mu m)^2$ pixel dimension. The upper image was obtained using a gradient echo and the lower with a spin echo. In each case $T_E = 3$ ms and $T_R = 1$ s. The inhomogeneous broadening arising from susceptibility variations leads to pronounced dark regions in the gradient echo image. (From D. Gross and V. Lehman, personal communication.)

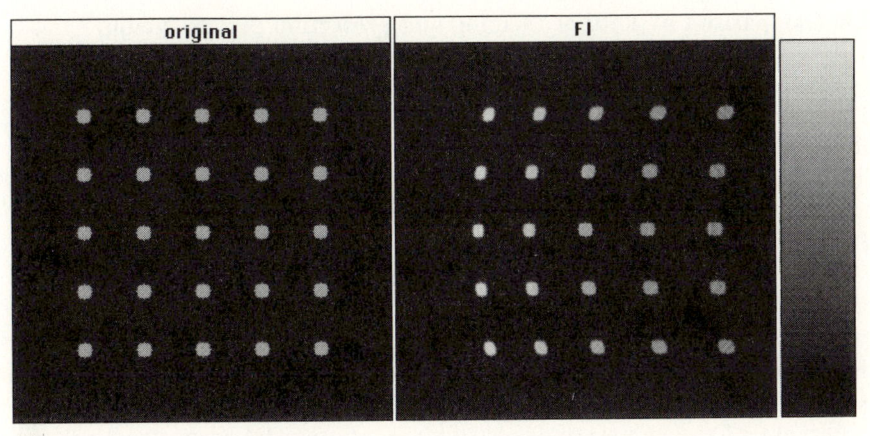

(a)

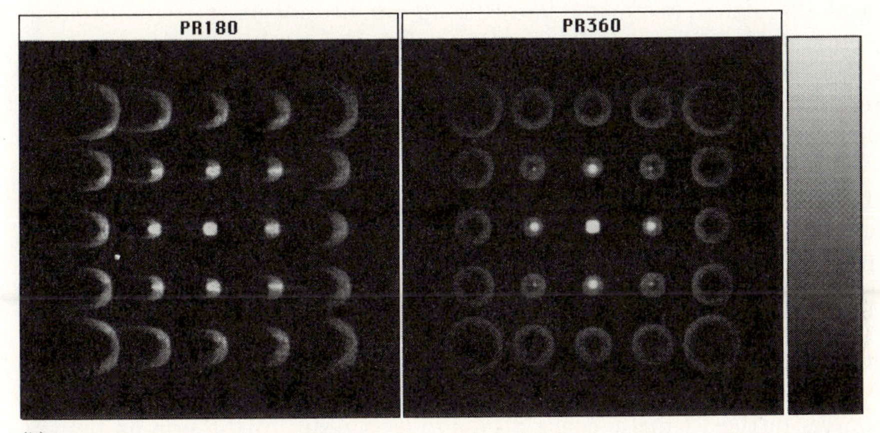

(b)

Fig. 4.22 Simulated FI and PR reconstruction of an object consisting of a matrix of small cylinders normal to the slice, under a quadratic B_0 field inhomogeneity. (a) shows the original object along with the FI reconstruction in which distortion occurs only along the read (horizontal) axis. Note that this FI image is obtained with two-quadrant gradient orientation although this may correspond to four quadrants of **k**-space provided full echo sampling is used. (b) PR reconstruction using two and four quadrants of gradient orientation.

4.8 Comparison of Fourier imaging and projection reconstruction

The analysis here has shown that PR offers a very slight sensitivity advantage over FI, an advantage which is enhanced by the use of smoothing filters. The PR method also obtains a slightly greater advantage from echo summation, yielding close-to-optimum resolution at variable bandwidth. The underlying reason for this superiority lies in noise power attenuation which results from the use of a polar raster in PR. This raster concentrates data sampling at the centre of **k**-space where the local signal-to-noise is at its highest. Additional sensitivity advantages exist when the specimen T_2 is very short since FI requires that the phase gradient displacement in **k**-space is established before sampling can begin.

Because of the correspondence of the Cartesian FI reconstruction grid with the image matrix grid it is possible to compute the images more rapidly with this technique. However, the speed of reconstruction is not so important in microscopy where the time bottleneck is more often related to the need for signal averaging in order to obtain adequate signal-to-noise. Furthermore PR offers the possibility of obtaining images more rapidly by using fewer projections. In FI a similar procedure would involve taking fewer phase steps, thus incurring the problem of performing an $N \times M$ point digital FFT where $N \ll M$.

Despite the apparent signal-to-noise advantages, PR is less satisfactory in the matter of image quality. Obviously the uneven **k**-space sampling over-emphasizes low spatial frequencies. Thus fine details are better represented in FI. PR is also susceptible to distortion[39] due to sample movement or gradient non-orthogonality or gradient non-uniformity and the point spread function varies across the image plane.[33] A comparison of PR and FI reconstructions for a phantom sample suffering quadratic field inhomogeneity is shown in Fig. 4.22. It is clear that, whereas the positions of the sources in the FI method are displaced as a result of the non-uniformity in the magnetic field, the intensity of each source is correctly represented. One interesting feature of image reconstruction under B_0 non-uniformity concerns the distinction between gradient and time reversal in setting the sign of **k**. Normally FI is performed with positive-only read gradients and four-quadrant **k**-space sampling is achieved by means of full echo acquisition. This leads to an asymmetric shift of image elements in the direction of the read gradient. For PR the same asymmetric shift phenomenon is observed, as apparent in Fig. 4.22(b), provided that only two quadrants of gradient vector orientation are used (180° reconstruction). When four quadrants are used (360° reconstruction) the image element displacements lead to circularly symmetric fringe effects.

Clearly the effect of field inhomogeneity is more severe in PR than in FI.

However the choice of which method to use depends on whether the actual limit to resolution arises from intrinsic distortion rather than intrinsic sensitivity and this will depend on specific features of the pulsed gradient system to be used.

4.9 Further resolution enhancement

In optical microscopy the image is obtained from a section with longitudinal dimensions $\approx$ 100 μm. While the transverse resolution is potentially 0.2 μm, this is possible only when the specimen has longitudinal symmetry. Taking the smallest depth of field as 30 μm the corresponding voxel resolution in optical microscopy is $\sim$ 1 μm. Compare this with the voxel resolution of 10 μm obtainable under ideal conditions in proton NMR microscopy at 400 MHz. This latter resolution is obtained in an experiment lasting about 15 minutes and it is still one order of magnitude poorer than can be obtained in optical microscopy in the 10 millisecond time it takes to expose one photograph.

Can the resolution in NMR microscopy be further improved? Given a sample which is sufficiently stable to permit an acquisition time of several hours, the effect of further signal averaging could reduce the limit by a factor of 2. Another method of improving the spectral signal-to-noise ratio is the Maximum Entropy Method (MEM). This technique has been applied with some success to one-dimensional spectroscopy and to two-dimensional reconstruction.[40–42] The MEM produces a signal-to-noise enhancement equivalent to that gained by the use of optimal filters but with no additional broadening introduced. One disadvantage with the method is the large computing time required. It cannot, for example, be applied until the final image is computed since MEM depends on the magnitude spectrum whereas both FI and PR reconstruction require full phase-dependent amplitudes. One-dimensional spectra require about 10^2 additional Fourier transformations to 'smooth' the spectra whereas two-dimensional MEM processing would require about 10^4. More efficient algorithms are possible where the image shape is already known but such a feature is not particularly attractive!

One of the most hopeful prospects for fundamental signal-to-noise enhancement is via improved receiver coil and preamplifier technology. In particular, if the receiver coil is operated at liquid helium tempertures then the Johnson noise may be significantly reduced through both the T_c and $\rho(T_c)$ terms, by a factor of about 20. To take advantage of such an improvement the spectrometer noise figure would need to be kept low so that liquid helium immersion would be necessary for the preamplifier as well. Styles *et al.* (1984) have reported the development of a receiver coil assembly with an 11 mm diameter room temperature space surrounded by a glass cryostat containing a 19 mm diameter receiver coil.[43] Their system, which also used a liquid helium cooled preamplifier, gave an improvement in signal-to-noise of only half that predicted. This gain was, of course, offset by the need to work with a larger receiver coil. Assuming a need to double receiver coil dimensions it seems that this method realistically offers the prospect of enhancing signal-to-noise by around 6, so improving the resolution by a factor of 2.

All these advances suggest that NMR microscopy may achieve a voxel resolution of around 2 μm but only with great difficulty and for specially favoured samples. Despite this limitation the NMR microscope offers significant unique features for it to be of major importance in biology, medicine, and materials science. These features centre on the myriad contrast qualities with which the NMR image may be endowed.

4.10 References

1. Brunner, P. and Ernst, R. R. (1979). *J. Magn. Reson.* **33**, 83.
2. Mansfield, P. and Morris, P. G. (1982). *NMR imaging in biomedicine*, Academic Press, New York.
3. Callaghan, P. T. and Eccles, C. D. (1987). *J. Magn. Reson.* **71**, 426.
4. Callaghan, P. T. and Eccles, C. D. (1988). *J. Magn. Reson.* **78**, 1.
5. Abragam, A. (1961). *Principles of nuclear magnetism*, Clarendon Press, Oxford.
6. Slichter, C. P. (1963). *Principles, of magnetic resonance*, Harper and Row, New York.
7. *Bruker Almanac*, Bruker Analytische Messtechnik GmbH, Karlsruhe (1988).
8. Hoult, D. I. and Richards, R. E. (1976). *J. Magn. Reson.* **24**, 71.
9. Nyquist, H. (1928). *Phys. Rev.* **32**, 110.
10. Kraus, J. D. and Carver, K. R. (1973). *Electromagnetism*, McGraw Hill, New York.
11. Ernst, R. R. (1966). *Adv. Magn. Reson.* **2**, 1.
12. Lindon, J. C. and Ferrige, A. G. (1980). *Progr. NMR Spectr.* **14**, 27.
13. Ernst, R. R., Bodenhausen, G. and Wokaun, A. (1987). *Principles of nuclear magnetic resonance in one and two dimensions*, Clarendon Press, Oxford.
14. Brigham, E. O. (1974). *The fast Fourier transform*, Prentice Hall, Englewood Cliffs.
15. Rabiner, L. R. and Gold, B. (1975). *Theory and application of digital signal processing*, Prentice Hall, Englewood Cliffs.
16. Oppenheim, A. V. and Schaefer, R. W. (1975). *Digital signal processing*, Prentice Hall, Englewood Cliffs.
17. Lord Rayleigh, (1879). *Phil. Mag.* **8**, 261.
18. Waugh, J. S. (1970). *J. Mol. Spectr.* **35**, 298.
19. Maki, H., Johnson, G. A., Cofer, G. P. and Macfall, J. R. (1988). *J. Magn. Reson.* **80**, 482.
20. Shoup, R. R., Becker, E. D. and Farrar, T. C. (1972). *J. Magn. Reson.* **8**, 298.
21. Ahn, C. B. and Cho, Z. H. (1989). *Med. Phys.* **16**, 22.
22. Bleaney, B. I. and Bleaney, B. (1976). *Electricity and magnetism*, Oxford University Press.

23. Ludeke, K. M., Roschmann, P. and Tischler, R. (1985). *Magn. Reson. Imaging* **3**, 329.
24. McKinnon, G., Roschmann, P. and Tischler, R. (1985). *Magn. Reson. Imaging* **3**, 417.
25. Lauterbur, P. C., Bernado, M. L., Mendonca Dias, M. H. and Hedges, L. K. (1986). *Works in Progress Abstract, Fifth Annual Meeting of the Society of Magnetic Resonance in Medicine*, Montreal, pp. 292–230.
26. Packer, K. J. (1973). *J. Magn. Reson.* **9**, 438.
27. Case, T. A., Durney, C. H., Ailion, D. C., Cutilo, A. C. and Morris, A. H. (1987). *J. Magn. Reson.* **73**, 304.
28. Villringer, A., Rosen, B. R., Belliveau, J. W., Ackerman, J. L., Lauffer, R. B., Wedeen, V. J., Buxton, R. B. and Brady, T. J. (1988). *Magn. Reson. Med.* **6**, 164.
29. Wismer, G. L., Buxton, R. B., Rosen, B. R., Fisel, C. R., Oot, R. F., Brady, T. J. and Davis, K. R. (1988). *J. Comput. Assisted Tomography* **12** 259.
30. Belliveau, J. W., Villringer, A., Rosen, B. R., Lauffer, R. B., Ackerman, J. L., Buxton, R. B., Chao, Y. S., Frazer, J., Johnson, K. A., Moore, J., Wedeen, V. J. and Brady, T. J. (1986). In *Proceedings, Fifth Annual Meeting of the Society of Magnetic Resonance in Medicine*, Montreal.
31. Majunder, S. and Gore, J. C. (1988). *J. Magn. Reson.* **78**, 41.
32. Lai, C. M. (1982). *J. Phys. E* **15**, 1093.
33. O'Donnell, M. and Edelstein, W. A. (1985). *Med. Phys.* **12**, 20.
34. Callaghan, P. T. (1990). *J. Magn. Reson.* **87**, 304.
35. Edmonds, D. T. and Wormald, M. R. (1988). *J. Magn. Reson.* **77**, 223.
36. Miller, J. B. and Garroway, A. N. (1986). *J. Magn. Reson.* **67**, 575.
37. Thulborn, K. R., Waterton, J. C., Matthews, P. M. and Radda, G. K. (1982). *Biochim. Biophys. Acta* **714**, 265.
38. Gutowsky, H. S., McCall, D. W. and Slichter, C. P. (1953). *J. Chem. Phys.* **21**, 279.
39. Sekihara, K., Kuroda, M. and Kohno, H. (1984). *Magn. Reson. Med.* **1**, 247.
40. Gull, S. I. and Daniell, G. J. (1978). *Nature* **272**, 686.
41. Sibisi, S. (1984). *Nature* **301**, 134.
42. Sibisi, S., Skilling, J., Brereton, R. G., Lane, E. D. and Staunton, J. (1984). *Nature* **311**, 446.
43. Styles, P., Soffe, N. F., Scott, C. A., Cragg, D. A., Row, F., White, D. J. and White, P. C. J. (1984). *J. Magn. Reson.* **60**, 397.

5

k-SPACE MICROSCOPY IN BIOLOGY AND MATERIALS SCIENCE

5.1 Proton NMR in biological, synthetic, and mineral materials

Because of its high gyromagnetic ratio and its abundance in natural materials, the hydrogen proton is the predominant nucleus of NMR microscopy. In biological tissue it is present in the major molecular species, water, lipids, proteins, and polysaccharides. Hydrogen is abundant in synthetic polymers and in natural oils, waxes, and tars. It is significant in some minerals such as coal, and in others it may be introduced by adsorption of water.

Since all hydrogen protons are identical the amplitude of the NMR signal should, in principle, give a direct, quantitative measure of hydrogen content. However, the process of acquiring the signal involves finite time-scales over which relaxation effects are important in determining the actual signal amplitude. As we have seen in Chapter 2 the nuclear spin relaxation time depends on molecular motion. The proton NMR signal is therefore highly sensitive to the physical environment of the host molecule and this sensitivity will be reflected in the NMR image. In attempting to obtain spin density images one must be aware of the role of spin relaxation in determining relative amplitudes, since the time delay between selective excitation and the origin of **k**-space at the centre of the acquired echo exposes the nuclear magnetization to transverse relaxation. The dependence of transverse relaxation on molecular motion is extreme. Unless special imaging methods are employed the signal will be dominated by protons whose molecular hosts undergo rapid reorientational motion. The signal from protons in solid state environments will normally be entirely absent.

While presenting some complications in the quantitative interpretation of signal amplitudes, the sensitivity of proton NMR to molecular environment is also very helpful because of the information with which it endows the image. In principle, any NMR property can be spatially localized by gradient encoding but it is beyond the scope of this text to review the vast and extraordinary literature of general NMR spectroscopy. Instead, in this chapter we will present examples of the application of imaging to the study of microstructure, so that the role of various contrast schemes can be assessed by the reader. Because of its importance in determining signal amplitude, spin relaxation is central to our understanding of imaging. Indeed its existence is crucial to the observation of structural features in medical MRI, since the proton density exhibits only a small variation between the various types of human tissue. Before discussing applications of imaging we therefore begin by briefly summarizing the elements of proton relaxation in differing molecular environments.

5.1.1 *Water*

Free water

In the water molecule a proton experiences an intramolecular dipolar interaction with the other proton on the same molecule and an intermolecular interaction from protons in neighbouring water molecules. Both interactions fluctuate as the water molecule diffuses in rotational and translational motion. For free water the rotational correlation time is very much faster than the Larmor period and, as indicated in Section 2.5.1, this leads to 'extreme narrowing' of the resonance with $T_1 \approx T_2$ and independent of Larmor frequency. The combined effect of rotational and translational diffusion in this limit leads to [1]

$$\frac{1}{T_1} = \left(\frac{\mu_0}{4\pi}\right)^2 \left(\frac{a}{b}\right)^2 \left(\frac{\gamma^4 \hbar^2}{3Db^4}\right) \left(1 + \frac{3\pi}{5} \frac{Nb^6}{a^3}\right) \tag{5.1}$$

where a is the molecular Stokes radius, b is the interproton distance, D is the self-diffusion coefficient and N is the number of protons per unit volume. The rotational contribution (the unity term in the final factor) is about three times that due to translation and the calculated value of around 3 seconds agrees well with the measured value of 2.72 s.[2]

In fact the extreme-narrowed dipolar limit value can be obtained only when the water is free from paramagnetic impurities and, in particular, oxygen. Impurity molecules which are paramagnetic in solution, act as relaxation centres[3] as the dipolar interaction between the proton and the impurity ionic moment is modulated by the relative motion of the water and the ion. Assuming that the translational diffusion of the water and ion are similar, the impurity-dominated relaxation limit is[1]

$$\frac{1}{T_1} = \left(\frac{\mu_0}{4\pi}\right)^2 \left(\frac{16}{15}\right) \pi^2 \gamma^2 \langle \mu_i^2 \rangle N_i \frac{D}{6\pi a} \tag{5.2}$$

where N_i is the ionic number density and $\langle \mu_i^2 \rangle$ is the mean square ionic moment.

Fig 5.1 shows the dependence of T_2 and T_1 on concentration for Mn^{2+} ions. The reduction of T_1 by the use of paramagnetic agents enables more rapid signal recovery between acquisitions and hence improved signal-to-noise and resolution in NMR microscopy. The method is useful in non-biological materials but, because of the quite high doping levels required to substantially reduce T_1, it is not strictly non-invasive when employed in

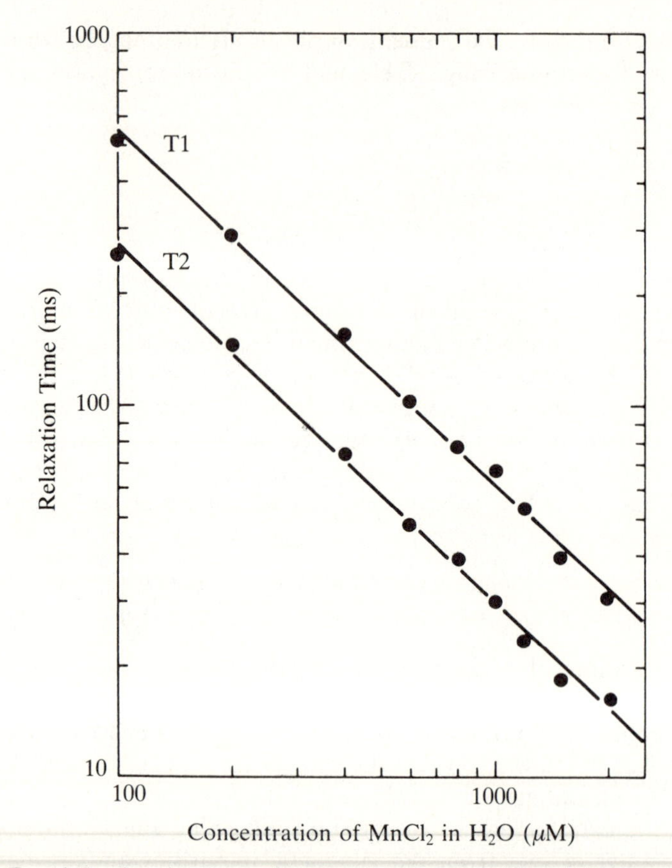

Fig. 5.1 T_1 and T_2 relaxation times of protons in water containing different concentrations of $MnCl_2$, from Kroeker and Henkelman[4]. Note that for the entire concentration range $T_1 \approx 2T_2$.

biological tissue. None the less, the use of paramagnetic agents in biological systems can give an indication of tissue or cellular compartmention via the reduction in T_1 and T_2 in specific regions of an image.

Bound water and exchange

The rotational correlation time for free water is around 10^{-12} seconds.[5] Water closely associated with larger molecules or with solid surfaces will in general have much slower tumbling rates. The slowing of rotational motion leads to a reduction in both T_1 and T_2 until the correlation time for dipolar fluctuation is of order the Larmor period of around 10^{-9} seconds. This is the characteristic T_1 minimum apparent in Fig. 2.12, and for a proton pair undergoing isotropic motion the relaxation values at this minimum

($\omega_0\tau_c \sim 1$) are of order 10–100 ms for $\omega_0 \sim 100$ MHz. At slower tumbling rates T_2 continues to fall while T_1 increases.

When water molecules are in close proximity with solid surfaces or with slowly moving macromolecules, rotational correlation times are typically 10^{-8} s[6] and proton relaxation rates are therefore dramatically affected. One indication that 'bound water' is structurally modified comes from the depression of the freezing point, an effect apparent in an NMR experiment by the persistence of a liquid-like signal at temperatures below 0°C.[7-9] The question of 'structure' in bound water is a contentious topic[10] which need not concern us. None the less, it is clear that at least two phases of water, bound and free, exist in biological tissue, as well as in mineral systems where there are solid surfaces with which the water can interact.

The slowing of reorientational motion in the water molecule will inevitably lead to altered proton relaxation for this phase. However, translational diffusion leads to exchange of molecules between the bound and free phases and, in principle, all relaxation rates can be affected. The basis of this exchange description is the model of Zimmerman and Brittin,[11] of which only a brief outline is given here, but the reader is referred to the excellent reviews on the subject of water relaxation.[12-15]

The mean time taken for a proton to 'sample' the differing water phases is termed the exchange correlation time, τ_e. (This transfer may take place due to molecular diffusion, or via chemical exchange of protons between adjacent water molecules. While chemical exchange can be important for short-range processes it is quite slow, taking the order of milliseconds per jump, so that the dominant process determining the rate of translation between spatially separated regions is molecular self-diffusion.) Exchange may be classified as slow or fast depending on the relative speed of exchange and relaxation. Suppose that the relaxation process is described by an exponential change of the magnetization, $M(t)/M_0 = \exp(-t/T)$, where T is the relaxation time such as T_1 (with $M(t)$ the longitudinal magnetization difference from equilibrium) or T_2 (with $M(t)$ the transverse magnetization). Labelling the fraction of water in the ith phase as P_i and the relaxation time in that phase as T_i then, in the slow-exchange case ($T_i \ll \tau_e$), the total magnetization relaxation is multi-exponential and given by

$$M(t)/M_0 = \sum_i P_i \exp(-t/T_i) \tag{5.3}$$

while in fast exchange ($T_i \gg \tau_e$) a common relaxation is observed for the whole system with

$$M(t)/M_0 = \exp(-t/T) \tag{5.4}$$

$$\frac{1}{T} = \sum_i P_i \left(\frac{1}{T_i}\right). \tag{5.5}$$

For intermediate exchange rates non-exponential relaxation is observed but the simple subdivision of eqn (5.3) no longer applies.[11]

Because rotational motion in the bound phase can be quite slow, with correlation times much less than the Larmor period, divergence of T_1 and T_2 is apparent as shown in Fig. 2.12. It is a common feature of biological systems that T_1 relaxation is single exponential while T_2 is multi-exponential. This arises because the exchange times can lie between the bound-water T_1 and T_2 values leading to fast and slow exchange, respectively.

Eqn (5.5) leads to a fast-exchange result for bound and free water,[2]

$$\frac{1}{T_1} = \left(\frac{1}{T_1}\right)_f + P_b\left\{\left(\frac{1}{T_1}\right)_b - \left(\frac{1}{T_1}\right)_f\right\}. \tag{5.6}$$

The second term in eqn (5.6) is generally dominant when the bound-water correlation time is small[16,17] leading to a simple proportionality between the T_1 relaxation rate and the bound fraction, P_b.[18] This is a very useful relationship since it indicates that, over a narrow composition range, T_1 should be approximately proportional to water content.[2]

It should be noted that the fast-exchange description is too simplistic when τ_e becomes comparable with the correlation time governing the relaxation process and relaxation and exchange processes become coupled. This problem has been discussed in detail by Wennerstrom.[19]

It is clear that in biological samples the transition between the bound and free water is not sharp so that a two-phase description is simplistic. This means that there is seldom a single characteristic exchange time but rather a continuous spectrum.[20,5] A variety of correlation time distribution models, applicable to biological tissues, is available.[21–23] Because of the role of diffusion in determining transfer between phases, the exchange process will also be strongly influenced by geometry. This problem has been treated in detail by Brownstein and Tarr[24] who find the relaxation 'normal modes' which arise due to diffusion between bulk water and a surrounding surface on which 'relaxation sinks' are present. This model can be used to explain relaxation of water in plant cells and animal cells but is particularly important in porous mineral structures such as clays and sandstones. Because of the characteristic way in which they restrict molecular diffusion, porous materials are very effectively examined using pulsed gradient spin echo NMR, as discussed in detail in Chapter 7. The echo amplitude can be strongly influenced by surface effects and a detailed discussion of the Brownstein–Tarr model is contained in that chapter.

Susceptibility effects
In Chapter 4 attention was drawn to the role of susceptibility inhomogeneity in influencing the local magnetic fields in heterogeneous systems. These fields not only perturb the image but may also induce spin relaxation, an effect which is discussed in detail in Chapter 7. The broad lines which result from susceptibility effects are partly homogeneous and partly inhomogeneous in character, depending on the time taken for a water molecule to diffuse across the local field fluctuations. This means that it is possible to recover part of the transverse magnetization using spin echo methods. A comparison of T_2 and T_2^* relaxation and their dependence on correlation time is shown in Fig. 7.20

Because susceptibility gradients depend on local surface symmetry, a distinct difference can sometimes arise in the water proton signals from intracellular and extra-cellular water. This had led to some ingenious indications of compartmentation as well as quantitative studies on membrane permeability in erythrocytes.[27,28]

5.1.2 *Plant and animal tissue*

The spectrum of proton NMR relaxation behaviour in biological tissue depends on the specific system under study. However, broad similarities arise in the division of water into intracellular and extra-cellular components, and the division of the remaining, non-water proton signal into lipid and macromolecule. Of course, one major difference between plant and animal structures concerns the macromolecular component, and the respective role of polysaccharide and protein. The non-aqueous proton signal in plant and animal tissue can be revealed by exchanging all water with D_2O. Part of this signal will be due to rigid macromolecules with characteristic solid state relaxation times of order a few microseconds, while others may be sufficiently flexible to have T_2 relaxation of order a few milliseconds. In Tables 5.1–5.3 we present typical examples taken from plants and animals, respectively.[29-31] Great similarity is found in the component distribution of rat and barnacle muscle, although in the latter case, exchange with D_2O showed that the millisecond fraction could not be regarded as entirely bound water. The most remarkable features of these tables is the similarly wide spectrum of proton relaxation times.

Plant tissues exhibit much greater variation in water content than animal tissue. Flibotte *et al.* have followed the T_2 variation in cedar sapwood as a function of moisture content and shown that while the amplitudes of the two longer T_2 components (non-bound water) decrease with decreasing moisture content, the T_2 values remain relatively unchanged.

The independence of water proton relaxation on water content is

Table 5.1. *Proton relaxation times for different molecular components in cedar sapwood with moisture content 140% w/w, from S. G. Flibotte, R. S. Menon, A. L. Mackay, and J. R. T. Hailey.*[29]

Proton fraction	T_2	relative amplitude	tentative proton source
Cellulose and lignin	20 μs	0.2	cell wall polymer
bound water	4 ms	0.2	cell wall water
small-diameter intracellular water	30 ms	0.2	latewood lumen
large-diameter cell water	100 ms	0.4	earlywood lumen

Table 5.2. *Proton relaxation times for different molecular components in rat muscle from C. F. Hazelwood, D. C. Chang, B. L. Nichols, and D. E. Woesser.*[30]

Proton fraction	T_2	relative amplitude	tentative proton source
Protein	20 μs	0.2	tissue protein
millisecond	0.4 ms	0.06	bound water
major	45 ms	0.65	intracellular water
extra-cellular	200 ms	0.08	extra-cellular water

Table 5.3. *Proton relaxation times for different molecular components in single barnacle cell muscle from K. R. Foster, H. A. Resing, and A. N. Garroway.*[31]

Proton fraction	T_2	relative amplitude	tentative proton source
Protein	20 μs	0.2	tissue protein
millisecond	0.73 to 4 ms	0.02 to 0.05	three components: mobile protons in tissue protein and lipid
major	35 ms	0.7	intracellular water
extra-cellular	400 ms	0.02	extra-cellular water

consistent with exchange times being slow in comparison with T_2. By contrast, and as indicated earlier, T_1 relaxation is often sufficiently slow that exchange effects can be significant. In most plant and animal tissue, therefore, proton T_1 behaviour is typically single exponential with values in the range 0.2 to 1 second, the major factor determining the precise value being the water content of the tissue.[32,33] One of the more remarkable aspects of T_1 times in animal tissue is the apparent sensitivity to malignancy.[34,35]

5.1.3 *Polymers*

Spin relaxation in polymers exhibits behaviour ranging from that of the rigid solid to the extreme-narrowing conditions usually associated with rapidly tumbling small molecules. Because proton spin relaxation is largely determined by proton dipolar interactions whose effect varies as r_{ij}^{-6}, where r_{ij} is the interproton distance, it is therefore sensitive to local motion. In consequence, although the host macromolecule may be large and cumbersome, the relaxation behaviour will generally be characteristic of the motion of the segment containing the proton of interest. For random coil polymers in the solution state, this generally means that the correlation times are sufficiently rapid for proton relaxation to fall on the 'fast side' of the T_1 minimum with specific values determined by the local stereochemistry.[36] Relaxation behaviour for such systems will exhibit a weak molar mass and concentration dependence in dilute solution but will be strongly influenced by the effects of entanglements at high concentration.[37] By contrast with random coils the local dipole–dipole correlation times in rigid rod polymers are characteristic of macromolecular rotational diffusion. At all concentrations both T_1 and T_2 will be strongly influenced by molar mass and for rods longer than a few 1000 daltons, relaxation behaviour will be solid-like. Proton signals from such polymers will generally only be visible in NMR imaging experiments if there are locally mobile side chains attached to the rods. Most biological macromolecules are stabilized by internal hydrogen bonding which hinders local motions. In general, therefore, aqueous solutions of proteins, polypeptides, and carbohydrates exhibit proton relaxation times somewhat shorter than those of synthetic polymers in organic solvents, with the longest T_2 components typically of order or less than 1 ms.

In the molten phase polymer proton transverse relaxation times are generally strongly molar mass dependant. This is result of entanglements forming above a critical molar mass,[38] M_c, which is generally of order 1000–10 000 daltons for synthetic polymers. Above this mass the tube formed by entanglements prevents isotropic reorientation of local segments and the small residual anisotropy results in a finite dipolar linewidth[39] which is slowly averaged to zero by longer range motions of the entire polymer molecule. This tube motion is strongly dependent on chain length so that T_2 may shorten by several orders of magnitude for increasing polymer lengths above M_c.[39-42] By contrast T_1, which is sensitive to spectral density only near the Larmor frequency, is unaffected by small anisotropic effects in the local motion so that its value will be characteristic of the rapid segmental correlation time and therefore largely molar mass independent.[40]

As the temperature of a polymer melt is lowered below the 'glass transition temperature', T_g,[38] the molecules condense into a solid phase. Polymers are generally unable to form simple crystals with long-range order because of complex defects which arise from chain topology. In those polymers with highly ordered chains (such as polyethylene), local crystallization can occur but the material subdivides into amorphous and crystalline regions in which the local motions relevant to spin relaxation may be strongly differentiated. A comprehensive review of spin relaxation in solid polymers has been given by McBrierty.[43]

Another class of polymer 'solid' is that formed by physically or chemically cross-linking chains. In many systems this leads to elastomeric properties in which large elastic deformations are possible, the modulus of elasticity being determined by the change in chain entropy that results as the random coils are straightened. Because of the space existing between chains in elastomers the segments are free to reorient, leading to relaxation behaviour characteristic of the solution phase. Elastomers are therefore a peculiar class of solid which appear as 'liquids in the NMR sense'. Examples of polymer proton relaxation times in elastomeric solids are given in Table 5.4. A review of proton relaxation in synthetic polymer networks is available in the recent article by Andreis and Koenig.[44]

Polymer proton NMR spectra are often complex because of a variety of hydrogen chemical environments in the monomer subunits. Furthermore, the monomers do not pack uniformly, so that unlike a simple liquid there exists local void spaces which contribute susceptibility defects. As a consequence the proton NMR linewidth is partly inhomogeneously broadened due to chemical shift variation and susceptibility inhomogeneity. This is manifested in the difference between T_2 and T_2^* in Table 5.4.

5.2 Proton density studies in the 'liquid' state

One of the most straightforward applications of imaging is the measurement of spin density maps for spins in molecules sufficiently mobile that T_2 relaxation does not significantly attenuate the image. In plant and animal

Table 5.4. *Proton relaxation times for various elastomers, from Chang and Komorowski.*[45] *Note that the T_2 and T_1 values are approximate averages obtained by a single exponential fit. $\Delta\nu_{1/2}$ is the proton NMR linewidth.*

	T_g (°C)	$10^9 \tau_c$ (s)	T_2 (ms)	T_2^* (ms)	$\Delta\nu_{1/2}$ (Hz)	T_1 (ms)
cis-polybutadiene	−102	0.01	13	2.1	150	300
cis-polyisoprene	−70	0.4	9	1.0	310	380
polyisobutylene	−70	20	<1	0.2	1650	
cured, filled *cis*-polybutadiene			4	0.4	740	250

tissue, such proton maps will be dominated by the signal from non-bound water. In porous solids the void spaces may be imaged using absorbed liquid probe molecules. In elastomeric polymers, the macromolecular proton T_2 may be sufficiently long that it is possible to obtain an image of the polymer matrix without the need to employ the special line-narrowing methods discussed in Section 5.6.

The idea that proton density maps can be obtained using NMR imaging is a gross simplification. Because of the finite time between the selective excitation and the **k**-space origin during the subsequent read gradient, T_2 contrast is always implicitly present in NMR microscopy. At the simplest level this is the basis of the exclusion of the proton signal associated with molecules in the solid state but in the 'liquid signal' remaining at the start of acquisition, a spectrum of T_2 relaxation times will lead to intensity contrast dependent on T_2. Furthermore, unless very long repetition times are used, T_1 contrast effects will also be present. These effects mean that relative intensities in the image must be interpreted with care but the picture is by no means gloomy since the effect of this implicit contrasting is to endow the image with clearly defined features which relate to molecular properties. Quantitative interpretation requires deliberate contrasting pulse schemes which yield NMR parameter maps and we shall discuss examples of these in later sections. For the moment, however, we examine some simple applications of **k**-space microscopy in plane, animal, and polymer systems, aware of the role that relaxation will play in influencing the intensity maps.

5.2.1 *Plant tissue images*

Living plants

The application of NMR imaging to plant systems is quite recent. Since 1986, however, a number of groups have reported studies of water distribution, indicating that NMR microscopy not only gives useful information about morphology but is particularly valuable in following slow physiological changes associated with water movement. Some of the highest resolution micrographs reported in the literature have been obtained on plants, an example being the leaf image of *Bryophyllum tubiflore*[46] shown in Fig. 5.2(a) in which the individual cells are clearly visible.

NMR microscopy is particularly suited to plant stems where a degree of longitudinal symmetry means that high-resolution images can be obtained in quite thick slices. Plant stems can be broadly divided into parenchymous and vascular tissue, with cell sizes typically of order 50 μm. Fig. 5.3(a) shows an image of the stem of a 15-day-old runner bean plant obtained by Connelly *et al.*[47] in which the parenchyma cells in the central part of the image are visible. This region is clearly distinguished from the surrounding vascular tissue and epidermal layer. The contrast between different parts of the image

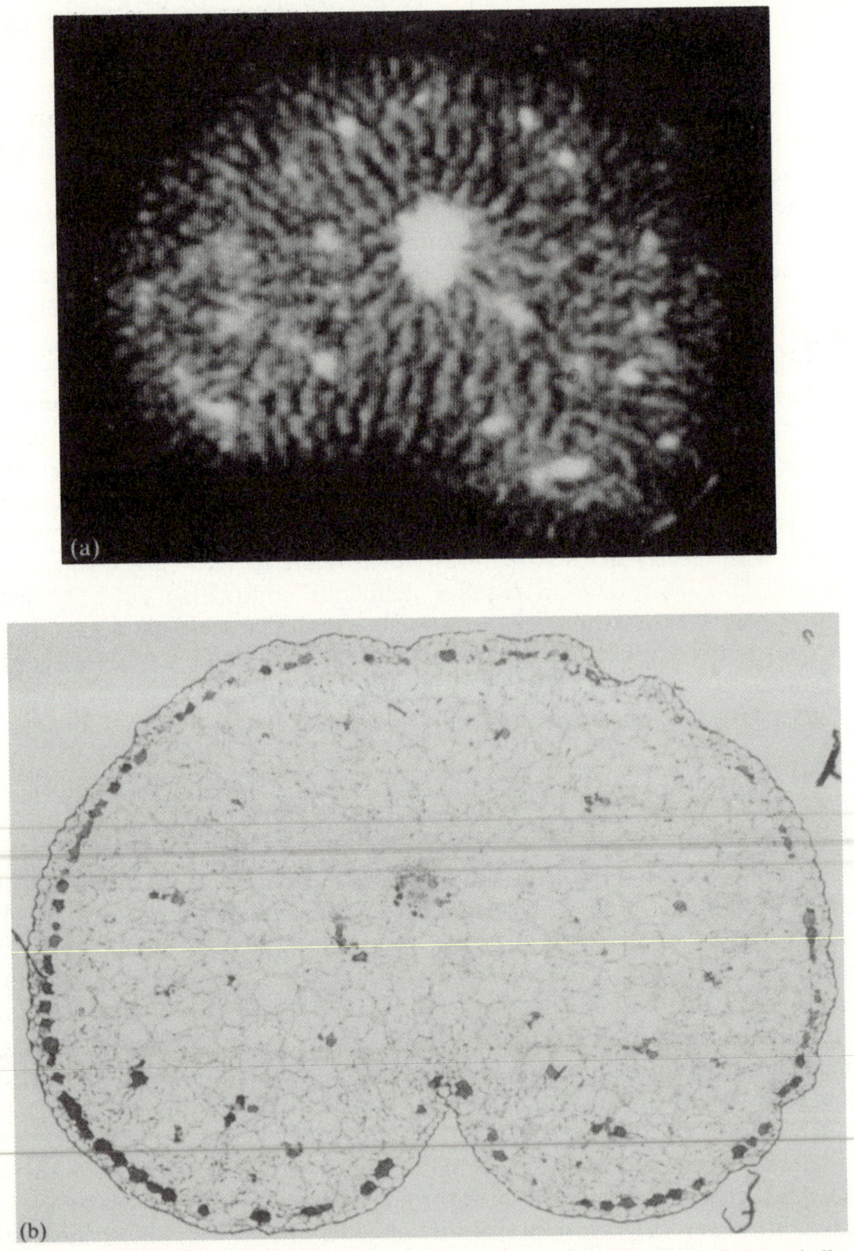

Fig. 5.2 (a) 300 MHz ^{1}H NMR microscope cross-section through a leaf of *Bryophyllum tubiflore* with a slice thickness of 250 μm and with pixel dimension of (15 μm)^2 and (b) light microscope image. Note that the cell diameter is approximately 40 μm (from Kuhn[46]).

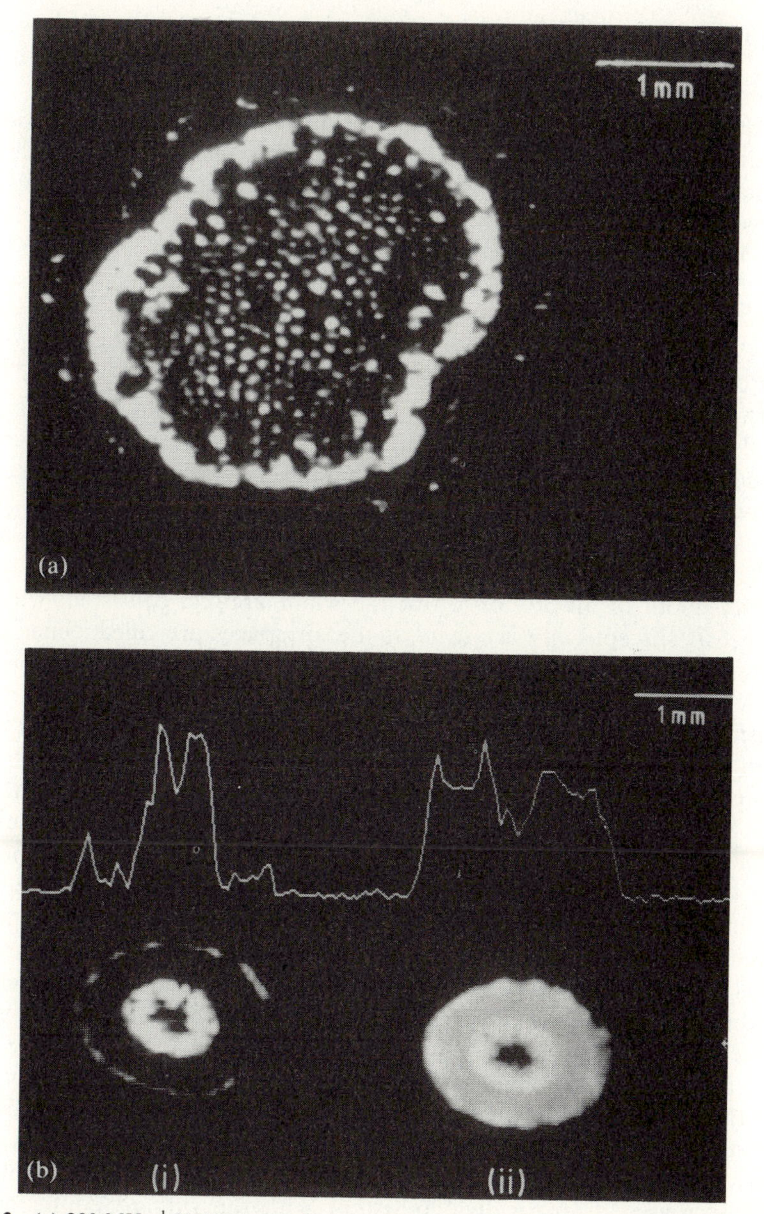

Fig. 5.3 (a) 200 MHz [1] NMR image from a transverse section of the stem of a 15-day-old runner bean using a slice thickness of 560 μm and pixel dimension (60 μm)^2. The total acquisition time was 40 minutes. the parenchyma cells inside the stem are clearly visible. (b) Transverse images of an excised root section of a 4-day-old maize seedling (i) before and (ii) after vacuum infiltration with water. (From Connelly *et al.*[47])

arises partly from differences in total water content, the vascular bundles clearly having a higher water content due to the presence of xylem and phloem vessels. However, considerable care is required in assigning water content on the basis of the image because of the influence of relaxation in affecting signal intensity. In cellular systems this relaxation may often be dominated by the effect of susceptibility inhomogeneity in influencing spin dephasing.

Fig 5.3(b) shows two images obtained in maize root as part of the same study by Connelly *et al.* Image (i) is of normal tissue and shows a strong signal from the stele, a much weaker signal from the cortex, and a relatively strong signal from water absorbed on the surface of the root. The authors point out that the relative image intensities are quite inconsistent with the known relative water contents since on a tissue dry-weight basis, the stele actually contains less water than the cortex.[48] Their explanation for the attenuation of the cortex water signal is that the cortical air spaces are a cause of susceptibility boundaries which create large local field gradients, thus attenuating the water signal. Evidence for this is clearly provided by vacuum infiltration of water as shown in (ii) where the dramatic increase in signal intensity is out of all proportion to the water uptake, consistent with the removal of susceptibility artefacts as the air spaces are filled. Since these images were obtained using spin echo methods, any decay of the signal must be due to homogeneous broadening, such as may arise from molecular diffusion in the presence of the local gradient.

The influence of intercellular air spaces is particularly clear in the very high-resolution NMR micrograph shown in Fig. 5.4. This image, obtained by G. D. Mateescu (personal communication) of Case Western Reserve University, is of an african violet petiole (the slender stalk joining the leaf to the stem). Also shown in Fig. 5.4(c) is an optical micrograph taken from the same stem section in which the air spaces between parenchyma cells can be seen. The dark regions in the optical micrograph correspond to vascular tissue. Around this, water has interpenetrated the spaces between cells so that the area surrounding dark spots in the NMR image are considerably brighter.

An interesting example of this phenomenon is seen in the work of Mansfield and co-workers (personal communication) who imaged the stem of *Pelargonium* (geranium) at 11 tesla where susceptibility effects were clearly apparent. On bruising the stem the intensity of the water proton image increased dramatically, consistent with the filling of intercellular spaces with water as the plant cells were damaged, thus reducing some of the local susceptibility variation.

These experiments highlight some of the difficulties experienced in interpreting image intensity data, especially in high-field instruments. However, the susceptibility effect clearly provides useful contrast in the image. For

(a)
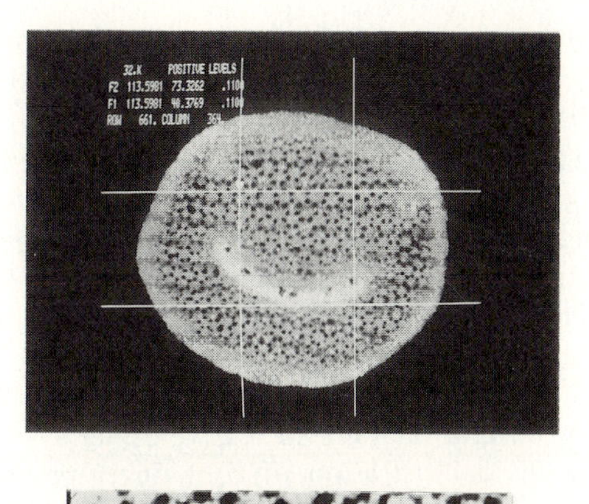

(b)

(c)
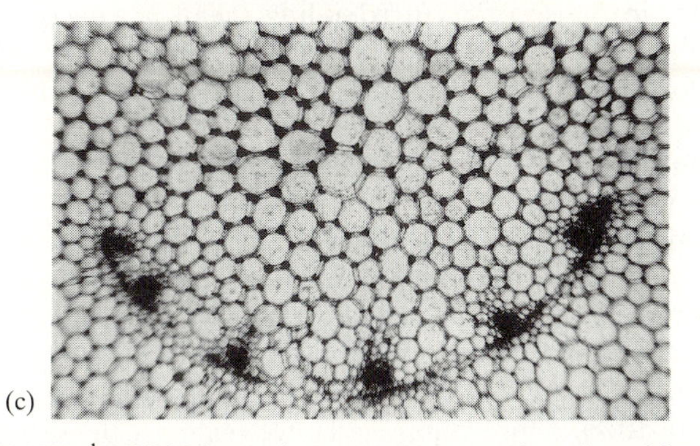

Fig. 5.4 (a) 400 MHz ^{1}H NMR image from a transverse section of an african violet petiole of 3.3 mm diameter. The slice thickness was 270 μm and pixel dimension $(4\ \mu m)^2$ with a total acquisition time of 900 minutes. The image was obtained using a spin echo with $T_E = 11$ ms. (b) Enlargement of the image which can be directly compared with the optical micrograph in (c). The parenchyma cells inside the stem are clearly visible as are the intercellular air spaces. These spaces cause enlarged artefacts in the NMR image probably due to susceptibility inhomogeneity. Note the bright NMR image in the vicinity of vascular tissue. (From G. Mateescu (personal communication.))

example, it enables one to distinguish between water in cells whose symmetry results in uniformity of the interior magnetic fields, and extra-cellular water whose image intensity is severely perturbed by local surfaces. One interesting possibility concerns the facility to alter the influence of susceptibility artefacts, thus giving information about such surface symmetry effects by examining changes in the image intensity. This can be performed by obtaining images using a much lower polarizing field, by shortening the time between the selective excitation and the centre of the echo, or by 'susceptibility matching' using an inert filler such as D_2O.

Another application of proton intensity imaging in plants concerns the monitoring of structural changes in a developing system. Connelly *et al.* have provided a vivid illustration of root development in a germinating mung bean seed[47] while Harrison *et al.*[49] have observed the maturation process in a single-celled marine organism, *Acetabularia*. These authors point out that, despite the limited resolution of NMR microscopy, the method is a powerful tool in cytoplasmic studies, especially if cytoplasmic reorganization during development involves the appearance or disappearance of extensive binding surfaces for water. In the *Acetabularia* cap, the transient appearance of large numbers of microtubules induce changes in relaxation time which effectively contrast the image by influencing the T_1 values of water in their vicinity. Fig. 5.5 compares optical micrographs and NMR micrographs for mature and immature *Acetabularia* caps. The similar appearance of the corresponding micrographs is striking, although certain visual contrasts, such as between dark bundles of cysts in mature rays and the free water in the solution surrounding them, are absent in the NMR images. The physical separation of intracellular cyst water and that in the surrounding medium is clearly demonstrated by soaking the *Acetabularia* cap for 5 minutes in a D_2O medium, thus allowing all free water to be exchanged with D_2O. This is shown in the comparative optical and NMR micrographs in Fig. 5.6.

The bulk movement of water in plants can be very effectively monitored using NMR microscopy methods. Where this movement occurs more slowly than the time taken to acquire an image, a simple real-time monitoring of the image can prove effective, as demonstrated by Tamiya *et al.*[50] in a study of water redistribution in the pulvinius of *Mimosa* which results from external stimulation. More rapid water movement can be observed using dynamic imaging techniques which have been used to measure vascular water velocity in living wheat grains.[51] These methods are discussed in detail in Chapter 8. Even where there is no net water transport, the measurement of local water self-diffusion coefficients can be a useful indicator of water binding, tissue porosity, or cellular compartmentation. Applications at microscopic resolution include studies of water mobility in corn roots[52] and wheat grains.[53]

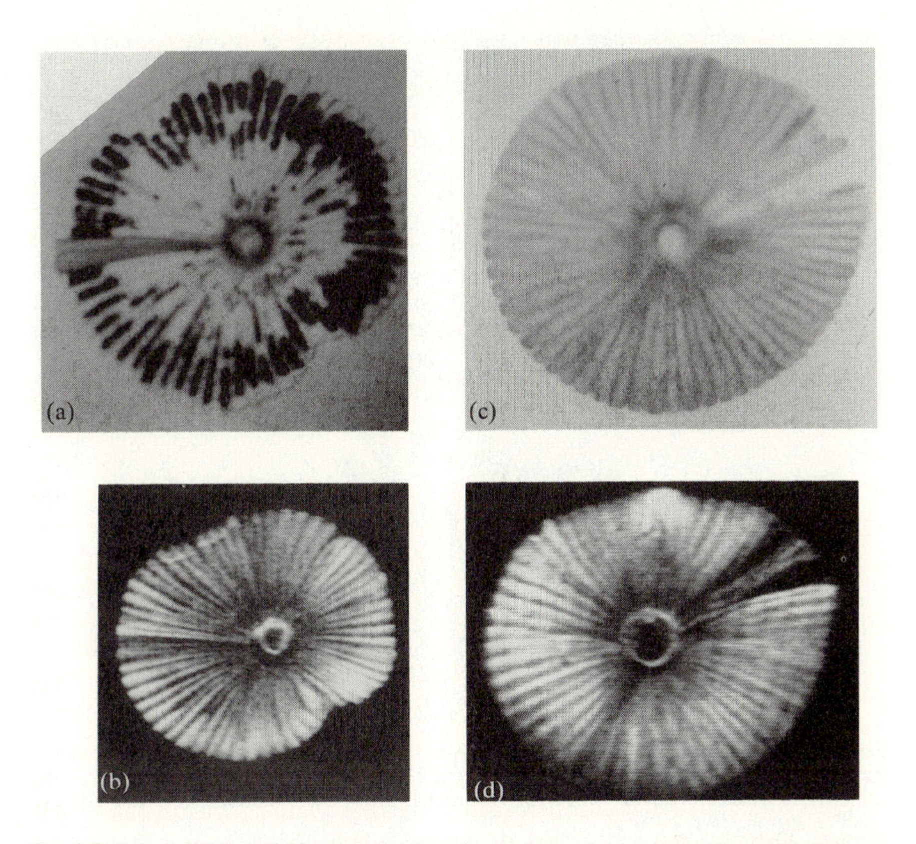

Fig. 5.5 (a) and (c) optical micrographs of mature and immature *Acetabularia* caps, respectively. The caps are approximately 7 mm in diameter. Immature rays are apparent in (a) at '9 o'clock'. (b) and (d) are 270 MHz ^{1}H NMR micrographs corresponding to (a) and (c). The rays are resolved down to 50 μm. Because the caps are flat no slice selection was employed. (From Harrison *et al.*[49])

Wood

A number of publications have appeared in which NMR imaging is used to examine the distribution of water in wood.[54-58] The work of Menon *et al.*[58] indicates that by careful analysis of T_2 relaxation it is possible to obtain separate images of water associated with earlywood and latewood components. An example of this separated component imaging is shown in Section 5.3.

Table 5.5 summarizes applications of **k**-space NMR microscopy studies in plant systems. Some of these studies were performed with voxel sizes too large to be considered microscopic but they none the less represent

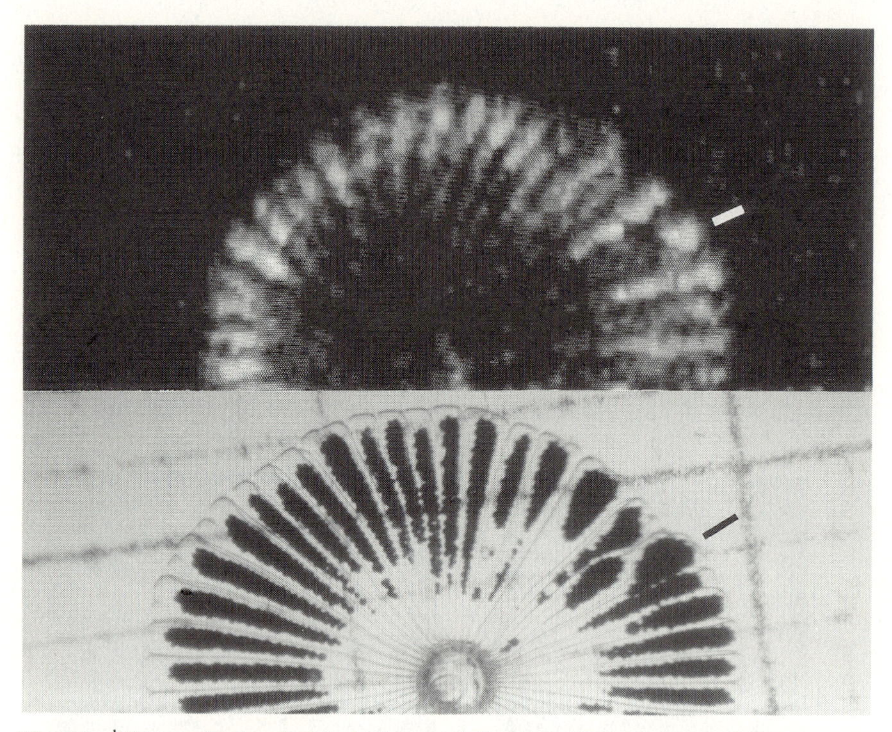

Fig. 5.6 ^{1}H image (top) and optical micrograph (bottom) of a mature *Acetabularia* cap with a millimetre grid in the background. The proton NMR image was obtained after soaking the cap in a culture medium made with D_2O. The marker shows corresponding features on the cyst bundles. (From Harrison *et al.*[49]).

interesting applications of the method. In most of this work the authors have considered the role of relaxation in the resulting image and in some cases special contrast schemes have been used, as indicated in the right-hand column. Many of the applications listed here will be referred to in later sections of this chapter.

5.2.2 *Animal tissue images*

Animal tissue applications of NMR microscopy fall into three broad categories which successively scale in resolution requirements. These concern the radiology of model mammalian animals such as rats and mice in the testing of new treatments for human disease, the developmental biology of very small animals and insects, and the study of biochemical processes at the cellular level.

Foremost among the groups working on rat NMR microscopy has been that of Johnson *et al.*[66–70] at Duke University medical centre. Johnson and

Table 5.5. *Summary of applications of NMR microscopy to plants.*

sample	molecule	nucleus/ frequency	pixel (μm)/ slice (μm)	contrast	reference
alyssum tenium stem	water	^{1}H 60 MHz	12 1500		59
Vicia faba L. root	water	^{1}H 63 MHz	560	paramagnetic agent	60
Equisetum arvense stem	water	^{1}H 270 MHz	80 80		61
P. coccineus (runner bean) stem	water	^{1}H 200 MHz	70 590	T_2	47
Z. Mays (maize) root	water	^{1}H 200 MHz	30 2000	T_2 paramagnetic agent	47
Phaseolus aureus L. (mung bean)	water	^{1}H 200 MHz	70 590	real-time evolution	47
Pelargonium hortorum stem	water	^{1}H 64 MHz	100 1250	T_1	62
Triticum aestivum L. (wheat) grain	water	^{1}H 60 MHz	30 1300	diffusion	53
Triticum aestivum L. (wheat) grain	water	^{1}H 60 MHz	30 1000	velocity diffusion	51

Table 5.5. *Continued*

sample	molecule	nucleus/ frequency	pixel (μm)/ slice (μm)	contrast	reference
Acetabularia cap	water	^{1}H 270 MHz	35	T_1 D_2O exchange paramagnetic agent	49
Mimosa pudica L.	water	^{1}H 85 2MHz	250 5000	real-time evolution	50
asparagus	water	^{1}H 300 MHz	70 200		63
Bryophyllum tubiflore	water	^{1}H 300 MHz	15 250		63
seed	water oil	^{1}H 300 MHz		chemical shift	64
Thuja plicata (cedar)	water	^{1}H 90 MHz	100	T_2	58
stem	water	^{17}O 54 MHz	50 3000		65
Z. mays (corn stem)	water	^{1}H 85.5 MHz	20	diffusion	52
Pelargonium (geranium stem)	water	^{1}H 500 MHz	4.5 100	susceptibility	(P. Mansfield personal communication)
Bryophyllum	water	^{1}H 300 MHz	15 250		46

co-workers point out[66] that the rat is a model for studies in mammalian anatomy, physiology, and pathology[71–73] and have carried out extensive MRI investigations using both close-wound r.f. coils as well as inductively coupled[74] implanted coils in the region of specific organs such as the kidney.[69] Human MRI systems commonly employ 1000 μm × 1000 μm × 5000 μm pixels. The linear dimensions of a rat are some 20 times smaller so that the comparable resolution would be 50 μm × 50 μm × 250 μm. Clearly the gain in sensitivity resulting from the use of smaller r.f. coils ($\sim$dimension^{-1}) will never compensate for the loss associated with restoring an equivalent resolution ($\sim$dimension^3) in studies of increasingly smaller animals. However, the ability, in small-animal studies, to use much higher polarizing fields than in human MRI goes a long way towards restoring comparability.

The Duke University group have correlated pathology specimens at various slices in the rat brain with MRI images obtained at 115 μm × 115 μm × 1200 μm, an example of which is shown in Fig. 5.7.[68] By synchronizing scan acquisition with breathing they have also demonstrated the possibility of obtaining high-resolution images in the rat thorax and abdomen.[66] One of their more remarkable images (Fig. 5.8) is that taken through the upper thorax, with motional artefacts reduced by cardiac gating. This image, with a transverse resolution of 200 μm shows the rat cardiac chambers and peripheral vasculature.

In the images shown here some potential resolution was sacrificed because the data set was limited to 256^2 pixels and the pixel size had to exceed 100 μm in order to cover the field of view. In principle, given a sufficiently high polarizing field, there is no reason why equivalent images cannot be obtained with 50 μm transverse resolution in a 250 μm slice and, using small fixed specimens of rat kidney and rat brain, Suddarth and Johnson (private communication) have recently obtained three-dimensional 256^3 array images at 300 MHz with a cubic voxel size of $(70\ \mu\text{m})^3$. The field of view limitation in intact animals can be avoided by the use of a surface coil so that only the region close to the coil is imaged. Using this method Rudin[75] has achieved a pixel size of 60 μm in 800 μm slices, imaging the head of a rat at 200 MHz.

By working at 9.2 T Mattingley and co-workers[76,77] have obtained very high-resolution NMR micrographs in laboratory mice, detecting metastatic deposits in a study that mimics several aspects of the human disease. Given that it is possible to image any area of either rats or mice with a scaled resolution at least comparable with that used in human studies, well-characterized rodent models are now available for the study of a wide range of fundamental diseases.

Among the first applications of NMR microscopy to insect development has been the *in vivo* monitoring of embryo development. Organisms studied

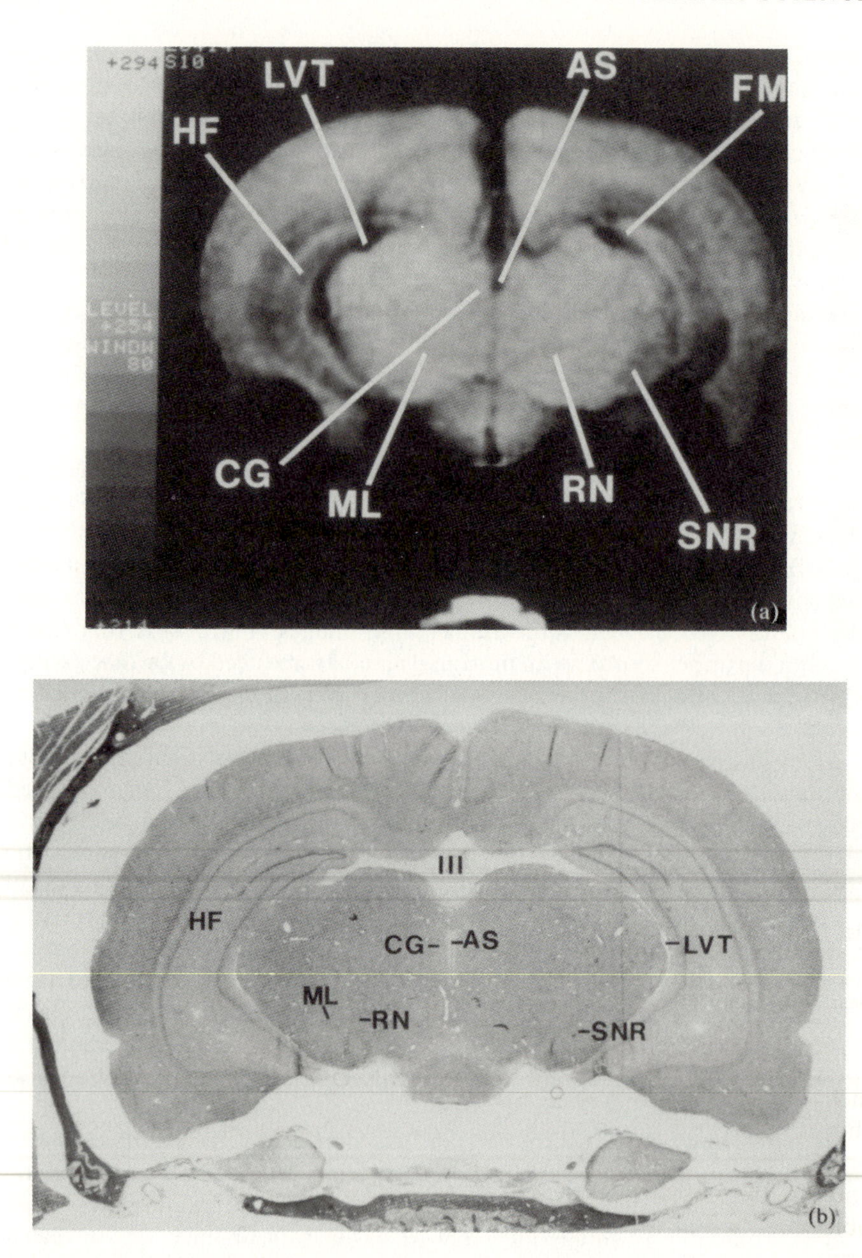

Fig. 5.7 (a) 64 MHz ^{1}H NMR image of a rat brain using a slice thickness of 1200 μm and an inplane pixel of (115 μm)^2. The labelled features are AS: aqueduct of sylvius; FM: forceps major of corpus callosum; RN: red nucleus; SNR: reticular substantia nigra; ML: medial lemniscus; CG: central tegmental gray; and LVT: lateral ventrical, temporal horn. (b) correlated pathology specimen of rat brain shown in (a). (From Johnson *et al.* in Investigative Radiology published by J. B. Lippincott.[68])

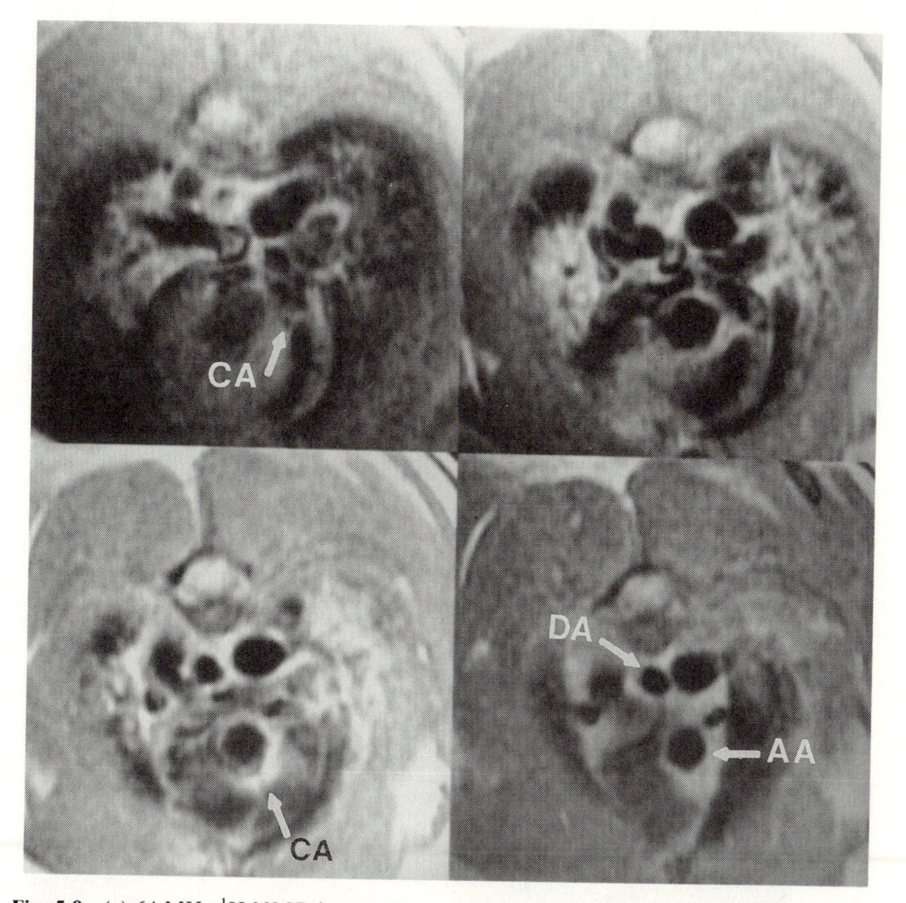

Fig. 5.8 (a) 64 MHz ^{1}H NMR image of a rat upper thorax showing four contiguous slices from an eight-slice cardiac-gated acquisition. The labels refer to AA: ascending aorta; DA: descending aorta; and CA: coronary artery. Cardiac chambers and pulmonary vasculature are clearly visible. (From Hedlund *et al.*[66])

include the tobacco horn worm[78–80] and the locust embryo.[81,82] Fig. 5.9 shows a series of slices obtained by Gassner and Lohman in the locust *Schistocerca gregaria* from a few hours after fertilization to the formation and hatching of the first nymphal stage. The proton NMR images were obtained at 200 MHz using a 3 mm diameter solenoidal r.f. coil. The transverse pixel resolution was 100 μm in a 500 μm slice thickness. Each image took 27 minutes to acquire using a multislice STEAM technique[83,84] (discussed in Section 3.5.2) to acquire three adjacent slices simultaneously. A remarkable feature of this work was the separate imaging of water and lipid distributions in the same object. The images show that the locust

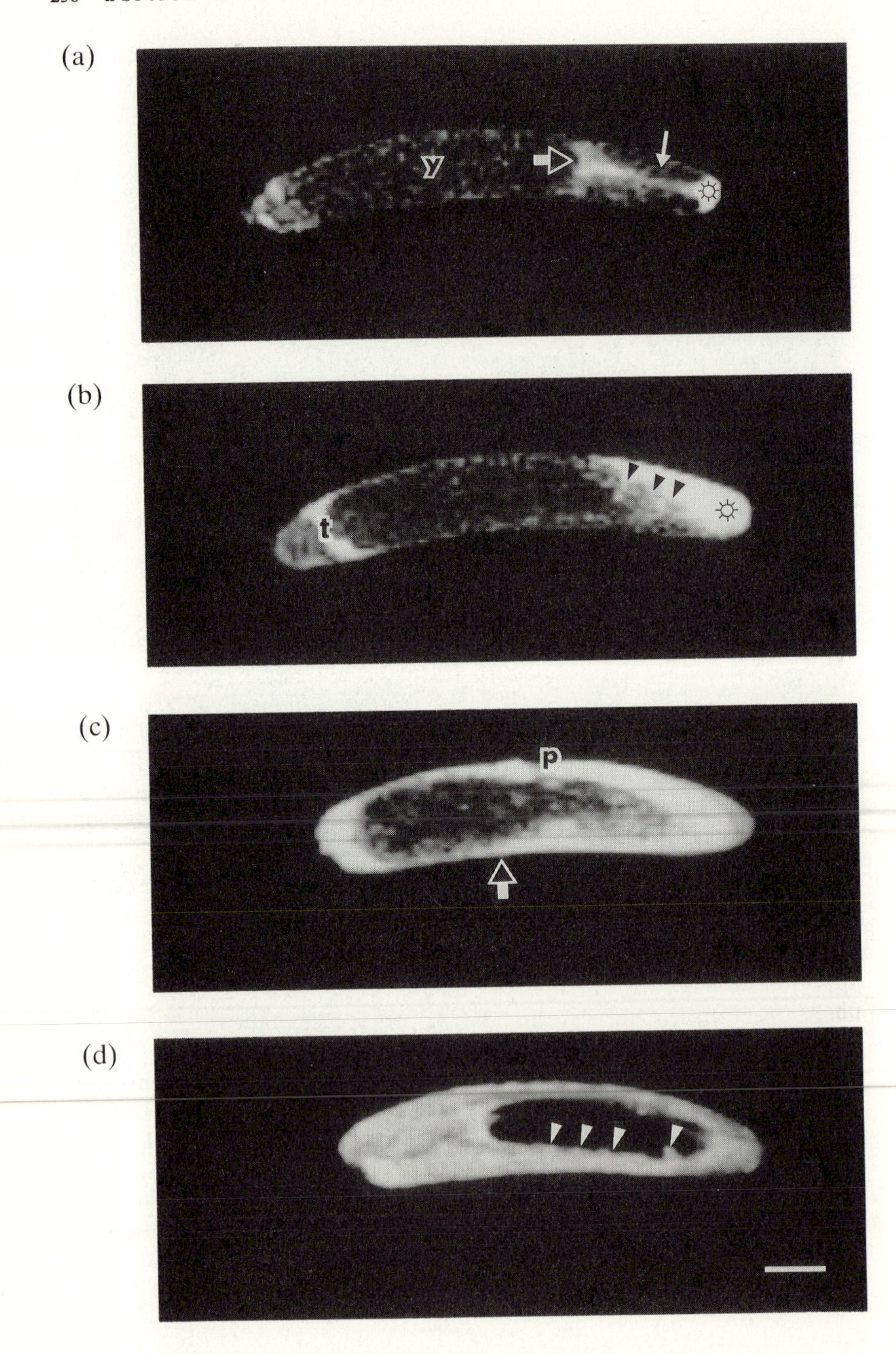
(a)
y
(b)
t
(c)
p
(d)

embryo undergoes positional changes during the course of its development. Delineation of anatomical features by the water and lipid images show that the water is concentrated in the ganglia, musculature, and integuments while the lipid resides in the body cavity.

Fig. 5.10(a) and (b) shows the time dependence of the total lipid and water signals as well as the water signal intensities for the yolk. The proton NMR signals associated with different chemical environments were separated using chemical shift contrast methods discussed in Section 5.3. These graphs illustrate the remarkable process of water uptake and consumption of egg lipid over the development process, demonstrating how the dynamics of embrogenesis can be visualized *in vivo*.

The study of tissue at the cellular level presents one of the most exciting challenges for NMR microscopy. In an early demonstration of NMR microscopy, Aguayo *et al.*[85] obtained a water proton density image at 400 MHz for a single toad ovum cell, with 250 μm slice thickness and a transverse pixel dimension of 15 μm. Cleary visible in the image, shown in Fig. 5.11, is the cell nucleus. This particular image was obtained by placing the cell in a glass tube inside a 5 mm diameter r.f.coil.

The toad ovum has a diameter of around 1 mm, a rather large object compared with most animal cells. While no investigation of isolated mammalian cells has yet been reported, there are some very high-resolution NMR microscopy studies on cell aggregates. Sillerud *et al.*[86] have followed the development of necrotic tissue in multicellular tumour spheroids (MTS). Human and mouse carcinomas were cultured at 37°C and 1500 μm diameter MTS samples were placed in 2 mm thin-walled glass tubes filled with phosphate buffered saline. Using a 5 mm r.f. coil at 400 MHz they obtained images with 8 μm transverse resolution in a slice thickness of 125 μm. The necrotic core showed a distinctly lower intensity and careful analysis showed that this was due to a shorter T_2. By comparison, T_1-weighted images showed a uniform signal intensity across the spheroid, as apparent in Fig. 5.12. As part of this study the authors obtained proton T_2 contour maps in the MTS and used ^{31}P NMR to monitor the characteristic disappearance of ATP peaks and pH change (visible in the inorganic phosphorous chemical shift) associated with the development of necrosis. One of

---

Fig. 5.9 200 MHz ^{1}H NMR images of the water distribution in midsaggital slices through locust eggs (anterior end on left). Four different development stages are shown (a): 14%; (b): 25%; (c): 47%; and (d): 86%. The bar indicates 1 mm. Structural features in (a) are as follows; y: yolk; small arrow: water column; asterisk: embryo during anatrepis; large arrowhead: transverse water compartment. Structural features in (b) (embryo during katatrepsis) are; darts: abdominal somites; t: transverse water compartment. Features in (c) (embryo undergoing katatrepsis) are; p: posterior end of embryo; arrowhead: protocephalon. In (d) katatrepsis is completed. The darts indicate somites. (From Gassner and Lohman.[81])

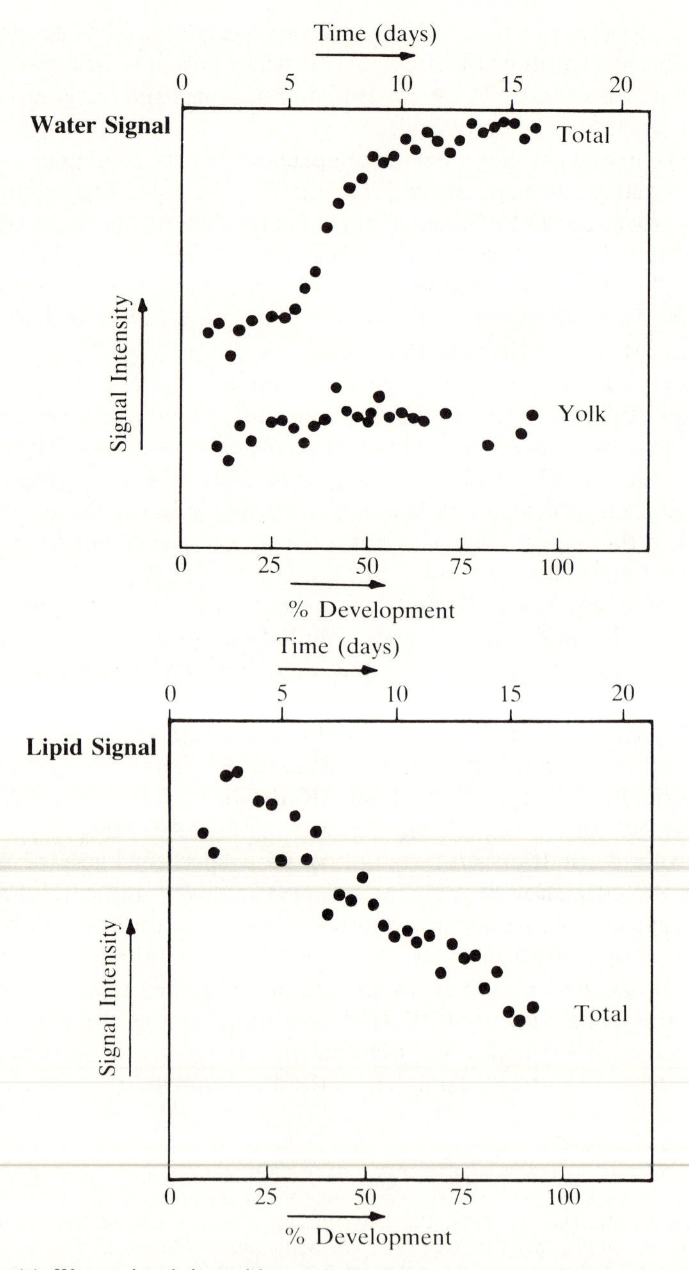

Fig. 5.10 (a) Water signal intensities and (b) lipid signal intensities as a function of development in the locust eggs. Chemical shift contrast is used to distinguish the water and aliphatic proton NMR signals. The water content of the whole eggs increases markedly during development while that of the yolk remains constant. By contrast the lipid content of the egg decreases uniformly over the development period. (From Gassner and Lohman.[81])

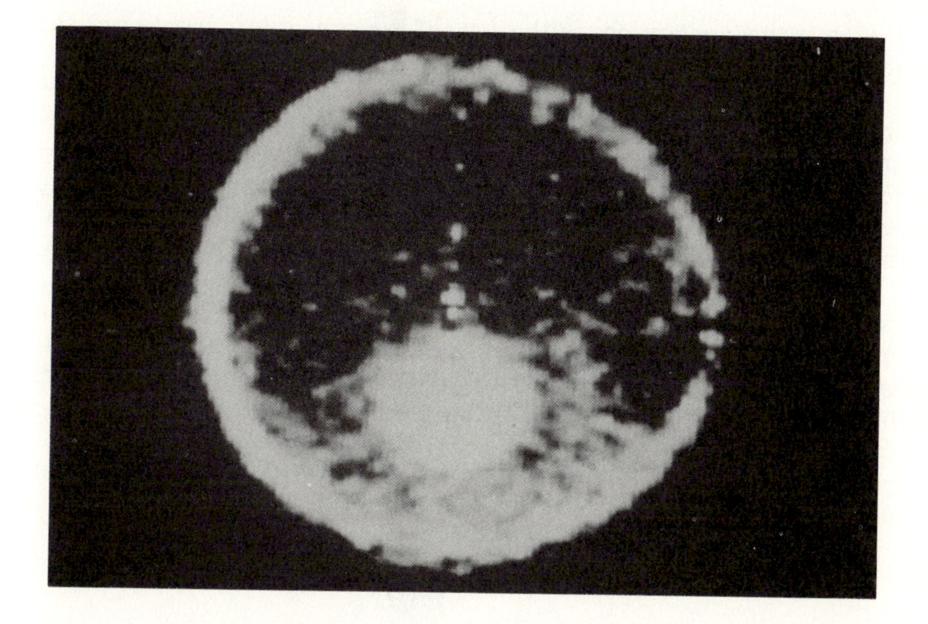

Fig. 5.11 400 MHz ^{1}H NMR image of a transverse slice through the ovum (approximately 1 mm in diameter) of *Xenopus laevis*. The slice thickness is 250 μm with an in-plane pixel resolution of (10 μm $\times$ 13 μm). The total acquisition time was 32 minutes. (From Aguayo *et al.*[85])

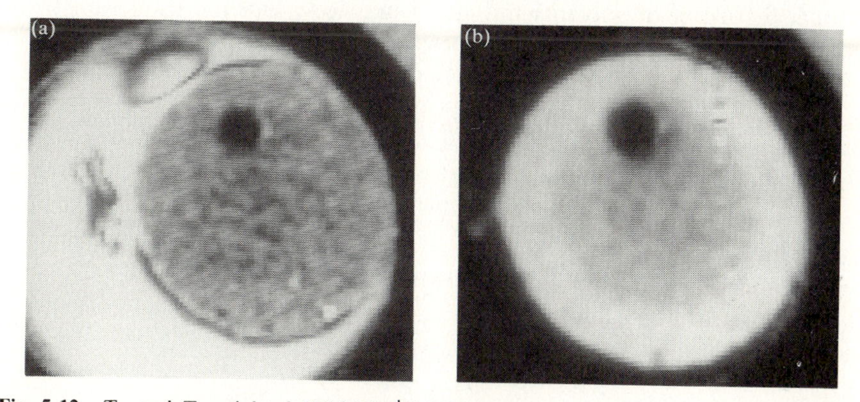

Fig. 5.12 T_1- and T_2-weighted 400 MHz ^{1}H NMR images of a transverse slice through an EMT6/Ro multicellular tumour spheroid of 1.6 mm diameter. The slice thickness was 125 μm with in-plane pixel resolution of 30 μm. The T_1-weighted image (a) is obtained using $T_E = 17.6$ ms and $T_R = 320$ ms while the T_2 weighting (b) arises from $T_E = 25.6$ ms and $T_R = 3020$ ms. The total acquisition time was 2 hours. In the T_2-weighted image the viable cells in the outer rim (thickness 200 μm) have a signal 40% brighter than the necrotic core. (From Sillerud *et al.*[86])

the interesting applications of this type of experiment is in monitoring the response of tissue to drug treatment or other biochemical stresses.

Other recent applications of NMR microscopy in mammalian cells include a study of human skin cells[87] using a 20 mm surface coil in a standard medical imager, and the imaging of a selenite cataract in a rat pup lens.[88] Unlike plant stems, animal tissue seldom possesses symmetry along one axis so that there is no real justification for the common practice of employing slice thicknesses much larger than the transverse pixel dimension. Lauterbur and co-workers have emphasized the importance of obtaining cubic voxels and employ a three-dimensional projection reconstruction method for this purpose.[89] They report a 200 MHz study of tissue contrasting using dextran magnetite in which the voxel size is $(13\,\mu\text{m})^3$, close to the theoretical optimum. This work[90] demonstrates the selective uptake of the contrast agent by macrophages in spleen red pulp, showing the usefulness of particulate super-paramagnetic contrast agents in studying the effects of developing pathology on local NMR tissue properties.

A summary of NMR microscopic studies in animal tissue is given in Table 5.6.

5.2.3 *Imaging in non-biological materials and in food products*

While the non-invasive character of NMR microscopy is less important in materials science than in the study of living organisms, the application of imaging is no less interesting. It offers the possibility of studying physical and chemical heterogeneity in composite materials, porosity in minerals, mutual diffusion of solvents into solid matrices, and the transition from liquid to solid state in ceramics, glues, and expoxies. It is also possible, in

Table 5.6. *Summary of applications of NMR microscopy to animals.*

sample	molecule	nucleus/ frequency	pixel (μm)/ slice (μm)	contrast	reference
X. laevis toad ovum	water	^{1}H 400 MHz	15 250		85
rat brain and abdomen	water	^{1}H 64 MHz	50 1000		67
snail	water	^{1}H 90 MHz	20 70		89
rat thorax and abdomen	water	^{1}H 64 MHz	200 1200		66
quail egg	water/ lipid	^{1}H 300 MHz	115 275	chemical shift	63

sample	molecule	nucleus/ frequency	pixel (μm)/ slice (μm)	contrast	reference
rat brain	water	^{1}H 64 MHz	115 1200		68
rat kidney	water	^{1}H 64 MHz	115 1200		69
Schistocerca gregaria (locust embryo)	water lipid	^{1}H 200 MHz	100 500	chemical shift	82
multicellular tumour spheroids	water	^{1}H 400 MHz	30 125	T_2, T_1	86
rat head and eyes	water	^{1}H 200 MHz	60 800		75
mouse brain	water	^{1}H 400 MHz	25 500	diffusion	91
Manduca sexta tobacco horn worm (moth pupa)	water	^{1}H 90 MHz	100 1250	T_1	78
Rana pipiens eggs	water	^{1}H 360 MHz	15 200		79
capillary	PCr/ ATP	^{31}P	200 1000	chemical exchange	79
Manduca sexta tobacco horn worm (moth pupa)	water	^{1}H 200 MHz	35 1000		80
mouse	water	^{17}O 54 MHz	50 3000		65
rat spleen	water	^{1}H 200 MHz	13 13	$T_1 T_2$	90
rat kidney	water	^{1}H 85.5 MHz 300 MHz	60 1250 350	$T_1 T_2$	70
mouse tumour	sodium ions	^{23}Na 105.8 MHz	200		92
mouse colon tumour	water	^{1}H 400 MHz	100 700		76
mouse kidney	water	^{1}H 400 MHz	36 320		77
rat kidney/brain	water	^{1}H 300 MHz	70 70		70

principle, to study the fundamental physics of liquid polymer systems by measuring NMR parameters sensitive to local dynamics in spatially resolved velocity gradients. Some of these latter applications are given in Chapter 8. One potentially interesting application of microscopy is in the study of food products, and especially in monitoring the distribution of oil, water, and other molecular species during drying, ripening, or cooking. A number of groups have begun work in this area but as yet few results have been reported in the literature. One such is a recent proton imaging study by Mateescu[93] involving a comparison of deep frying and microwave cooking uniformity in potatoes, where clear differences were apparent.

One of the major challenges in applying imaging to solid materials is the problem of the broad NMR linewidth for protons in rigid molecular environments. Here internuclear dipolar interactions can reduce the transverse magnetization coherence time to a few tens of microseconds, making conventional **k**-space imaging methods impossible. In recent years some powerful methods have been developed to obtain images for protons in such environments. These involve the combination of line narrowing pulse techniques, and in some cases magic angle spinning, with NMR imaging. This highly specialized form of imaging is discussed separately in Section 5.6.

For the moment, however, we concentrate on those solid state systems in which it is possible to obtain proton NMR signals with T_2 relaxation times greater than or of order a few milliseconds. We may group these into two broad categories. First, it is possible to use the NMR signal from small molecules in the liquid state which have been absorbed into a solid matrix, a class of material we could label 'heterogeneous solid/liquid systems'. This form of imaging is useful for investigating porosity and for following mutual diffusion processes. Second, we can obtain relatively long T_2 proton NMR signals from certain classes of elastomeric polymer molecules. These macromolecules can exhibit 'liquid' NMR properties because of high local mobility and rapid isotropic reorientation of segments while, on the macroscopic scale, showing 'solid' properties because of cross-linking or impeded translational diffusion because of entanglements coupling. The local mobilities of polymer segments are especially sensitive to chemical structure so that T_2 values, which depend on rate of rotational motion as well as the anisotropy of this motion, will vary considerably from one material to another. A key parameter in determining segment reorientation is the proximity to the glass transition temperature of the polymer, with typical proton NMR linewidths for rubbery polymers being in the range 100 to 2000 Hz at a temperature of 100°C above T_g.[45] Examples of such systems are given in Table 5.4.

While the inhomogeneous broadening of the line can be refocused by echo methods, it still contributes to the decay of the transverse magnetization during the read gradient and so limits the resolution attainable. T_2^* relaxation times of a few milliseconds mean that its is extremely difficult to achieve a resolution below 100 μm in elastomer imaging.

Heterogeneous solid/liquid systems

The imaging of small molecules introduced as markers in porous materials provides a useful means of investigating morphology of pores with diameters >10 μm and for detecting defects and voids. This type of study has been performed on sponges by J. Gelan (personal communication), foams,[94] rubbers,[95] and polymer composites.[96] A particularly interesting application of this imaging in which a liquid component is observed in a solid matrix concerns the evaluation of adhesives where the process of curing and solidification leads to reduction in the liquid component relaxation times and changes in chemical shift. Studies of adhesive penetration and curing at high spatial resolution have been reported for PV Ac-emulsions in wood[97] and ceramic,[98] and polyester polyol/isocyanate in plastic.[99]

Much of the contrast in water proton images obtained in biological systems arises from the sensitivity of T_1 and T_2 to the state of the water. This same idea is apparent in a study of slip casting of ceramics by Hayashi *et al.*[100] The formation of solid deposit from Al_2O_3 slip was monitored over a period of one hour, taking an image every 10 minutes. While the water density is very similar in the slip and deposit, the difference of T_1 and T_2 was sufficient to provide a clear contrast so that the deposit thickness could be accurately measured. The authors suggest the possibility of obtaining T_1, T_2, and diffusion maps across the cast in order to determine the physical state of the water accurately. Other ceramics applications include a study of the pore distribution[101] in unsintered ceramic via the imaging of absorbed fluids and a study of the distribution of low molecular weight polymer binding agents.[102]

The observation of signal intensity changes during the curing of a two-part resin[103] provides another example in which a transition in the state of molecular rotational mobility is apparent. Rothwell and co-workers have used NMR imaging to observe the penetration of water at high temperatures into different epoxy resin composites.[104] Their study shows the importance of thermal shrinkage cracks in determining the rate of water ingress. This group has also monitored the mutual diffusion of solvents into solid polymer matrices, with a series of images obtained successively for a 19 mm diameter polystyrene rod immersed in toluene.[105] The polystyrene proton T_2 is sufficiently short that the solid polymer contributes effectively zero image intensity and the proton density maps result from the toluene molecules which have a very much longer T_2. A similar set of proton density maps has been obtained by Samoilenko *et al.* for the diffusion of acetone into polymethylmethacrylate.[106] Other examples of interdiffusion studies include the

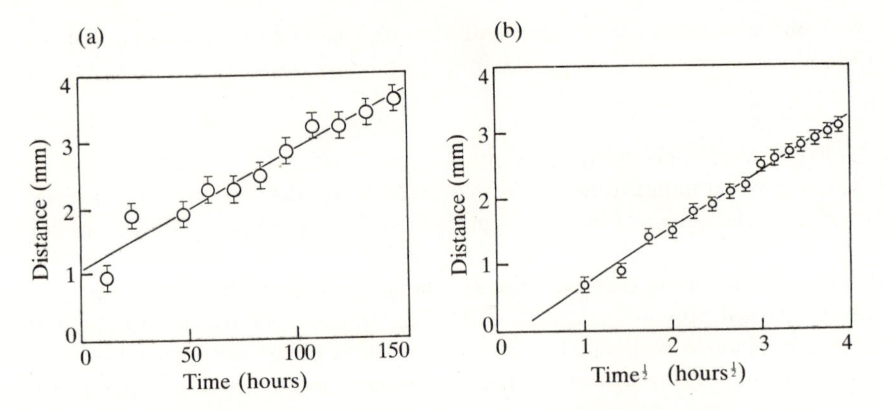

Fig. 5.13 Change in the diameter of the glassy core of a PMMA rod placed in methanol. (a) shows case II mutual diffusion at 30°C, while (b) shows case I diffusion at 60°C. (From Weisenberger and Koenig.[111])

proton imaging of benzene in graphite[107] and the ^{19}F imaging of AsF_5 in normal[107] and highly oriented pyrolytic graphite.[108]

Precise measurements of mutual diffusion are possible using imaging methods.[109,110] Koenig and co-workers[111,112] have used FLASH imaging to study case II diffusion of methanol into cylindrical rods of polymethylmethacrylate in which a Fickian concentration profile precedes a sharp front that advances through the glass at constant velocity. As the temperature is raised from 30°C to 60°C the diffusion profile changes to normal Fick's law behaviour with the front advancing as time$^{\frac{1}{2}}$. These data are shown in Fig. 5.13. Fig. 5.14 shows an example of such an experiment in which water is observed to diffuse into the face of a nylon block at 100°C. In this study by Blackband and Mansfield the diffusion was described by a one dimensional Fick's equation with a concentration-dependant diffusion coefficient D(c)[113] as

$$\frac{\partial C}{\partial t} = \frac{\partial}{\partial x}\left(D(C)\frac{\partial C}{\partial x}\right) \tag{5.7}$$

Eqn (5.7) may be integrated[114] to give the diffusion coefficient at concentration C_1 as

$$D(C_1) = -\frac{1}{2}\left[\frac{\partial(x/t^{\frac{1}{2}})}{\partial C}\right]_{C_1} \int_0^{C_1} (x/t^{\frac{1}{2}})\,\mathrm{d}C. \tag{5.8}$$

The dependence of the mutual diffusion coefficient on concentration obtained in this experiment is shown in Fig. 5.14(b), indicating the precision possible. It should be noted that the diffusion coefficients measured are close to the lower limit obtainable using pulsed gradient spin echo NMR.

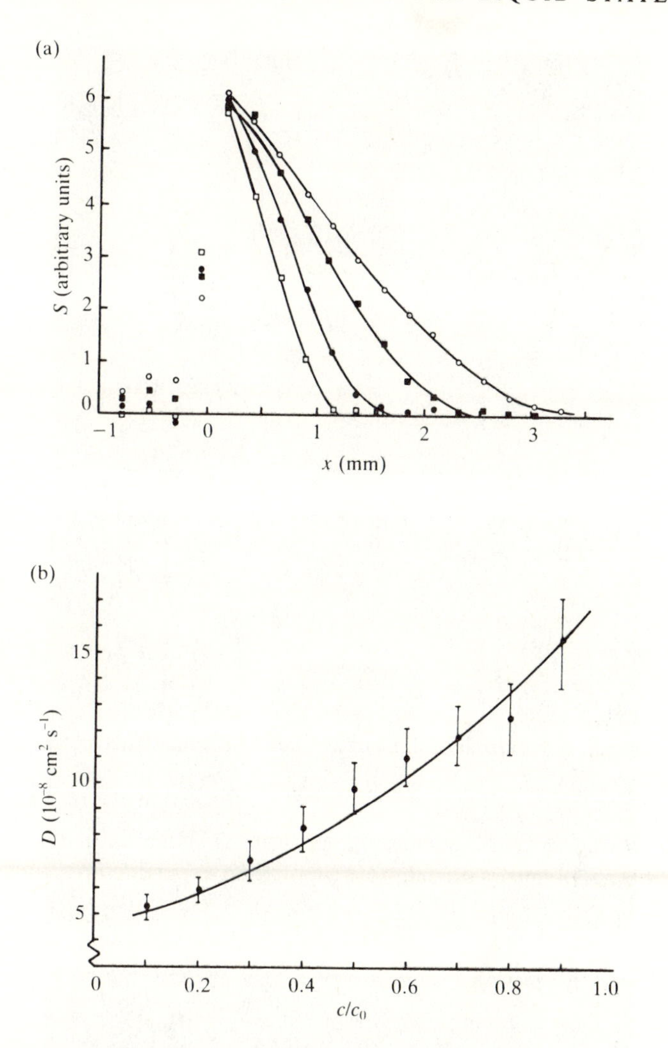

Fig. 5.14 (a) Water signal profiles as a function of distance into the face of a nylon block at 100°C recorded at time intervals of 6, 9, 18, and 33 hours, respectively. The resolution is 240 μm. (b) Calculated mutual diffusion coefficient versus concentration. (From Blackband and Mansfield.[110])

One of the effects of solvent adsorption in amorphous polymers is to induce a swelling disordering transition. Marechi *et al.*[115] have obtained a series of images following the diffusion of chloroform into a bar of polymethylmethacrylate. In one series of measurements they used normal chloroform and in the other deutero-chloroform. These images are shown

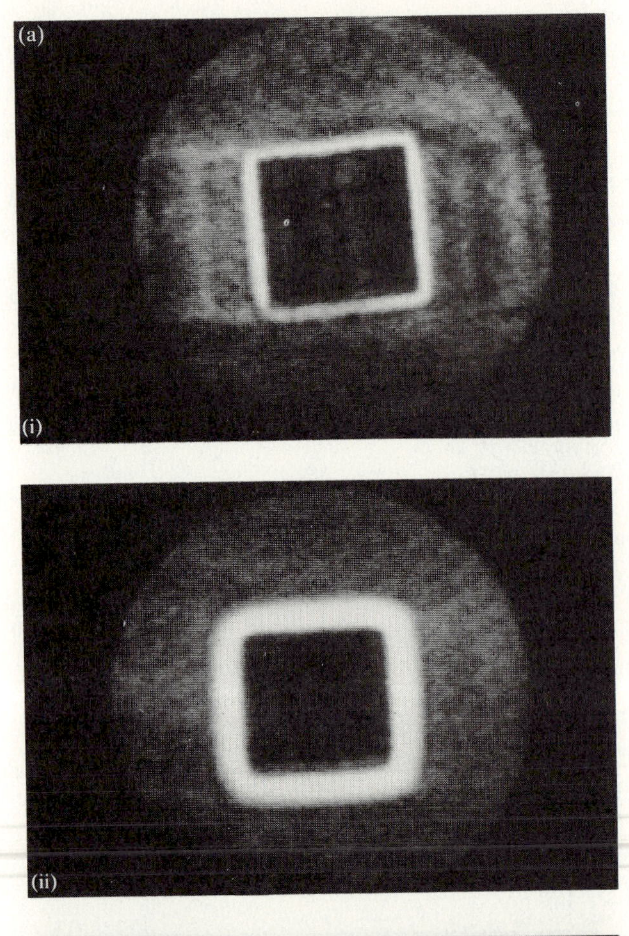
(a)
(i)
(ii)

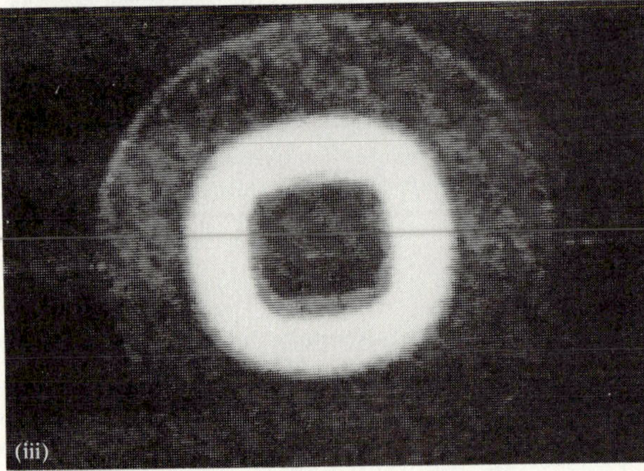
(iii)

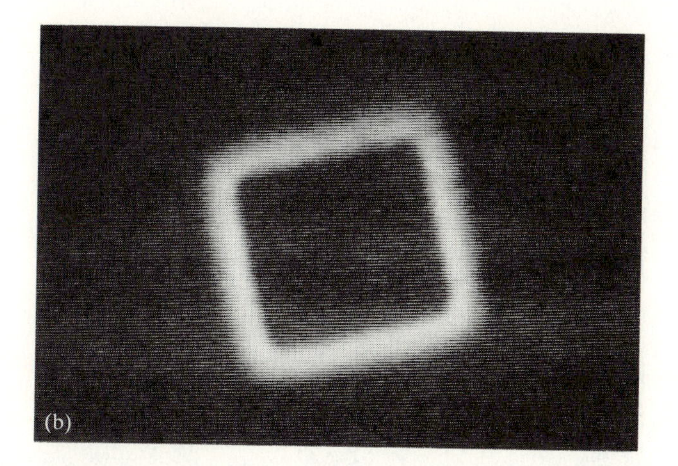

Fig. 5.15 (a) ^{1}H NMR images of a bar of PMMA in $CHCl_3$ as a function of time after the beginning of swelling. Each image covers a field of view of 80 mm × 80 mm. The images are (i) $t = 1$ hour, (ii) $t = 3.5$ hours, and (iii) $t = 20$ hours. (b) As in (a) but where the PMMA bar is swollen by deuterated chloroform ($t = 3$ hours). (From Marechi *et al.*[115])

in Fig. 5.15. This second experiment is particularly interesting since the deutero-chloroform contributes no proton NMR signal and the regions of high intensity result from increased segmental mobility in PMMA molecules swollen by the solvent. A remarkable feature of the swelling process is its geometric uniformity. On the basis of these images and localized T_1 measurements, the authors postulate that the swelling process occurs as a first-order transition, as a form of 'surface melting'.

A particularly beautiful example of solvent imaging in a swollen polymer has been provided by Kuhn and Mattingley[63] in obtaining water proton images for hygroscopic polymer beads. This very high-resolution micrograph is shown in Fig. 5.16. The bright edges in the polymer gel particles are probably due to susceptibility artefacts. The four images shown were obtained at differing T_E values in a spin echo imaging sequence, thus showing the influence of T_2 relaxation.

Among mineral rocks studied by NMR imaging are the variety of sandstones so important to the petrochemical industry. NMR microscopy has the potential to provide information about the relative distribution of oil and water in the multiple fluid phases contained in the pores, of considerable significance in understanding oil recovery. Applications using water and oil proton imaging have been reported for Lion Mountain,[116] Berea,[116,117] and Bentheim sandstones,[117,118] and Bakers and San Andreas dolomite.[117] The water signals exhibited short T_2 relaxation times of a few tens of milliseconds, the reduction in T_2 from the free-water value being characteristic for water molecules in small pores (see Section 7.6). In order to obtain a

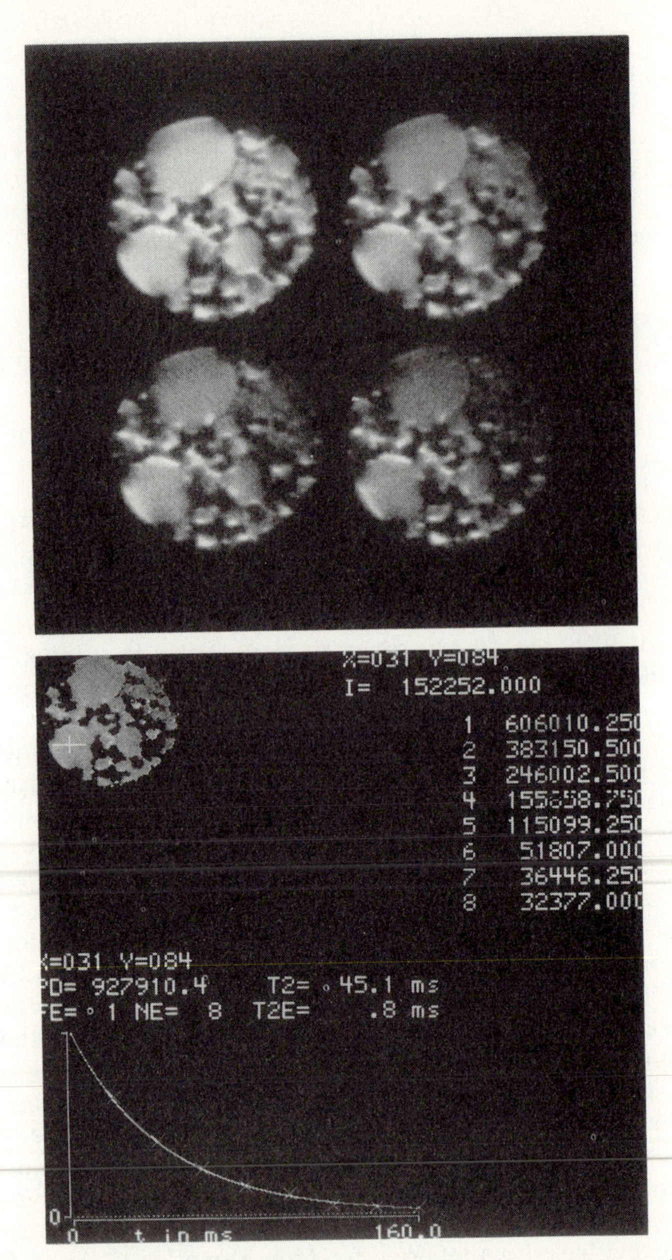

Fig. 5.16 Multi-echo 300 MHz ^{1}H NMR images of hygroscopic polymer beads (diameter approximately 1.5 mm). The slice thickness is 200 μm with an in-plane pixel resolution of (18 μm)^2. In the lower part of the diagram the exponential T_2 decay for a specified point in the image is shown. (From Kuhn and Mattingley.[63])

water density map free of relaxation attenuation a series of images was obtained in a CPMG train and the signal in each pixel extrapolated back to the time origin at the initial 90° pulse. This enabled a map of $M(0)$ to be calculated and these measurements gave a porosity value in good agreement with conventional hydrostatic weighing methods, but indicated a surprising degree of heterogeneity in the pore distribution.

Elastomers

Imaging in elastomers provides an effective means of measuring void and defect distribution. One of the consequences of such inhomogeneity is the existence of bulk susceptibility variations which induce additional inhomogeneous line-broadening. This broadening can be distinguished from true relaxation by echo refocusing. Chang and Komorowski[45] have carried out extensive imaging study in polymer elastomers in which two different pulse sequences were used in order to distinguish the effects of T_2 and T_2^* relaxation. In the first method (spin echo) a 180° r.f. pulse is used to refocus dephasing caused by the selective excitation pulse and the precursor read pulse, G_y. The echo produced by this sequence will have all inhomogeneous broadening effects refocused so that the image will be attenuated by T_2 relaxation only. In the second sequence (gradient echo) the gradients are reversed in sign but no 180° pulse is employed. This means that inhomogeneous broadening, whether of chemical shift or susceptibility origin, will continue to cause dephasing and the image will be attenuated by T_2^*.

Fig. 5.17 compares the images obtained at 200 MHz from a block of solid *cis*-polybutadiene using the two methods. This material has a glass temperature of $-102°C$ with highly mobile polymer segments leading to a T_2 of 13 ms. Several defects, including two larger voids (upper left and lower right), are clearly visible. The gradient echo image shows more structure with air bubbles more obvious and with voids appearing larger. This effect is also observed in medical imaging at tissue–air boundaries when gradient echo refocusing is used.[119] The inhomogeneous broadening component of T_2^* scales with the polarizing field strength. This means that the influence of susceptibility inhomogeneity is less severe in images obtained at lower fields, an effect which is clearly apparent in images obtained on the same elastomer but at 1.5 T rather than 4.7 T.

The same method was also used to obtain images of natural rubber at 200 MHz although these required considerably longer signal averaging times because of the shorter T_2. Again the gradient echo image showed enhanced defect size.

The effect of curing and filling is to reduce T_2 due to the slowing of chain motion and T_2^* due to bulk susceptibility effects.[120] The T_2^* decay for protons in cured, carbon black-filled *cis*-polybutadiene was so rapid that gradient echo images at the shortest available echo time of 12 ms, were

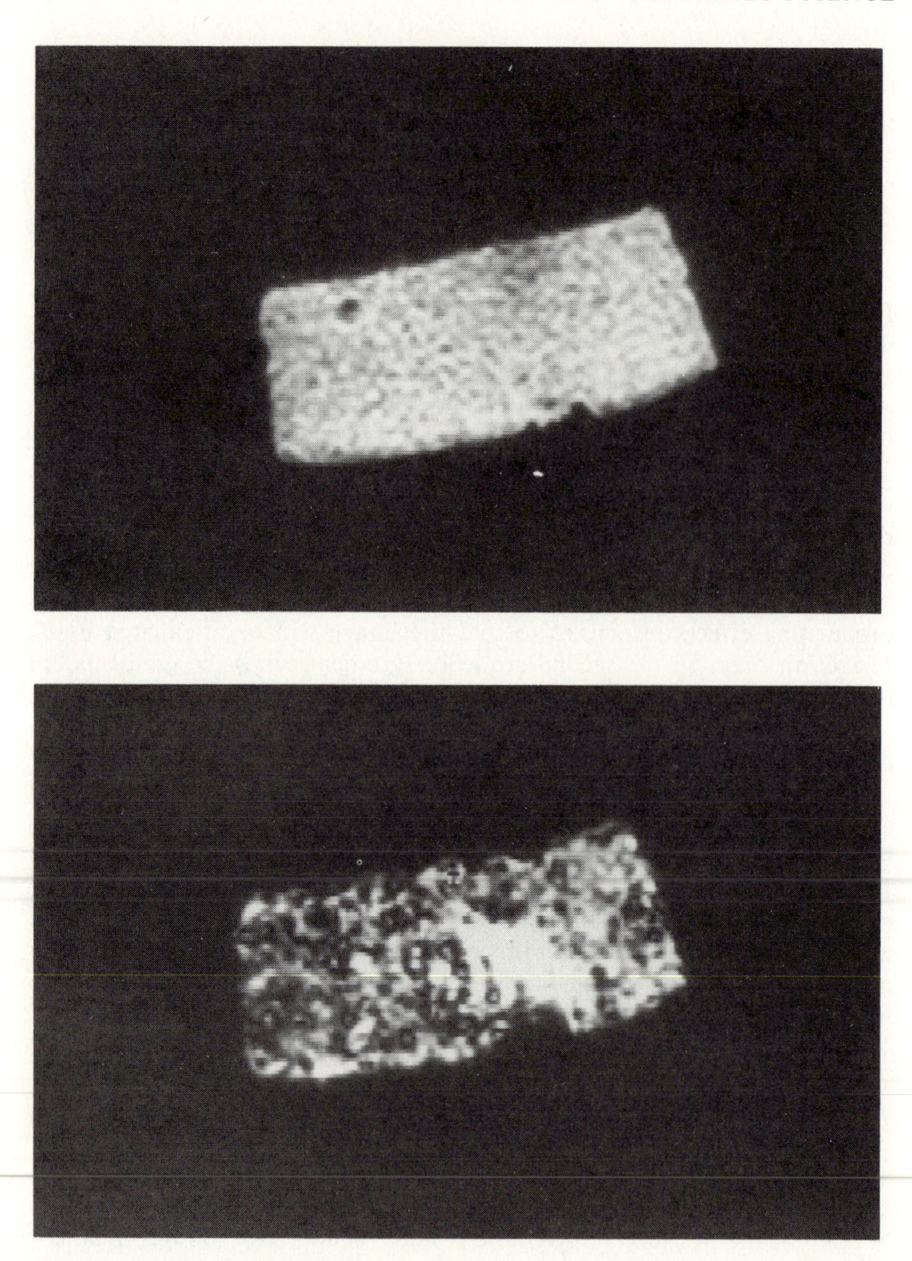

Fig. 5.17 200 MHz ^{1}H NMR images of a portion of a block of solid *cis*-polybutadiene. The face shown is 20 mm × 55 mm and the slice thickness is 1 mm. The top image is obtained using a spin echo with $T_E = 12$ ms, while the lower image was obtained using a gradient echo with $T_E = 6.4$ ms. (From Chang and Komorowski.[45])

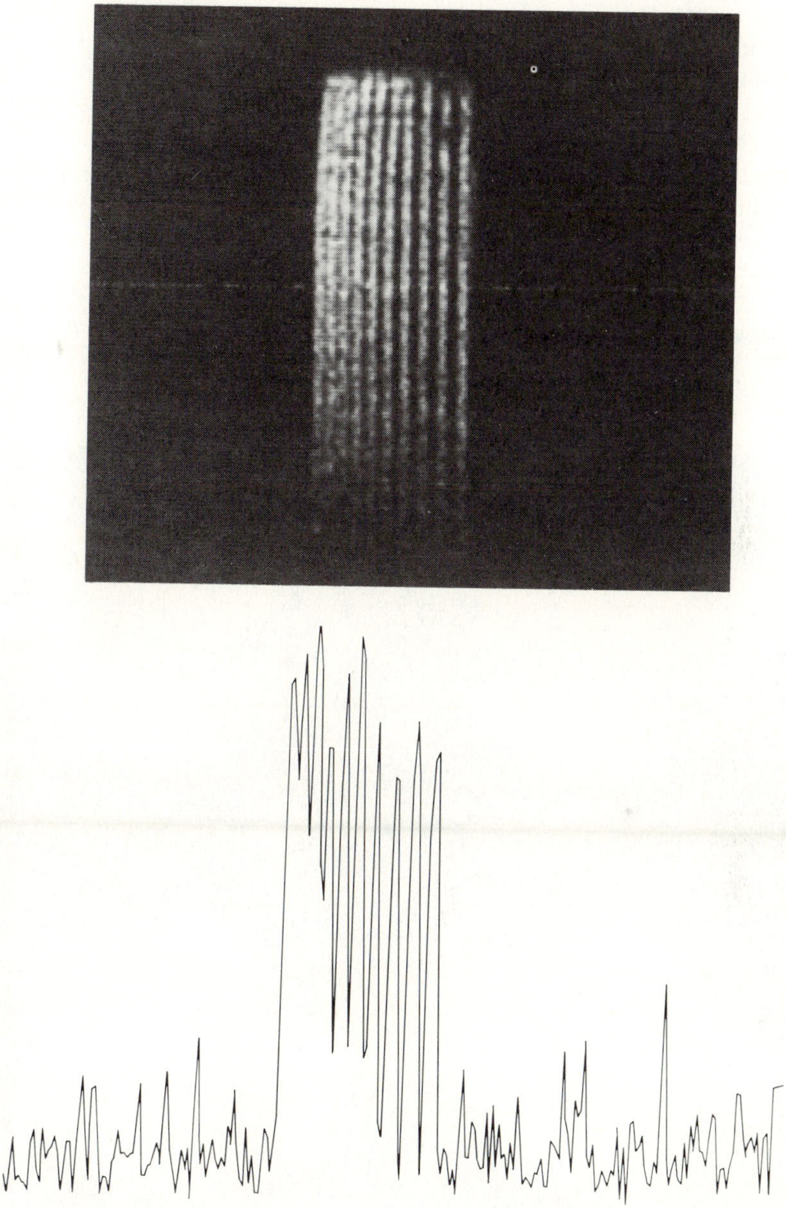

Fig. 5.18 Spin echo image of ten sheets of cured carbon black filled *cis*-polybutadiene (2 mm sheet thickness) separated by spacers of various thicknesses: from left, 0.07, 0.15, 0.30, 0.45, 0.60, 0.80, 1.0, 1.15, and 1.30 mm. The slice thickness is 1.6 mm while the field of view is 120 mm × 123 mm. $T_E = 12$ ms. The intensity fall-off at the edge is due to the sample extending beyond the edge of the r.f. coil. The lower part of the diagram shows a horizontal profile through the image. (From Chang and Komorowski.[45])

severely attenuated. Fig. 5.18 shows an r.f. echo image and corresponding line profile of ten sheets of 2.0 mm thick cured, carbon black filled *cis*-polybutadiene separated by glass spacers of thicknesses ranging from 0.07 mm to 1.3 mm, along with a horizontal intensity profile through the image. One remarkable feature is the apparent resolution of the 0.07 spacer, despite the fact that this dimension is smaller than the pixel size. One possibility is that the polymer mobility is reduced near the surface leading to a shorter T_2. This hypothesis is supported by the continued existence of a dark line between the 2 mm thick sheets even when the spacer was removed.

An example of NMR microscopy in a natural rubber is shown in Fig. 5.19. Here multislicing reveals the connectivity of internal void spaces. Once

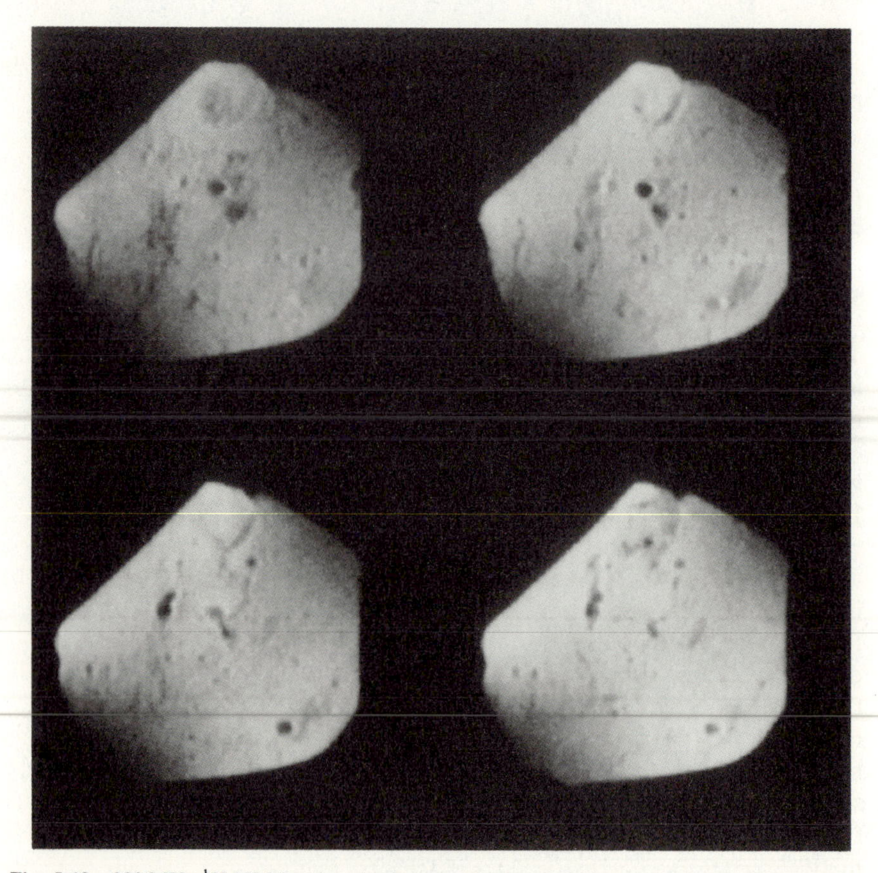

Fig. 5.19 300 MHz ^{1}H NMR spin echo images of natural rubber. A series of slices is shown each with a thickness of 500 μm. The pixel dimension is $(45\ \mu m)^2$. Note the bright edges close to voids caused by susceptibility artefacts. (From Kuhn and Mattingley.[63])

again, susceptibility effects are seen as bright edges at the boundary between the rubber and the air pockets.

The susceptibility artefacts in elastomer imaging, far from presenting a severe problem, can be regarded as useful. Comparison of spin echo and gradient echo imaging can reveal the existence of these effects. In the spin echo image the amplitude loss due to T_2^* can be avoided but the resolution loss will still be effective. However, by using modulated-gradient or high-gradient echo summation methods, as discussed in Section 4.7.2, it is possible to obtain images with even the resolution artefacts removed. Furthermore, unlike the imaging of small molecules in porous solids the attenuation of the echo due to diffusion in the local gradients will be absent. Because of the facility to generate three types of image, namely T_2^*-weighted and distorted, T_2-weighted and distorted, and T_2-weighted and undistorted, it should prove possible to calculate maps of the local susceptibility field profiles. These could provide an effective indication of structural morphology. A summary of NMR imaging applications in polymeric and mineral materials is given in Table 5.7.

Finally, it should be noted that NMR microscopy can prove particularly useful in studying viscous liquids. In particular the method lends itself to profiling slowly developing droplets or aqueous foams,[121] and the measurement of molecular properties during flow in specific geometries.[122,123]

5.3 Contrast techniques in imaging

The process of image reconstruction using **k**-space magnetic field gradients is a relatively recent development in the long history of nuclear magnetic resonance. Indeed the subject of NMR imaging represents, even now, but a fraction of the general field of NMR spectroscopy. The diversity of this wider area of research arises from a variety of sources. First, the nuclear spin Hamiltonian contains several well-characterized terms which are sensitive to the chemical and physical environment of the nucleus. Specifically, these are the chemical shift, the spin–spin scalar coupling interaction, and the internuclear dipolar interaction. Second, the nuclear magnetization is sensitive to molecular dynamics. Spin relaxation gives a measure of the spectral density of this motion via the fluctuating terms in the Hamiltonian, while the magnitude of residual terms indicates motional anisotropy and local order. Finally, because the coherences generated in NMR experiments are relatively long lived, the method is ideally suited to time domain methods, so that a seemingly limitless combination of r.f. pulses of differing amplitudes and relative phases, may be used to manipulate the spin system. These methods include contrast schemes designed to elucidate molecular translational motion. This particular contrast scheme is dealt with in detail in Chapter 8.

Table 5.7. *Summary of applications of NMR microscopy in materials science. Note that these applications involve the use of conventional, as opposed to solid state line-narrowing, imaging sequences.*

sample	molecule	nucleus/ frequency	pixel (μm)/ slice (μm)	contrast	reference
polymer composites	water	^{1}H 35 MHz	500 15000		104
sandstone	water	^{1}H 35 MHz	500 60000	T_2	116
polystyrene rod	toluene	^{1}H 20 MHz	200 2000	T_2 real-time evolution	105
Polymer rods	polypropylene/ polymethylmethacrylate	^{1}H 200 MHz	100	T_2 real-time evolution	106
polymer beads	water	^{1}H 300 MHz	18 250	T_2	63
charcoal	benzene	^{1}H 60 MHz	300	real-time evolution	107
charcoal	AsF_5	^{19}F 56 MHz	300	real-time evolution	107

capillary tube	water	^{1}H 60 MHz	20 1000	velocity/ diffusion	122
Al_2O_3 slip	water	^{1}H 42 MHz	250 8000	real-time evolution	100
polymethylmethacrylate	chloroform/ PMMA	^{1}H 85 MHz	600 2000	real-time evolution	115
polymethylmethacrylate	methanol/ PMMA	^{1}H 300 MHz	140 700	real-time evolution	111
rubber/ polybutadiene	polymer	^{1}H 200 MHz	400 2000	T_2	45
polyethylene oxide	water/ polymer	^{1}H 60 MHz	20 1000	T_2 velocity/diffusion	123

This panoply of complexity can be grafted on to the spatial localization process in the manner indicated in Fig. 5.20[124] in which imaging is viewed as the final phase of the experiment. In the precursor phase the nuclear transverse magnetization is 'preconditioned' by a pulse sequence designed to impart a contrast to the image which is then sensitive to some chosen property of the spin system. A simple example of preconditioning might involve a simple time delay after excitation, so that signals from those spins with longer T_2^* relaxation times will be relatively enhanced. Alternatively, the contrast phase may involve the selection of spins associated with a specific chemical environmental, for example, by selective excitation at a specific chemical shift region of the NMR spectrum.

In our previous discussion of imaging, the role of contrast was never far from the surface. The image signal intensity is always subject to T_2 or T_2^* relaxation because of the finite time delay from the initial excitation pulse to the particular acquisition time which represents the origin of **k**-space, and the degree of image attenuation depending on the relative magnitudes of T_E and T_2. Similarly, T_1 relaxation will impose contrast if the pulse repetition time is short, with T_1 weighting depending on the ratio of T_R to T_1. A nice example of this effect is shown in Fig. 5.12 where adjustment of T_E and T_R respectively, provided T_1 and T_2 weighting in images of the multicellular tumour spheroids.

In this section, however, we explore the possibility of using contrast in a quantitative or precisely determined manner in order to obtain maps of specific molecular species or of specific NMR parameters.

5.3.1 *T_1 and T_2 contrast*

In Section 2.6.1 we outlined the derivation of the equilibrium transverse magnetization for a repetitive single r.f. pulse sequence. The area under the NMR spectrum is determined by the signal amplitude at the origin of sampling. In **k**-space imaging the area in one pixel (i.e. the pixel amplitude) is determined by the signal amplitude at the origin of **k**-space for spins at the location corresponding to that pixel. Because the gradient and r.f. echo

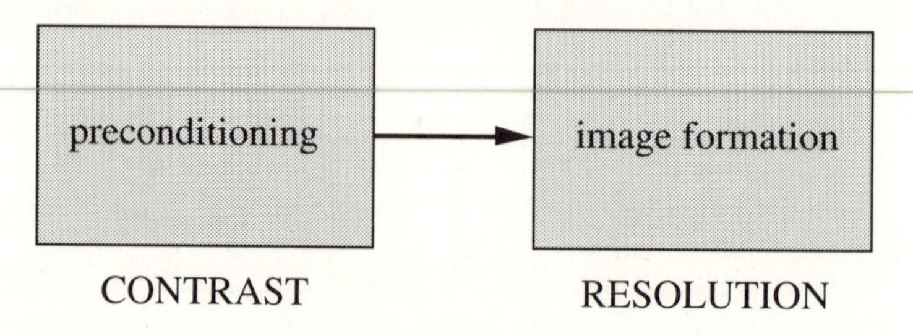

Fig. 5.20 The two phases of an NMR imaging experiment (after Ernst[124]).

methods discussed in Chapter 3 displace the origin of **k**-space to a time T_E after the selective excitation, this simple relationship expressed by eqn. (2.8) is no longer valid. This effect can, however, be accounted for by an additional factor $\exp(-T_E/T_2)$ or $\exp(-T_E/T_2^*)$ for the spin and gradient echo methods, respectively. A further difference arises in the spin echo sequence (Fig. 3.17) where the 180° pulse inverts the residual z-magnetization. This effect is accounted for by inverting the sign of $\cos\theta$ in the denominator of eqn. (2.81). Assuming total transverse magnetization dephasing between cycles and using the symbol M_0 for the equilibrium slice magnetization, we can write, for the signal amplitude at the echo centre, the following expressions.

gradient echo

$$M_y = M_0 \frac{\{1 - \exp(-T_R/T_1)\}\exp(-T_E/T_2^*)}{1 - \cos\theta\exp(-T_R/T_1)} \sin\theta \tag{5.9}$$

spin echo

$$M_y = M_0 \frac{[1 - 2\exp\{-(T_R - T_E/2)/T_1\} + \exp(-T_R/T_1)]\exp(-T_E/T_2)}{1 + \cos\theta\exp(-T_R/T_1)} \sin\theta \tag{5.10}$$

Where transverse coherence is not entirely destroyed in the imaging cycle, decaying by a factor $f > 0$, it can be shown[2] that the spin echo sequence leads to a signal

$$M_y \approx M_0 \frac{\{1 - \exp(-T_R/T_1)\}\exp(-T_E/T_2^*)}{1 - f\exp(-T_R/T_1) + \cos\theta\{\exp(-T_R/T_1) + f\}}. \tag{5.11}$$

Generally, however, the dephasing of the tranverse magnetization by the read gradient results in f being close to zero, in which case eqn (5.11) reduces to the simpler result, eqn (5.10).

T_1 contrasting

The calculation of T_1 values across the image can be performed by obtained a sequence of images in which either θ or T_R is varied. Using either eqn. (5.9) or (5.10), and some appropriate fitting algorithm, T_1 values can be computed in each pixel and a T_1 map displayed. θ variation leads to a more complicated dependence of image intensity on T_1 and is conveniently and rapidly performed using the Partial Saturation sequence. This method is appropriate for providing qualitative contrast but is less suited to T_1 mapping, although recent work using snapshot FLASH[125,126] with a precursor inversion recovery pulse has proved effective.[127]

A simpler fitting procedure results when 90° pulses are used, and eqns (5.9) and (5.10) reduce to:

gradient echo

$$M_y = M_0\{1 - \exp(-T_R/T_1)\}\exp(-T_E/T_2^*) \tag{5.12}$$

spin echo

$$M_y = M_0[1 - 2\exp\{-(T_R - T_E/2)/T_1\} + \exp(-T_R/T_1)]\exp(-T_E/T_2) \tag{5.13}$$

Note that the gradient echo refocuses only those phase shifts which arise from the applied magnetic field gradient pulse, so that the resultant transverse relaxation depends on T_2^* rather than T_2 as in the case of the spin echo.

Eqns (5.12) and (5.13) represent single exponential recovery, in most cases a reasonable assumption for T_1 relaxation. The existence of the terms involving T_E complicates the fitting process although specific algorithms for this purpose are available.[128] The simplest procedure is to choose T_E sufficiently short that transverse relaxation effects can be neglected, the approach adopted by Johnson *et al.*[62] in obtaining T_1 profiles across a plant stem (*Pelargonium hortorum*). For $T_R \gg T_1$ the image signal amplitude is proportional to the number of protons. By obtaining signal amplitudes for various $T_R \lesssim T_1$ eqn (5.13) can be used to yield T_1 profiles. Fig. 5.21 shows both the M_0 and T_1 profiles across the stem for two different levels of transpiration. The correlation between T_1 and M_0 values is remarkable, indicating the well-known dependence of T_1 on water content in biological tissue, exemplified by eqn (5.6).

Because proton T_1 relaxation rates depend on the spectral density of dipolar fluctuations at the Larmor frequency, increasing the polarizing field strength generally leads to an increase in T_1, one consequence of which can be a reduction in contrast. However, this loss is often more than compensated for by the gain in sensitivity as B_0 increases. The comparative micro-imaging experiments of Johnson and co-workers[70] on rat kidneys at 85 and 300 MHz shows a remarkably improved image structure at the higher field despite some convergence in T_1 values, the improved spatial resolution revealing microstructural variations in proton density which had been previously averaged at lower field.

T_2 contrasting

By keeping T_R constant and obtaining a sequence of images with varying T_E, a map of T_2 values can be obtained. An example of such an applica-

Fig. 5.21 (a) Plot of M_0 (relative hydrogen density) across the stem of *Pelargonium hortorum*) at two different levels of transpiration. Areas labelled are a: pith parenchyma; b: vascular tissue; c: fibrous sheath; d: cortical parenchyma; and e: epidermal tissue. (b) as in (a) but showing the variation in T_1 across the stem. (From Johnson *et al.*[62])

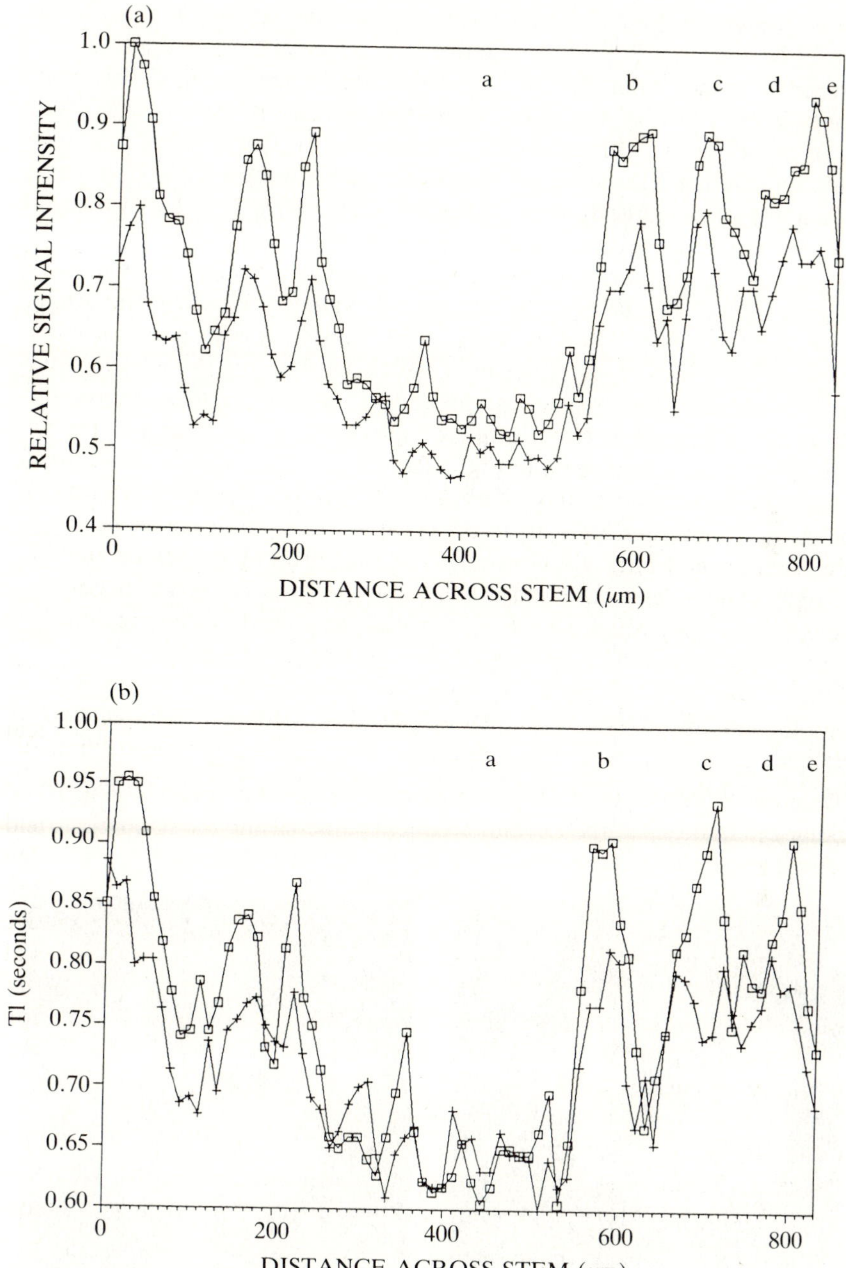

(a)
RELATIVE SIGNAL INTENSITY
1.0
0.9
0.8
0.7
0.6
0.5
0.4
a
b
c
d
e
0
200
400
600
800
DISTANCE ACROSS STEM (μm)
(b)
Tl (seconds)
1.00
0.95
0.90
0.85
0.80
0.75
0.70
0.65
0.60
a
b
c
d
e
0
200
400
600
800
DISTANCE ACROSS STEM (μm)

tion is shown in Fig. 5.16. Because of the spin–lattice recovery term, $\exp\{-(T_R - T_E/2)/T_1\}$, in eqn (5.13), it is important to ensure that T_R is sufficiently larger than T_E for the dominant dependence of the image amplitude on T_E to be due to the spin–spin relaxation, $\exp(-T_E/T_2)$. One way of avoiding the influence of T_1 relaxation contrast in the T_2 distribution is to generate a sequence of images in a multi-echo CPMG sequence. Of course such a method, while efficient, allows only T_E values which are multiples of the acquisition time for each echo.

Weighted least squares fitting to the exponential decay of the CPMG sequence images can be used to compute T_2 values for each pixel. Again, the fits yield not only T_2 values but also M_0 values for each pixel, provided T_1 contrast is removed by choosing T_R sufficiently long. As an additional benefit, therefore, the T_2 mapping procedure leads to a correct representation of the spin density. An example of spin density and T_2 maps obtained using this procedure is shown in Fig. 5.22[116] Fig. 5.22(a) shows an M_0 image of water in Berea sandstone while Fig. 5.22(b) shows the corresponding T_2 plot. At first sight there appears to be a simple correspondence between high water content and long T_2. However, the thin bedding planes on the left- and right-hand sides of the image exhibit long T_2 but lower water content. One possible explanation is that the porosity is reduced in these regions but the pore diameters may be larger or the clay content higher.[129-131]

Single exponential fitting to T_2 relaxation must be used with caution. In many systems multi-exponentiality is a feature. Provided enough data is obtained, however, it is quite feasible to separate different exponentials[132,133] and hence the three M_0 maps associated with each component in

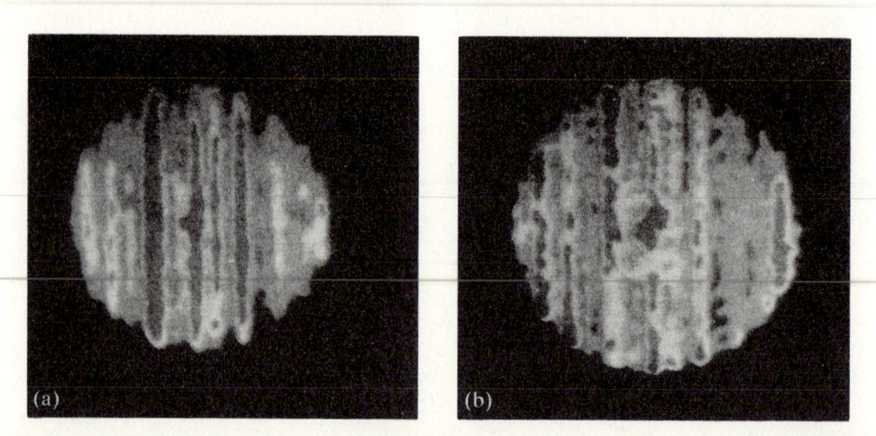

Fig. 5.22 (a) M_0 image and (b) T_2 image of Berea sandstone. These maps were obtained by exponential decay analysis for each pixel in a series of T_2-weighted images acquired successively in a CPMG train. (From Rothwell and Vinegar.[116])

the relaxation. A beautiful example of this is shown in Fig. 5.23, where, in a one-dimensional imaging experiment on green cedar wood, separated images of earlywood tracheid lumen water, latewood tracheid and ray water, and cell wall water were obtained.[133] An interesting feature of this data is

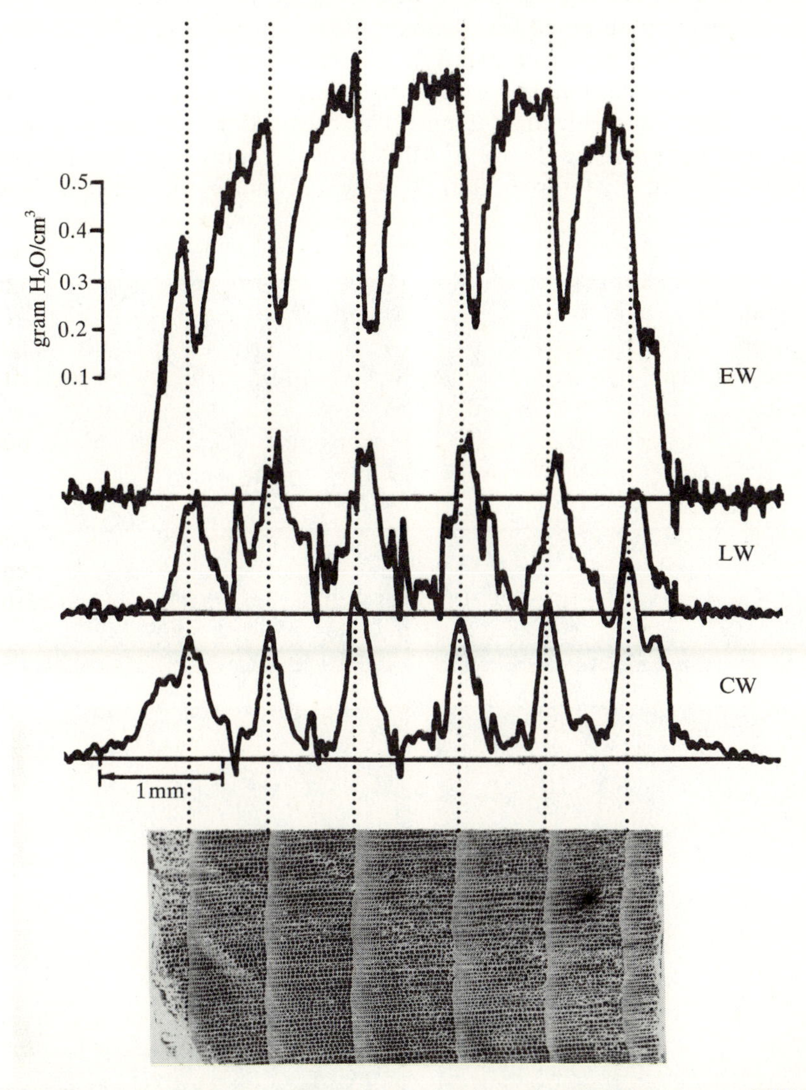

Fig. 5.23 Separated one-dimensional images of earlywood tracheid lumen water (EW), latewood tracheid and ray lumen water (LW), and cell water (CW) compared with the scanning electron micrograph from the same sample. The different water component images have been independently resolved by careful T_2 analysis. (From Menon *et al.*[133])

the close correspondence of the distribution of bound water with the growth rings apparent in the scanning electron microscope and the solid wood density as observed using X-ray densitometry.

A remarkable application of T_2 contrast has been recently reported by Armstrong *et al.*[134] in which chemical waves and their time evolution in the Belosov–Zhabotinsky reaction[135] were detected by NMR imaging. The particular reaction observed was the oxidation of malonic acid by bromate ions in which Mn^{2+} acts as a catalyst. The sample consisted of a mixture of bromate, bromide, malonic acid, Mn^{2+}, sulphuric acid, and agar, with diameter 40 mm and thickness 2 mm. The chemical waves occur spontaneously and represent the change of Mn^{2+} to Mn^{3+}, thus leading to a strong T_2 weighting. Fig. 5.24 shows four T_2-weighted images obtained using a

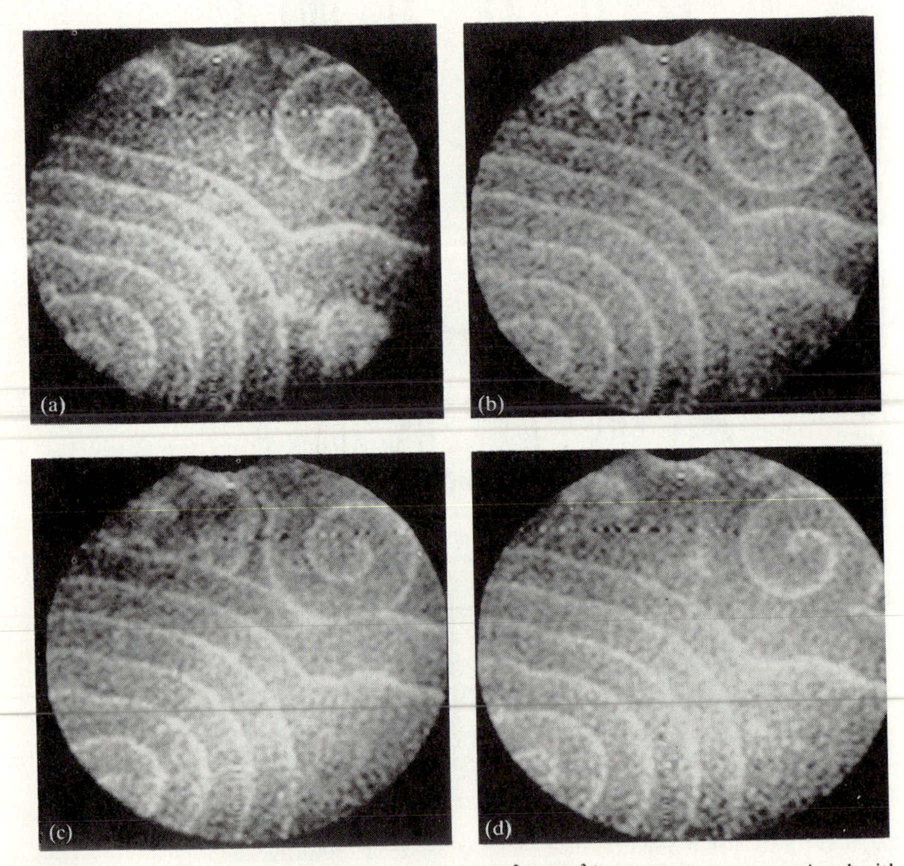

Fig. 5.24 T_2-weighted proton images showing Mn^{2+}/Mn^{3+} chemical waves associated with the Belosov–Zhabotinsky reaction. Times are separated by one minute with the top left image corresponding to the time origin. (From Armstrong *et al.*[134])

spin echo pulse sequence with $T_E = 15$ ms at times separated by 1 minute. Complete image acquisition took 12.8 s (128 phase-encoding steps with $T_R = 100$ ms). Chemical wavefronts are clearly apparent, emanating from a point of origin at the bottom left. The wavefront velocity is approximately 2 mm min^{-1}. A spiral wave can be seen at the top right and interference effects can be seen as the waves collide. This example graphically illustrates the potential use of NMR imaging in the study of reaction kinetics.

One potentially very interesting application of T_2 contrast concerns the study of water and cryofluid freezing in which the fluid phase changes cause a dramatic signal loss due to the significant reduction in transverse relaxation. Preliminary work using NMR microscopy to map heterogeneous freezing processes in biological tissue and in soils and composite materials has been reported by Fyfe.[136]

$T_{1\rho}$ contrasting

While T_2 contrast provides a high degree of sensitivity to slow motions its multi-exponential relaxation makes for complexity in analysis. As an alternative $T_{1\rho}$ relaxation is similarly dependent on the low-frequency spectrum of molecular motion but is generally single exponential in character. Rommel and Kimmich[137] have applied $T_{1\rho}$ contrast methods in the imaging of food products and have demonstrated that the method is especially useful in highlighting water molecules which are in slow exchange with slowly tumbling macromolecules.

Contrast enhancement

The introduction of substances which cause proton relaxation enhancement represents a useful mechanism for identifying structure. Where the contrast enhancing molecular species binds to or is concentrated in a specific region of a sample because of known chemical affinity or because of sample compartmentation, the method is particularly effective. Contrast agents usually contain paramagnetic electrons, examples being in stable free radicals such as nitroxides or transition metal/lanthanide cations.[138,139] As indicated in Section 5.1.2, the much larger electron magnetic moment has the effect of enhancing the local magnetic fields which fluctuate (due to thermal motion) in the vicinity of a proton. Table 5.8 lists common paramagnetic substances with unbalanced electron spins.[138]

There have been few reported uses of paramagnetic contrasts agents in NMR microscopy although their potential is considerable. However, in medical imaging, chelated paramagnetic species with tissue targeting functions have been used extensively.[140-144] By comparison with bare cations, the slower rotational correlation times for these larger complexes assist the relaxation process, partially compensating for the reduction in dipolar field at the proton resulting from the larger electron–proton distances.

Table 5.8. *Relaxation contrast agents commonly used in NMR imaging.*

Cations containing unpaired electrons:	
Transition metals:	
Mn^{2+}, Mn^{3+}, Fe^{2+}, Fe^{3+}, Ni^{2+}, Cr^{2+}, Cu^{2+}	
Lanthanides:	
Gd^{3+}, Eu^{3+}	
Paramagnetic molecules	
Nitric Oxide (NO)	-single unpaired electron
Nitrogen Dioxide (NO_2)	-single unpaired electron
Molecular Oxygen (O_2)	-two unpaired electrons but with parallel spins
Spin labels:	
Nitroxides	-stable free radical

Furthermore some complexes allow water molecules access to the paramagnetic core. For example the paramagnetic relaxation effect of Gadolinium-diethylene-triaminepenta-acetic acid (Gd-DTPA), one of the best known medical contrast agents, is nearly half that of the free Gd^{3+} ion.[139]

5.3.2 *Chemical shift contrast*

The description of image reconstruction given in Chapter 3 is based on the premise that the Larmor frequency of a nuclear spin is simply related to its location in the sample because of the imposition of magnetic field gradients. This is precisely true only when the NMR spectrum, in the absence of an applied gradient, consists of a single infinitely narrow line. T_2 relaxation leads to a finite linewidth and, because we generally wish to confine this width to less than one pixel spacing in the image, a minimum value for the bandwidth and imaging gradient is implied. Where more than one chemical environment is available for spins contributing significantly to the signal, the chemical shift leads to differing Larmor frequencies. If the imaging gradients are sufficiently small that this separation is larger than one pixel, the simple correspondence between resonant frequency and nuclear position no longer applies and image artefacts appear. In the simplest case of a pair of resonances, the chemical shift leads to two displaced images for the two chemical species of spin. A graphic example of this effect is apparent in the clove stem image obtained by J. M. Pope and shown in. Fig. 5.25.[64] Fig. 5.25(b) shows the chemical shift spectrum associated with the essential oils in the clove. As a consequence of this spectral complexity the upper image in Fig. 5.25(a), which is not chemical shift selective exhibits four superimposed oil images chemically shifted from each other. The lower image was obtained by exciting only the largest oil peak using the pulse sequence shown in Fig. 5.27

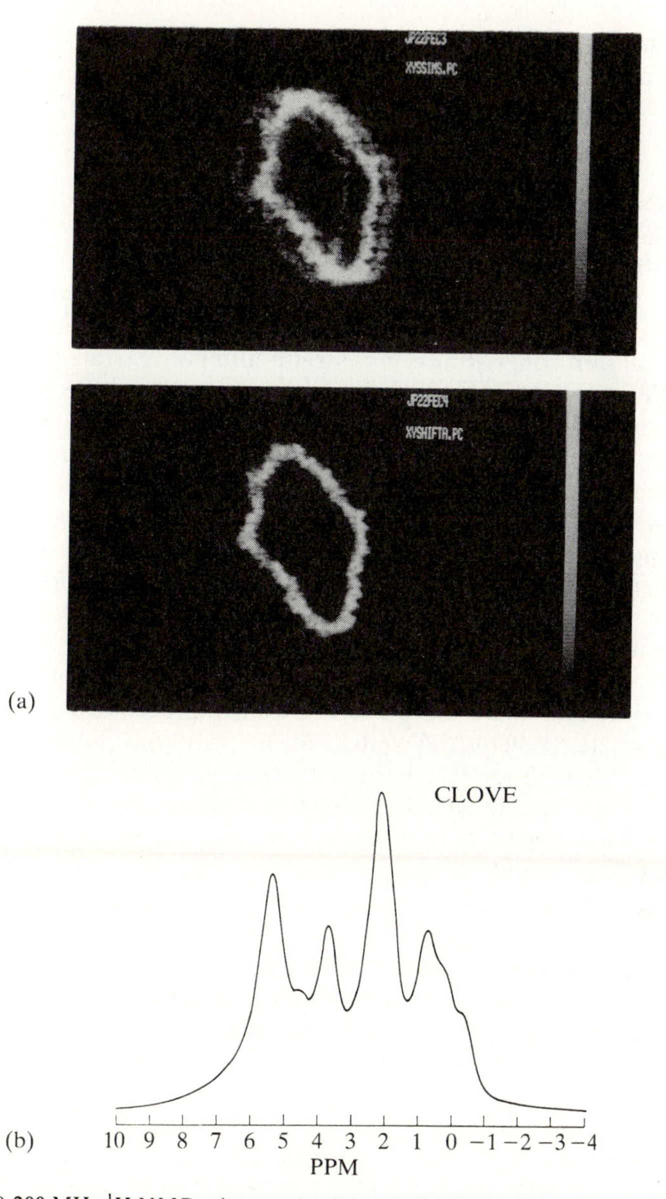

Fig. 5.25 (a) 200 MHz ^{1}H NMR micrograph of the oil distribution in a clove stem without chemical shift selectivity (upper image) and using a chemical shift selective excitation (lower image). The slice thickness is 1 mm and the field of view is 6.8 mm with a pixel dimension of $(26\ \mu m)^2$. The image acquisition times were 15 minutes and 120 minutes, respectively. (From J. M. Pope personal communication.[64]) (b) The ^{1}H NMR spectrum from the clove oils.

The removal of chemical shift artefacts can be achieved by further increasing the imaging gradients in order to compress the frequency separation within a single pixel. The use of larger gradients implies larger acquisition bandwidths and hence reduced read times. This enables image signal-to-noise, and hence pixel resolution, to be restored by multiple echo co-addition, as discussed in Section 3.5.5. Alternatively, the spectrum can be deconvoluted so as to remove the chemical shift effect.[145, 146] This deconvolution filtering process is performed most easily as a division in Fourier space by normalizing the time domain signal obtained during the read gradient to the time domain signal obtained under zero gradient conditions.

Chemical shift contrasting represents the alternative strategy. Rather than submerging the chemical shift it is possible to use it to selectively image a specific species of molecule. One very simple approach which is possible when the chemical shift is very large, is to employ imaging gradients sufficiently weak that the different spectral lines do not overlap.[147–149] Unfortunately, except in the case of nuclei such as ^{13}C, where the lines are narrow and well-separated, gradients this small will invariably result in the inhomogeneous or T_2 linewidth exceeding the pixel separation. A 'cleaner' approach to contrasting is to use pulse sequences which deliberately separate the imaging and spectral dimensions.

The most whole-hearted approach to chemical shift contrasting is to increase the dimensionality of the acquisition process, so that frequency spread associated with the zero-gradient spectrum becomes a fourth dimension[150] in addition to the three spatial dimensions of imaging. This is performed by spatially encoding the NMR signal prior to reading in the absence of a magnetic field gradient a method known as Chemical Shift Imaging (CSI). A pulse sequence which achieves this result using a slice-selective r.f. pulse followed by two phase-encoding gradients is shown in Fig. 5.26a.

The disadvantage of this higher-dimensional strategy is twofold. First, the signal is acquired soon after the switching of magnetic field gradient pulses associated with the phase encoding. The residual eddy gradients associated with these are slow to decay entirely so that the magnetic field will not have recovered to the uniformity associated with steady-state conditions, an effect

Fig. 5.26 (a) Four-dimensional (CSI) imaging scheme in which the three dimensions of spatial encoding are imposed before acquiring the signal in the absence of a magnetic field gradient. Each NMR spectrum so obtained corresponds to a single point in *k*-space. The 180° r.f. pulse is needed to refocuss the chemical shift induced dephasing which arises during the phase-encoding period. (b) Prior chemical shift encoding in which the spin system is first allowed to evolve for a period τ, the resulting magnetization then being stored along the *z*-axis so that the differences in Larmor precession due to chemical shift differences result in an amplitude modulation. This modulated *z*-magnetization is then recalled using a conventional imaging pulse sequence.

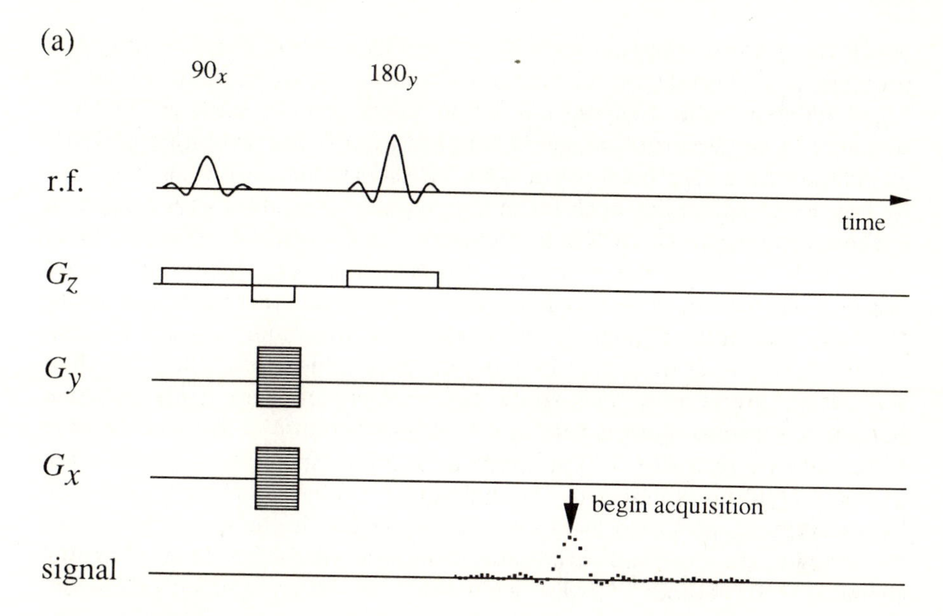
(a)
90x
180y
r.f.
time
Gz
Gy
Gx
begin acquisition
signal

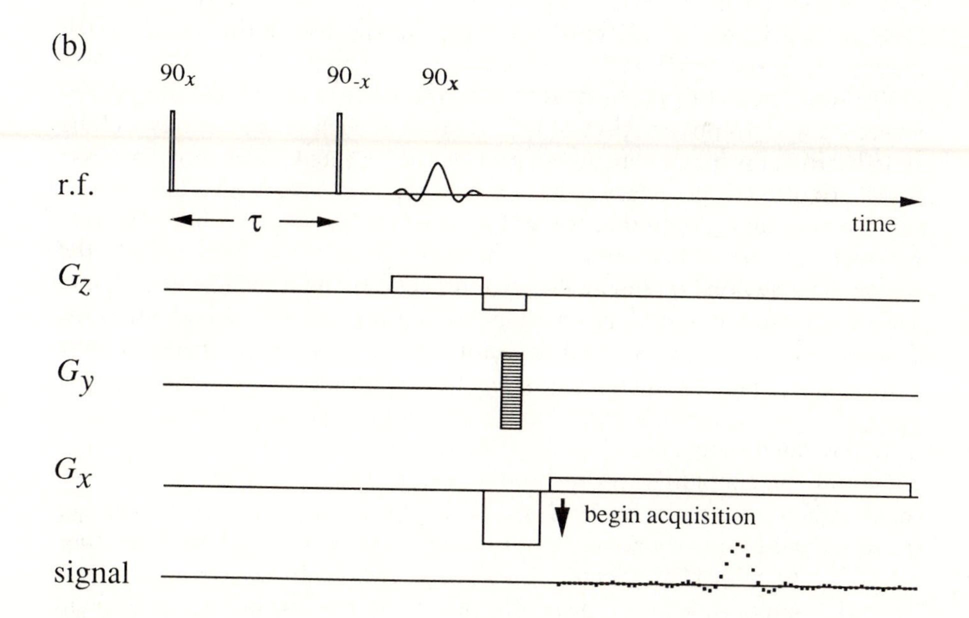
(b)
90x
90-x
90x
r.f.
τ
time
Gz
Gy
Gx
begin acquisition
signal

which can lead to spectral distortion. One nice method[151] of avoiding this problem is to encode the chemical shift information first, as shown in Fig. 5.26(b), by using an evolution period based on a two-pulse HCORR[152] sequence. A succession of images is obtained for different evolution periods, τ, and the chemical shift spectrum at each pixel in the image is then recovered by Fourier transforming each set of single pixel amplitudes with respect to τ. From the perspective of NMR microscopy, however, the second problem associated with four-dimensional imaging is more serious. It consumes more imaging time and so must sacrifice signal-to-noise and therefore resolution. Consider the pulse sequences for spatial encoding and spectral reading shown in Fig. 5.26(a). Unlike imaging in three dimensions where the read acquisition covers a line in **k**-space, the read process here yields only one point in **k**-space so that the total imaging time for an $N \times N$ pixel image is increased by a factor of N. The spectral encoding/spatial reading sequence of Fig. 5.26(b) takes longer by the number of τ values employed and hence the total time is governed by the resolution desired in the spectral domain. More rapid data acquisition is possible using combined echo planar imaging/chemical shift methods [153-155] but while these have been used with considerable success in medical imaging with large voxel sizes, the wide bandwidths and lower signal-to-noise ratios associated with EPI methods mean that these are not well suited to microscopy. Unless one of the imaging dimensions is sacrificed[156-158] the addition of a full spectral dimension to each pixel of the image is too costly in time or signal-to-noise ratios to be reasonably applied in NMR microscopy.

The solution to this problem is to adopt a selective approach. Generally, where we wish to obtain spectral information at each pixel, we may well be satisfied to know that we are imaging a specific, selected molecule of interest. It will not often be necessary to have a full spectrum at each position. Where full spectral data is required, we seldom need this at all pixels in the image, in which case we will be satisfied to achieve this result in some preselected region. The obvious spatial analogy to this reduced data set approach is our use of preselected slices rather than attempting to fully image all three dimensions simultaneously. These conjugate approaches to chemical shift contrasting represent chemically selective excitation and volume-selective spectroscopy respectively. Both have been applied with considerable success in NMR microscopy.

One of the simplest ways to obtain chemically selective excitation is to make either the 90° or 180° r.f. pulses in the r.f. echo imaging sequence narrow-band frequency-selective pulses, while using the other pulse for slice selection. As with slice selection pulses, the chemically selective pulse will have a low-amplitude long time profile in the time domain but will be applied at some chosen frequency in the NMR spectrum and in the absence of a magnetic field gradient. It will therefore excite only the nuclear spins whose chemical shift places them at this position in the frequency spectrum.

In practical terms the use of such pulses implies the facility to rapidly switch the oscillator frequency, a fairly common feature in most modern spectrometers.

The usual r.f. echo sequence, shown in Fig. 3.17, has a non-selective 180° pulse which at first sight might seem the obvious candidate for conversion to the chemical selection role. But such a change will produce partial selectivity only. So long as the initial 90° pulse excites all spins in the slice irrespective of chemical shift, the unwanted spins will make a contribution to the signal. The relative magnitudes of the 'selected' and 'non-selected' chemical species will be determined by the respective decays by T_2 and T_2^* relaxation. To avoid this problem either a CHESS stimulated echo sequence as shown in Fig. 3.20 may be used or, if a spin echo is preferred, the initial 90° excitation should be made chemical shift selective while the 180° pulse is slice selective as shown in Fig. 5.27. Note that the absence of the rephasing slice gradient after the 180° pulse ensures that there is no contribution to the image from spins of the desired chemical species but outside the chosen slice. An example of separate fat and water images obtained using this method is shown in Fig. 5.28.[64] Here the oil-bearing channels around the outside of the fennel seed are clearly separated from the water in the interior tissue. Chemical shift selective imaging of the very weak aromatic proton signal indicates that the aromatic oil anethole responsible for the fennel flavour reside exclusively in these channels. This remarkable result is shown in Fig. 5.28(c).

Because of the problems in obtaining good slice definition with 180°

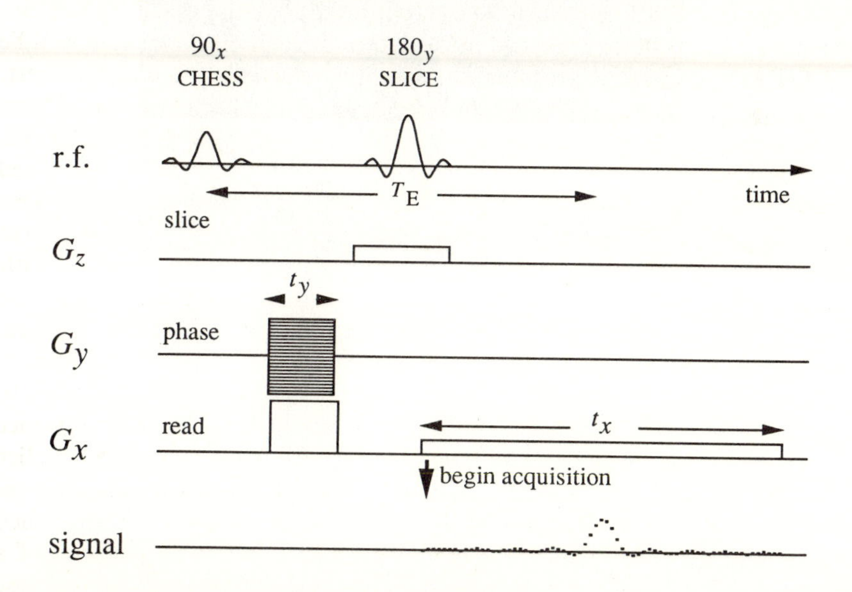

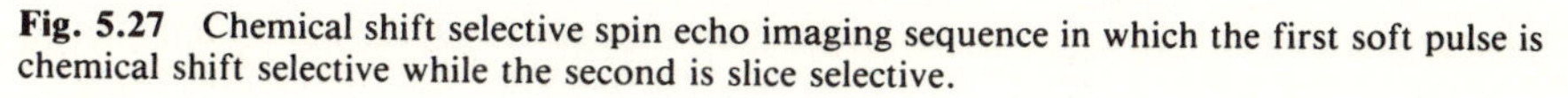

Fig. 5.27 Chemical shift selective spin echo imaging sequence in which the first soft pulse is chemical shift selective while the second is slice selective.

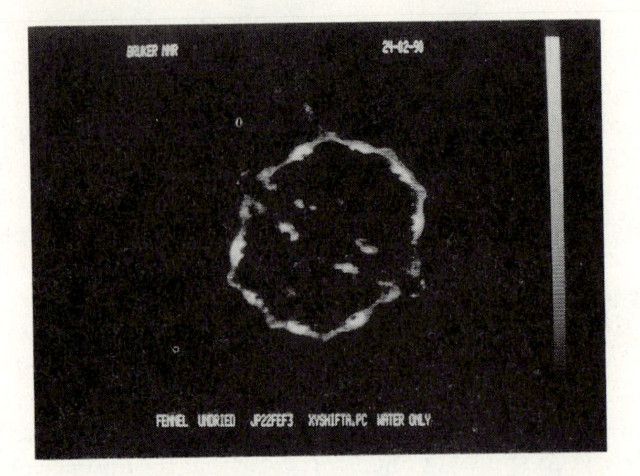
BRUKER NMR
24-02-90
FENNEL UNDRIED JP22FEF3 XYSHIFTN,PC WATER ONLY

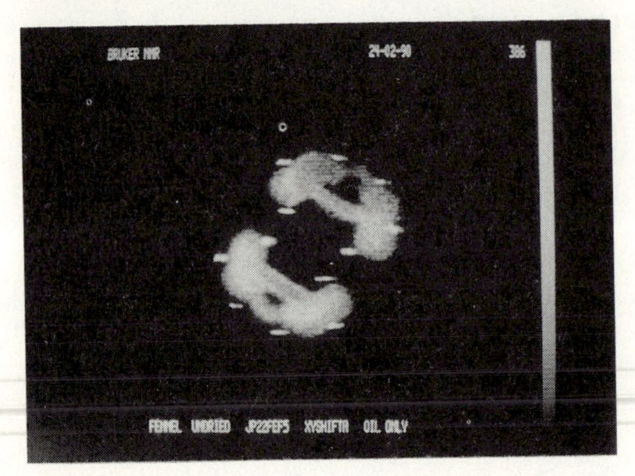
BRUKER NMR
24-02-90
386
FENNEL UNDRIED JP22FEF5 XYSHIFTN OIL ONLY

(a)

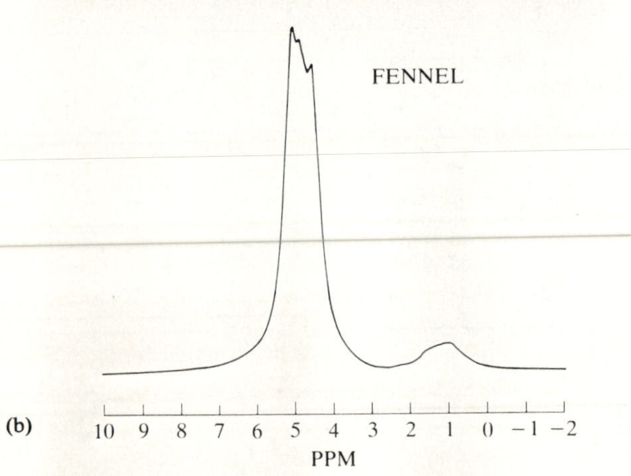
FENNEL
10 9 8 7 6 5 4 3 2 1 0 −1 −2
PPM

(b)

(c)

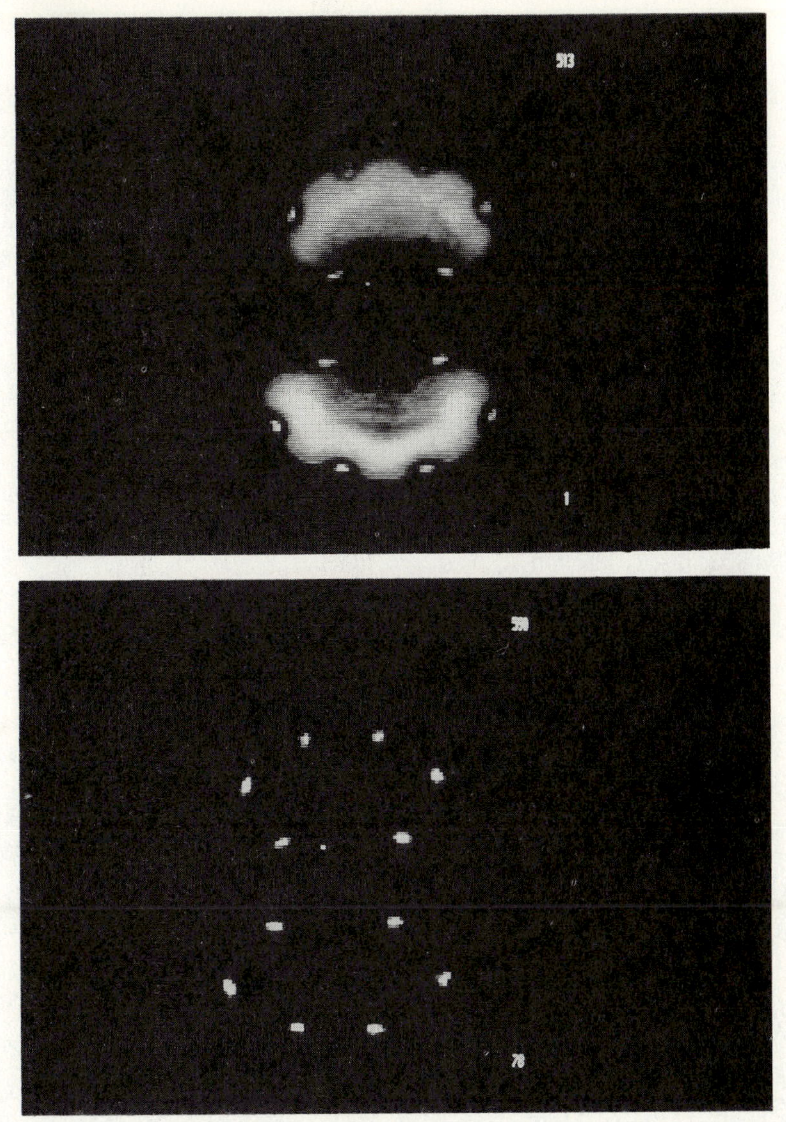

Fig. 5.28 (a) 200 MHz ^{1}H NMR micrograph of the water distribution (upper image) and the oil distribution (lower image) in a fennel seed. The slice thickness is 260 μm with a pixel resolution of (15 μm)^2. These images were obtained using the chemical shift selective imaging scheme shown in Fig. 5.27. Oil and water are separated by 3.5 p.p.m. corresponding to 700 Hz in this experiment. The bandwidth of the CHESS pulse was 500 Hz. The slice thickness is 1 mm and the field of view is 6.8 mm with a pixel dimension of (26 μm)^2. The image acquisition times were 8 minutes and 270 minutes, respectively, in the water and oil images. The oil canals around the outside of the seed are clearly visible. The slight blurring in these canals is due partly to the thick slice employed coupled with a small degree of sample tilt but also because of a chemical shift artefact. Close inspection of the oil canals reveals that each is a double image. (From Pope and Sarafis.[64]) (b) The ^{1}H NMR spectrum from the fennel, (from Pope and Sarafis.[64]) (c) 200 MHz ^{1}H NMR micrograph of oil distribution (upper image) and anethole distribution (lower image) in a Fennel seed. The images were obtained using a CSI sequence with two phase encoding gradients similar to that shown in Fig. 5.26(a). (From J. M. Pope, H. Rumpel, and V. Sarafis, personal communication)

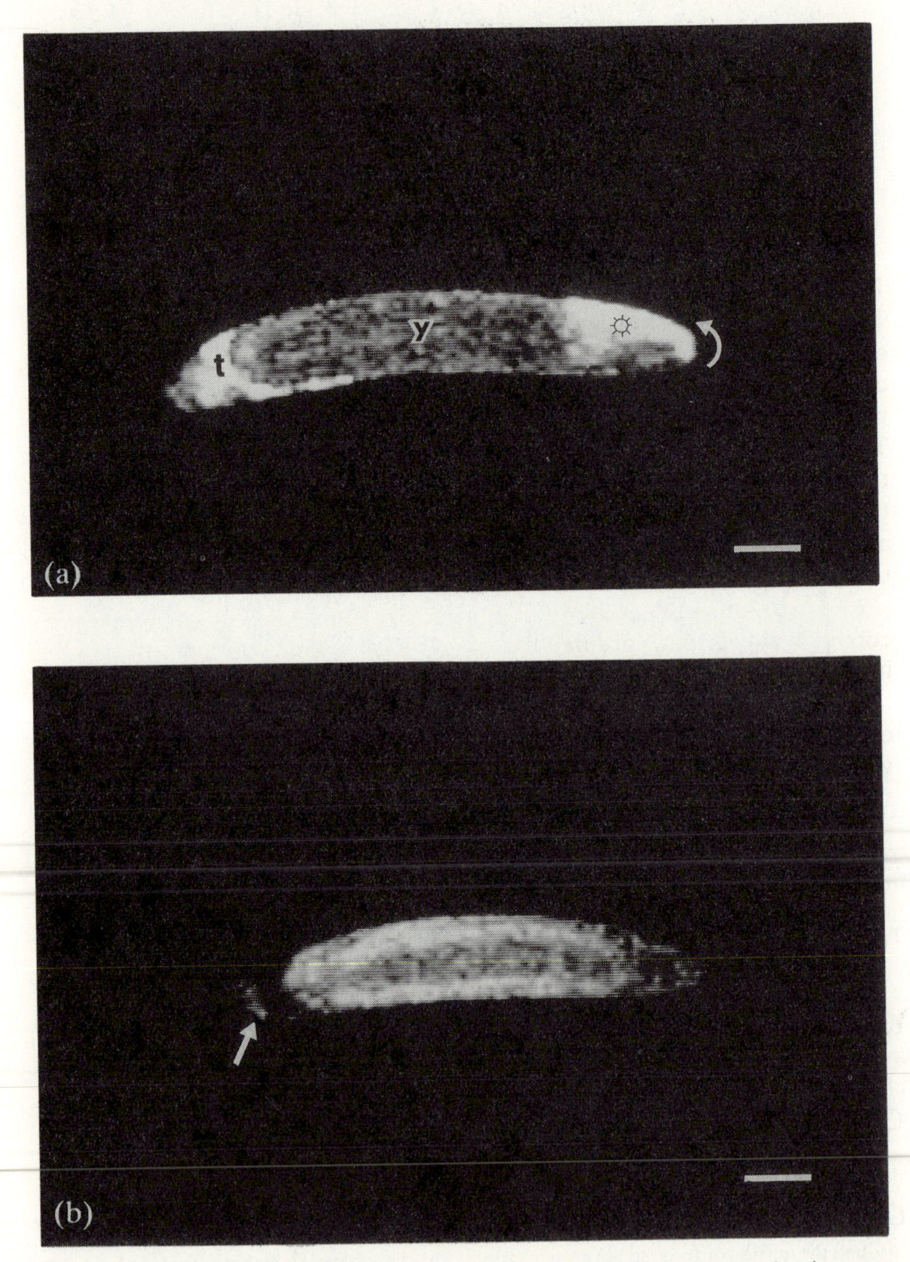

Fig. 5.29 Midsaggital slice through a locust embryo at 22% development showing water proton image. (a) and lipid proton image (b). Lipid protons are more intense around the periphery of the yolk than in the central core. The bar corresponds to 1 mm. The asterisk shows the embryo at completion of anatrepsis, t is the transverse boundary and y is the yolk. Note that the anterior compartment contains a small amount of lipid (arrow). (From Gassner and Lohman.[81])

pulses, a preferred alternative to this sequence is the CHESS variant of STEAM imaging shown in Fig. 3.20. The use of stimulated echoes involves 90° pulses only but also results in the sacrifice of a factor of 2 in the image signal-to-noise ratio. An interesting *in vivo* application of CHESS imaging is the mapping of water and lipid distributions in the developing locust embryo by Gassner and Lohman.[81,82] An example of the separated water and fat images is shown in Fig. 5.29.

Volume-selective spectroscopy

The use of volume-selective spectroscopy is widespread in medical imaging and a variety of methods exist for this purpose. These include the use of direct excitation of a specific voxel[158,159-161] as well as the localization of maxima in the r.f. field strength, the so-called 'depth-selective' pulses.[162-167] The latter methods produce volumes which are irregular in shape while the former require good r.f. homogeneity and can be adversely affected by residual gradients caused by prior gradient pulse switching. The SPACE and DIGGER schemes discussed in Section 3.2.5 use destruction of the extra-slice magnetization followed by a gradient settling period before slice excitation using a 90° r.f. pulse. Another method which also overcomes most of these problems, and which has been successfully applied in NMR microscopy, is ISIS[168] (or Image Selected *In vivo* Spectroscopy) and a modification of the method (OSIRIS) which involves the addition of selective noise pulses for the suppression of signal from the outer volume.[169]

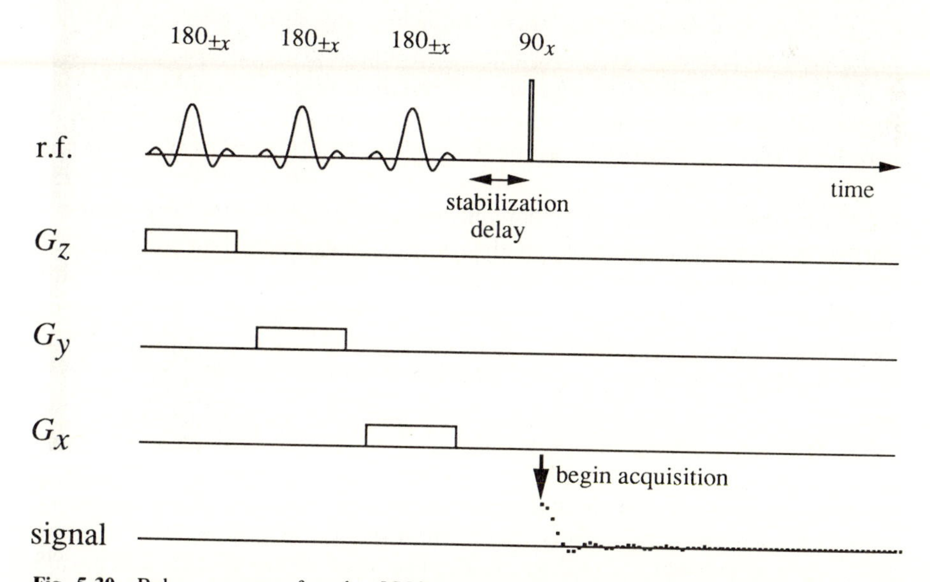

Fig. 5.30 Pulse sequence for the ISIS pulse experiment. Selective inversion of three orthogonal slices takes place during the preparation period. (After Ordidge *et al.*[168])

The pulse sequence for the ISIS method is shown in Fig. 5.30. It relies on three successive variable-frequency 180° selective inversions which define a cube of material with inverted magnetization. Because this prepared magnetization is 'stored' along the z-axis and is subject only to T_1 relaxation, a reasonably long delay can be used to allow gradients induced by the preparation pulses to decay. The experiment uses a specific pattern of phase cycling to achieve accurate volume selection with minimal dependence on r.f. field amplitude. It is therefore very tolerant of pulse turn angle errors. An additional advantage of the method is that the use of the ISIS pulse sequence as a precursor to a standard image sequence produces a distinct null in the image at the selected volume element, thus enabling the user to accurately observe this volume and shift it as desired by altering the spectrometer frequency offsets.

Fig. 5.31 shows the application of the ISIS method in the locust embryo study of Gassner and Lohman. The spectra obtained, respectively, in the stomatodeum and hemocoel, not only give information about relative water and lipid ratios, but, through the linewidths, also indicate the relative transverse relaxation times of the spins. Remarkably, these spectra were obtained in $(500\,\mu m)^3$ voxels!

The SPACE sequence of Doddrell *et al.* (Fig. 3.12) has been used to obtain high-resolution spectra from selected microvolumes in a rat brain.[170,171] Because of the use of a surface coil in this work, the selected volumes were considerably larger than those used in the ISIS study quoted previously. The authors report high-quality water-suppressed spectra from 0.2 ml voxels and estimate that this volume could be reduced to 0.05 ml with some improvements in gradient and r.f. transmitter hardware.

One intriguing development in the methodology of volume-selective spectroscopy has been the suggestion by Lauterbur[172,173] that spectra can be computed from any desired volume within the sample using post-processing of a set of total sample signals, provided that a sufficient number of such signals is obtained using different phase-encoding gradients. The method, known as SLIM for Spectral Localization by Image Modulation, works by solving a set of simultaneous linear equations in which the selected volume coordinates enter as a set of chosen parameters. Among the many advantages of such a method are its operational simplicity and the potential to define a volume shape which matches the region of interest.

5.3.3 *Multiple quantum filters*

In principle, any term in the Hamiltonian common to a particular group of spins can be used to selectively excite those nuclei for subsequent imaging. In the case of the scalar coupling interaction, one of the most effective ways of discriminating spins experiencing this interaction is via the coherence transfer processes of the multiple quantum filter. These filters are particularly effective in suppressing the signal from two of the major proton

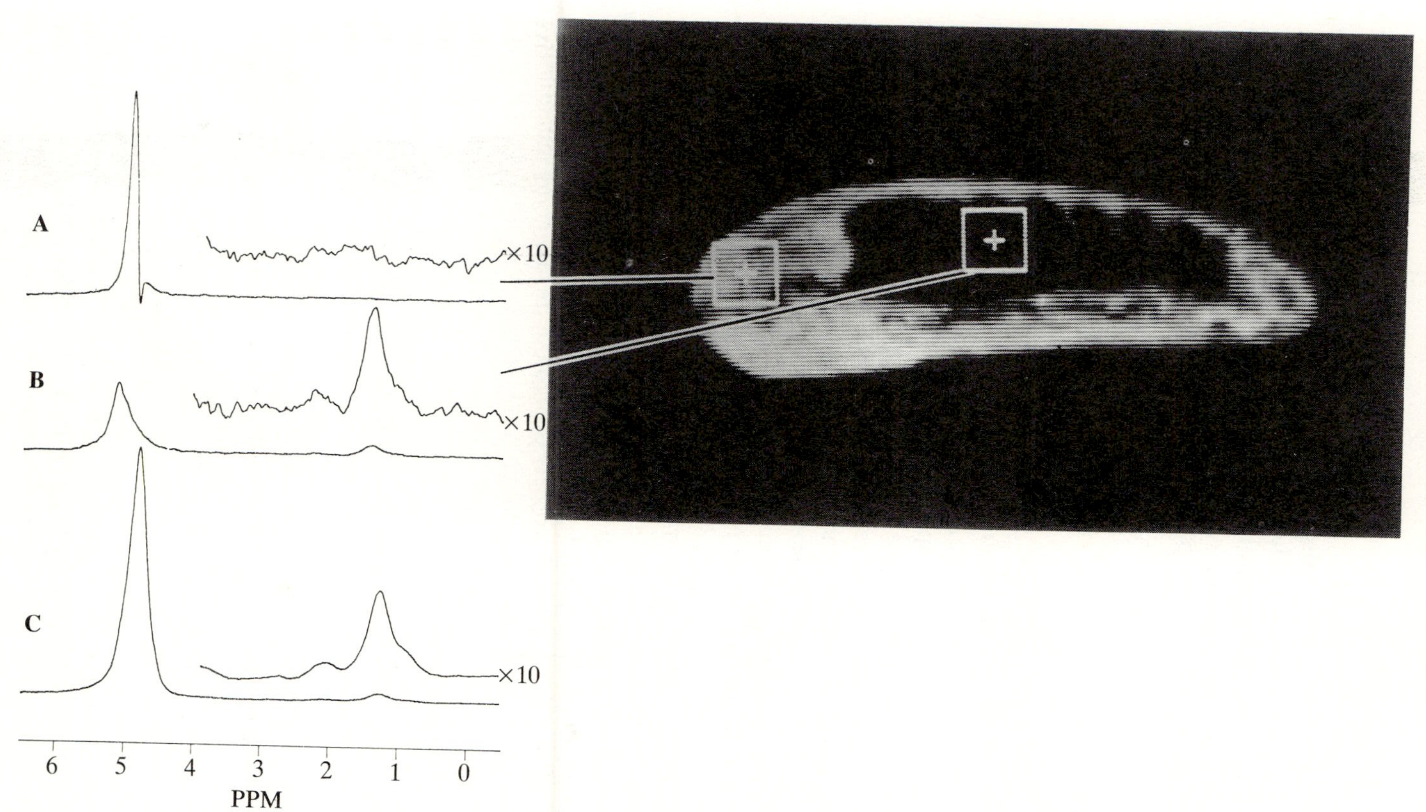

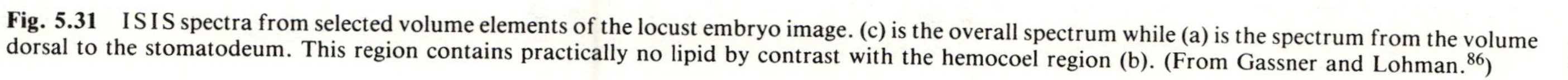

Fig. 5.31 ISIS spectra from selected volume elements of the locust embryo image. (c) is the overall spectrum while (a) is the spectrum from the volume dorsal to the stomatodeum. This region contains practically no lipid by contrast with the hemocoel region (b). (From Gassner and Lohman.[86])

groups in biological tissue, namely those in water and fat, where the identical chemical shift between coupled nuclei leads to a singlet resonance. Such methods have been used to obtain water- and fat-suppressed proton NMR signals from metabolites *in vivo*.[174–177]

Two simple multiple quantum filters are discussed in Section 2.6.7, one working via double quantum coherence (2QC) and the other through zero-quantum coherence (ZQC). Fig. 5.32 shows two pulse sequences in which each of these is incorporated as the initial excitation in an imaging sequence. The multiple quantum filter plays the same role as the chemically selective excitation in the CHESS–STEAM pulse train. Note that the point in time equivalent to the initial CHESS pulse is the instant when selectively excited 1QC is generated, as shown by the arrow in the diagram. The use of a homospoil pulse in conjunction with the ZQC filter leads to particularly effective filtering since all higher quantum coherences are destroyed. An example of a simple imaging application using this approach has been given by Hall and Norwood.[178] Because at least p-coupled spins are required to produce p-quantum coherence, the coherence transfer sequences can, in principle, be used to select signals from spins with a coupling multiplicity above any desired value. However, one disadvantage of multiple quantum filters is their low signal efficiency due to 'leakage pathways', a factor which is clearly disadvantageous in NMR microscopy. In this regard the ZQC filter, which has only half the final signal amplitude of the 2QC, is clearly handicapped. Another problem is the dependence of coherence transfer on the time of evolution under the scalar coupling since multiplets with differing J values will have differing amplitudes in the final single-quantum coherence. While this can be used to advantage in enhancing one particular multiplet, it will in general lead to difficulties in assigning quantitative significance to the image amplitude.

5.3.4 *Signal suppression*

One approach to molecular specificity in imaging was seen in the CHESS excitation where a narrow-band selective r.f. pulse was used to excite only those spins with a chosen chemical shift. This method is very effective provided that the particular proton resonance we wish to observe is a significant component in the total NMR spectrum. In biological systems this spectrum often contains an overwhelmingly dominant water proton peak, a feature which forms the basis of our judgement that, in most cases, we image the distribution of water in biological tissue. Selective excitation of the minority species of interest may not lead to sufficient enhancement with respect to the water. What is required is some additional water resonance suppression. Another circumstance in which signal suppression is called for is when we wish to observe all resonances in the spectrum except that of one

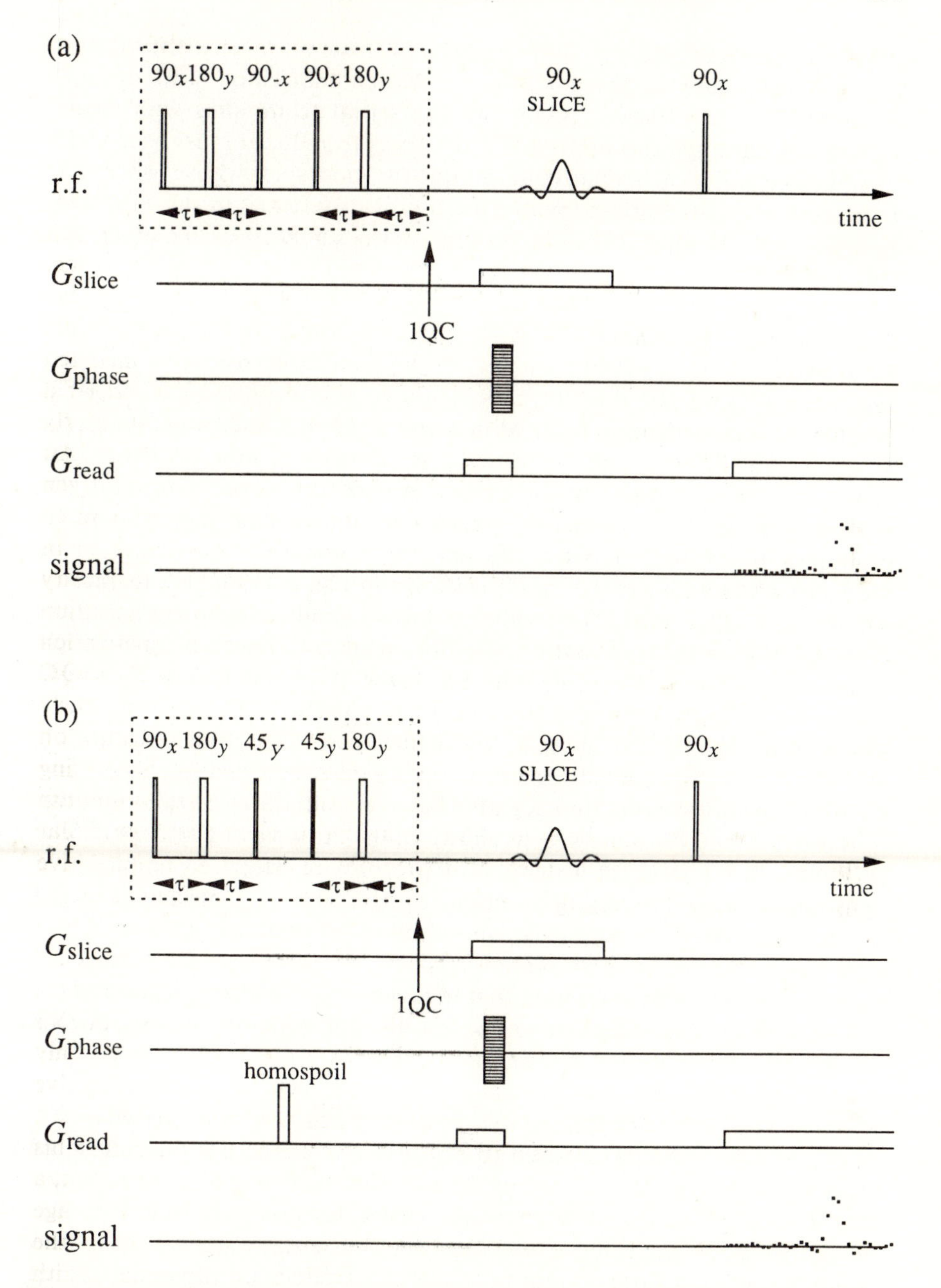

Fig. 5.32 Stimulated echo imaging sequences in which the initial 90_x pulse is replaced by a multiple quantum-selective excitation. The part of the pulse sequence surrounded by the box plays the same role as the CHESS pulse in Fig. 3.20. In (a) the image is double quantum filtered while in (b) it is zero-quantum filtered.

particular proton. Selective suppression represents an inverse approach to selective excitation in the imaging of chosen molecules. The multiple quantum filter is effective at suppressing signal from spins with singlet resonances although this approach often leads to reduced sensitivity and a complicated variation in relative amplitude for those spins whose coherences pass the filter. Alternatives involve the use of differences in the T_1 relaxation time and the chemical shift to suppress the signal from one molecular species of proton.

Inversion recovery nulling

In discussing the measurement of T_1 by inversion recovery methods (Section 2.6.2) we saw that the application of a 180° pulse to the thermal equilibrium magnetization resulted in a null in M_z at a later time, 0.693 T_1. If there is a difference in T_1 values for differing protons in the NMR spectrum, we can use a prior inversion recovery process to null the z-magnetization of a particular nucleus at the instant that the initial excitation is applied to begin the imaging sequence.[179] A simple pulse sequence which achieves this result is shown in Fig. 5.33(a). One feature of this method is that other spins, while not having a null in their magnetization at the point of imaging excitation, will not, in general, have a magnetization equal in amplitude to their equilibrium value, M_0, unless their T_1 is very much shorter (in which case their magnetization is $\approx M_0$) or very much longer than the nulled spin T_1 (in which case their magnetization is $\approx -M_0$). While this partial relaxation does lead to reduced signal-to-noise ratios the facility to generate a negative signal amplitude for certain spins can also provide a useful signature, as shown in the example of Fig. 5.34.[180] The difference in T_1 values for water and oil protons are effectively employed in contrasting. One interesting example of this use in materials imaging involves the selective imaging of water and oil in porous sandstone.[181]

A more efficient use of inversion recovery properties can be achieved if narrow-band r.f. pulses are used first to invert only the magnetization of the spins to be nulled, and then to excite only the spins of interest. Such a sequence is shown in Fig. 5.33b.

Frequency-selective destruction and frequency-selective non-excitation

The time delay of 0.693 T_1 used to produce the magnetization null in the inversion recovery method is sometimes sufficiently long to permit significant proton exchange between water molecules and, for example, NH groups in solute molecules. Furthermore, where the water signal is very large, the method suffers from severe phase artefacts in the remainder of the spectrum. Two elegant methods which avoid this problem and achieve suppression ratios of 1000:1 are the binomial selective non-excitation pulse sequence proposed by Hore[182,183] and the SUBMERGE selective

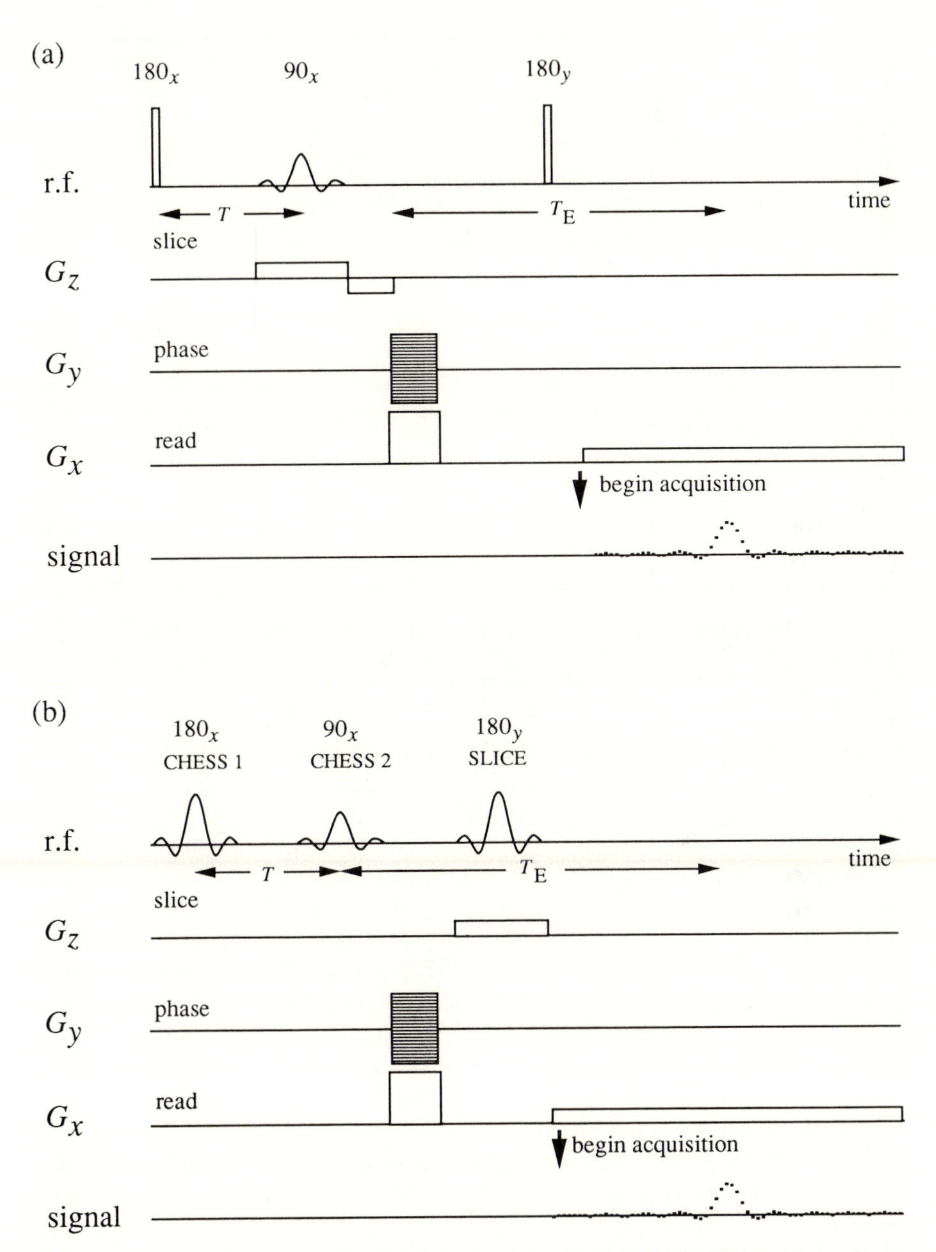

Fig. 5.33 Inversion recovery imaging pulse sequences using an initial 180_x inversion pulse to prepare the z-magnetization with a magnitude $M_0\{1-2\exp(-T/T_1)\}$ prior to excitation with a 90_x r.f. pulse. In (a) hard 180_x and 90_x pulses are used while in (b) they are shown as frequency selective with the first pulse being used to suppress the undesired region of chemical shift while the second is used to excite only those spins in the desired region.

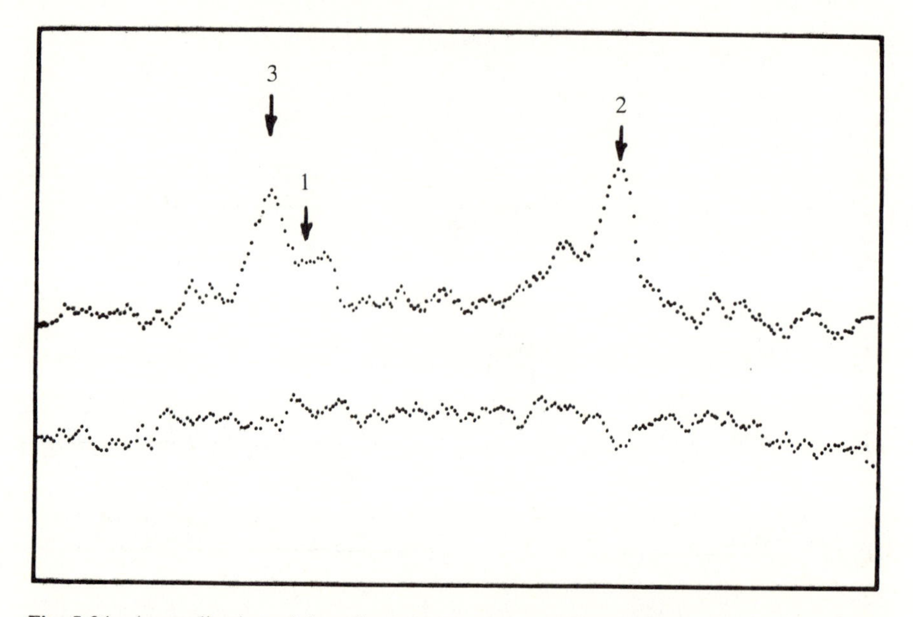

Fig. 5.34 An application of the pulse sequence shown in Fig. 5.33 (a) in which a line profile across an image of a rabbit trachea is shown. The upper trace shows the result without T_1 contrast while, in the lower trace, signal from the cartilage (2) and trachealis muscle (3) has been suppressed whereas the more slowly relaxing connective tissue signal (1) is still apparent. Note that a small amount of negative signal from 2 is apparent, consistent with a slight difference in relaxation time between the cartilage and muscle. (From Lambert *et al.*[180])

destruction method of Doddrell *et al.*[184] The binomial sequence produces a response in the region of the transmitter frequency which varies as $\sin^n(\Delta\omega\tau/2)$ where $\Delta\omega$ is the offset frequency and τ is the pulse spacing. This particular function has a flat null at $\Delta\omega = 0$. The most effective sequence is the $1\ \bar{3}\ 3\ \bar{1}$ train where the numerals refer to relative pulse length while the bar indicates the relative phase. Thus $1\ \bar{3}\ 3\ \bar{1}$ represents $\theta_x - \tau - 3\theta_{-x} - \tau - 3\theta_x - \tau - \theta_{-x}$. This sequence suffers from a 'phase jump' at the spectral null, a problem that can be corrected by an adjustment of the pulse spacings and durations.[185] The SUBMERGE method involves a successive excitation and dephasing of the unwanted resonance by the use of three Gaussians and three orthogonal gradients as shown in Fig. 5.35(a). Even higher levels of suppression are possible using combinations of these different methods.

An alternative approach, proposed by Blondet *et al.*,[186] uses spin-lock pulses to preserve the magnetization of some components while destroying that of the unwanted signal. This method has the advantage that signal suppression is independent of r.f. field strength, making the sequence ideal for surface coil applications where the r.f. field distribution is very

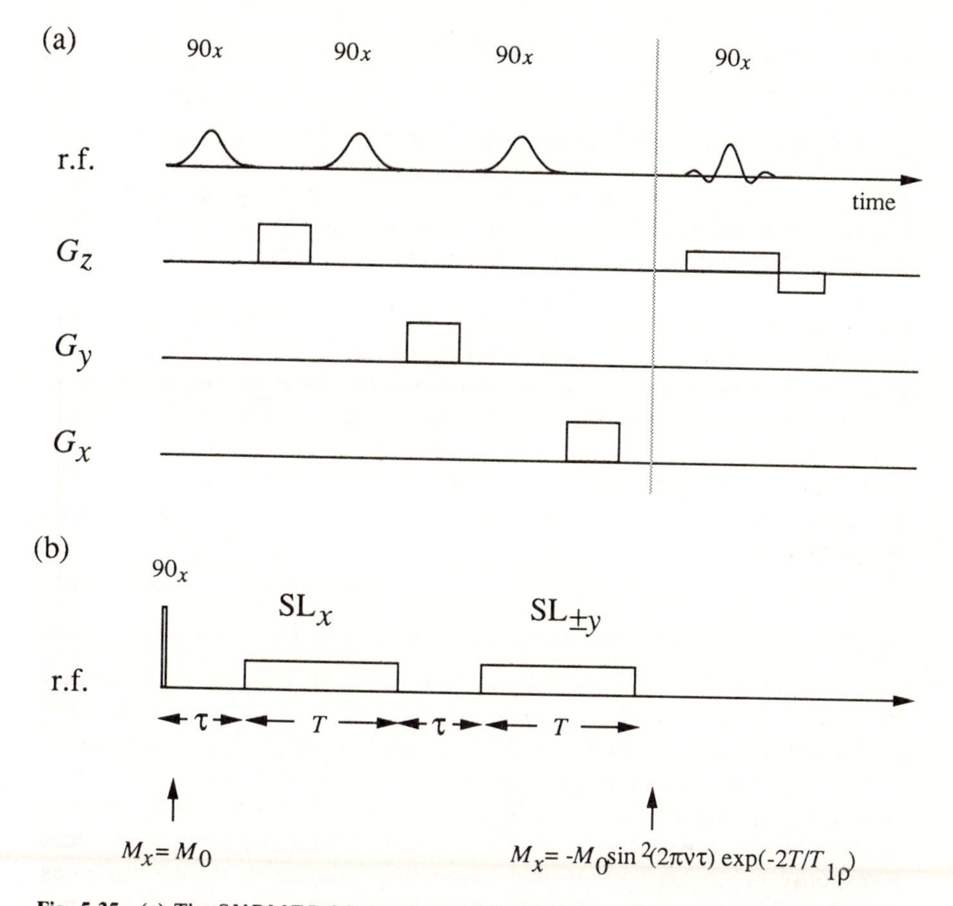

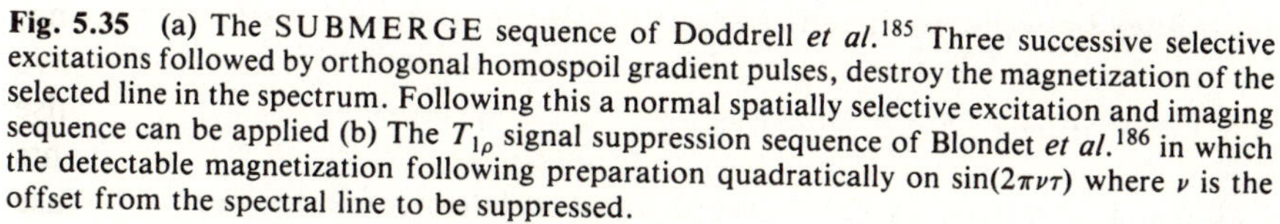

Fig. 5.35 (a) The SUBMERGE sequence of Doddrell *et al.*[185] Three successive selective excitations followed by orthogonal homospoil gradient pulses, destroy the magnetization of the selected line in the spectrum. Following this a normal spatially selective excitation and imaging sequence can be applied (b) The $T_{1\rho}$ signal suppression sequence of Blondet *et al.*[186] in which the detectable magnetization following preparation quadratically on $\sin(2\pi\nu\tau)$ where ν is the offset from the spectral line to be suppressed.

inhomogeneous. The spin-lock pulse sequence is shown in Fig. 5.35(b). After the first θ_x pulse the transverse magnetization is $M_x = 0$, $M_y = M_0 \sin \theta_x$. As a result of the evolution period τ, only the component which remains along the x-spin-lock pulse field direction is preserved leaving a magnitude $M_x = M_0 \sin \theta_x \sin \Delta\omega\tau \exp(-T_{SL}/T_{1\rho})$ at the end of SL_x. A second-order suppression is achieved by using a second evolution period τ and a subsequent spin-lock field directed along the y-axis in the rotating frame. The resulting magnetization is $M_x = -M_0 \sin \theta_x \sin^2 \Delta\omega\tau \exp(-2T_{SL}/T_{1\rho})$ where $\Delta\omega$ is the frequency offset from the signal component being suppressed.

Further improvement is obtained by adding a third SL_x pulse or a selective spin echo pulse.

5.3.5 *Magnetization transfer via molecular exchange*

Where molecules are in exchange between two environments *i* and *j* this can often be detected via an exchange of longitudinal magnetization. Two different chemical sites will generally be characterized by different NMR chemical shifts while differing physical environments will generally be characterized by different T_2 relaxation. As a result of exchange, a disturbance of the longitudinal magnetization of the *j* spins will produce a response in the *i* spins in excess of that expected on the basis of cross-relaxation via dipolar interactions. The usual method employed is to saturate the magnetization of the *j* spins prior to excitation of the *i* spins.

In the case of slow exchange following a steady-state saturation of M_j, the magnetization of M_i is reduced from its equilibrium value M_{i0} to a value M_{isat} given by[187]

$$\frac{M_{\mathrm{isat}}}{M_{\mathrm{i0}}} = (1 + k_{ij} T_{1i})^{-1} \tag{5.14}$$

where k_{ij} is the exchange rate constant and T_{1i} is the spin–lattice relaxation time of spin *i*. One method appropriate when spins *i* and *j* have different chemical shifts is termed Chemical Exchange Magnetic Resonance Imaging (CHEMI).[79] Here two images are acquired, one with and one without selective presaturation of the *j* spins. The two images are then subtracted to reveal where spin transfer has taken place. Another example of magnetization transfer contrast[188] has been developed to indicate the transfer of water molecules between environments with different local relaxation times.

5.4 Other nuclei

Table 4.1 summarizes the properties of other nuclei relevant to NMR imaging. In the very wide literature of this subject there are only a few examples in which nuclei other than protons have been used for imaging. Most notable among these are the nuclei whose natural abundance is close to 100% where imaging applications have been reported for ^{7}Li,[189,190] ^{14}N,[191] ^{19}F,[192-196] ^{23}Na,[197-206] ^{29}Si,[207] and ^{31}P.[79,150,208,209] In addition, some work has been reported on the imaging of isotopically enriched probe molecules for ^{2}H[210-212] and ^{13}C.[213-216]. In seeking candidates for NMR microscopy, the three crucial factors to be considered are: the relative sensitivity of the nucleus in comparison with the proton; the intrinsic linewidth due to T_2 relaxation; and the minimum repetition time which is allowed by spin–lattice relaxation. Apart from ^{19}F (relative sensitivity 0.833) and ^{7}Li (relative sensitivity 0.291) the potential signal-to-noise ratios for the remaining nuclei in Table 4.1 appear, at first sight, fairly dismal. Unfortunately, neither fluorine nor lithium are significant components in biological systems although they are present in some drug compounds,[190,196] and water soluble, non-toxic, fluorinated compounds are available which are suitable as markers in medical imaging.[195] Fluorine micro-imaging may prove useful in materials science in applications using fluorinated solvents or polymers.

The next most sensitive nuclei, ^{31}P and ^{23}Na, have only 10% of the proton NMR signal amplitude but both have high isotopic abundance and both are of considerable biological significance. Through the chemical shifts of inorganic phosphate resonances, ^{31}P NMR is able to give information about pH and temperature,[217,218] an effect which has been used to map these parameters using chemical shift resolved imaging.[209] The intensity of phosphocreatine (PCr) and adenosinetriphosphate (ATP) peaks gives an indication of metabolic activity. A ^{31}P study at near-microscopic resolution has been reported by McFarland *et al.*[79] who obtained a transverse resolution for ^{31}P spins of around 200 μm in a slice thickness of 1 mm. Given the concentrations of PCr (80 mM) and ATP (20 mM) in their sample this is a significant achievement. Fig. 5.36 shows the chemical shift selective ^{31}P image from the phosphocreatine peak in the mixture of PCr and ATP in two concentric tubes. In the central tube the enzyme creatine kinase is present. This enzyme catalyses slow phosphorous exchange between PCr and ATP and, in the lower part of the figure, the profile across the line *a* in the image shows the effect of a presaturation pulse applied selectively to the ATP ^{31}P resonance. While the spatial resolution is larger than the dimension of most single cells, the experiment demonstrates how NMR imaging can be used to characterize kinetic heterogeneity on an intercellular level.

There are few reports of NMR imaging performed at microscopic resolution with ^{23}Na. Moseley *et al.*[219] have obtained transaxial ^{23}Na images in a rat brain using a surface coil without slice selection and with a transverse resolution of around 200 μm. The brain sodium at a level of 25 mM gave little signal in the image, obtained over 10 minutes, but the high levels of sodium in the eyes (150 mM) were clearly visible. In another study Burstein and Mattingley have imaged nude mice tumours and a perfused frog heart to reveal the distribution of intracellular sodium.[205]

Because ^{23}Na is a quadrupolar nucleus its T_2 values are quite short, ranging from 1 to 50 ms in biological tissue[203,204] and generally with biexponential character.[206] While rapid T_2 relaxation is clearly disadvantageous in attempting to achieve high spatial resolution, this is compensated by shorter T_1 values which allow more rapid signal averaging. In Chapter 4 we saw that the transverse spatial resolution in a selectively excited slice varies as $(T_1/T_2)^{-\frac{1}{4}}$. Because T_1 values for ^{23}Na in biological tissues are typically in the range 30 to 50 ms and therefore of the same order of magnitude as T_2,

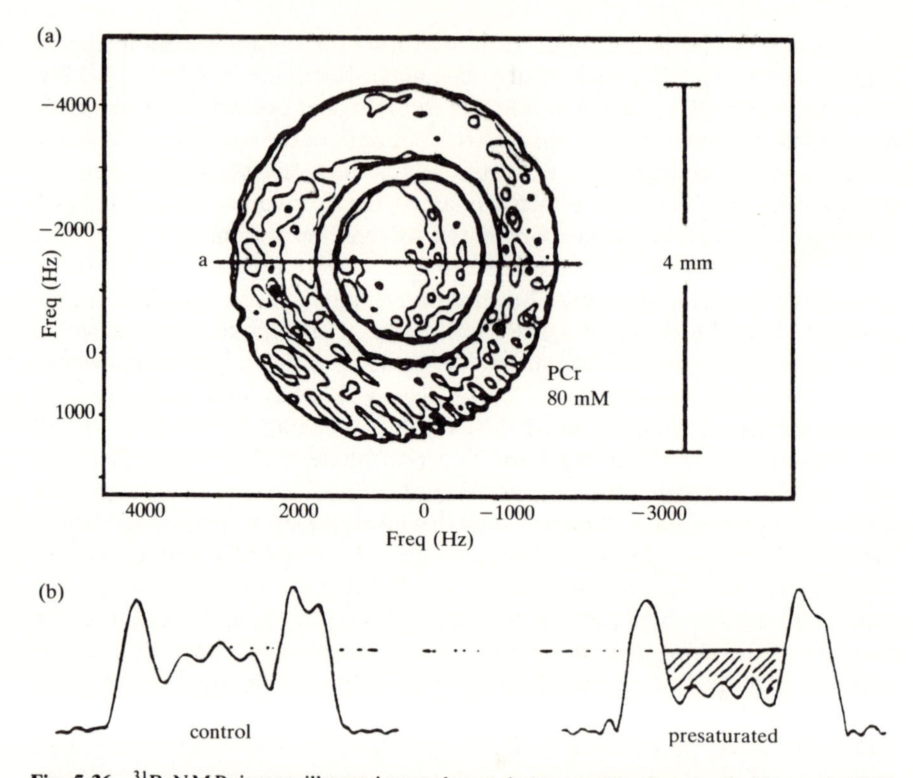

Fig. 5.36 ^{31}P NMR image illustrating exchange between phosphorous in PCr and ATP catalysed by creatine kinase (CK). (a) shows the chemical selective image of PCr in a phantom containing 80 mM PCr, 20 mM ATP, and 5 mM Pi in the outer annulus and the same solution but with creatine kinase present in the central tube. The reduction in PCr signal in the central tube is due to equilibrium between reactants and products of the CK reaction. When the ATP ^{31}P signal is presaturated the PCr ^{31}P signal in the central tube is reduced, thus showing the effect of magnetization transfer due to chemical exchange. (From McFarland *et al.*[79])

the rapid T_2 relaxation presents no particular handicap in comparison with protons in water where T_1 is usually somewhat longer than T_2. It should be noted that the T_1 advantage can apply only if rapid r.f. pulsing is permissible. This is not a problem in microscopic imaging but can be a serious problem in medical applications.

In materials science applications, sodium imaging has been used (at low spatial resolution) to monitor the presence of defect densities in ionic crystals.[220] Dislocations disrupt the octahedral symmetry in cubic crystals and result in large local electric field gradients. Since ^{23}Na is a quadrupolar nucleus ($I > \frac{1}{2}$) this results in a shifting of the $\pm\frac{3}{2}$ to $\pm\frac{1}{2}$ transitions outside the receiver bandwidth, leaving only the $-\frac{1}{2}$ to $+\frac{1}{2}$ transition which is unaffected by the quadrupole interaction. A reduced image intensity results, indicating the presence of crystal defects.

Of the other nuclei listed in Table 4.1, none is a possible candidate for microscopy studies unless isotopic enrichment is used. While isotopic enrichment can be expensive it offers an effective means of introducing a well-defined label on to the system being studied. The three major species in this regard are ^{2}H, ^{13}C and ^{17}O. Deuterium has long been used as a label in biophysical studies, and in the investigation of structural anisotropy in polymers and liquid crystals where the deuterium quadrupole interaction provides a measure of local order. Even at $\approx$ 100% enrichment, the deuterium sensitivity is poor in comparison with the proton. However, because of quadrupolar relaxation, its T_1 value is short (100 ms for D_2O in tissue) and its T_1/T_2 ratio close to unity. Muller and Seelig[210] have obtained low-resolution 12.2 MHz ^{2}H images of a rat head *in vivo*, having previously fed the animal with a 5% deuterium oxide solution for five weeks while Ewy *et al.*[211] have performed a similar experiment at 27.8 MHz in a rabbit heart.

As pointed out by Link and Seelig[212], the deuterium sensitivity problem is exacerbated when using heavy water in *in vivo* tracer studies because of the need to work with reduced D_2O concentration. This dilution is required because of changes in reaction and diffusion rates of deuterium compared with hydrogen due to the significant relative mass difference. In general a 10% concentration of D_2O in H_2O is considered safe for animals with around 20% being tolerated by plants. The sensitivity reduction for 10% D_2O deuterium NMR, in comparison with water proton NMR, is around 1000. Link and Seelig suggest that this can be restored by taking advantage of the following: typically one tenth T_1 values and hence faster repetition rates; typically three times T_2^* values because of reduced γ and hence reduced sensitivity to local field inhomogeneity. (This allows a correspondingly smaller acquisition bandwidth); sacrifice in resolution by a factor of three in all dimensions; an increase in imaging time by a factor of 40. The reduction in imaging bandwidth and in voxel resolution helpfully compensates for the need to otherwise work with higher gradients because of the lower value of γ for deuterium.

While the prospects for NMR microscopy using deuterium are poor the picture is much brighter for isotopically enriched ^{17}O labels. The high spin of ^{17}O leads to a relative sensitivity some ten times that of deuterium despite a similar gyromagnetic ratio and very high-resolution images have been obtained by Mateescu *et al.* for ^{17}O-labelled water molecules in capillaries, plant stems, and mice.[65] T_1 and T_2 are around 6ms for ^{17}O in pure water and shorter for labelled water in biological tissue. In order to reduce signal loss by T_2 relaxation in the delay between excitation and sampling, these authors used a projection reconstruction pulse sequence,

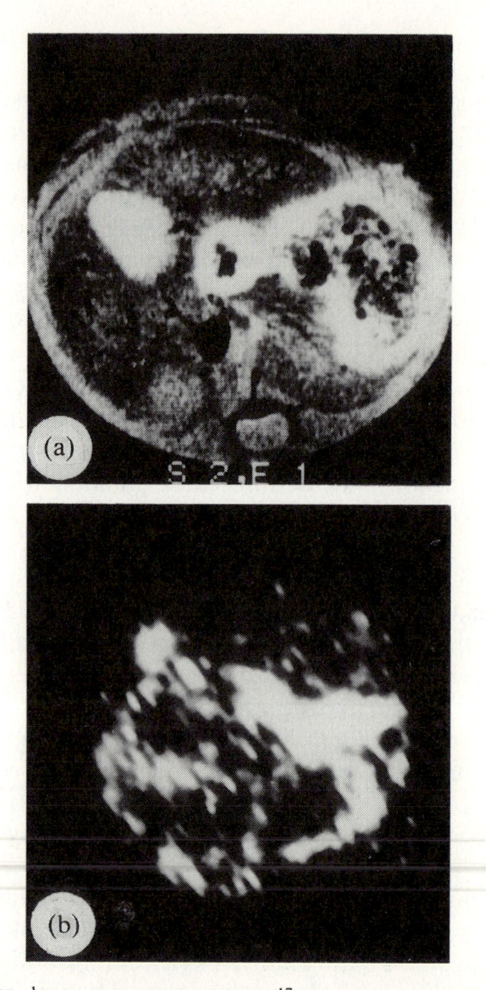

Fig. 5.37 (a) 400 MHz ^{1}H image and (b) 54 MHz ^{17}O image of mouse stomach cross-section 40 minutes after ingestion of ^{17}O-enriched water. The oxygen image was acquired in 8.6 minutes with a slice thickness of 10 mm, $T_0 = 3.9$ ms and $T_R = 120$ ms. For the proton image the acquisition time was 17 minutes, slice thickness 0.5 mm, $T_0 = 12$ ms, and $T_R = 15$. (From Mateescu[65])

thus avoiding the need for a phase gradient delay. Fig. 5.37 shows an ^{17}O image obtained in a 2.5 mm slice encompassing the stomach of a mouse 40 minutes after ingestion of ^{17}O-enriched water, along with a proton image obtained on the same slice. This method not only enables the labelled water to be directly observed but, through the enhancement of proton relaxation in the ^{1}H image, shows the interaction of the labelled water with the stomach and oesophagus walls.

The intrinsic sensitivity of ^{13}C is only 50% greater than that of ^{2}H and the sensitivity of ^{15}N is one order of magnitude smaller. Despite this, ^{13}C- and ^{15}N-labelled molecules may, in principle, be imaged with the full sensitivity of the proton provided that these spins are in close proximity to a hydrogen nucleus, a common occurrence in organic molecules. This remarkable possibility arises because of recent developments in heteronuclear polarization transfer methods.[221] Four types of heteronuclear coherence transfer are possible as illustrated schematically in Fig. 5.38 where I stands for the high gyromagnetic ratio spin (the proton) and S for the low γ spin (^{13}C or ^{15}N). The coherence transfer can be used to enhance the initial magnetization of the S-spin to that of the proton (example (b)), or to indirectly detect the presence of the S-spin via the much stronger proton NMR signal (example (c)). Where both processes take place (example (d)) the effective sensitivity is the same as that of an experiment performed

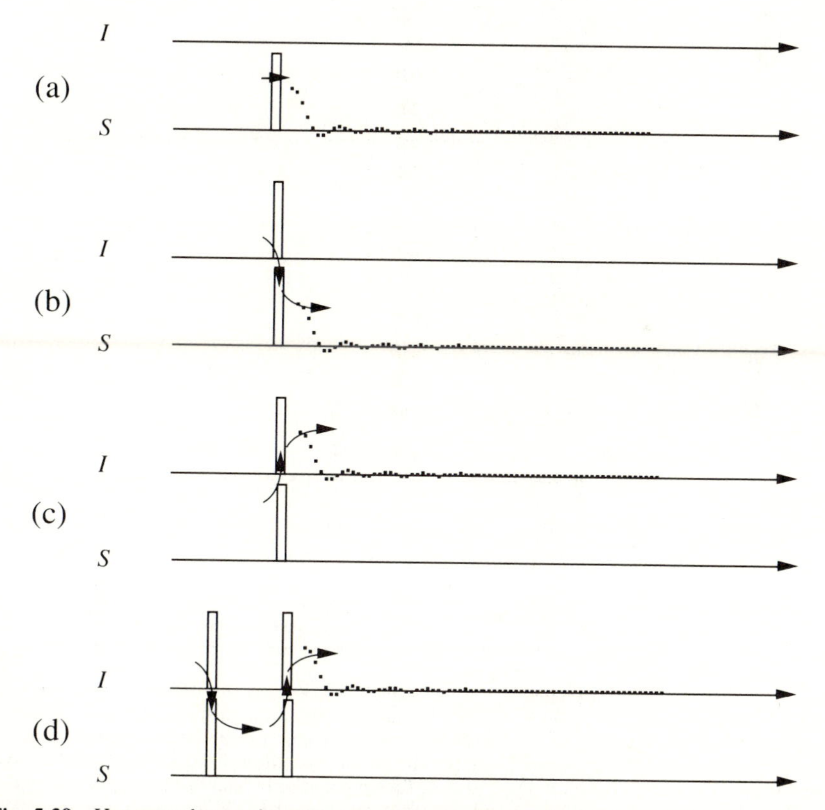

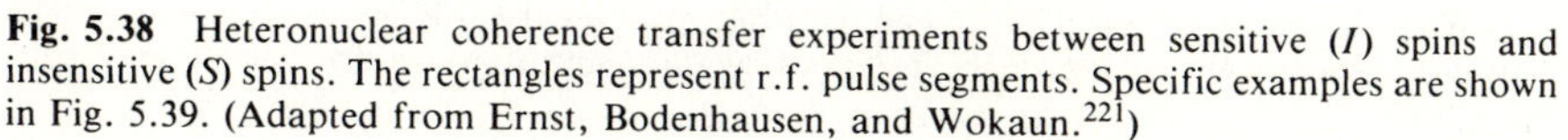
Fig. 5.38 Heteronuclear coherence transfer experiments between sensitive (I) spins and insensitive (S) spins. The rectangles represent r.f. pulse segments. Specific examples are shown in Fig. 5.39. (Adapted from Ernst, Bodenhausen, and Wokaun.[221])

entirely with protons. Furthermore, the relevant T_1 recovery between acquisitions is that of the proton from which the enhanced S-spin polarization is derived. Note that the equilibrium polarization is proportional to γ whereas the detection amplitude is proportional to $\gamma^{7/4}$ (see Section 4.2). The ratio of the overall sensitivity factors for the experiments (a), (b), (c), (d) shown in Fig. 5.38 are therefore $\gamma_S^{11/4}: \gamma_I \gamma_S^{7/4}: \gamma_I^{7/4} \gamma_S: \gamma_I^{11/4}$ which, for ^{13}C, gives 1:4:11:44, while, for ^{15}N, the ratio is 1:10:55:540.

Coherence transfer using r.f. pulses normally takes place by using the J-coupling between the I- and S-spins. Fig. 5.39(a) shows the INEPT

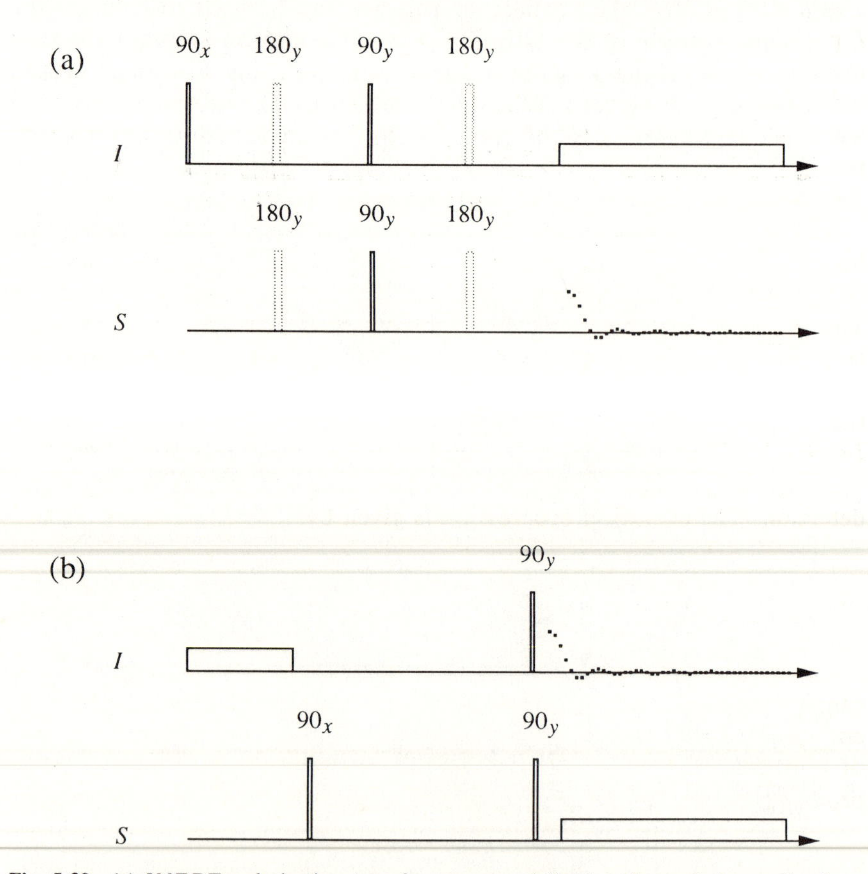

Fig. 5.39 (a) INEPT polarization transfer sequence following the evolution path of eqn (5.15). This sequence achieves the transfer of I-spin polarization to the S-spins as indicated schematically in Fig. 5.38(b). (b) Incoherent transfer of I-spin polarization to the S-spins by means of the Overhauser effect (resulting from the initial continuous irradiation of the I-spins). After S-spin evolution their polarization is transferred coherently back to the I-spins using the mechanism indicated by eqn (5.16). The sequence achieves the full sensitivity enhancement indicated schematically in Fig. 5.38(d). Note that the I-spin signal is detected under S-spin decoupling.

(Insensitive Nuclei Enhanced by Polarization Transfer) sequence which achieves the result of experiment (b) in Fig. 5.38. The 180_y pulses (indicated by dashed lines) can be inserted to refocus chemical shift or B_0 inhomogeneity dephasing. The remaining pulses achieve the following sequence of transformations:

$$I_z \xrightarrow{-(\pi/2)I_x} I_y \xrightarrow{(\pi/2)2I_zS_z} -2I_xS_z \xrightarrow{-(\pi/2)I_y} -2I_zS_z \xrightarrow{-(\pi/2)S_y} 2I_zS_x \xrightarrow{(\pi/2)2I_zS_z} S_y. \tag{5.15}$$

The final transverse magnetization of the S-spin, S_y, derives from the initial I_z polarization and therefore acquires its equilibrium magnitude. While this method of polarization transfer is highly effective for ^{15}N spins, a simpler procedure, which works well for ^{13}C, employs the Overhauser effect in which the I-spins are continuously irradiated on resonance to destroy their polarization. Because of cross-relaxation via I–S dipolar interactions, the S-spin equilibrium polarization is enhanced by $(1 + \gamma_I/2\gamma_S)$.

The reverse process of S-spin polarization transfer to the I-spin via the heteronuclear J-coupling is shown in Fig. 5.39(b). The initial irradiation of the I-spins causes Overhauser enhancement of the S-spin magnetization. The initial 90_x pulse applied to the S-spins produces the single-quantum coherence S_y which evolves under the Heteronuclear scalar coupling into I_zS_x. 90_y pulses applied to both the I- and S-spins convert this to I_xS_z which then subsequently evolves to I_y, enabling detection on the more sensitive I-spin. Note that the I-spin resonance will have a J-splitting due to the scalar coupling which can be removed by irradiating the S-spins during signal detection. The evolution progression is given by:

$$S_z \xrightarrow{-(\pi/2)S_x} S_y \xrightarrow{(\pi/2)2I_zS_z} -2I_zS_x \xrightarrow{-(\pi/2)I_y} 2I_xS_x \xrightarrow{-(\pi/2)S_y} 2I_xS_z \xrightarrow{(\pi/2)2I_zS_z} I_y. \tag{5.16}$$

It should be noted than the overall sensitivity of coherence transfer methods can be reduced because of transverse relaxation during the evolution periods and because of 'leakage' through other pathways. One of the difficulties of indirect detection is the need to suppress direct I-spin magnetization produced by the I_y pulse, especially that of the water proton signal *in vivo*. This is normally performed by some type of phase cycling. A very simple scheme for detecting ^{13}C via ^1H precession has been used by Sillerud *et al.*[216] to image enriched ^{13}C spins with practically the full sensitivity of the proton. Their imaging pulse scheme is shown in Fig. 5.40. Note that the protons are irradiated before the I-spin excitation in order to enhance the S-spin magnetization via the Overhauser effect, and the ^{13}C spins are broadband decoupled during observation of the echo on the I-spins. The initial proton pulses form a binomial $1\ \bar{3}\ 3\ \bar{1}$ sequence centred on water so as to suppress the water proton signal and excite the protons coupled to the ^{13}C spins. The evolution uses a simple phase cycling to subtract all I_y magnetization not derived from coupling to ^{13}C. In successive acquisitions the echoes are added and then subtracted while the phase of the second S_y pulse is alternated between 0° and 180°. The evolution is therefore:

$$\text{add:} \quad I_z \xrightarrow{-(\pi/2)I_x} I_y \xrightarrow{(\pi/2)2I_zS_z} -2I_xS_z \xrightarrow{-(\pi/2)S_y} 2I_xS_x \xrightarrow{-(\pi/2)S_y} 2I_xS_z \xrightarrow{(\pi/2)2I_zS_z} I_y$$

$$\text{subtract:} \quad I_z \xrightarrow{-(\pi/2)I_x} I_y \xrightarrow{(\pi/2)2I_zS_z} -2I_xS_z \xrightarrow{-(\pi/2)S_y} 2I_xS_x \xrightarrow{(\pi/2)S_y} -2I_xS_z \xrightarrow{(\pi/2)2I_zS_z} -I_y. \tag{5.17}$$

The $\pm I_y$ states add coherently in the alternating add/subtract acquisitions. By contrast a directly excited I-spin which has not evolved through the I_xS_x pathway will have a polarization I_y for both phases and will cancel in the add/subtract phase cycling.

One major problem with the methods outlined here concerns the large chemical shifts associated with the ^{13}C spins which results in the need for very large gradients if spectral–spatial overlap is to be avoided. One solution to this problem is to ensure that phase encoding involves only the proton transverse magnetization. Recent use of coherence transfer methods employing this approach has been demonstrated by Yeung and Swanson,[222] and Knüttel *et al.*[223,224] In the method of Knüttel, Kimmich, and Spohn, (MQF-VOSING) a heteronuclear two-quantum filter is used to destroy unwanted magnetization and to ensure that the final image derives only from protons coupled to ^{13}C nuclei. The MQF-VOSING pulse sequence is shown in Fig. 5.40(b) and the density matrix evolution is as follows:

$$I_z \xrightarrow{-(\pi/2)I_x} I_y \xrightarrow{(\pi/2)2I_zS_z} -2I_xS_z \xrightarrow{-(\pi/2)S_x} -2I_xS_y \xrightarrow{-\Delta\omega_0\tau(I_z+S_z)-\pi I_y-\Delta\omega_0\tau(I_z+S_z)}$$

$$-2I_x(S_y\cos\phi_2 - S_x\sin\phi_2) \xrightarrow{-(\pi/2)S_x} 2I_x(S_z\cos\phi_2 - S_x\sin\phi_2) \xrightarrow{(\pi/2)2I_zS_z}$$

$$I_y\cos\phi_2 \xrightarrow{-(1/r)\Delta\omega_0\tau I_z} \overline{[I_y\cos^2\phi_2 + I_x\cos\phi_2\sin\phi_2]} = \tfrac{1}{2}I_y \tag{5.18}$$

Fig. 5.40 (a) Imaging pulse sequence used by Sillerud *et al.*[216] in which polarization is transferred from ^1H to ^{13}C spins by Overhauser enhancement and, following evolution in the phase-encoding gradient, the ^{13}C polarization is transferred to the protons via the evolution progression indicated in eqn (5.17). Note that no slice selection was employed so that a projected plane image was acquired. (b) Imaging pulse sequence similar to the MQF-VOSING method of Knüttel *et al.*[224] The 90_x and 180_y proton r.f. pulses are slice selective while the ^{13}C r.f. pulses are chemical shift selective. The gradient pulses shown shaded produce a dephasing by ϕ_2 of ^{13}C magnetization in the DQ coherence state. This is refocused once the polarization is transformed back to proton single-quantum coherence, and thus requires a gradient with only 1/4 of the net time integral. Clearly protons not coupled to the selected ^{13}C nuclei will be suppressed by this filter. Note that all imaging gradients relate to proton coherences.

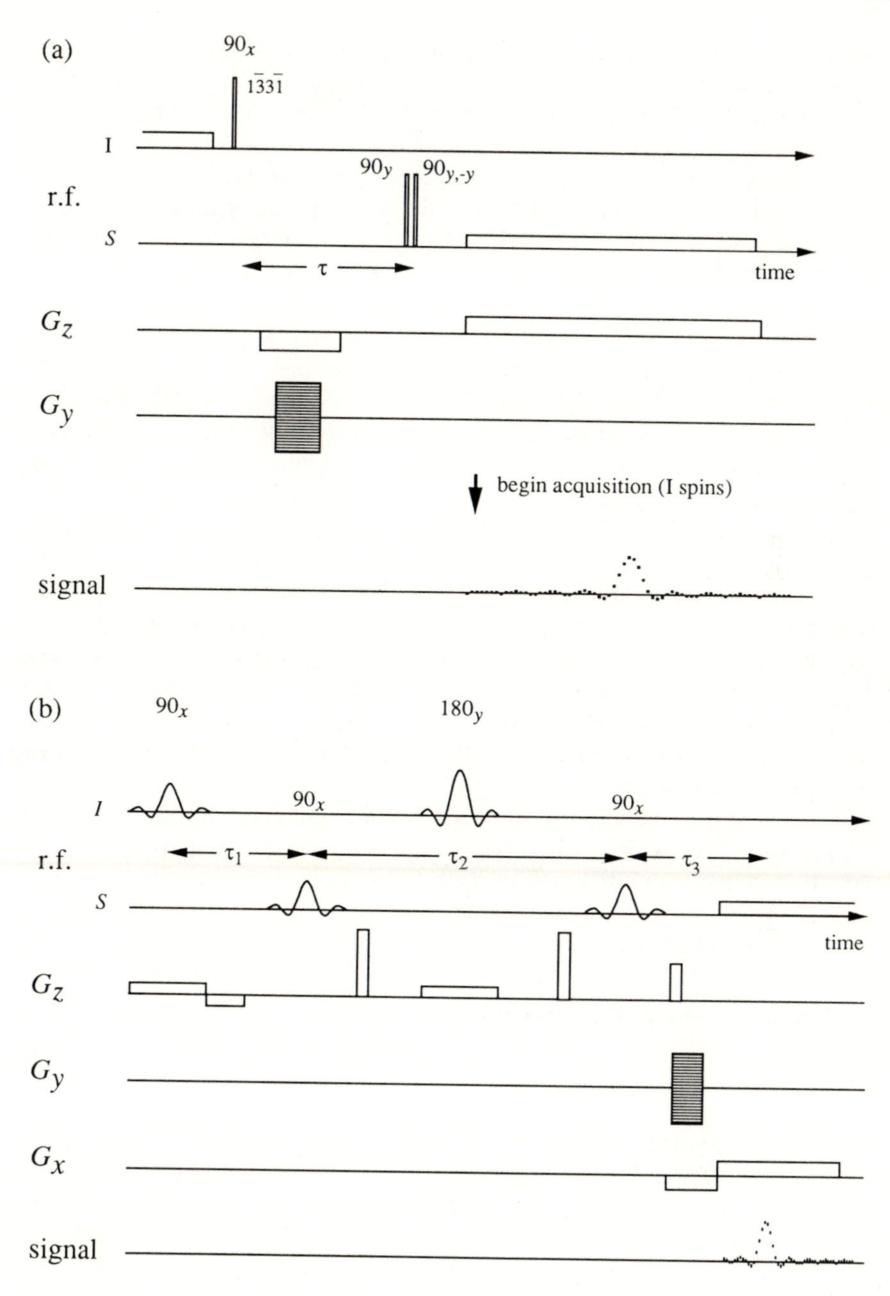
(a)
90x
1̄33̄1
I
r.f.
90y
90y,-y
S
τ
time
Gz
Gy
begin acquisition (I spins)
signal
(b)
90x
180y
I
90x
90x
r.f.
τ1
τ2
τ3
S
time
Gz
Gy
Gx
signal

where ϕ_2 is the ^{13}C dephasing angle due to the combined effect of both gradient pulses applied during the period τ_2, denoted τ in eqn. (5.18). The final echo amplitude is, of course, derived by averaging over the distribution of ϕ_2 values.

The period τ_1 is chosen to optimize coherence transfer, i.e., $\tau_1 = 1/2J$. Proton magnetization phase shifts arising from applied gradient, chemical shift, and field inhomogeneity effects during the period τ_2 are refocused by means of the 180_y I-spin r.f. pulse, whereas the ^{13}C phase shifts (ϕ_2) remain. These latter shifts are, however, refocused during the τ_3 period using the proton single-quantum coherence, I_y. In consequence, the refocusing gradient must be smaller by the quotient of the ^{13}C and ^{1}H gyromagnetic ratios, namely $(1/r)$, where r is very close to 4. This results in a coherence transfer echo at τ_3 in which signals from uncoupled protons are suppressed. Note that the chemical shift and field inhomogeneity dephasing suffered by the ^{13}C nuclei during τ_2 must also be refocused using the proton single-quantum coherence so that the echo formation time, τ_3, is equal to $\tau_1 + (1/r)\,\tau_2$.

The slice selection, phase encoding, and read gradients are all applied while proton pulses or coherences are relevant. Furthermore, for each ^{13}C nucleus, the full proton sensitivity is achieved since the initial polarization, and detection sensitivity derives from the protons coupled to the ^{13}C. Knüttel *et al.* have demonstrated that the method achieves approximately the same spatial resolution and sensitivity as proton imaging, provided ^{13}C-enriched molecular species are imaged. However, they have also demonstrated that the method is effective in the imaging samples at natural ^{13}C abundance provided that larger voxels are used to compensate for the reduction in nuclear numbers. The method has not yet been applied in NMR microscopy but promises to be highly effective in tracing the distribution of ^{13}C-labelled molecules.

5.5 NMR imaging in the solid state

5.5.1 *The sensitivity problem*

In NMR terminology the words 'solid' and 'liquid' are used to describe the local motion of the nuclear environment. In that sense spins belonging to liquid molecules absorbed in a solid state matrix or to mobile polymer segments in a rubbery solid are said to be in a liquid state. The locality of the motional state arises because of the very short range (a few Angstroms) of the through-space dipolar interaction between spins. Nuclei residing in locally solid molecular environments experience static spin–spin dipolar interactions. In Chapter 2 we saw that these internuclear dipolar interactions in solids cause a rapid dephasing of the FID, the so-called Bloch decay. This presents a formidable problem in NMR imaging. In principle, it would be possible to carry out imaging experiments in solids provided that these could be performed in a time much shorter that the decay time of the FID. To produce a spectral spread across a specimen of a few mm, which is much larger than the intrinsic proton dipolar linewidth of order 20 to 100 kHz, requires enormous magnetic field gradients. While this presents a daunting practical limitation, even more serious is the need to sample the signal with very short dwell times, and hence very large acquisition bandwidths, in order to obtain a sufficient number of pixels in the image. The signal-to-noise penalty inherent in such an approach is so severe that imaging at high resolution becomes a practical impossibility.

In the imaging of spins in the solid state, achieving microscopic resolution is particularly difficult. Because of its high gyromagnetic ratio the most sensitive nucleus, the proton, inevitably experiences the largest dipolar interactions. While the use of lower γ nuclei relaxes the gradient strength requirement, the associated loss of sensitivity represents a severe trade-off. The most promising approaches seem to lie in proton imaging using pulse techniques designed to suppress the dipolar interaction. These are extremely difficult to perform and are only a realistic proposition for researchers experienced in the use of multiple pulse line-narrowing methods.

Our review of solid state imaging includes few examples in which the resolution approaches voxel sizes below 1 mm^3 and none in which the microscopic voxel of $(100\,\mu m)^3$ is obtained. None the less the development of solid state methods has progressed significantly over the last few years and some of the techniques developed are potentially useful for improving resolution in more conventional liquid state microscopy.

5.5.2 *Dilute and low γ spins*

The troublesome internuclear dipolar interactions are much lower if the spins to be imaged are sufficiently dilute or their gyromagnetic ratios small. This has allowed the mapping of ^{23}Na ions in β-alumina[225] where quite small gradients have been used.

^{13}C, at natural abundance in organic materials and polymers, has a small probability of having another ^{13}C spin close enough to produce a significant dipolar interaction but will experience dipolar coupling from attached protons. Because of the reduced ^{13}C gyromagnetic ratio this is usually an order of magnitude less than the 20 kHz dipolar linewidths commonly experienced for protons in rigid polymers. Furthermore the ^{1}H–^{13}C dipolar broadening can be substantially removed by continuous irradiation of the proton resonances during detection of the ^{13}C FID. This decoupling process requires a proton r.f. field strength sufficiently large that $\gamma_H B_1$ exceeds the dipolar interaction strength, a criterion which can be relatively easily satisfied given a proton irradiation transmitter of a few hundred watts. The remaining linewidth of a few hundred p.p.m. arises from the isotropic and anisotropic chemical shifts. The anisotropic shift can normally be removed by employing magic angle spinning (MAS) methods, although this requires special gradient manipulation where imaging is intended. An alternative scheme for removing all chemical shift effects by employing a Carr–Purcell pulse train has been demonstrated in one-dimensional imaging of ^{13}C by Szeverenyi and Maciel[226] and by Miller and Garroway.[227]

The low sensitivity of ^{13}C associated with the small gyromagnetic ratio can be partially compensated by the use of magnetization transfer from the protons, an effect which is normally performed by Overhauser enhancement in liquids and Hartmann–Hahn cross-polarization[228] in solids. In certain favourable cases where paramagnetic centres are present, a further enhancement is possible by pre-irradiating the paramagnetic electron spin resonance, thus enabling polarization transfer from the much larger electron spin moments. This effect, known as Dynamic Nuclear Polarization (DNP) has been used to obtain ^{13}C images in pitch[229] at a transverse resolution below 1 mm.

5.5.3 *Proton imaging in the solid state*

In the selective excitation of slices a narrow-band r.f. pulse is applied in the presence of a gradient which spreads the spectrum. When imaging protons in liquids the r.f. pulse bandwidth can be made much larger than the intrinsic linewidth while still representing a narrow fraction of the gradient-induced spread. In solids this is impossible. While it is feasible to produce gradients large enough to spread the spectrum beyond the dipolar-broadened linewidth, it is difficult to produce a hard r.f. pulse sufficient to encompass this width, let alone a shaped, soft pulse. One method (LOSY) of avoiding the problem of shaped pulses is to spin-lock the transverse magnetization, in the presence of a magnetic field gradient, following a 90_x hard pulse.[230] The only transverse magnetization remaining at the end of the spin-lock pulse is that within a frequency γB_1 of the transmitter frequency, where B_1 is the amplitude of the spin-lock field. This slice magnetization can then be stored along the z-axis for subsequent use by applying a 90_{-x} pulse. Another approach is to use a selective r.f. pulse train which uses only hard pulses. The DANTE sequence is ideal for this and, by amalgamating the DANTE train and the selection gradient (applied as very short pulses) into appropriate windows of a dipolar line-narrowing MREV-8 sequence, Cory *et al.*[231] have demonstrated very efficient slice selection for protons in a sample of ferrocene.

The development of these promising solid state slice selection methods is very new and in nearly all applications of imaging to solids so far reported, no slice selection has been employed, and either cut slices or samples with longitudinal symmetry have been investigated.

Phase encoding without line-narrowing

Initial attempts to obtain proton images in solids have employed a sensitive volume approach in which a signal was acquired from a small region of the sample[232] or a phase encoding of the FID[233,234], and subsequent Fourier transformation with respect to the phase-encoding time. The FID is detected in the absence of magnetic field gradient so that the use of gradient spreads and reception bandwidths well in excess of the dipolar linewidth is avoided. One problem with this method is the need to employ gradients sufficiently large that phase encoding occurs in a time comparable with the Bloch decay of the proton magnetization. A variant of this method which avoids this, works by using solid echoes to prolong the transverse magnetization.[235] In this approach the phase gradient is applied throughout the experiment, making rapid switching unnecessary. Because this gradient is present both before and after the refocusing 90_y pulse, an unusual phase cycling scheme is needed to achieve the appropriate **k**-space modulation. This method has produced low-resolution one-dimensional images for 8 mm diameter samples of powdered hexamethyl benzene, for which the Bloch decay time was around 40 μs.[235]

Another approach to phase encoding without the use of read gradients has been taken by Rommel *et al.*[236] Their pulse sequence is shown in Fig. 5.41 and employs a LOSY slice selection followed by a Jeener–Broekhart echo. The problem of large gradient switching is avoided by storing the slice magnetization along the z-axis during switch-on, and as dipolar order during switch-off. Alternating the phase of the dipolar recall pulse between 45_x and 45_y enables both quadrature phases of the signal to be obtained. Note that the pulse sequence of Fig. 5.41, while requiring high gradients, does permit the simultaneous measurement of spectroscopic parameters associated with the natural linewidth, since the FID is sampled once the phase gradients have been removed. A one-dimensional version of the method has been applied to KEVLAR with a resultant transverse spatial resolution of 1.7 mm in a slice of 2.5 mm selected with a spin-lock B_1 amplitude of 1 Gauss and duration 100 μs.

One of the major problems associated with these phase-encoding methods is the need for large gradients if high spatial resolution is to be achieved. For protons in solids the local field due to dipolar interactions is of order 5 Gauss so that gradients of 500 Gauss cm^{-1} are needed to achieve 0.1 mm transverse resolution. By coherence transfer to states of multiple quantum coherence[237] the effect of the gradient can be enhanced although such methods are inherently inefficient in signal-to-noise terms. It is the sensitivity problem which lies at the heart of the difficulty in achieving high resolution

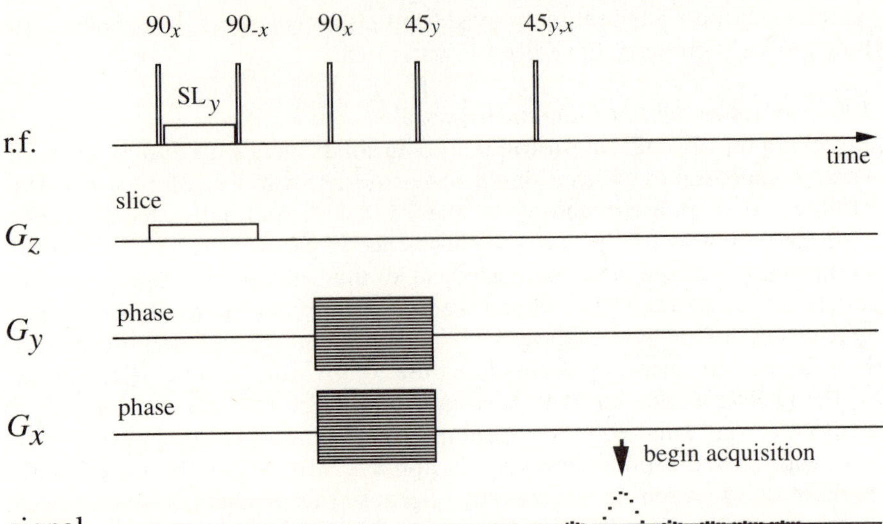

Fig. 5.41 The LOSY pulse sequence of Rommel *et al.*[236] in which a hard 90_x pulse is applied in the presence of the slice gradient, the on-resonant magnetization being subsequently spin-locked by a continuous phase-shifted r.f. pulse and then stored along the z-axis. This slice magnetization is later recalled in the presence of the phase-encoding gradients and, after a period of evolution, stored as dipolar order while the gradients are switched off, to be recalled for reading at a later time. The image can then be reconstructed by PR.

images for protons in the solid state. Because no read gradient is applied in the methods discussed, only one point in **k**-space is obtained for each acquisition, so that such a method, while avoiding the dipolar broadening problem, is inherently inefficient and therefore unsuited to microscopy. The appropriate solution to the problem is found in the application of line-narrowing methods which permit modest read gradients and reception bandwidths to be used.

Line-narrowing methods

Applications of line-narrowing methods to imaging include the use of multiple pulse coherent averaging (MP)[238–242], magic angle spinning (MAS),[243,244] and combinations of these (CRAMPS).[245,246] The MP approach is effective in reducing the dipolar linewidth by at least two orders of magnitude but does not remove broadening arising from chemical shift anisotropy.

Chingas *et al.*[238] have used MREV-8 cycles to produce line-narrowing during both the evolution (phase) and detection (read) periods with full, symmetric echo detection. These authors used alternate acquisition storage of the in-phase (M_y) or quadrature (M_x) signal along the z-axis to allow time for gradient settling. Their pulse sequence is shown in Fig. 5.42. The

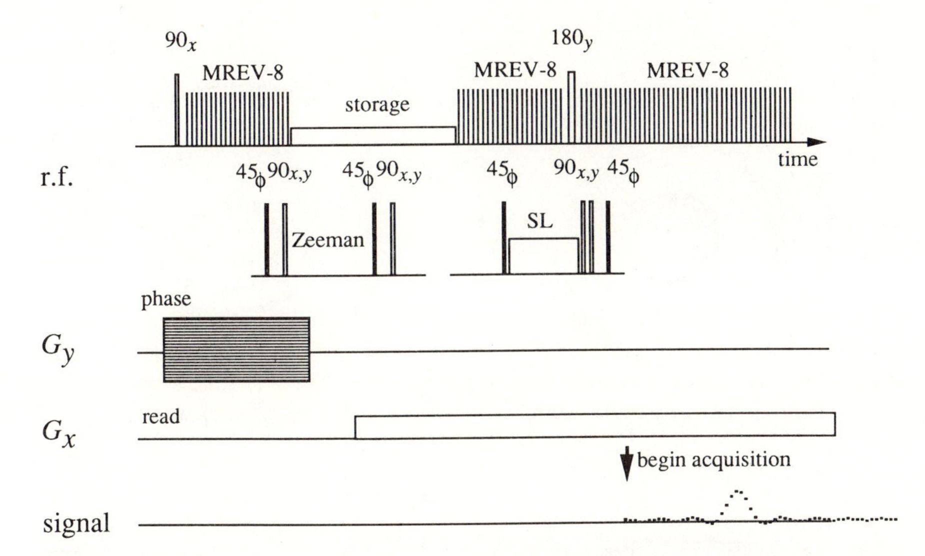

Fig. 5.42 Solid state image pulse sequence of Chingas *et al.*[238] The storage sequences (both Zeeman and spin-lock storage pulse sequences are shown) are used to protect the phase-encoded magnetization while the phase gradient is removed and the real gradient is switched on. An MREV-8 sequence is used to produce line-narrowing during the phase-encoding and read periods.

45_y pulses are required at the beginning and end of the storage intervals to rotate the magnetization from the MREV-8 frame direction, $(-1, 0, 1)$ in the xz-plane, to the x-axis in the xy-plane. Fig. 5.43 shows an image, obtained by this method, of a 12 mm outer diameter neoprene hose surrounding a 5 mm tube of adamantane. One of the curious features of multiple pulse imaging is that the gradient-induced frequency offsets are scaled down by the same factor as the chemical shift, $\sqrt{2}/3$, for MREV-8 as discussed in Section 2.6.9.

For large samples it is difficult to generate the r.f. field strengths required for MP line-narrowing if surrounding r.f. coils are used. One satisfactory alternative is to use surface coils.[239,240] Another problem associated with MP methods concerns the anisotropic chemical shift which is reduced but not eliminated by the MREV-8 cycle. One ingenious approach developed by Miller and Garroway[241,242] involved the use of two MREV-8 sequences preceded by a four-pulse frame reference rotation sequence.[248] These have the effect of refocusing the chemical shift or susceptibility broadening Zeeman interactions in a series of spin echoes (in the manner of a CPMG train), while at the same time maintaining the line-narrowing. So that the desired applied gradient dephasing is not refocused, the phase and read gradients are applied as sinusoids synchronous with the 24-pulse cycle. The

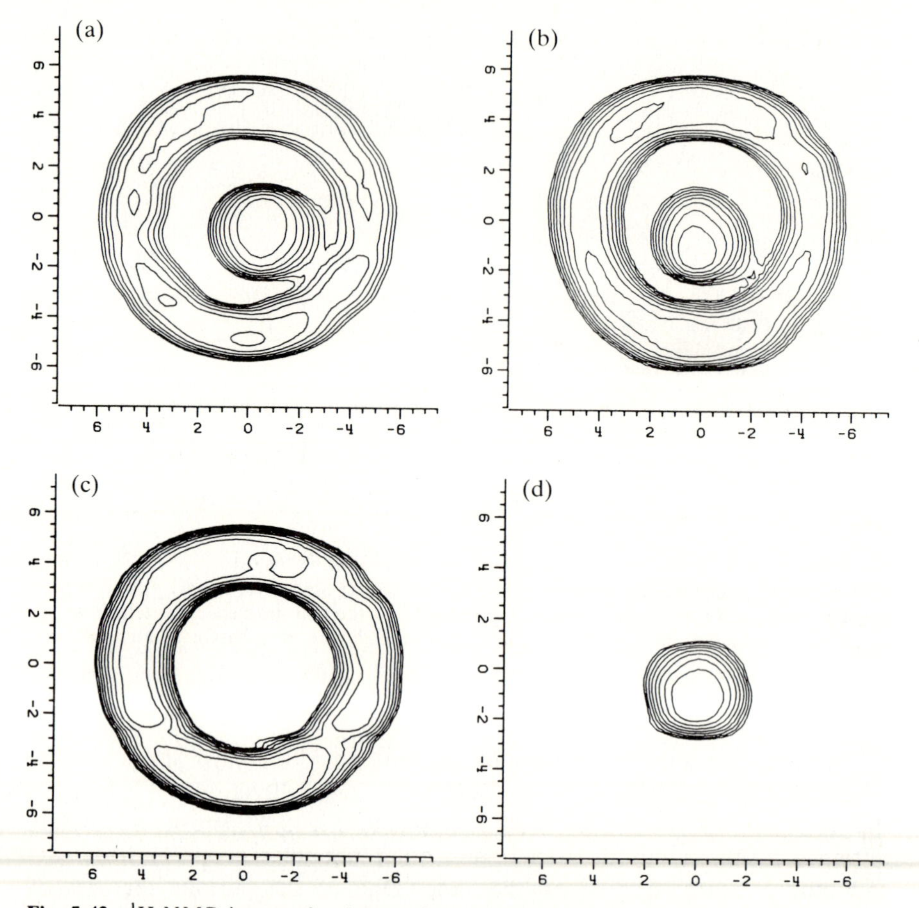

Fig. 5.43 ^{1}H NMR image of a 5 mm tube of adamantane surrounded by a 12 mm outer diameter neoprene hose section. In (a) 5 ms Zeeman storage is used while in (b) a 5 ms spin-locked pulse is employed. In (c) T_R is reduced from 2 s to 0.25 s and the adamantane signal saturates. In (d) $T_R = 2s$ but the spin-lock time is increased to 50 ms. Here the neoprene signal has decayed because of the relatively short $T_{1\rho}$. (From Chingas *et al.*[238])

sequence is, in effect, a solid state line-narrowing version of the alternating gradient, CPMG method[249] which can be used to remove susceptibility artefacts in liquid state imaging (see Section 4.7.2). This highly sophisticated imaging method was used to produce a two-dimensional image of protons in a polyacrylic sample at a transverse resolution (without slice selection) of around 200 μm. The removal of chemical shift and susceptibility broadening by the 24-pulse cycle resulted in a linewidth of 200 Hz in comparison with the MREV-8 linewidth of 1 kHz and the unnarrowed linewidth of 25 kHz.

The first application of MAS methods to imaging was independently suggested by Wind and Yannoni[243] and by de Luca and Maraviglia[244] using an imaginative rotating frame method whereby the spins precess about an off-resonant effective field inclined at the magic angle to the z-axis. By using a large enough r.f. field the rotation rate can be made sufficiently high to remove the full dipolar linewidth. While this approach enables two-dimensional reconstruction via PR it suffers from the need to detect the signal at the rotation frequency of a few kHz using a z-axis coil. In principle this leads to much lower signal-to-noise ratios although de Luca and Maraviglia suggest that this could be restored by using a greater number of turns in the receiver coil.[244]

The use of MAS sample rotation in imaging at speeds of up to 5 kHz has been pioneered by Cory, Veeman, and co-workers.[245–247] The immediate challenge facing the experimenter is how to rotate the imaging gradients with the sample. This is achieved by a quite ingenious trick, as shown in Fig. 5.44(a). The coils G_x and G_y produce gradients normal to the rotation axis. They are driven by currents applied as two sinusoids in quadrature phase whose frequencies are synchronous with the rotor speed. The frequency locking is achieved by means of an optical sensor which measures the sample rotation. PR is achieved by successively stepping the sinusoid phase. This has the effect of rotating the gradients with respect to the sample. FI can also be executed as shown in Fig. 5.44(b) although this is more difficult since rotating the gradient orientation between the phase and read in the sample reference frame requires a 90° phase switch of the gradients.

MAS imaging has been applied with considerable success to the imaging of elastomers in which the dipolar linewidth is already motionally narrowed (see Section 2.6.9). It has the particular advantage over MP methods that anisotropic chemical shifts are also removed. One particular problem with elastomers is rotor-induced distortion, an effect which is apparent in the imaging of silicone rubber[245] where an image with 200 μm transverse resolution was obtained in a period of 35 minutes. For rigid solids where dipolar narrowing cannot be achieved by MAS, a combination of MREV-8 and MAS (CRAMPS) has led to the measurement of high-resolution proton images in a variety of polymeric materials.[246,247] Figure 5.45 shows a CRAMPS image of two 1 mm slices of low-density polyethylene. Where MAS narrowing alone is employed the image is strongly T_2-weighted. In a MAS imaging study of a 750 μm thick disc of polybutadiene/polystyrene by MAS narrowing alone, Cory *et al.*[247] found that the blend image amplitude is dominated by the more mobile polybutadiene indicating the non-uniformity of the mixture.

The great difficulty associated with the use of these line-narrowing methods has dissuaded all but the practised specialist from attempting this field of imaging. Furthermore, it appears that the MP approach suffers from

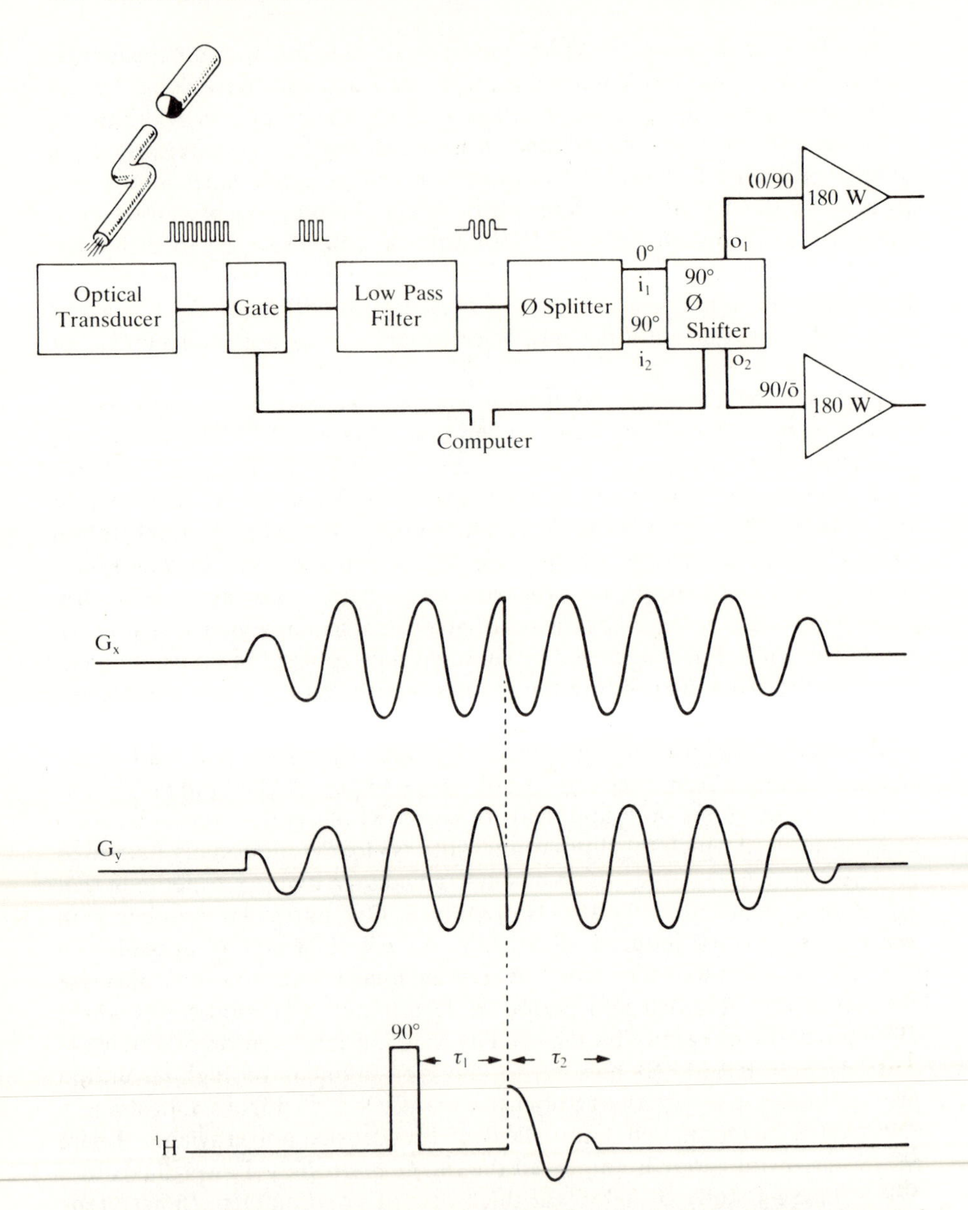

Fig. 5.44 (a) Schematic of control circuit used to synchronize the gradient to the spinner and to gate and phase-shift it. (From Cory *et al.*[245]) (b) Pulse sequence used to acquire images of rotating solids. The gradient is made to rotate synchronously with the spinner by applying two quadrature audio signals to a pair of orthogonal gradient coils. Phase shifting the gradient by 90° is equivalent to a gradient rotation by 90° in the sample frame. (From Cory *et al.*[245])

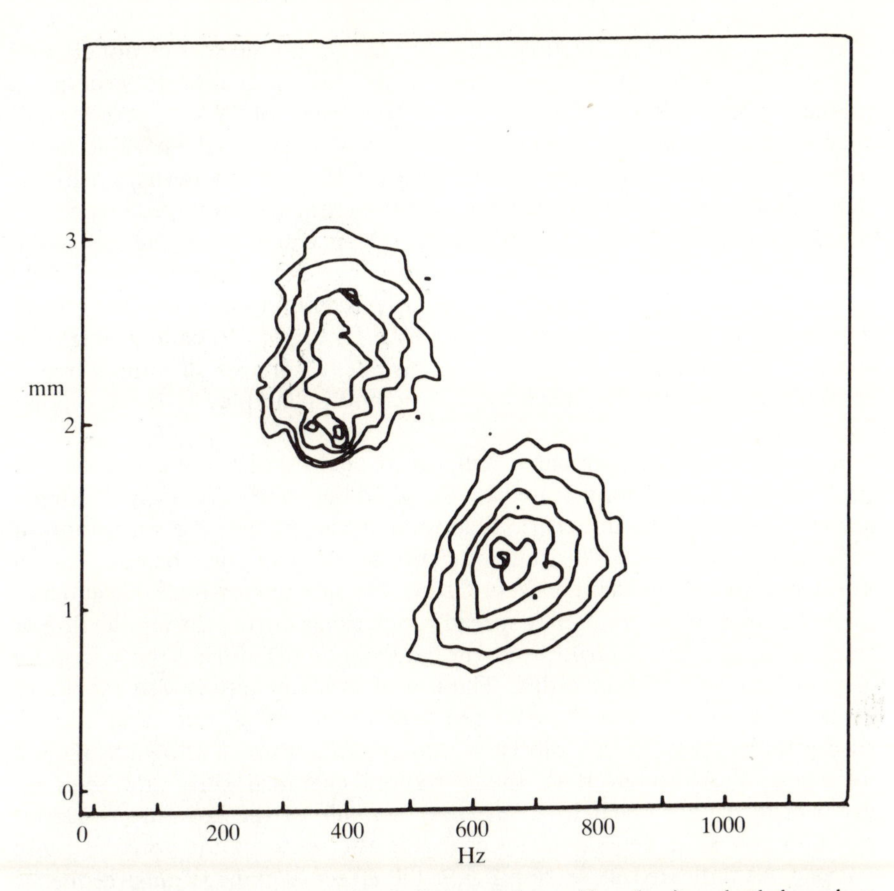

Fig. 5.45 Solid state image of two irregularly shaped pieces of low-density polyethylene where the spinning rate was 1850 Hz. The transverse resolution is of order 100 μm. (From Cory *et al.*[246])

a fundamental difficulty, namely the breakdown in line-narrowing for off-resonant spins. The effective frequency width for MREV-8 is approximately 1.5 kHz[250] which limits the maximum scaled gradients which can be used to of order $2\pi \times 1.5 \times 10^3/\gamma l_s$ where l_s is the transverse sample dimension. A frequency spread of only 1.5 kHz demands that the line-narrowing be particularly effective. To place this in perspective, a reduction of the dipolar linewidth to 200 Hz (no mean achievement) allows only 8 transverse pixel elements across the image! Furthermore the line-narrowing by MP does not allow a similar reduction in the acquisition bandwidth since the magnetization must be sampled in a few microseconds after the r.f. pulse.

One solution to the off-resonant spins problem is to apply the magnetic field gradients as very high-frequency sinusoids with all r.f. pulses occurring at the zero cross-overs. Such an approach has been used by Macdonald and Tokarczuk[251] who, by using a series tuned coil, managed to produce a gradient amplitude of 2.2 G cm^{-1} at a frequency of 25 kHz, this amplitude being considerably reduced from the static field gradient value at the same current amplitude because of severe eddy current induction in the magnet pole pieces. The line-narrowing pulse sequence employed is specially modified to allow for cumulative phase advancement from the sinusoidal gradient but cancellation of phase shifts due to susceptibility of field inhomogeneity offsets. The particular sequence employed by McDonald and Tokarczuk produced a line-narrowing by a factor of 50, comparable with the best achievable by MREV-8, but as yet no images of submillimetre resolution have been produced by this method in samples with full dipolar broadening.

Recently a very sophisticated method of dealing with the off-resonance problem has been demonstrated by Cory, Miller, and Garroway[252] using a sequence in which the magnetic field gradient is applied as a succession of very short (μs) pulses. This method relies on the fact that the spin system sensitivity to off-resonant effects during the line-narrowing r.f. sequence arises because of the frequency spread which exists during the application of the r.f. pulses. By ensuring that the gradient is off during the r.f. pulse this problem is largely avoided. The use of gradient pulses also facilitates an alternative approach to the removal of chemical shift and sample susceptibility terms in the spin Hamiltonian, an approach known as second averaging.[253] Although both the unwanted chemical shift and required gradient terms are linear in I_z, use can be made of the fact that the latter can be made time dependent by changing the sense of the gradient pulse current. As indicated in Section 2.6.9, the effect of the r.f. pulse sequence is to transform the Hamiltonian into the 'toggling frame'. This causes the dipolar interaction to vanish but leaves Zeeman terms transformed into effective magnetic fields along some particular axis in the toggling frame. Using a basic MREV-8 sequence with specially selected r.f. pulse phases and taking into account the fact that the effect of the Zeeman Hamiltonian terms differs in the various 'windows of evolution', the authors ensure that the effective direction of the chemical shift field is normal to the much larger gradient field, thus resulting in a second averaging to zero. (This vanishing of small orthogonal terms is akin to the insignificant influence of magnetic fields applied normal to B_0, as discussed in Section 3.1.) This approach has led to the polymethylmethacrylate image shown in Fig. 5.46 in which the transverse resolution is 100 μm.

Provided that the dipolar linewidth is sufficiently small to permit MAS narrowing, this approach to imaging appears to be the most straightforward.[254] It is likely, therefore, that very high-resolution proton images will be obtained only for elastomeric polymers with small

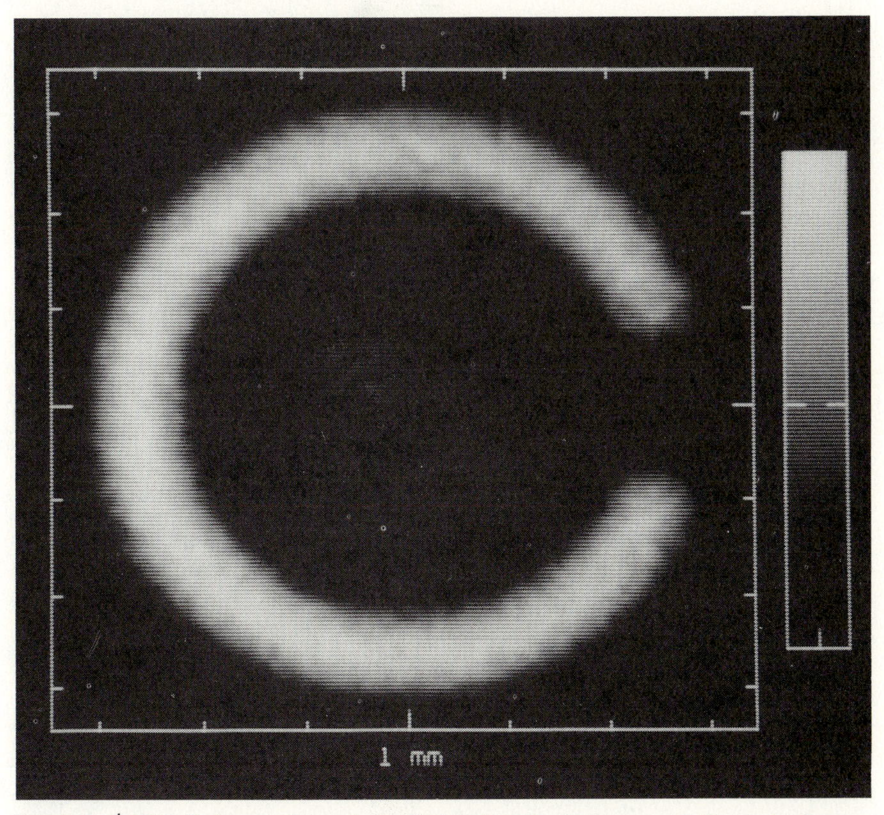

Fig. 5.46 ^{1}H NMR image of a poly (methylmethacrylate) phantom, 6.35 mm in diameter with a wall thickness of 0.79 mm. The image was acquired using the second averaging multiple pulse line-narrowing method with a 2.5 μs 90° pulse time and a cycle time of 60 μs. (From Cory *et al.*[252])

residual dipolar interactions. Magic angle spinning in conjunction with ^{1}H–^{13}C cross-polarization (CPMAS)[255] is, however, particularly effective in producing high-resolution ^{13}C spectra from rigid solids. One interesting possibility concerns the use of CPMAS methods combined with gradient rotation to obtain images at microscopic resolution for samples enriched with ^{13}C, although no such experiments have been reported at this time. For fully dipolar-broadened samples the pulsed gradient, multiple pulse line-narrowing method of Cory, Miller, and Garroway looks the most promising, although the technical problems associated with its implementation are extremely daunting. None the less, the early results are impressive and suggest that proton NMR microscopy in rigid solids is a real possibility.

5.5.4 *Deuteron imaging and two-quantum coherence*

At first sight the prospects for deuterium imaging in the solid state would seem very poor. Not only are quadrupole-broadened ^{2}H NMR spectra similar in width to those of dipolar-broadened protons, but the deuteron suffers from a much smaller gyromagnetic ratio, thus causing a severe sensitivity disadvantage. There is, however, a significant difference in the nature of the line-broadening in deuteron NMR in that it arises almost entirely from a rank-two tensorial interaction, whereas in dipolar coupling where multispin interactions are possible, higher rank tensorial terms will be present. This Hamiltonian 'purity' offers the possibility of transforming the deuteron spin magnetization into a variety of well-characterized coherences each with special properties in the presence of the two Hamiltonian terms of interest to us, namely the Zeeman interaction associated with the magnetic field gradient, and the quadrupolar interaction responsible for the line-broadening.

An elegant demonstration of the usefulness of coherence transfer in deuteron imaging has been provided by Günter, Blumich, and Speiss[256] who employ a pulse sequence in which the spin polarization is first stored as quadrupolar order (T_2^0), invariant under both H_Q and the Zeeman interaction, while the magnetic field gradient is turned on and allowed to settle. A 90_x pulse then transforms T_2^0 into the double quantum coherence state, $T_2^{\pm 2}$ in which precession in the magnetic field gradient occurs at twice the usual Larmor frequency, but the system is protected from quadrupolar precession. The spins are then returned to a state of spatially phase-encoded transverse magnetization via the T_2^0 polarization, where the gradients are removed. Because no read gradient is used, the full spectral information of the quadrupolar precession is available as an additional dimension.

Günter *et al.* have used this sequence to provide spectral–spatial images of drawn deuterated polyethylene with segments at different orientations, and for a sample consisting of adjacent ^{2}H-polycarbonate and ^{2}H-polystyrene segments. Although only one spatial dimension was employed, a sub-millimetre resolution was achieved, the spectral components of these two experiments showing features in the quadrupole lineshape associated with orientation and dynamics, respectively. This preliminary study indicates the potential of more sophisticated contrast schemes in the NMR microscopy of model polymer structures and, in particular, for the imaging of molecular order and dynamics in polymer composites.

5.6 References

1. Abragam, A. (1961). *Principles of nuclear magnetism*, Clarendon Press, Oxford.
2. Mansfield, P. and Morris, P. G. (1982). *NMR imaging in biomedicine*, Academic Press, New York.

3. Bloembergen, N. (1949). *Physica* **15**, 386.
4. Kroeker, R. M. and Henkelman, R. M. (1986). *J. Magn. Reson.* **69**, 218.
5. Bottomley, P. A., Foster, T. H., Argersinger, R. E. and Pfeifer, L. M. (1984). *Med. Phys.* **11**, 425.
6. Burnell, E. E., Clark, M. E., Chapman, N. R. and Hinke, J. A. M. (1981). *Biophys. J.* **33**, 1.
7. Belton, P. S. Jackson, R. R. and Packer, K. J. (1972). *Biochim. Biophys. Acta.* **286**, 16.
8. Fung, B. M. and McGaughy, T. W. (1974). *Biochim. Biophys. Acta.* **343**, 663.
9. Rustgi, S. N., Peemoeller, H., Thompson, R. T., Kydon, D. W. and Pintar, M. M. (1978). *Biophys. J.* **22**, 439.
10. Kolata, G. B. (1976). *Science* **192**, 1220.
11. Zimmerman, J. R. and Brittin, W. E. (1957). *J. Phys. Chem.* **61**, 1328.
12. Franks, F. (1972). *Water: a comprehensive treatise*, Vols 1–4, Plenum, New York.
13. Derbyshire, W. (1976). *Chem. Soc. Spec. Period. Rep*; *Nuclear Magnetic Resonance* **5**, 264.
14. Derbyshire, W. (1978). *Chem. Soc. Spec. Period. Rep*; *Nuclear Magnetic Resonance* **7**, 193.
15. Packer, K. J. (1977). *Phil. Trans. Roy. Soc. Lond. Ser. B* **278**, 59.
16. Knipsel, R. R., Thomson, R. T. and Pintar, M. M. (1974). *J. Magn. Reson.* **14**, 44.
17. Koenig, S. H. and Shillinger, W. E. (1969). *J. Biol. Chem.* **244**, 3283.
18. Daszkiewicz, O. K., Hennel, J. W., Lubas, B. and Szczepkowski, T. W. (1963). *Nature* **200**, 1006.
19. Wennerstrom, H. (1972). *Mol. Phys.* **24**, 69.
20. Lillford, P. J., Clark, A. M. and Jones, D. V. (1980). *ACS Symp. Ser.* **127**, 177.
21. Held, G., Noack, F., Pollack, V. and Melton, B. (1973). *Z. Naturforsch.* **28c**, 59.
22. Escaynye, J. M., Canet D. and Robert, J. (1982). *Biochim. Biophys. Acta.* **721**, 305.
23. Koenig, S. H., Brown, R. D., Adams, D., Emerson, D. and Harrison, C. G. (1983). *IBM Research Report* **RC 10116**, No. 44807.
24. Brownstein, K. R. and Tarr, C. E. (1979). *Phys. Rev.* **19**, 2446.
25. Brindle, K. M., Brown, F. F., Campbell, I. D., Grathwohl, C. and Kuchel, P. W. (1979). *Biochem. J.* **180**, 37.
26. Brown, F. F. (1983). *J. Magn. Reson.* **54**, 385.
27. King, G. F., York, M. J., Chapman, B. E. and Kuchel, P. W. (1983). *Biochem. Biophys. Res. Commun.* **110**, 305.
28. Endre, Z. H., Kuchel P. W. and Chapman, B. E. (1984). *Biochim. Biophys. Acta.* **803**, 137.
29. Flibotte, S. G., Menon, R. S., Mackay, A. L. and Hailey, J. R. T. (1990). *Wood and Fiber Science,* **22**, 362.
30. Hazelwood, C. F., Chang, D. C., Nichols, B. L. and Woessner, D. E. (1974). *Biophys. J.* **14**, 583.
31. Foster, K. R., Resing, H. A. and Garroway, A. N. (1976). *Science* **194**, 324.
32. Hollis, D. P., Saryan, L. A., Eggleston, J. C. and Morris, H. P. (1975). *J. Natl. Cancer Institute* **54**, 1469.

33. Lauterbur, P. C., Frank, J. A. and Jacobson, M. J. (1976). *4th Dig. Int. Conf. Med. Phys., Physics in Canada*, **32**, Abstract 33.9.
34. Damadian, R., Zaner, K., Hor, D. and DiMaio, R. (1974). *Proc. Natl. Acad. Sci. U.S.A.* **71**, 1471.
35. Cameron, I. L., Ord, V. A. and Fullerton, G. D. (1984). *Magn. Reson. Imaging* **2**, 97.
36. Bovey, F. A. and Jelinski, L. W. (1985). *J. Phys. Chem.* **89**, 571.
37. Heatly, P. (1980). *Progr. NMR Spectr.* **13**, 47.
38. Ferry, J. D. (1980). *Viscoelastic properties of polymers*, (3rd ed), Wiley, New York.
39. Kimmich, R. (1984). *Polymer* **25**, 187.
40. Callaghan, P. T. (1988). *Polymer* **29**, 1951.
41. Huirua, T. M., Wang, R. and Callaghan, P. T. (1990). *Macromolecules* **23**, 1658.
42. Charlesby, A. and Jaroszkiewicz, E. M. (1985). *Eur. Polym. J.* **21**, 55.
43. McBrierty, V. J. (1983). *Magn. Reson. Rev.* **8**, 165.
44. Andreis, M. and Koenig, J. L. (1989). *Adv. Polym. Sci.* **89**, 69.
45. Chang, C. and Komorowski, R. A. (1989). *Macromolecules* **22**, 600.
46. Kuhn, W. (1990). *Angewandte Chemie* **29**, 1.
47. Connelly, A., Lohman, J. A. B., Loughman, B. C., Quiquampoix, H. and Ratcliffe, R. G. (1987). *J. Exp. Bot.* **38**, 1713.
48. Bacic, G. and Ratkovic, S. (1984). *Biophys. J.* **45**, 767.
49. Harrison, L. G., Luck, S. D., Munasinghe, B. D. J. P. and Hall, L. D. (1988). *J. Cell Sci.* **91**, 379.
50. Tamiya, T., Miyazaki, T., Ishikawa, H., Iriguchi, A., Maki, T., Matsumoto, J. and Tsuchiya, T. (1988). *J. Biochem.* **104**, 5.
51. Jenner, C. F., Xia, Y., Eccles, C. D. and Callaghan, P. T. (1988). *Nature* **336**, 399.
52. Cofer, G. P., Brown, J. M. and Johnson, G. A. (1989). *J. Magn. Reson.* **83**, 608.
53. Eccles, C. D., Callaghan, P. T. and Jenner, C. F. (1988). *Biophys. J.* **53**, 77.
54. Wang, P. C. and Chang, S. J. (1986). *Wood Fiber Sci.* **18**, 308.
55. Hall, L. D. and Rajanayagam, V. (1986). *Wood Sci. Technology* **20**, 329.
56. Hall, L. D., Rajanayagam, V., Stewart W. A. and Steiner, P. R. (1986). *Can. J. For. Res.* **16**, 423.
57. Hall, L. D., Rajanayagam, V., Stewart, W. A., Steiner, P. R. and Chow, S. (1986). *Can. J. For. Res.* **16**, 684.
58. Menon, R. S., Mackay, A. L., Flibotte, S. and Hailey, J. R. T. (1989). *J. Magn. Reson.* **82**, 205.
59. Eccles, C. D. and Callaghan, P. T. (1986). *J. Magn. Reson.* **68**, 393.
60. Bottomley, P. A., Rogers, H. H. and Foster, T. H. (1986). *Proc. Natl. Acad. Sci. U.S.A.* **83**, 87.
61. Kamei H. and Katayama, Y. (1986). *Proc. 8th Ann. Conf. IEEE Engin. in Med. and Biol. Soc.*, Forth Worth. p. 1159.
62. Johnson, G. A., Brown, J. and Kramer, P. J. (1987). *Proc. Natl. Acad. Sci. U.S.A.* **84**, 2752.
63. Kuhn, W. and Mattingley, M. (1988). *Special Bruker report on NMR microscopy*, Bruker GmbH, Karlsruhe.

64. Pope, J. M. and Sarafis, V. (1990). *Chemistry in Australia*. Royal Australian Chemical Institute, Melbourne. **57**, 221.
65. Mateescu, G. D., Yvars, G. M., Pazara, D. I., Alldridge, N. A., Lamanna, J. C., Lust, W. D., Mattingley, M. and Kuhn, W. (1989). In *Synthesis and applications of isotopically labelled compounds*, (eds. T. A. Bailey and J. R. Jones), Elsevier, Amsterdam.
66. Hedlund, L. W., Johnson, G. A. and Mills, G. I. (1986). *Investigative Radiology* **21**, 843.
67. Johnson, G. A., Thompson, M. B., Gewalt, S. L. and Hayes, C. E. (1986). *J. Magn. Reson*. **68**, 129.
68. Johnson, G. A., Thompson, M. B. and Drayer, B. P. (1987) *Magn. Reson. Med*. **4**, 351.
69. Hollet, M. D., Cofer, G. P. and Johnson, G. A. (1987). *Investigative Radiology* **22**, 965.
70. Dockery, S. E., Suddarth, S. A. and Johnson, G. A. (1989). *Magn. Reson. Med*. **11**, 182.
71. Baker, H., Lindsey, J. and Weibroth, S. *The laboratory rat: research applications*, (1980). Academic Press, New York.
72. Fox, J., Cohen, B. and Loew, F. (1984). *Laboratory animal medicine*, Academic Press, New York.
73. Andrews, E., Ward, B. and Altman, N. (1979). *Spontaneous animal models of human disease*, Academic Press, New York.
74. Schnall, M. D., Barlow, C., Subramanian, V. H. and Leigh, J. S. (1986). *J. Magn. Reson*. **68**, 161.
75. Rudin, M. (1987). *Magn. Reson. Med*. **5**, 443.
76. Sarkar, S. K., Clark, R. C., Rycyna, R. E., Mattingley, M. M. and Greig, R. (1989). *Magn. Reson. Med*. **12**, 268.
77. Kuhn, W. and Mattingley, M. (1989). *Special Bruker report on NMR microscopy*, Bruker GmbH, Karlsruhe.
78. Conner, W. E., Johnson, G. A., Cofer, G. P. and Dittrich, K. (1988). *Experimentia* **44**, 11.
79. McFarland, E. W., Neuringer, L. J. and Kushmerick, M. J. (1988). *Magn. Reson. Imaging* **5**, 507.
80. Listerud, J. M., Sinton, S. W. and Drobny, G. P. (1989). *Analytical Chemistry* **61**, 23A.
81. Lohman, J. A. and Gassner, G. (1988). *Ann. of N.Y. Acad. Sci*. **508**, 435.
82. Gassner, G. A. and Lohman, J. A. B. (1987). *Proc. Natl. Acad. Sci. U.S.A*. **84**, 5297.
83. Frahm, J., Merboldt, K. D., Hanicke, W. and Haase, A. (1985). *J. Magn. Reson*. **64**, 81.
84. Haase, A. and Frahm, J. (1985). *J. Magn. Reson*. **64**, 81.
85. Aguayo, J. B., Blackband, S. J., Shoeniger, J., Mattingley, M. M. and Hinterman, M. (1986). *Nature* **322**, 198.
86. Sillerud, L. O., Freyer, J. P., Neeman, M. and Mattingley, M. M. *Magn. Reson. Med*. In press.
87. Bittoun, J., Saint-Jalmes, H., Querleux, B., Jolivet, O., Darrasse, L., Idy-Peretti, I., Leroy-Willig, A. and Leveque, J. L. (1989). *Abstracts of 8th Annual Meeting, SMRM*, Amsterdam, p. 289.

88. Listerud, J., Beaulieu, C., Drobny, G. and Clark, J. (1989). *Abstracts of 8th Annual Meeting, SMRM*, Amsterdam, p. 288.
89. Lauterbur, P.C. (1986). In *NMR in biology and medicine* (eds. S. Chien and C. Ho), Raven Press, New York.
90. Zhou, X., Alameda Jr, J. C., Magin, R. L. and Lauterbur, P. C. (1989). *Abstracts of 8th Annual Meeting, SMRM*, Amsterdam, p. 1121.
91. Meyer, R. A. and Brown, T. R. (1988). *J. Magn. Reson.* **76**, 393.
92. Burstein, D., Fox, B. D. and Mattingley, M. M. (1989). *Bruker Reports* **2**, 12.
93. Tewari, M. (1989). *Eng. Sci. Rev.*, **XXXI**, No. 1, Fall 1989, p. 12. Student Media Board, Case Western Reserve University, (describing work by G. Mateescu)
94. Perry, B. C. and Koenig, J. L. (1989). *J. Polym. Sci., Polym. Chem.* **27**, 3429.
95. Clough, R. S. and Koenig, J. L. (1989). *J. Polym. Sci., Polym. Lett.* **27**, 451.
96. Hoh, K. P., Perry, B., Rotter, G., Ishida, H. and Koenig, J. L. (1989). *J. Adhesion* **27**, 245.
97. Nieminen, A. O. K. and Koenig, J. L., (1988). *J. Adhesion Sci. Technol.* **2**, 407.
98. Nieminen, A. O. K. and Koenig, J. L. (1989). *Appl. Spectr.* **43**, 1358.
99. Nieminen, A. O. K., Liu, J. and Koenig, J. L. (1989). *J. Adhesion Sci. Technol.* **3**, 455.
100. Hayashi, K., Kawashima, K., Kose, K. and Inouye, T. (1988). *J. Phys. D* **21**, 1037.
101. Ellingson, W. A., Ackerman, J. L., Garrido, L., Weyand, J. D. and deMilia, R. A. (1987). *Ceram. Eng. Sci. Proc.* **8**, 503.
102. Garrido, L., Ackerman, J. L., Ellingson, W. A. and Weyand, J. D. (1988). *Proc. Conf. Compos. Adv. Ceram.*, Cocoa Beach, Florida.
103. Vander Heiden, B. and Hurd, R. E. (1986). *CSI 2T: Non-medical applications*, General Electric Company publication, Fremont, California.
104. Rothwell, W. P., Holecek, D. R. and Kershaw, J. A. (1984). *J. Polym. Sci.* **22**, 241.
105. Rothwell, W. P. and Gentempo, P. P. (1985). *Bruker Reports* **1**, 46.
106. Samoilenko, A. A., Artemov, D. Yu. and Sobeldina, L. A. (1987). *Bruker Reports* **2**, 30.
107. Resing, H. A. and Miller, J. B. *Proc. 1987 US Army CRDEC Conf. on Chem. Defence Research*,
108. Chingas, G. C., Milliken, J., Resing, H. A. and Tsang, T. (1985). *Synthetic Metals* **12**, 131.
109. Pickup, S. and Blum, F. D. In *Solid state NMR of polymers*, (ed L. Mathias). In press.
110. Blackband, S. and Mansfield, P. (1986). *J. Phys. C* **19**, L49.
111. Weisenberger, L. A. and Koenig, J. L. (1989). *Appl. Spectr.* **43**, 1117.
112. Weisenberger, L. A. and Koenig, J. L. (1989). *J. Polym. Sci, Polym. Lett.* **27**, 55.
113. Crank, J. (1975). *The mathematics of diffusion*, Oxford University Press.
114. Matano, C. (1932). *Japan J. Phys.* **8**, 109.
115. Marechi, T. H., Donstrup, S. and Rigamonti, A. (1988). *J. Mol. Liquids* **38**, 185.
116. Rothwell, W. P. and Vinegar, H. J. (1985). *Applied Optics* **24**, 3969.
117. Edelstein, W. A., Vinegar, H. J., Tutunjian, P. N., Roemer, P. B. and

Mueller, O. M. (1988). *Soc. Petrol. Eng. 63 Ann. Conf. Proceedings*, Houston, p. 101.
118. Vinegar, H. J. (1986). *J. Petrol. Techn.* (March) p. 257.
119. Haacke, E. M. and Bellon, E. M. (1988). In *Magnetic resonance imaging* (ed. D. D. Stark and W. G. Bradley), C. V. Mosby, St Louis, p. 138.
120. Komorowski, R. A. (1986). In *High resolution NMR spectroscopy of synthetic polymers in bulk*, (ed. R. A. Komorowski), VCH publishers, Deerfield Beach, Fla. p. 121.
121. Assink, R. A., Caprihan, A. and Fukushima, E. (1988). *AIChE Journal* **34**, 2077.
122. Callaghan, P. T., Eccles, C. D. and Xia, Y. (1988). *J. Phys. E* **21**, 820.
123. Xia, Y. and Callaghan, P. T. (1990). *Die. Makrom, Chemie. Macromol. Symp.* **34**, 277.
124. Ernst, R. R. (1987). *Q. Rev. Biophysics*, **19**, 183.
125. Haase, A., Frahm, J., Matthei, D., Hanicke, W. and Merboldt, K. D. (1985). *Abstracts of 4th Annual Meeting, SMRM*, London, p. 980.
126. Haase, A., Frahm, J., Matthei, D., Hanicke, W. and Merboldt, K. D. (1986). *J. Magn. Reson.* **67**, 258.
127. Haase, A. (1990). *Magn. Reson. Med.* **13**, 77.
128. Liu, J., Nieminen, A. O. K. and Koenig, J. L. (1989). *J. Magn. Reson.* **85**, 95.
129. Glasel, J. A. and Lee, K. H. (1974). *J. Am. Chem. Soc.* **96**, 970.
130. Bull, T. E. and Tiddy, G. J. T. (1977). *J. Am. Chem. Soc.* **97**, 236.
131. Pickett, A. G. and Lemcoe, M. M. (1959). *J. Geophys. Res.* **64**, 1579.
132. James, F. and Roos, M. (1975). *Comput. Phys. Commun.* **10**, 343.
133. Menon, R. S., Mackay, A. L., Flibotte, S. G. and Hailey, J. R. T. (1989). *J. Magn. Reson.* **82**, 205.
134. Armstrong, R. L., Tzalmona, A., Menzinger, M., Cross, A. and Lemaire, C. (1990). *Abstracts, XXV Congress Ampere*, Stüttgart.
135. Field, R. J. and Burger M. (eds.) (1984). *Oscillations and travelling waves in chemical systems*, Wiley, New York.
136. Fyfe, C. A. (1990). *Abstracts, 31st ENC Meeting*, Asilomar, California.
137. Rommel, E. and Kimmich, R. (1989). *Magn. Reson. Med.* **12**, 209.
138. Brasch, R. C. (1983). *Radiology* **147**, 781.
139. Niendorf, H. P., Semmler, W., Laniado, M. and Felix, R. (1986). *Diagn. Imag. Clin. Med.* **55**, 25.
140. Huberty, J., Engelstad, B., Wesbey, G., Moseley, M., Young, G., Tuck, D., Brito, A., Hattner, R. and Brasch, R. (1983). *Abstracts for Society of Magnetic Resonance in Medicine Meeting*, San Francisco, p. 175.
141. Runge, V. M., Stewart, R. G., Clanton, J. A., Jones, M. M., Lukehart, C. M., Partain, C. L. and James, A. E. (1983). *Radiology* **147**, 789.
142. Carr, D. H., Brown, J., Bydder, G. M., Steiner, R. E., Weinmann, H. J., Speck, U., Hall, A. S. and Young, I. R. (1984). *Am. J. Roentg.* **143**, 215.
143. Weinmann, H. J., Brasch, R. C., Brasch, W. R. and Wesby, G. E. (1984). *Am. J. Roentg.* **142**, 619.
144. Brown, M. A. (1985) *Magn. Reson. Imaging* **3**, 3.
145. Cory, D. G., Reichwein, A. M. and Veeman, W. S. (1988). *J. Magn. Reson.* **80**, 259.
146. Liu, J., Nieminen, A.O.K. and Koenig, J. L. (1989). *Appl. Spectr.* **43**, 1260.
147. Bendel, P., Lai, C. M. and Lauterbur, P. C. (1980). *J. Magn. Reson.* **38**, 343.

148. Hall, L. D. and Sukumar, S. (1982). *J. Magn. Reson.* **50**, 161.
149. Hall, L. D. and Sukumar, S. (1984). *J. Magn. Reson.* **56**, 179.
150. Maudsley, A. A., Hilal, S. K., Perman, W. H. and Simon, H. E. (1983). *J. Magn. Reson.* **51**, 147.
151. Martin, J. F. and Wade, C. G. (1985). *J. Magn. Reson.* **61**, 153.
152. Jeener, J. (1981). *Ampère International Summer School*, Basko Poljie, Yugoslavia.
153. Doyle, M. and Mansfield, P. (1987). *Magn. Reson. Med.* **5**, 255.
154. Bowtell, R., Cawley, M. G. and Mansfield, P. (1989). *J. Magn. Reson.* **82**, 634.
155. Guilfoyle, D. N., Blamire, A., Chapman, B., Ordidge, R. J. and Mansfield, P. (1989). *Magn. Reson. Med.* **10**, 282.
156. Lauterbur, P. C., Levin, D. N. and Marr, R. B. (1984). *J. Magn. Reson.* **59**, 536.
157. Hall, L. D. and Sukumar, S. (1984). *J. Magn. Reson.* **56**, 314.
158. Gordon, R. E., Hanley, P. E., Shaw, D., Gadian, D. G., Radda, G. K., Styles, P., Bore, P. J. and Chan, L. (1980). *Nature* **287**, 367.
159. Gordon, R. E., Hanley, P. E. and Shaw, D. (1982). *Progr. NMR Spectr.* **15**, 1.
160. Doddrell, D. M., Bulsing, J. M., Galloway, G. J., Brooks, W. M., Field, J., Irving, M. G. and Baddeley, H. (1986). *J. Magn. Reson.* **70**, 319.
161. Aue, W. P., Mueller, S., Cross, T. A. and Seelig, J. (1984). *J. Magn. Reson.* **56**, 350.
162. Haase, A., Hanicke, W. and Frahm, J. (1984). *J. Magn. Reson.* **56**, 401.
163. Bottomley, P. A., Foster, T. B. and Darrow, R. D. (1984). *J. Magn. Reson.* **59**, 338.
164. Bendall, M. R. and Gordon, R. E. (1983). *J. Magn. Reson.* **53**, 365.
165. Bendall, M. R. (1984). *J. Magn. Reson.* **59**, 406.
166. Bendall, M. R., McKendry, J. M., Cresshull, I. D. and Ordidge, R. J. (1984). *J. Magn. Reson.* **60**, 473.
167. Shaka, A. J., Keeler, J., Smith, M. B. and Freeman, R. (1985). *J. Magn. Reson.* **61**, 175.
168. Ordidge, R. J., Connelly, A. and Lohman, J. A. B. (1986). *J. Magn. Reson.* **66**, 283.
169. Connelly, A., Counsell, C. and Lohman, J. A. B. (1988). *J. Magn. Reson.* **78**, 519.
170. Doddrell, D. M., Field, J., Crozier, S., Brereton, I. M., Galloway, G. J. and Rose, S. E. (1990). *Magn. Reson. Med.* **13**, 518.
171. Brereton, I. M., Rose, S. E., Galloway, G. J., Nixon, L. N. and Doddrell, D. M. *Magn. Reson. Med.* In press.
172. Hu, X., Levin, D. N., Lauterbur, P. C. and Spraggins, T. (1988). *Magn. Reson. Med.* **8**, 314.
173. Lauterbur, P. C. and Lee, H. (1989). *International Society of Magnetic Resonance Meeting*, Morzine, Abstracts, pp. 11–14.
174. Prestegard, J. H. and Scarsdale, J. N. (1985). *J. Magn. Reson.* **62**, 136.
175. Sotak, C., Freeman, D. M. and Hurd, R. E. (1988). *J. Magn. Reson.* **78**, 355.
176. Thomas, M. A., Meyerhof, D. J., Hetherington, H. P. and Weiner, M. W. (1989). *Abstracts of 8th Annual Meeting, SMRM*, Amsterdam, p. 1115.
177. Dumoulin, C. L. (1985). *J. Magn. Reson.* **64**, 38.
178. Hall, L. D. and Norwood, T. J. (1986). *J. Magn. Reson.* **67**, 382.

179. Maerefat, N. L., Palmer, I., Yamanashi, W. S. and Lester, P. D. (1986). *Magn. Reson. Imaging* **4**, 122.
180. Lambert, R. K., Pack, R. J., Xia, Y., Eccles, C. D. and Callaghan, P. T. (1988). *Journal of Applied Physiology* **65**, 1872.
181. Hall, L. D. and Rajanayagam, V. (1987). *J. Magn. Reson.* **74**, 139.
182. Hore, P. J. (1983). *J. Magn. Reson.* **54**, 539.
183. Hore, P. J. (1983). *J. Magn. Reson.* **55**, 283.
184. Doddrell, D. M., Galloway, G. J., Brooks, W. M., Field, J., Bulsing, J. M., Irving, M. G. and Baddeley, H. (1986). *J. Magn. Reson.* **70**, 176.
185. Galloway, G. J., Haseler, L. J., Marshman, M. F., Williams, D. H. and Doddrell, D. M. (1987). *J. Magn. Reson.* **74**, 184.
186. Blondet, P., Decorps, M., Albrand, J. A., Benabid, A. L. and Remy, C. (1986). *J. Magn. Reson.* **69**, 403.
187. Forsen, S. and Hoffman, R. A. (1963). *J. Chem. Phys.* **39**, 2892.
188. Kelly, D. A. C., Graham, R., Kappler, F., Kowalyshyn, M., Keller, S. and Brown, T. R. (1989). *Abstracts of 8th Annual Meeting, SMRM*, Amsterdam, p. 1036.
189. Renshaw, P. F., Haelgrove, J. C., Leigh, J. S. and Chance, B. (1985). *Magn. Reson. Med.* **2**, 512.
190. Renshaw, P. F., Haelgrove, J. C., Bolinger, L., Chance, B. and Leigh, J. S. (1986). *Magn. Reson. Imaging* **4**, 193.
191. Haase, A. and Hopfel, D. (1988). *J. Magn. Reson.* **80**, 389.
192. Joseph, P. M., Fishman, J. E., Mukherji, B. and Sloviter, H. A. (1985). *J. Comput. Assisted Tomography.* **9**, 1012.
193. Thomas, S. R., Clark, L. C., Ackerman, J. L., Pratt, R. G., Hoffman, R. E., Busse, L. J., Kinsey, R. A. and Samaratunga, R. C. (1986). *J. Comput. Assisted Tomography* **10**, 1.
194. Koutcher, J. A., Rosen, B. R., Metarland, E. W., Telcher, B. A. and Brady, T. J. (1985). *Abstracts of 4th Annual Meeting, SMRM*, London, p. 435.
195. Nunnally, R. L., Babock, E. E., Horner, S. D. and Peshok, R. M. (1985). *Magn. Reson. Imaging* **3**, 399.
196. Chew, W. M., Moseley, M. E., Mills, P. A., Sessler, D., Gonzalez-Mendez, R., James, T. L. and Litt, L. (1987). *Magn. Reson. Imaging* **5**, 51.
197. De Layre, J. L., Ingwall, J. S., Malloy, C. and Fossel, E. T. (1981). *Science* **212**, 935.
198. Hilal, S. K., Maudsley, A. A., Simon, H. E., Perman, W. H., Bonn, J., Mawad, M. E. and Silver, A. J. (1983). *Am. J. Neurol. Res.* **4**, 245.
199. Feinberg, D. A., Crooks, L. A., Kaufman, L., Zawadski, M. B., Posin, J. P., Arakawa, M., Watts, J. C. and Hoenniger, J. H. (1985). *Radiology* **156**, 133.
200. Hilal, S. K., Maudsley, A. A., Ra, J. B., Simon, H. E., Roschmann, P., Wittkoek, S., Cho, Z. N. and Mun, S. K. (1985). *J. Comp. Assisted Tomography* **9**, 1.
201. Moseley, M. E., Chew, W. M., Nishimura, M. C., Richards, T. L., Murphy-Boesch, J., Young, G. B., Pitts, L. H. and James, T. L. (1986). *Magn. Reson. Imaging* **3**, 383.
202. Joseph, P. M. and Summers, R. M. (1986). *J. Magn. Reson.* **69**, 198.
203. Granot, J. (1986). *J. Magn. Reson.* **68**, 578.
204. Burstein, D., Fox, B. D. and Mattingley, M. M. (1989). *Bruker Reports* **2**, 12.
205. Burstein, D. and Mattingley, M. M. (1989). *J. Magn. Reson.* **83**, 197.

206. Maudsley, A. A. and Hilal, S. K. (1984). *Br. Med. Bull.* **40**, 165.
207. Hall, L. D. and Webb, A. G. (1989). *J. Magn. Reson.* **83**, 371.
208. Hall, L. D. and Talagala, S. L. (1985). *J. Magn. Reson.* **65**, 501.
209. Hsieh, P. S. and Balaban, R. S. (1987). *J. Magn. Reson.* **74**, 574.
210. Mueller, S. and Seelig, J. (1987). *J. Magn. Reson.* **72**, 456.
211. Ewy, C. S., Babcock, E. E. and Ackerman, J. L. (1986). *Magn. Reson. Imaging* **4**, 407.
212. Link, J. and Seelig, J. (1990). *J. Magn. Reson.* **89**, 310.
213. Kormos, D. W., Yeung, H. N. and Gauss, R. C. (1987). *J. Magn. Reson.* **71**, 159.
214. Kormos, D. W. and Yeung, H. N. (1987). *Magn. Reson. Med.* **4**, 500.
215. Sillerud, L. O., Alger, J. R. and Shulman, R. G. (1981). *J. Magn. Reson.* **45**, 142.
216. Sillerud, L. O., Van Hulsteyn, D. B. and Griffey, R. H. (1988). *J. Magn. Reson.* **76**, 380.
217. Moon, R. B. and Richards, J. H. (1973). *J. Biol. Chem.* **248**, 7276.
218. Pople, J. A., Schneider, W. G. and Bernstein, H. J. (1959). *High resolution nuclear magnetic resonance*, McGraw Hill, New York.
219. Moseley, M. (1990). GEC applications note.
220. Suits, B. H. and Lutz, J. L. (1989). *J. Appl. Phys.* **65**, 3728.
221. Ernst, R. R., Bodenhausen, G. and Wokaun, A. (1987). *Principles of nuclear magnetic resonance in one and two dimensions*, Clarendon Press, Oxford.
222. Yeung, H. N. and Swanson, S. D. (1989). *J. Magn. Reson.* **83**, 183.
223. Knüttel, A., Kimmich, R. and Spohn, K. H. (1990). *J. Magn. Reson.* **86**, 526.
224. Knüttel, A., Spohn, K. H. and Kimmich, R. (1990). *J. Magn. Reson.* **86**, 542.
225. Suits, B. H. and White, D. (1984). *Solid State Communications* **50**, 291.
226. Szeverenyi, N. M. and Maciel, G. E. (1984). *J. Magn. Reson.* **60**, 460.
227. Miller, J. B. and Garroway, A. N. (1989). *J. Magn. Reson.* **85**, 255.
228. Hartmann, S. R. and Hahn, E. L. (1962). *Phys. Rev.* **128**, 2042.
229. Maciel, G. E. and Davis, M. F. (1985). *J. Magn. Reson.* **64**, 356.
230. Rommel, E. and Kimmich, R. (1989). *J. Magn. Reson.* **83**, 299.
231. Cory, D. *Abstracts, 31st ENC Meeting*, (1990). Asilomar, California.
232. Wind, R. A. and Yannoni, C. S. (1979). *J. Magn. Reson.* **36**, 269.
233. Emid, S. (1985). *Physica* **128B**, 79.
234. Emid, S. and Creyghton, J. H. N. (1985). *Physica* **128B**, 81.
235. McDonald, P. J., Attard, J. J. and Taylor, D. G. (1987). *J. Magn. Reson.* **72**, 224.
236. Rommel, E., Hafner, S. and Kimmich, R. (1990). *J. Magn. Reson.* **86**, 264.
237. Garroway, A. N., Baum, J., Munowitz, M. G. and Pines, A. (1984). *J. Magn. Reson.* **60**, 337.
238. Chingas, G. C., Miller, J. B. and Garroway, A. N. (1986). *J. Magn. Reson.* **66**, 530.
239. Miller, J. B. and Garroway, A. N. (1988). *J. Magn. Reson.* **77**, 187.
240. Miller, J. B. and Garroway, A. N. (1988). *Review of progress in quantitative nondestructive evaluation*, Vol 7A. (ed. D. O. Thompson and D. E. Chimenti), Plenum, New York.
241. Miller, J. B. and Garroway, A. N. (1988). *ACS Polym. Preprints* **29**, 90.
242. Miller, J. B. and Garroway, A. N. (1989). *J. Magn. Reson.* **82**, 529.
243. Wind, R. A. and Yannoni, C. S. (1981). *U.S. Patent No. 4.*, 301, 401, Nov. 17.

244. de Luca, F. and Maraviglia, B. (1986). *J. Magn. Reson.* **67**, 169.
245. Cory, D. G., Van Os, J. W. M. and Veeman, W. S. (1988). *J. Magn. Reson.* **76**, 543.
246. Cory, D. G., Reichwein, A. M., Van Os, J. W. M. and Veeman, W. S. (1988). *Chem. Phys. Lett.* **143**, 467.
247. Cory, D. G., de Boer, J. C. and Veeman, W. S. (1989). *Macromolecules* **22**, 1618.
248. Mansfield, P. and Grannell, P. K. (1975). *Phys. Rev.* **B12**, 3618.
249. Miller, J. B. and Garroway, A. N. (1986). *J. Magn. Reson.* **67**, 575.
250. Burum, D. P., Linder, M. and Ernst, R. R. (1981). *J. Magn. Reson.* **44**, 173.
251. McDonald, P. J. and Tokarczuk, P. F. (1989). *J. Phys. E* **22**, 948.
252. Cory, D. G., Miller, J. B., Turner, R. and Garroway, A. N. (1990) *Mol. Phys.* **70**, 331.
253. Pines, A. and Waugh, J. S. (1972). *J. Magn. Reson.* **8**, 354.
254. Cory, D. G. and Veeman W. S. (1989). *J. Magn. Reson.* **84**, 392.
255. Mehring, M. (1983). *Principles of high resolution NMR in solids*, Springer-Verlag, Berlin.
256. Gunter, E., Blumich, B. and Speiss, H. W. *Mol. Phys.* In press.

6

THE MEASUREMENT OF MOTION USING SPIN ECHOES

6.1 Motional contrast and microstructure

In Chapter 5 the concept of image contrast was introduced. The nuclear magnetic resonance spectrum is rich in details which reveal the molecular environment of the nucleus. Chemical shift contrast and other more sophisticated methods of spectral editing can provide molecularly selective images while other schemes take advantage of the various characteristic time-scales in NMR, giving effects which are highly sensitive to the motion of the surrounding molecule. For example, the use of relaxation time contrast can differentiate molecules via their characteristic rotational motions. The suggestion that nuclear magnetic resonance could be used to measure molecular translational motion dates from the earliest days of this field of research.[1,2]

In this chapter we shall be exclusively concerned with translational motion. We have already seen how important translational diffusion is in affecting the intrinsic resolution of NMR microscopy. This is because random molecular displacements in the presence of magnetic field gradients impart a phase spread to the spin transverse magnetization. While phase effects due to motion can sometimes degrade the image quality, when carefully employed they can provide a useful signature. A powerful tool in the measurement process is the spin echo. The formation of an echo arises because precessional phase shifts associated with the position of a molecule in a magnetic field gradient are cancelled by reversing the phase evolution in some manner. Any movement of a molecule over this time-scale results in residual phase offsets and the effect on the echo signal can be used to deduce the form of the motion. The sensitivity of spin echoes to both random and directed translational motion was pointed out by Hahn[2,3] and by Carr and Purcell[4] in their classic papers on time domain NMR.

Since NMR is capable of giving information about molecular translational motion, we are led to ponder two questions. The first concerns the imaging of microstructure. In a heterogeneous structure the motion of molecules is interrupted by interfaces. For example, the intravascular water in a plant moves within the confines of the vascular geometry. In a food substance the lipids may be confined to isolated droplets. In a porous sandstone, the interpenetrating water is confined to the labyrinth of channels. Can a measurement of molecular motion reveal information about very small-scale structural geometry and, if so, is the spatial resolution limited by the same constraints as occur in imaging the static spin distribution? We shall see that certain geometries are indeed susceptible to such a probe and that in such cases the limits to resolution may be many orders of magnitude smaller than in static microscopy.

The second question concerns the motions themselves. Consider a sample in which there exists a 'large-scale' structural heterogeneity over a distance scale larger than a **k**-space imaging pixel but where the molecular translational motion differs from pixel to pixel. This might arise, for example, in a capillary which is much larger than one pixel but where the fluid experiences a continuous velocity shear in capillary flow. Alternatively, the microstructure may be smaller than the size of a pixel, for example, when dealing with restricted diffusion in biological cells. If the cell sizes vary across the image then a measurement of the local motions gives a useful clue concerning the microstructure. Can we obtain information about molecular translational motion in one pixel, even though the signal intensity may vary from place to place in the image? The answer again is yes. In particular we shall see that it is possible to obtain velocity and diffusion maps within a sample so that NMR microscopy can be used both to carry out velocimetry studies at a resolution of a few microns and to examine heterogeneity in local microstructure.

The examination of microstructure via the observation of restricted motion is the subject of Chapter 7 while the facility to spatially resolve motional behaviour by imaging methods is dealt with in Chapter 8. The combination of 'static' imaging methods with contrast schemes which reveal the spectrum of local translational motion represents an advanced stage of imaging. In effect we are attempting to separate the three dimensions of spatial location from the three dimensions of spin displacements over some predetermined time-scale. This sophisticated form of motional contrast is still in its infancy and as yet has been applied only to the simplest possible molecular motions, namely those which can be characterized by a local constant velocity and local self-diffusion coefficient.

In this chapter attention is focused on the determination of the motion itself. Thus we deal with a system for which the motions of the spins, while possibly complicated, are at least held in common by all spins. This may indeed be a uniform sample without heterogeneity on the pixel scale. Here the use of **k**-space imaging is unnecessary and the entire NMR signal may be analysed without the need for spatial reconstruction. Where the sample, or the motion, is non-uniform on the pixel scale we shall anticipate the experiments described in Chapter 8 and presume that we are dealing with the signal from one pixel.

6.2 Introduction to translational dynamics

6.2.1 *The conditional probability function and self-diffusion*

In NMR a signal is never detected from one spin alone but instead via a coherent superposition of signals from a very large number of spins. Because of this we need to adopt an 'ensemble-averaged' view in order to depict the overall behaviour. This section describes approaches for dealing with translational motions in molecular ensembles.

In general, the motion of a molecule i can be characterized by some time-dependent displacement $\mathbf{r}_i(t)$. This function will often vary in a random way from molecule to molecule so that one is forced to adopt the language of statistics. There are a variety of mathematical tools which can be helpful in this approach. For example, we can speak of the probability that a particle will have displacement $\mathbf{r}'$ at a time t. Usually, this probability function will depend not only on the time interval t but also on the starting position $\mathbf{r}$. One such function is well known in scattering theory and sometimes called the Van Hove correlation function,[5] $P(\mathbf{r}|\mathbf{r}', t)$. $P(\mathbf{r}|\mathbf{r}', t)$ gives the likelihood of finding *any* scattering centre 'positioned' at $(\mathbf{r}', t)$ if there was a scattering centre at $(\mathbf{r}, 0)$. The Van Hove function determines the nature of the inelastic scattering which will occur when X-rays, light, or neutrons are incident on an ensemble of moving scatterers.[6] The problem with the Van Hove function is that it correlates the positions of $\mathbf{r}_i'(t)$ and $\mathbf{r}_j(0)$ for $i \neq j$ as well as for $i = j$ and is therefore sensitive to relative motions.

There is another conditional probability which describes the correlations when $i = j$ only. This is the *self*-correlation function,[7] $P_s(\mathbf{r}|\mathbf{r}', t)$ which gives the chance that a molecule initially at $\mathbf{r}$ will have moved to $\mathbf{r}'$ after a time t. Obviously the facility to measure P_s would be very helpful since descriptions of self-motion are so much simpler. At the microscopic level only two techniques have the inherent labelling required to measure self-correlations. These are pulsed gradient spin echo (PGSE) NMR[8-11] and polarized neutron scattering[12-14] in which the polarized and depolarized scattering cross-sections separately reveal $P(\mathbf{r}|\mathbf{r}', t)$ and $P_s(\mathbf{r}|\mathbf{r}', t)$. In the case of neutron diffraction the depolarization of the neutrons is associated with spin flips in the scattering nuclei, an effect which makes it intrinsically possible to locate the scattering centre. The associated 'collapse of the wave-function' necessarily removes interference effects, a point which is delightfully described in the book by Feynman, Leighton, and Sands,[15] so removing all dynamical effects due to relative motion. An obvious analogy is possible in the case of nuclear magnetic resonance where the observed photon can be clearly identified with a specific nucleus via the characteristic Larmor frequency. Any sensitivity to motion in NMR necessarily reveals the self-motion rather than relative motion of the nuclei.

It is important to note that the total probability, $\Psi(\mathbf{r}', t)$ of finding a particle at position $\mathbf{r}'$ at time t is given by

$$\Psi(\mathbf{r}', t) = \int \Psi(\mathbf{r}, 0)\, P_s(\mathbf{r}|\mathbf{r}', t)\, d\mathbf{r} \tag{6.1}$$

where $\Psi(\mathbf{r}, 0)$ is, of course, just the particle density, $\rho(\mathbf{r})$.

A nice application of the function Ψ can be found in the example of self-diffusion. The classical description of diffusion is via Fick's law[16] where the particle flux (per unit area per unit time) is set proportional to the particle concentration gradient. For self-diffusion, there is no net concentration gradient. None the less, the Fick's law description is possible using $\Psi(\mathbf{r}', t)$ in place of the concentration, since, like a concentration function, $\Psi(\mathbf{r}', t)$ describes the likelihood of finding a particle in a certain place at a certain time. The difference in approach arises from the fact that, whereas concentration refers to the density of any particle of a particular type, $\Psi(\mathbf{r}', t)$ is a sort of ensemble-averaged probability concentration for a *single* particle. It is therefore appropriate in considering self-diffusion. Because the spatial derivatives in Fick's equations refer to the coordinate $\mathbf{r}'$, inspection of eqn (6.1) shows that they can be written in terms of P_s with the initial condition,

$$P_s(\mathbf{r}|\mathbf{r}', 0) = \delta(\mathbf{r}' - \mathbf{r}). \tag{6.2}$$

Thus

$$\mathbf{J} = -D\nabla' P_s \tag{6.3}$$

where $\mathbf{J}$ is the 'conditional probability flux'. Because the total conditional probability is conserved, the continuity theorem applies and

$$\nabla' \cdot \mathbf{J} = -\partial P_s / \partial t. \tag{6.4}$$

Combining eqns (6.3) and (6.4) we obtain the differential equation sometimes known as Fick's second law,

$$\partial P_s / \partial t = D\nabla'^2 P_s \tag{6.5}$$

where D is the molecular self-diffusion coefficient.

We will first obtain the solution to eqn (6.5) for the special boundary condition which applies for unrestricted self-diffusion. This is simply $P_s \to 0$ as $r' \to \infty$ and, combined with the initial condition of eqn (6.2), yields,

$$P_s(\mathbf{r}|\mathbf{r}', t) = (4\pi Dt)^{-3/2} \exp\{-(\mathbf{r}' - \mathbf{r})^2 / 4Dt\}. \tag{6.6}$$

The fact that P_s depends only on the net displacement, $\mathbf{r}' - \mathbf{r}$, and not the initial position, $\mathbf{r}$, reflects the Markov nature of Brownian motion statistics.[17] In a fundamental sense it is this net displacement which is measured in PGSE NMR. For the purpose of this book we will refer to the vector $\mathbf{r}' - \mathbf{r}$ moved over a time t as the dynamic displacement $\mathbf{R}$. This is illustrated in Fig. 6.1.

It is important to distinguish this dynamic displacement from the static quantity $\mathbf{r}$ measured in $\mathbf{k}$-space imaging. In neutron scattering the momentum transfer determines the dynamic displacement scale giving an upper limit of around 100 Å. The relevant time-scale of neutron scattering is determined by the energy resolution, the best result being achieved by the neutron spin echo method[18,19] for which the time-scales are a few microseconds or less. By contrast, PGSE NMR is limited, in practice, to dynamic displacements of between 100 Å and 100 μm and over time-scales of a few milliseconds to a few seconds. The two methods are therefore complementary. The dimensional scale of PGSE NMR corresponds to the 'organizational' domain. It includes structural features of macromolecular solutions, mesophase structure of liquid crystals, emulsions, porous solids, and highly disperse biological tissue.

Developing this concept of the dynamic displacement it is possible to use eqn (6.1) to define a new and very useful function, sometimes called the 'average propagator',[20] $\overline{P_s}(\mathbf{R}, t)$. This function gives the average probability for any particle to have a dynamic displacement $\mathbf{R}$ over a time t and is given by

$$\overline{P_s}(\mathbf{R}, t) = \int P_s(\mathbf{r}|\mathbf{r} + \mathbf{R}, t)\rho(\mathbf{r})\,\mathrm{d}\mathbf{r}. \tag{6.7}$$

In cases such as the free self-diffusion example just considered, where $P_s(\mathbf{r}|\mathbf{r} + \mathbf{R},t)$ is independent of starting position, the average propagator is common to all molecules in the ensemble so that the bar above may be dropped. We could then write

$$P_s(\mathbf{R}, t) = (4\pi Dt)^{-\frac{3}{2}} \exp(-\mathbf{R}^2/4Dt). \tag{6.8}$$

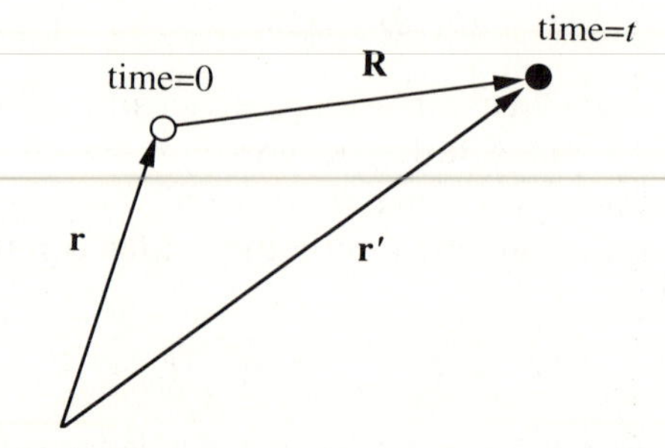

Fig. 6.1 Displacement of particle from $\mathbf{r}$ to $\mathbf{r}^1$ over time t.

For completeness we now make a simple extension to eqn (6.8) which applies when the diffusion is superposed on flow of velocity $\mathbf{v}$. In this case a term $\nabla' \cdot \mathbf{v}P_s$ must be added to the right-hand side of eqn (6.5) to give[10,21]

$$\partial P_s/\partial t = \nabla' \cdot \mathbf{v}P_s + D\nabla'^2 P_s. \tag{6.9}$$

If $\mathbf{v}$ is constant then the solution is

$$P_s(\mathbf{R}, t) = (4\pi Dt)^{-\frac{3}{2}} \exp\{-(\mathbf{R} - \mathbf{v}t)^2/4Dt\}. \tag{6.10}$$

Note that P_s is a normalized Gaussian function of dynamic displacement $\mathbf{R}$, with width increasing as time advances. Sometimes it is helpful to write $P_s(\mathbf{R}, t)$ in Cartesian component form as

$$P_s(\mathbf{R}, t) = (4\pi Dt)^{-\frac{3}{2}} \exp\{-(X - v_x t)^2/4Dt\} \exp\{-(Y - v_y t)^2/4Dt\} \times \exp\{-(Z - v_z t)^2/4Dt\} \tag{6.11}$$

and where we are concerned only with motions along one dimension (e.g. the z-axis) we can integrate over the remaining two dimensions to obtain

$$P_s(Z, t) = (4\pi Dt)^{-\frac{1}{2}} \exp\{-(Z - v_z t)^2/4Dt\}. \tag{6.12}$$

The behaviour of $P_s(Z, t)$ as a function of time is illustrated in Fig. 6.2 in the cases where the velocity is finite. It is a useful exercise in calculating ensemble-averaged properties to evaluate the mean square dynamic displacement for $v_z = 0$. Using a bar to represent an ensemble average,

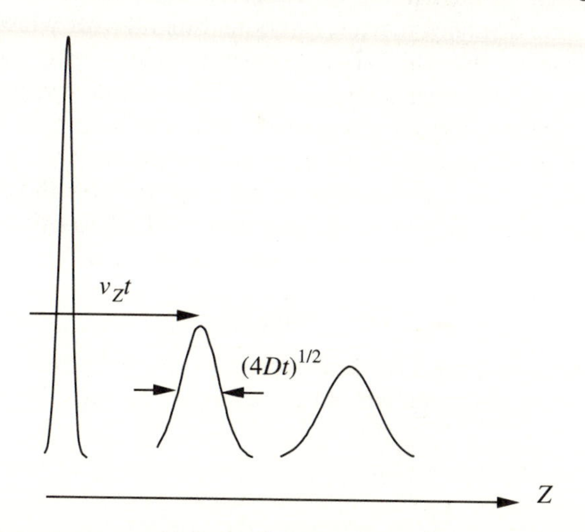

Fig. 6.2 Behaviour of the conditional probability function, P_S, for an ensemble of particles undergoing Brownian motion and flow. The successive Gaussians correspond to successively increasing time.

$$\overline{Z^2} = \int_{-\infty}^{\infty} Z^2 P_s(Z, t)\, \mathrm{d}Z$$
$$= 2Dt. \tag{6.13}$$

Once again it must be emphasized that the Markov statistics of self-diffusion are a special case and care must be exercised in using the propagator function $P_s(\mathbf{R}, t)$. Later we shall deal with specific examples of $P_s(\mathbf{r}|\mathbf{r}', t)$ for more complicated motion where no common propagator exists for all particles in the system. In such cases the averaging process appropriate to the NMR experiment must be carefully adopted.

6.2.2 *Velocity correlation, spectral density, and the self-diffusion tensor*

$P_s(\mathbf{r}|\mathbf{r}', t)$ helps us calculate averages across the ensemble of spins. Sometimes a problem naturally lends itself to a description in terms of this function but in other situations the connection is not so obvious. One alternative approach in dealing with behaviour which fluctuates with time is to define families of autocorrelation functions. Suppose we have some molecular quantity A which is a function of time. The autocorrelation function of A is[22,23]

$$G(t) = \int_{0}^{\infty} A(t')A(t' + t)\mathrm{d}t'.^{\dagger} \tag{6.14}$$

Now suppose that we are dealing with a *stationary ensemble*. This means that the origin of time does not matter, only the interval over which we measure. In effect this means that the NMR experiment can begin whenever we like and that the result will always be the same. This will be a very good assumption, except in those special cases dealt with in Chapter 8 where slow or periodic fluctuations occur. In a stationary ensemble the average over time implied by eqn (6.14) could equally be an average over the particles in the ensemble since one particle is representative of all particles over a long enough time interval. Incorporating these ideas we will rewrite $G(t)$ as

$$G(t) = \overline{A(0)A(t)}. \tag{6.15}$$

If we happen to know the function A in terms of position coordinates rather than time coordinates, then we can calculate $G(t)$ with the help of P_s since

$$\overline{A(0)A(t)} = \iint \rho(\mathbf{r}) P_s(\mathbf{r}|\mathbf{r}', t) A(\mathbf{r}') A(\mathbf{r})\, \mathrm{d}\mathbf{r}'\, \mathrm{d}\mathbf{r}. \tag{6.16}$$

In effect $G(t)$ measures the rate at which $A(t)$ 'loses memory' of its pre-

†Strictly speaking eqn (6.14) should read $A(t')A^*(t' + t)$ but we shall mostly be concerned with observable, and hence real, functions.

vious values. The time-scale for this loss of memory is called the correlation time, τ_c, and is defined by[22,23]

$$\tau_c = \frac{\int_0^\infty \overline{A(0)A(t)}\,dt}{\overline{A(0)^2}}. \tag{6.17}$$

In some experiments it is the spectrum, or Fourier transform, of $G(t)$ which is important. For example, where A corresponds to the dipolar interaction between spins which fluctuates as a molecule rotates, the spin relaxation times are sensitive to the spectrum of $G(t)$, and, in particular, to its value at certain characteristic frequencies. Notice that for rotational motion the treatment is similar to that shown in eqn (6.16) except for the use of angular displacement variables instead of translational displacement variables. In relaxation theory the spectrum of the dipolar Hamiltonian correlation function is known as the spectral density function[24] $J(\omega)$. In translational motion theory the spectrum of the velocity correlation function is known as the self-diffusion tensor,[25,26] $D_{\alpha\beta}(\omega)$, where α and β may take each of the Cartesian directions, x, y, z. Hence

$$D_{\alpha\beta}(\omega) = \tfrac{1}{2}\int_{-\infty}^{\infty} \overline{v_\alpha(0)v_\beta(t)}\exp(i\omega t)\,dt. \tag{6.18}$$

Generally we shall not be concerned with correlations between differing components of $\mathbf{v}(t)$ except where we deliberately apply magnetic field gradients which fluctuate in direction. We shall therefore focus on the diagonal elements of D and, using the even property of $G(t)$, we write

$$D_{zz}(\omega) = \int_0^{\infty} \overline{v_z(0)v_z(t)}\exp(i\omega t)\,dt. \tag{6.19}$$

Eqn (6.19) tells us that the zero-frequency component of D is simply the time integral of the velocity correlation function, i.e.,

$$D_{zz}(0) = \int_0^{\infty} \overline{v_z(0)\,v_z(t)}\,dt. \tag{6.20}$$

By definition, therefore, $D_{zz}(0)$ is $\overline{v_z^2}\,\tau_c$.

6.2.3 *The relationship between $\overline{v_z(0)v_z(t)}$ and P_s*

In order to find the relationship between $\overline{v_z(0)\,v_z(t)}$ and P_s we employ eqn (6.16). The correlation $\overline{z(t_1)z(t_2)}$ can be obtained from

$$\overline{\{z(t_1) - z(t_2)\}^2} = \overline{z^2(t_1)} + \overline{z^2(t_2)} - 2\,\overline{z(t_1)z(t_2)}$$

$$= -2\overline{[z(t_1)z(t_2) - \overline{z^2}]} \tag{6.21}$$

where the time independence, and hence identity, of $\overline{z^2(t_1)}$ and $\overline{z^2(t_2)}$ arises because of the stationary nature of the ensemble. The term in the square brackets is sometimes called the 'location correlation',[27] $\overline{z(t_1)z(t_2)}_{LC}$, and yields the velocity correlation function on applying the operator $\partial^2/\partial t_1 \partial t_2$. Thus

$$\overline{v_z(t_1)v_z(t_2)} = -\tfrac{1}{2}(\partial^2/\partial t_1 \partial t_2)\iint \rho(z_1)P_s(z_1|z_2, t_2-t_1)\{z(t_1)-z(t_2)\}^2 dz_2 dz_1. \tag{6.22}$$

For a stationary ensemble only the time difference $t = t_2 - t_1$ is important. Furthermore we can use eqn (6.7) to rewrite eqn (6.22) as

$$\overline{v_z(0)v_z(t)} = \tfrac{1}{2}\int (\partial^2/\partial t^2)\overline{P}_s(Z, t)(Z)^2 dZ. \tag{6.23}$$

Because $\overline{v_z(0)v_z(t)}$ is a very 'local' quantity, it makes little sense to employ an average propagator when that average is taken over a distance range where motions are entirely uncorrelated. This is a subtle point which we need not address in detail. For the moment we shall only consider the case where the propagator is common to all particles.

We are now in a position to evaluate $\overline{v_z(0)v_z(t)}$ for the case of self-diffusion where $P_s(Z, t)$ is described by eqn (6.12). After some manipulation we find

$$\overline{v_z(0)v_z(t)} = 2D\,\delta(t). \tag{6.24}$$

Since the time integral of $\delta(t)$ from 0 to ∞ is $\tfrac{1}{2}$, this expression is entirely compatible with eqn (6.20), where $D_{zz}(0) = D$.

The velocity correlation function and the self-diffusion spectrum implied by eqn (6.24) correspond to a particle executing a random walk consisting of uncorrelated jumps spaced infinitesimally in time. It is somewhat unrealistic. Suppose that the jump time is τ_c as described in Chapter 3. It is convenient to assume that the velocity correlation is now, as illustrated in Fig. 6.3,

$$\overline{v_z(0)v_z(t)} = \overline{v_z^2}\exp(-t/\tau_c). \tag{6.25}$$

Eqn (6.20) is clearly consistent with $D_{zz}(0) = \tau_c\,\overline{v_z^2}$.

It is important to note that the diffusion P_s function of eqn (6.12) applies only at low frequencies and D is taken to be $D_{zz}(0)$. When we deal with apparently random motion using the language of diffusion, we must be aware that such a description applies only for times longer than the velocity correlation time. For the unbounded diffusion of small molecules the velocity correlation time is so short that PGSE NMR will always be safely

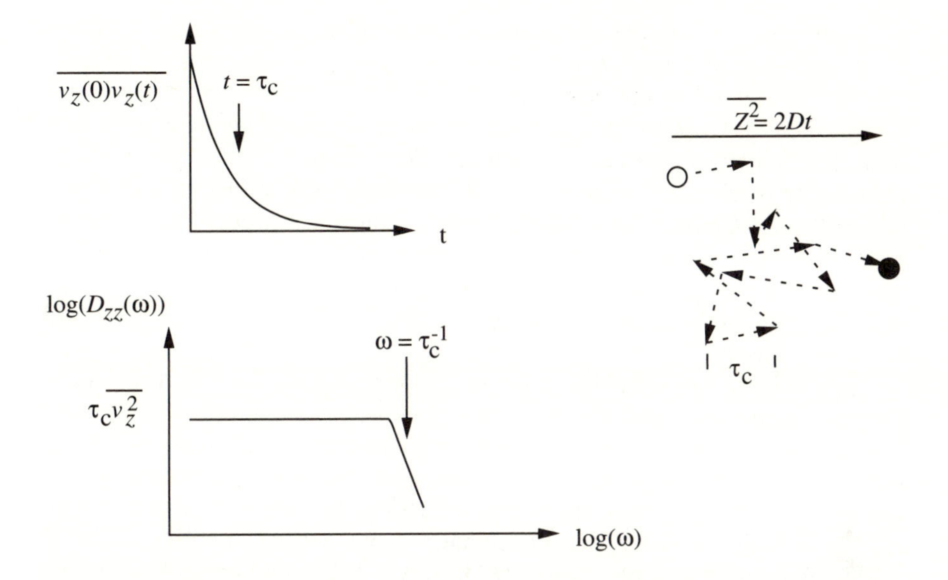

Fig. 6.3 Schematic representation of the velocity autocorrelation function, $\overline{v_z(0)v_z(t)}$, and associated spectral density, $D_{zz}(\omega)$, for a particle undergoing a Brownian random walk with velocity correlation time, τ_c. The r.m.s. jump length is of order $\tau_c(v_z^2)^{1/2}$.

in the low-frequency regime. For bounded diffusion and for turbulent motion it is possible that longer time correlations exist. In such circumstances great care is needed in probing the motional spectrum. The above analysis gives us the option of describing our motion in terms of the conditional probability, $P_s(\mathbf{r}|\mathbf{r}', t)$, or the diffusion tensor, $D_{\alpha\beta}(\omega)$. In the following sections of this chapter it will be shown how these parameters can be measured by NMR and how they can be used to characterize motion in specific examples.

6.3 PGSE: the scattering analogy and q-space

We have seen that nuclear magnetic resonance provides a molecular label via the characteristic Larmor frequencies of the component nuclei. This label is the phase of the transverse magnetization, an idea which lies at the heart of NMR imaging. In principle, it should therefore be possible to give a spatial label to nuclei at one instant of time, and then to check the labelling at a later time in order to see if they have moved. If we could measure the label shift then we can deduce the motion. In NMR imaging we measure, in effect, the absolute phase of the spins, relating this to the nuclear positions. In the measurement of motion what we require, therefore, is to measure phase differences. For this the spin echo is ideally suited.

Chapter 3 dealt with the attenuation of the Hahn echo caused by diffusion in the presence of a magnetic field gradient and with a more sophisticated method of measuring diffusion, the Stejskal–Tanner experiment[9] in which the gradient is applied in the form of pulses during the dephasing and rephasing parts of the echo sequence. This method avoids the need to transmit or receive in the presence of the gradient. The model used to depict this method in Chapter 3 involved calculating the phase accumulation during a random walk. Now we shall adopt a more analytical approach, showing how the PGSE method can be used to give information about general translational motion, and in particular the functions P_s and $\overline{v_z(0)\, v_z(t)}$ discussed above.

There are two key elements in the measurement of molecular translation. One concerns the use of magnetic field gradients to impose spatial labels. The other concerns the use of spin echoes to measure phase shifts. Applying our gradients in the form of two compensating pulses is indeed a nice technique. But sometimes, other approaches can be more effective in revealing the motion. A more suitable name for our general method might be 'Modulated Gradient Spin Echo' NMR or MGSE. However we shall retain the commonly used PGSE as a general label.

6.3.1 *The narrow-pulse approximation*

Figure 6.4 shows the r.f. and gradient pulse sequence used in the standard PGSE experiment. It is essentially the same as that shown in Fig. 3.21 except that now we have represented the gradient pulses as sufficiently narrow that we will neglect motion over their duration. In other words $\delta \ll \Delta$. This approximation can never be realized exactly in a real experiment but it does allow us powerful insight in cases where the motion is not so simple as the diffusive displacements considered in Chapter 3. As in Chapter 3 we use the symbol $\mathbf{g}$ to represent the magnitude (and direction) of the gradient pulse, distinguishing it from the upper case symbol used to describe the $\mathbf{k}$-space imaging gradient.

We will focus our attention on the evolution of the transverse magnetization in the period leading up to the echo. Note that, if there are small residual gradients due to polarizing magnet inhomogeneity, then the centre of this echo will be at a time τ after the 180_y pulse, the rephasing time being equal to the dephasing period separating the 90_x and 180_y pulses. The echo formation is shown in Fig. 6.4. Now consider the effect of the gradient pulses. Using the narrow-pulse approach we can see that the effect of the first gradient pulse is to impart a phase shift $\gamma\delta\mathbf{g}\cdot\mathbf{r}$ to a spin located at position $\mathbf{r}$ at the instant of the pulse. This phase shift is subsequently inverted by the 180_y r.f. pulse. Suppose that the molecule containing the spin has moved to $\mathbf{r}'$ at the time of the second gradient pulse. The net phase shift following this

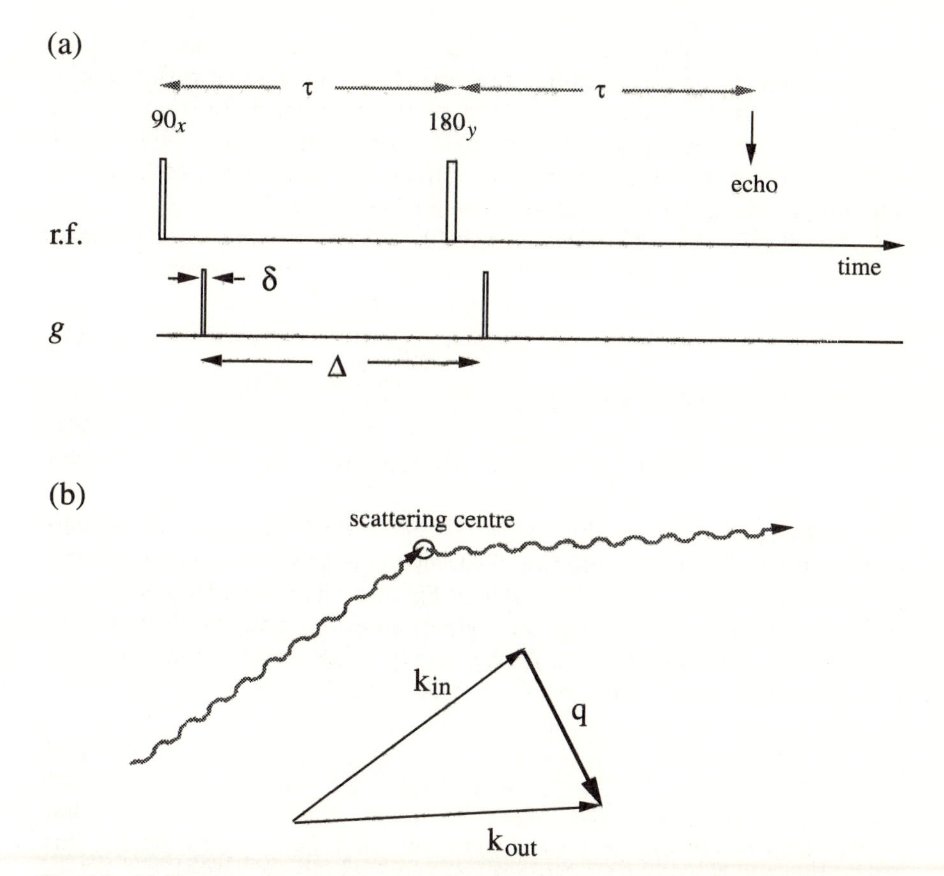

Fig. 6.4 (a) Narrow-pulse approximation PGSE, and (b) scattering analogy in which the scattering wavevector, has magnitude $\gamma\delta g$ and orientation given by the gradient direction.

pulse will be $\gamma\delta\mathbf{g}\cdot(\mathbf{r}'-\mathbf{r})$. If the spins are stationary then, of course, a perfectly refocused echo will occur. Any motion of the spins will cause phase shifts in their contribution to the echo. The size of this shift is a product of two vectors, the dynamic displacement $(\mathbf{r}'-\mathbf{r})$ and a vector $\gamma\delta\mathbf{g}$.

We will define our 'echo signal', $E_\Delta(\mathbf{g})$, as the amplitude of the echo at its centre. The total signal is a superposition of transverse magnetizations, an ensemble average in which each phase term $\exp[i\gamma\delta\mathbf{g}\cdot(\mathbf{r}'-\mathbf{r})]$ is weighted by the probability for a spin to begin at $\mathbf{r}$ and move to $\mathbf{r}'$. This probability is identically $\rho(\mathbf{r})P_s(\mathbf{r}|\mathbf{r}',\Delta)$. Thus

$$E_\Delta(\mathbf{g}) = \int \rho(\mathbf{r}) \int P_s(\mathbf{r}|\mathbf{r}',\Delta)\exp\left[i\gamma\delta\mathbf{g}\cdot(\mathbf{r}'-\mathbf{r})\right]\mathrm{d}\mathbf{r}'\,\mathrm{d}\mathbf{r}. \qquad (6.26)$$

As in Chapter 3 we shall not be concerned with absolute values nor with the effect of the attenuation of the echo due to T_2 relaxation, as important as these may be in determining the actual signal-to-noise ratio of our experiment. What we seek is the effect of the gradient pulses on the echo amplitude. This means that eqn (6.26) uses a normalized density, $\rho(\mathbf{r})$, so that $E_\Delta(\mathbf{g}) = 1$ when $\mathbf{g} = 0$. In practice $E_\Delta(\mathbf{g})$ can be obtained by dividing the echo amplitude under $\mathbf{g}$ by the amplitude when $\mathbf{g}$ is zero.

In fact eqn (6.26) bears a close resemblance to the expressions (such as eqn 3.4)) which were used to describe the signal in **k**-space imaging. We will pursue this analogy by defining a reciprocal space q where

$$\mathbf{q} = (2\pi)^{-1}\gamma\delta\mathbf{g}. \tag{6.27}$$

Now eqn (6.26) can be rewritten

$$E_\Delta(\mathbf{q}) = \int \rho(\mathbf{r}) \int P_s(\mathbf{r}|\mathbf{r}',\Delta) \exp\left[\mathrm{i}2\pi\mathbf{q}\cdot(\mathbf{r}' - \mathbf{r})\right] \mathrm{d}\mathbf{r}'\, \mathrm{d}\mathbf{r}. \tag{6.28}$$

This expression is akin to the scattering function which applies in neutron scattering where $\mathbf{q}$ is the scattering wavevector. PGSE and neutron scattering are closely analogous as illustrated in Fig. 6.4, the main differences being the scale of temporal and spatial regimes sampled, and the detection of $E_\Delta(\mathbf{q})$ in the time domain of Δ in the case of PGSE and in the frequency domain in the case of neutron scattering. Eqn (6.28) is equivalent to the neutron scattering function for the incoherent fraction,

$$S_{\text{incoherent}} = \overline{N^{-1} \sum_i \exp\left[\mathrm{i}2\pi\mathbf{q}\cdot\{\mathbf{r}_i(t) - \mathbf{r}_i(0)\}\right]} \tag{6.29}$$

where the sum is taken over all scattering centres. By contrast the coherent neutron scattering is described by

$$S_{\text{coherent}} = \overline{N^{-2} \sum_i \sum_j \exp\left[\mathrm{i}2\pi\mathbf{q}\cdot\{\mathbf{r}_j(t) - \mathbf{r}_i(0)\}\right]}. \tag{6.30}$$

This sensitivity to relative motion, $\mathbf{r}_j(t) - \mathbf{r}_i(0)$, makes the interpretation of coherent neutron scattering and quasi-elastic light scattering considerably more difficult. The direct measurement of self-motion is a major advantage in PGSE NMR.

Clearly the phase shifts appearing in the integrand of eqn (6.28) depend only on the dynamic displacement, $\mathbf{R}$. Making the substitution $\mathbf{r}' = \mathbf{r} + \mathbf{R}$, it is easy to see that this equation may be rewritten

$$E_\Delta(\mathbf{q}) = \int \overline{P}_s(\mathbf{R},\Delta) \exp(\mathrm{i}2\pi\mathbf{q}\cdot\mathbf{R})\, \mathrm{d}\mathbf{R}. \tag{6.31}$$

Eqn (6.31) expresses a simple Fourier relationship between $E_\Delta(\mathbf{q})$ and $\overline{P}_s(\mathbf{R},\Delta)$. The meaning of the narrow-pulse PGSE experiment is now transparent. Acquisition of the signal in **q**-space permits us to image $\overline{P}_s$ (**R**, Δ) just as acquisition in **k**-space permitted us to image $\rho(\mathbf{r})$! PGSE is an imaging experiment in its own right, probing the dynamic displacements, **R** rather than the static displacements, **r**. Note that while the echo attenuation, E, is implicitly a function of the PGSE pulse separation Δ, we shall henceforth omit the subscript Δ for simplicity.

Because PGSE is sensitive to the *averaged* propagator we must exercise care in using this imaging picture. $\overline{P}_s$ (**R**, Δ) will be equivalent to $P_s(\mathbf{r}|\mathbf{r}', \Delta)$ only when $P_s(\mathbf{r}|\mathbf{r}', \Delta)$ is independent of starting position **r** and depends only on the net displacement. This is clearly true for unbounded self-diffusion but it is not true where the molecules are moving in the vicinity of barriers. It is also untrue where $P_s(\mathbf{r}|\mathbf{r}', \Delta)$ is macroscopically inhomogeneous, such as may occur in capillary flow experiments where the molecular velocity varies across the sample. The word macroscopic here has the specific meaning that variations occur on a size scale larger than the resolution obtainable in (**k**-space) NMR microscopy. By combining PGSE and static imaging, it is possible to perform PGSE experiments in individual pixels of the image, thus avoiding the need, in eqn (6.28), to integrate **r** over the entire sample. In such a combined experiment the integration is confined to a single pixel. The combining of **k**-space and **q**-space imaging will be discussed in Chapter 8.

6.3.2 *Finite pulse widths, self-diffusion, and flow*

$E(\mathbf{q})$ is particularly simple to evaluate for the case of self-diffusion using the conditional probability function as first demonstrated for the PGSE experiment by Stejskal.[10] Suppose we define the z-axis as the direction of **g**. Then eqns (6.12) and (6.31) give

$$E(q) = \int_{-\infty}^{\infty} (4\pi D\Delta)^{-\frac{1}{2}} \exp\left(-Z^2/4D\Delta\right) \exp\left(\mathrm{i}2\pi qZ\right) \mathrm{d}Z. \qquad (6.32)$$

The Fourier transform of a Gaussian function $(2\pi\sigma^2)^{-\frac{1}{2}}\exp(-Z^2/2\sigma^2)$ is simply $\exp(-4\pi^2 q^2\sigma^2/2)$. In this example σ^2 is the mean square distance, $\overline{Z^2}$ travelled by the spins and eqn (6.32) gives

$$E(q) = \exp\left(-2\pi^2 q^2 \overline{Z^2}\right) \quad \text{or} \quad \exp\left(-\tfrac{1}{2}\gamma^2\delta^2 g^2 \overline{Z^2}\right) \qquad (6.33a)$$

$$= \exp\left(-4\pi^2 q^2 D\Delta\right) \quad \text{or} \quad \exp\left(-\gamma^2\delta^2 g^2 D\Delta\right). \qquad (6.33b)$$

Incorporation of flow is straightforward.[10] The conditional probability for diffusion superposed on flow is the convolution of P_s for $v = 0$ with the delta function, $\delta(Z - v\Delta)$. In consequence its Fourier transform is the product of eqn (6.33) and the Fourier transform of $\delta(Z - v_z\Delta)$, namely, $\exp(\mathrm{i}2\pi q v_z \Delta)$. In this case

$$E(q) = \exp(-4\pi^2 q^2 D\Delta + i2\pi q v_z \Delta) \tag{6.34a}$$

$$= \exp(-\gamma^2\delta^2 g^2 D\Delta + i\gamma\delta g v_z \Delta). \tag{6.34b}$$

Comparison of eqn (6.34b) with eqn (3.55) suggests that the narrow pulse evaluation of $E(q)$ is substantially correct but for finite δ the effective diffusion time takes the reduced value $\Delta - \delta/3$ while the effective flow time is unaltered. Remarkably, the case of self-diffusion superposed on flow is the only motion for which the exact finite pulse PGSE expression has been obtained analytically. For diffusion and flow, therefore, two-pulse PGSE gives us precise answers. For other motions it gives us useful, if inexact, insight by means of the scattering analogy. Some confidence in the scattering analogy may be obtained from the small nature of the deviation from ideality which is apparent in the two cases where it can be tested. Where exact answers are required, we must resort to numerical simulation.

One of the problems associated with the measurement of flow by pulsed gradient spin echo NMR concerns the spatial variation of velocity fields in most practical examples, except for the special case of plug flow in which the entire sample moves uniformly. Such variation can be resolved by static imaging methods and examples of spatially dependent flow measurements will be given in Chapter 8. However, in a pioneering experiment which predates imaging, Hayward *et al.*[28] measured the gradient dependence of the total echo signal for liquids moving in a tube, demonstrating that this dependence could be used to provide a quantitative signature for both lamellar and plug flow regimes.

Pulsed gradient spin echo NMR has been widely used in the precise measurement of self-diffusion in simple liquids.[29-32] An example of the precision possible in measuring diffusion is shown in Fig. 6.5 where the echo amplitude is plotted as $\log(E)$ versus $\delta^2 g^2 \Delta_r$, where Δ_r is the reduced observation time, $\Delta - \delta/3$. The experiments show the diffusion of water using proton and deuterium PGSE NMR, respectively. Note the linearity of the proton PGSE plot over two orders of magnitude in attenuation. PGSE measurements using deuterium NMR are doubly handicapped because of the poorer deuterium signal-to-noise (the $\omega_0^{7/4}$ effect) and the smaller phase shifts arising from diffusive motion (the γ^2 effect).

The molecular self-diffusion coefficient gives a great deal of information about molecular organization in the liquid state, especially in the case of macromolecules. A review of polymer diffusion is beyond the scope of this book and details can be found elsewhere.[33,34] However the reader should be aware of some of the possibilities. For example, at high dilution the molecular hydrodynamic radius, R_D, is available using the Stokes–Einstein law

$$D_0 = k_B T / 6\pi\eta R_D \tag{6.35}$$

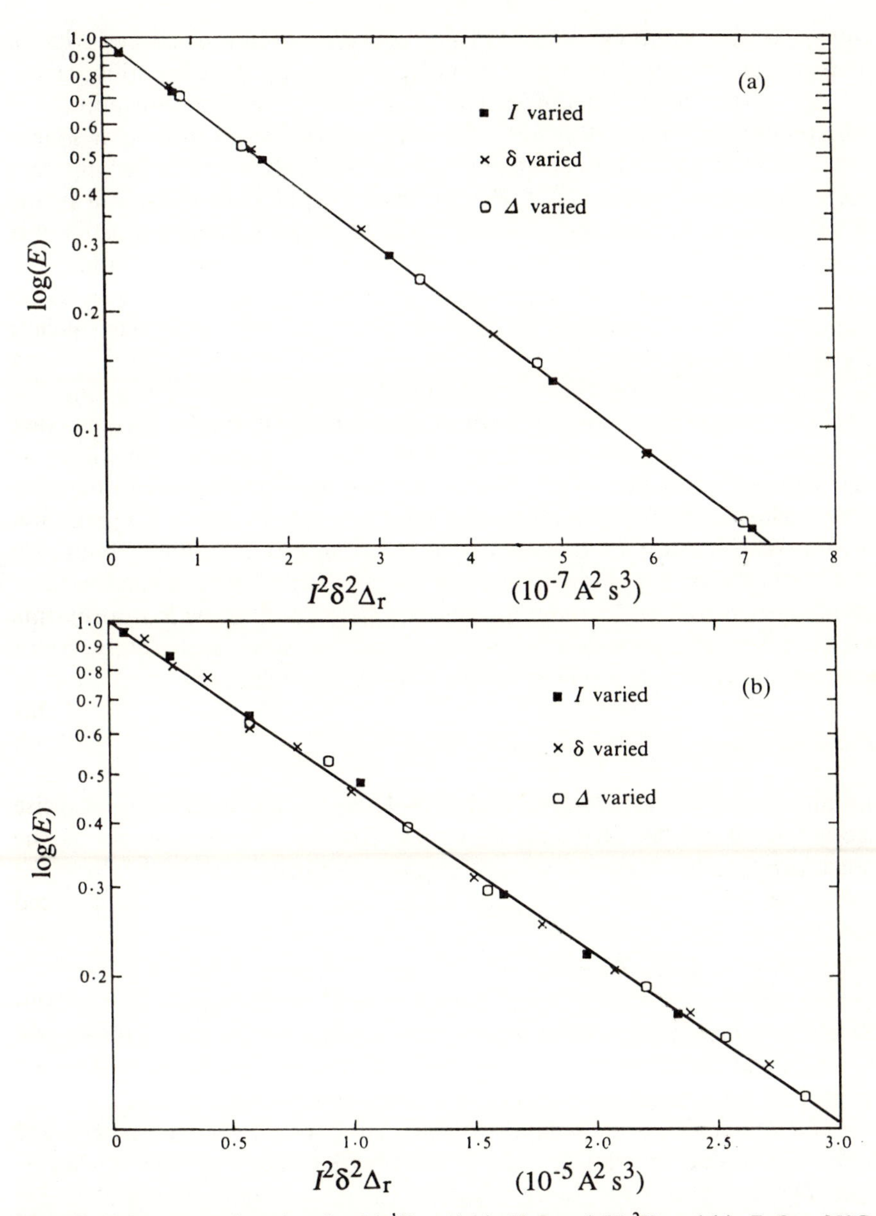

Fig. 6.5 Echo attenuation plots for (a) ^1H nuclei in H_2O and (b) ^2H nuclei in D_2O at 25°C. The abscissa is proportional to $q^2\Delta r$ where Δr is the effective diffusion time, $\Delta - \delta/3$. I refers to the current through the gradient coil and is proportional to the gradient amplitude, g. (From Callaghan *et al.*[67])

where η is the solvent viscosity. At finite concentrations the polymer friction increases due to collisions between molecules and the concentration dependence[35] of the diffusion coefficient may contain information about the macromolecular shape. Further information about macromolecular shapes can be obtained by observing the solvent self-diffusion as the polymer concentration is increased.[35-39] Finally, both in polymer melts, and in the solution phase 'semi-dilute' regime, the dependence of polymer self-diffusion on molar mass and concentration can reveal a great deal about the role of entanglements in determining polymer dynamics.[40-44] This is an important area of chemical physics in which PGSE NMR has made a significant contribution.[45-53]

In a sense, investigating the entanglement structure of random coil polymers can be regarded as a form of microscopy although the connection between this structure and $E(q)$ is somewhat remote. One exciting development made possible by NMR microscopy is the facility to measure polymer diffusion in inhomogeneous solutions. For example, the measurement of 'self-motion' in the presence of a concentration gradient could help elucidate relationships between mutual and self-diffusion. The measurement of polymer diffusion in a varying velocity field could provide information on the effect of shear on polymer conformation. Some preliminary results concerning this latter application are discussed in Chapter 8.

6.3.3 *Anisotropic self-diffusion*

Liquid crystals represent a class of materials exhibiting organizational anisotropy. In particular, the molecular self-diffusion coefficients are highly anisotropic and P_s obeys a self-diffusion equation of the form

$$\partial P_s/\partial t = \nabla \cdot \mathfrak{D} \nabla P_s \tag{6.36}$$

where $\mathfrak{D}$ is now the Cartesian tensor, $D_{\alpha\beta}$, defined as in eqn (6.18). The expression for the PGSE signal, $E(q)$, now has $g^2 D$ replaced by $\mathbf{g} \cdot \mathfrak{D} \cdot \mathbf{g}$. By alternating gradient orientation it is possible, in principle, to investigate off-diagonal elements of the self-diffusion tensor. In structures of right symmetry, $\mathfrak{D}$ is diagonal, although the three components D_{xx}, D_{yy}, and D_{zz} may differ. For example, cylindrical symmetry would exist in nematic and smectic A liquid crystals.[54,55] If the molecular director axis is labeled z' then $D_{z'z'} = D_{\parallel}$ and $D_{x'x'} = D_{y'y'} = D_{\perp}$. More generally, if the diffusion tensor is expressed in the natural reference frame of the molecular director as $\mathfrak{D}'$, then transformation to the laboratory frame gives

$$\mathfrak{D} = R^{-1}(\Theta, \Phi) \mathfrak{D}' R(\Theta, \Phi) \tag{6.37}$$

where Θ and Φ are the polar azimuthal angles between the director and gradient frames and R is a rotation matrix. PGSE methods have been used to examine diffusion anisotropy in thermotropic liquid crystals[56–59] and an example of the angular dependence of self-diffusion in smectic TBBA obtained by the Ljubljana group is shown in Fig. 6.6. One elegant feature of their work is the use of multiple-pulse line-narrowing techniques to extend the transverse coherence of the liquid crystal spin magnetization over a time-scale sufficient to measure diffusion. This need arises because the orientational order inherent in the liquid crystalline state causes the residual spin–spin dipolar interactions to persist, where in isotropically tumbling molecules they are averaged to zero. These are difficult experiments in which the gradient pulses are necessarily only a few microseconds long and applied in specific 'windows' of the multiple-pulse sequence.[26,60–62] NMR is uniquely

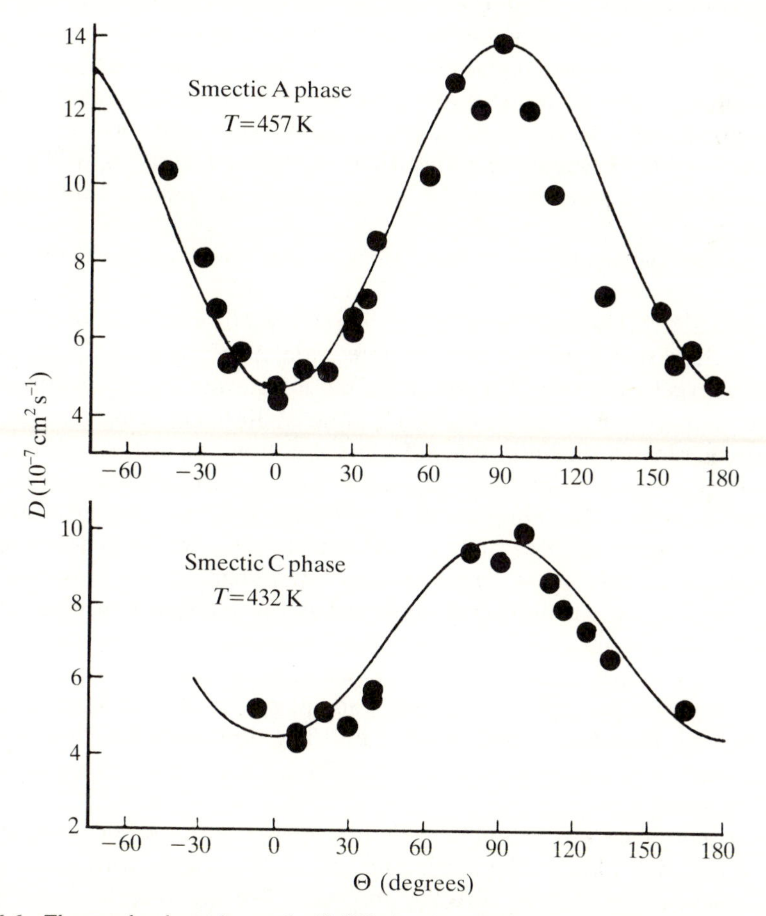

Fig. 6.6 The angular dependence of self-diffusion for TBBA in the smectic A and smectic C phases. Θ is the angle between the magnetic field gradient and the molecular directors. The theoretical fits give a measure of diffusion anisotropy. (From Blinc *et al.*[57])

equipped to demonstrate, and quantify, the solid-like orientational order of the molecules while at the same time revealing their liquid-like translational mobility.

The anisotropic diffusion apparent in Fig. 6.6 belongs to the amphiphiles of a thermotropic liquid crystal. In lyotropic liquid crystals, consisting of both amphiphile and solvent, both species of molecule can exhibit diffusional anisotropy. This behaviour is a useful aid in characterizing the structure and measurements of $D_{\perp}/D_{\parallel}$ by PGSE NMR and has proved effective in this regard.[63-66] For example, Fig. 6.7 shows the diffusive behaviour of heavy water in an oriented lamellar phase of potassium palmitate/D_2O.[67] Also shown is the ^2H NMR spectrum which shows a quadrupole splitting characteristic of the residual molecular order present in the water molecules. It exhibits a quadrupole splitting depending characteristically on the angle, θ, between the lipid director and the polarizing field as $\frac{1}{2}(3\cos^2\theta - 1)$. Despite the rapid tumbling motion of the water molecules, a small degree of orientational order remains because of the orientation of the lipid bilayers. In this experiment a quadrupolar $(90_x\text{–}90_y)$ echo[68] was used in the PGSE sequence rather than the usual Hahn $(90_x\text{–}180_y)$ echo. The use of solid echoes to measure diffusion has also been demonstrated by Kruger *et al.*[69]

Diffusion perpendicular to the bilayer director is unrestricted while diffusion parallel to the director is severely inhibited owing to the lipid bilayer barrier. The thickness of the water layer is only of order 100 Å, whereas the distance moved over the time-scale, Δ, of the PGSE experiment is much larger than this. The measured $D_{\parallel}$ is determined by the translayer permeation rate, a diffusion ratio, $D_{\perp}/D_{\parallel}$, of 30 being found.

The experiment shown in Fig. 6.7 illustrates how the diffusive behaviour of water can be used to probe the geometry of surrounding barriers, an important theme in this chapter. In thinking about the influence of diffusive barriers we must be clear about relative distances. At room temperature the water molecule has a free diffusion coefficient of $2.3 \times 10^{-9}\,\text{m}^2\,\text{s}^{-1}$. The time over which motion is observed is the interpulse separation, Δ, in the PGSE technique. Because of T_2 relaxation it is difficult to perform experiments with Δ larger than one or two seconds and because of the difficulty in providing rapidly switched gradient pulses of sufficient intensity, it is difficult to reduce Δ to below 0.5 ms. The time-scale $0.5\,\text{ms} < \Delta < 2\,\text{s}$ corresponds to a distance scale of 1 to 100 μm for the Brownian motion of water molecules.

In systems such as the amphiphilic bilayer where the water layers are at most a few 100 Å thick, the motion normal to the bilayer within one water layer thickness cannot be directly observed. We are therefore unable to probe this dimension. However, the angular dependence of self-diffusion can tell us a great deal about the orientation of barriers, as is immediately

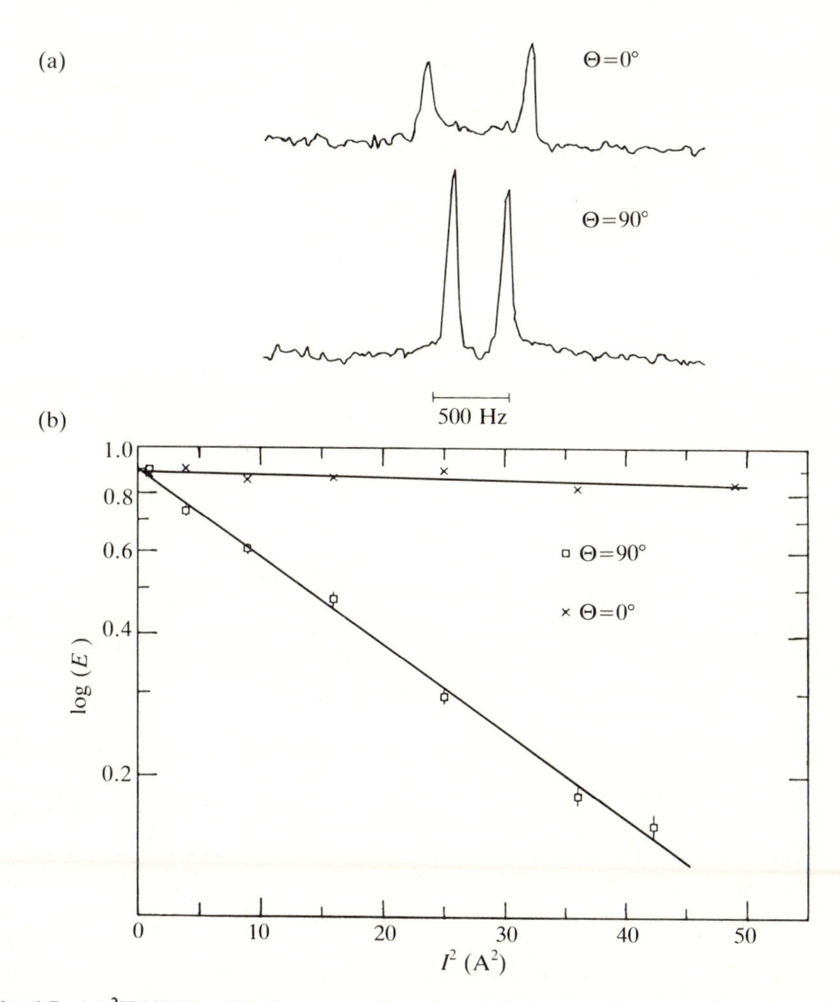

Fig. 6.7 (a) ^{2}H NMR solid echo spectra for oriented single crystalline potassium palmitate/ D_2O bilayers (50:50 w/w) at 65°C ($2\tau = 24$ ms). The splittings for orientation at 0° and 90° to the external field are in the predicted lamellar phase ratio of 2:1. (b) ^{2}H PGSE solid echo attenuation for the sample in (a) illustrating anisotropic water diffusion ($\Delta = 12$ ms and $\delta = 8$ ms). (From P. T. Callaghan *et al.*[67])

apparent in Fig. 6.7. Suppose, for example, we observe the diffusive motion of small molecules confined to water layers which are randomly oriented in the laboratory frame. Fig. 6.8 shows the ^{2}H NMR spectrum which results in the well-known 'powder pattern'. It is a direct consequence of co-adding spectra from all the randomly oriented directors in which each splitting, proportional to $\frac{1}{2}$ $(3\cos^2\theta - 1)$, is weighted by the usual polar angle factor, $\sin\theta \, d\theta$. This weighting factor reflects the greater amount of spherical

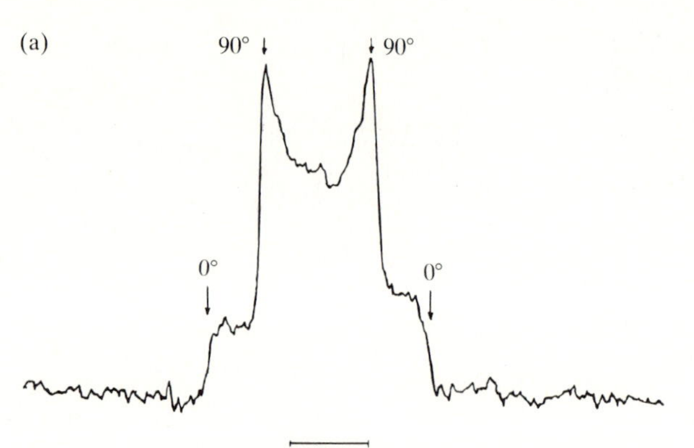

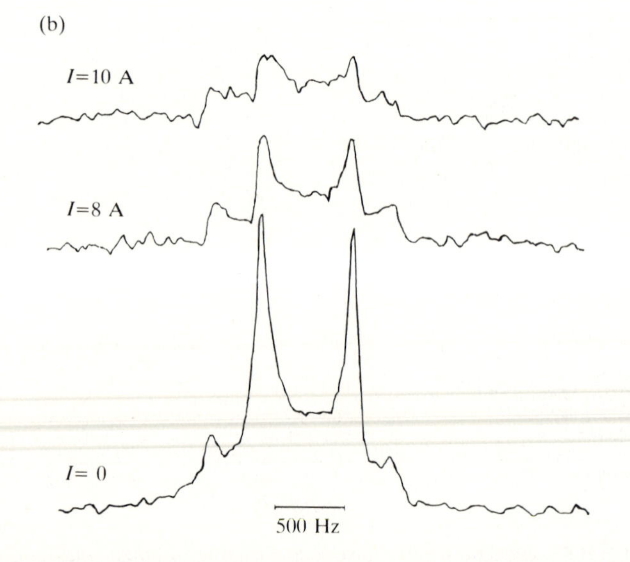

Fig 6.8 ^{2}H NMR solid echo spectra for polycrystalline potassium palmitate/D_2O (70:30 w/w) at 65°C for (a) $2\tau=4$ ms and (b) $2\tau=24$ ms. In (b) the spectrum is shown as a function of g and there exists a clear correspondence within the spectrum between the director orientations and the components of diffusion along the field gradient. (From Callaghan *et al.*[67])

surface area at equatorial rather than polar latitudes and therefore leads to prominent 90° edges.

The spectra in Fig. 6.8 were obtained using the solid echo sequence with gradient pulses inserted. As g increases the more rapid diffusion of water in the regions at 90° to $\mathbf{B}_0$ and $\mathbf{g}$ is apparent, consistent with the $D_\perp/D_\parallel$ anisotropy previously observed.

The facility to separate the signals from water in differently oriented regions is peculiar to the use of ^{2}H NMR with its quadrupolar,

$\frac{1}{2}(3\cos^2\theta - 1)$, label. In most experiments using ^{1}H, the signal is a simple superposition. We will, therefore, try to predict what the result of the PGSE experiment will be in this case, restricting ourselves to cylindrical symmetry ($D_{x'x'} = D_{y'y'} = D_\perp$) for convenience. Eqn (6.33b) states that the echo attenuation for spins with a mean square displacement $\overline{Z^2}$ is $\exp(-\frac{1}{2}\gamma^2\delta^2 g^2 \overline{Z^2})$. For a spin in a region where the axis of symmetry is aligned at θ to the gradient[70]

$$\overline{Z^2} = 2D_\parallel \Delta_r \cos^2\theta + 2D_\perp \Delta_r \sin^2\theta \tag{6.38}$$

where the reduced time has been used in order to give the exact echo attenuation factor. Hence the net $E(q)$ is the sum obtained by averaging element orientations over all angles, weighted by the sphere area element, $\sin\theta\, d\theta$. Thus[70]

$$E = \frac{\int_0^\pi \exp[-\gamma^2\delta^2 g^2 \Delta_r (D_\parallel \cos^2\theta + D_\perp \sin^2\theta)] \sin\theta\, d\theta}{\int_0^\pi \sin\theta\, d\theta}$$

$$= \exp(-\gamma^2\delta^2 g^2 \Delta_r D_\perp) \int_0^1 \exp[-\gamma^2\delta^2 g^2 \Delta_r (D_\parallel - D_\perp) x^2]\, dx. \tag{6.39}$$

Eqn (6.39) is very helpful. Suppose that we are dealing with the water/bilayer problem. Then the symmetry is 'two-dimensional' or 'lamellar' and $D_\perp \gg D_\parallel$. The echo attenuation varies as

$$E_{2D} = \exp(-\gamma^2\delta^2 g^2 \Delta_r D_\perp) \int_0^1 \exp[\gamma^2\delta^2 g^2 \Delta_r D_\perp x^2]\, dx. \tag{6.40}$$

If, on the other hand, the diffusing molecules are confined to narrow pipes then the symmetry is 'one-dimensional' or 'capillary' and $D_\perp \ll D_\parallel$ with

$$E_{1D} = \int_0^1 \exp[-\gamma^2\delta^2 g^2 \Delta_r D_\parallel x^2]\, dx. \tag{6.41}$$

Eqns (6.40) and (6.41) may be contrasted with the echo attenuation behaviour for unrestricted diffusion, E_{3D}, namely $\exp(-aD)$. The dependence of E_{1D}, E_{2D}, and E_{3D} on $\gamma^2\delta^2 g^2 \Delta_r$ gives a characteristic signature in the curvature of the $\log(E(q))$ versus $\gamma^2\delta^2 g^2 \Delta_r$ plot. Fig. 6.9 shows the echo attenuation plot for water protons in the poly-domain lamellar phase of aerosol OT/water.[71] As in Fig. 6.5 the data are plotted as $\log(E(q))$ versus $\gamma^2\delta^2 g^2 \Delta_r$. Also shown are the best fits using E_{1D}, E_{2D}, and E_{3D}, with the two-dimensional model representing the data well. Similar agreement using E_{2D} has been found for PGSE experiments on water in bilayers of the polycrystalline smectic liquid crystal, sodium 4-(1′-heptylnonyl)benzenesulphonate.[72]

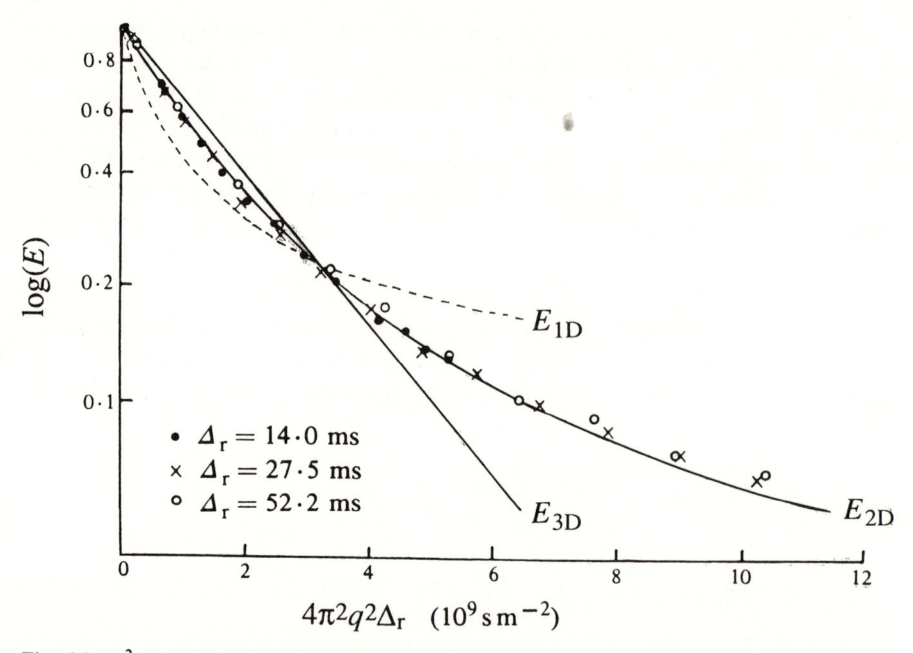

Fig. 6.9 $q^2\Delta_r$ dependence of water proton echo attenuation plot for 25% aerosol OT/water. The theoretical cuves labelled E_{1D}, E_{2D}, and E_{3D} refer to one-, two- and three-dimensional diffusion (eqns (6.41), (6.40), and (3.25)), respectively. (From Callaghan and Soderman.[71]) The data lie on a common curve characteristic of two-dimensional diffusion as Δ_r is varied, consistent with each molecule residing in a single lamellar domain.

Other dimensionalities have been observed in such experiments. Fig. 6.10 shows an echo attenuation plot for water in wheat grain endosperm where the one-dimensional model gives a good representation, consistent with the diffusion of water along the surface of extended protein fibres. In Fig. 6.11 the diffusion of water in alumina[73] is closer to two-dimensional. In all these experiments an important test is obeyed. The data obtained with differing Δ_r values exhibit a common attenuation, dependent on $\gamma^2\delta^2 g^2\Delta_r$, thus clearly indicating that the characteristic curvature in the echo attenuation plot arises from unbounded, but low-dimensional, diffusion.

6.3.4 *Comparison of the sensitivity resolution limits in* **q***-space and* **k***-space imaging*

In Chapter 4 the sensitivity limit to resolution in **k**-space imaging was examined in detail. The question arises as to whether a sensitivity limit also exists in **q**-space imaging. At first sight it may seem that, because the spin echo is acquired in the absence of a magnetic field gradient in the PGSE experiment, there is no spectral spreading and hence no sensitivity loss as g

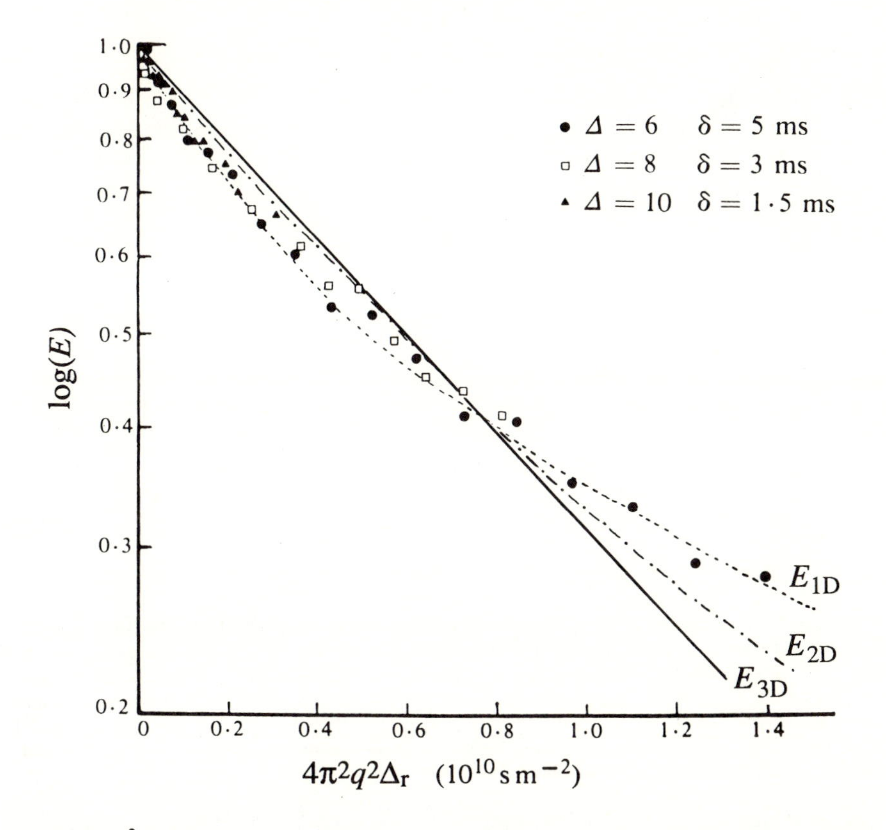

Fig. 6.10 $q^2\Delta_r$ dependence of water proton attenuation in wheat grain endosperm tissue equilibrated at 90% R. H. As in Fig. 6.9, the data lie on a common curve (this time characteristic of locally one-dimensional diffusion) as Δ_r is varied. (From Callaghan *et al.*[70])

is increased. However, it is also possible to perform **k**-space imaging with no read gradient by exclusively phase encoding so that such an argument is clearly simplistic.

Consider the simple two-dimensional circular slice of dimension r_0 and thickness Δz shown in Fig. 6.12. If there are N pixels across the image then the resolution, Δr, is of order r_0/N. We also allow that the time domain signal amplitude, S_0, is proportional to $r_0^2\Delta z$ while the noise power per unit bandwidth will be given by σ^2.

Suppose we perform **k**-space imaging in two dimensions without a read gradient using N^2 phase-encoding steps and detect the signal using the optimal bandwidth of order T_2^{-1}. The total noise power is therefore $\sigma^2 T_2^{-1} N^2$ giving $\sigma^2 T_2^{-1}$ per pixel. The time domain signal is S_0 giving a signal amplitude per frequency domain pixel proportional to $S_0 N^{-2}$. Now imagine that the same experiment is performed over the same total time-scale

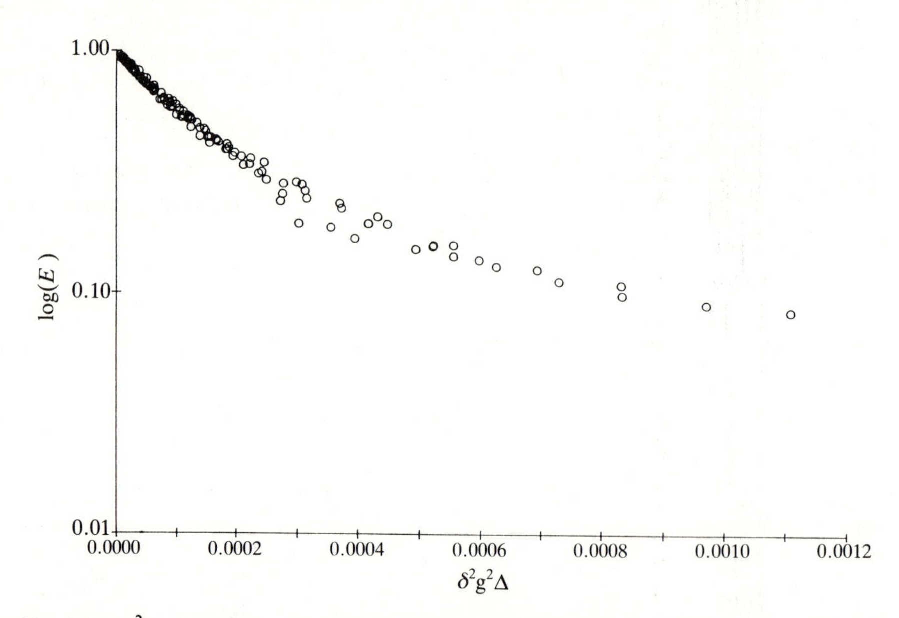

Fig. 6.11 $q^2\Delta_r$ dependence of water proton attenuation in alumina. The behaviour is intermediate between E_{1D} and E_{2D}. (From Packer and Zelaya.[73])

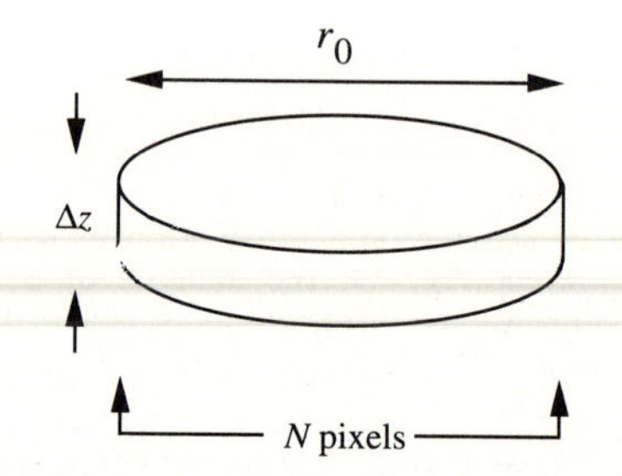

Fig. 6.12 Schematic diagram indicating slice of transverse dimension, r_0, thickness, Δz, with N^2 pixels in the plane.

but using phase encoding (with N steps) and a read gradient (with N acquisition points) at the new optimal bandwidth, NT_2^{-1}. We have sufficient time to repeat the experiment N times giving a total time domain signal NS_0 with S_0N^{-1} per pixel in the frequency domain. The noise power is now $N\sigma^2(NT_2^{-1})N^2$ giving $\sigma^2N^2T_2^{-1}$ per pixel. Both experiments yield an image signal-to-noise ratio of $S_0\sigma^{-1}N^{-2}T_2^{1/2}$. Given the dependence of S_0 and Δr on r_0 it is easy to show that

$$\Delta r \sim \left(\frac{\text{image signal}}{\text{noise amplitude}}\right)^{\frac{1}{2}} \sigma^{\frac{1}{2}} \Delta z^{-\frac{1}{2}} T_2^{-\frac{1}{4}}. \tag{6.42}$$

Eqn (6.42) is, as required, consistent with eqns (4.52) and (4.53). Clearly the use of pure phase encoding in **k**-space imaging yields the same intrinsic sensitivity-limited resolution as that which applies when a combination of phase and read gradients is employed.

In the case of **q**-space microscopy we now apply a similar reasoning to that used for pure phase encoding in **k**-space imaging. In **q**-space imaging the resolution, ΔR, is determined by the total range of dynamic displacements, ΔR_0, and is therefore given by $\Delta R_0/N$. By contrast, the time domain signal is proportional to $r_0^2 \Delta z$, where r_0 and Δz refer, as before, to sample dimensions. Following the same reasoning used previously, two-dimensional **q**-space microscopy has a resolution,

$$\Delta R \sim \left(\frac{\text{image signal}}{\text{noise amplitude}}\right)^{\frac{1}{2}} \sigma^{\frac{1}{2}} \Delta z^{-\frac{1}{2}} T_2^{-\frac{1}{4}} \left(\frac{\Delta R_0}{r_0}\right). \quad (6.43)$$

Note the scaling factor, $\Delta R_0/r_0$. This has the effect of improving the sensitivity-limited resolution over that of **k**-space microscopy by the ratio of the nuclear spin displacement range (0.1 to 100 μm) to the sample dimensions (1 to 10 mm). The resolution improves without limit as the sample dimension increases because in **q**-space imaging the spin displacement image, P_s, is common to the entire sample so that the sample itself corresponds to the 'sensitivity pixel'.

The limits to resolution in **q**-space imaging are therefore not determined by sensitivity but by artefacts which arise from sample movement and gradient pulse mismatch. A practical method of reducing these effects is discussed in Chapter 9.

6.4 General gradient modulation methods: the motional spectrum

The use of sharp gradient pulses allows precise definition of the time-scale of the motion. What is not clear is the sensitivity of the method to the motional spectrum, defined in the sense of the Fourier transform of the velocity correlation, $D_{\alpha\beta}(\omega)$. Furthermore, it is not necessarily the case that the sharp-pulse PGSE experiment provides the best probe of the motional spectrum. What is needed is some means of analysing the result of a spin echo experiment under any chosen gradient modulation. A nice method of tackling this problem has been outlined by Stepisnik.[26,27]

6.4.1 *The effective gradient*

The phase shift experienced at time t by a nuclear spin i following the path $\mathbf{r}_i(t')$ in a gradient $\mathbf{g}(t')$ is

$$\phi_i(t) = \gamma \int_0^t \mathbf{g}(t') \cdot \mathbf{r}_i(t')\, \mathrm{d}t'. \quad (6.44)$$

This simple relation takes no account of the influence of r.f. pulses on the spin phase. For example, a 180_y pulse inverts all prior phase shifts, an effect which is equivalent to that which would result if we had used a negative value of gradient up to the time of the r.f. pulse. This suggests that we can take account of r.f. pulses by defining an 'effective gradient', $\mathbf{g}^*(t)$ which has the actual sign at the current time t, but is historically inverted by all prior 180_y pulses. In the case of 90° r.f. pulses the value of $\mathbf{g}^*$ is transformed to zero. The effect is illustrated in Fig. 6.13. More generally $\mathbf{g}^*$ is defined by the equivalence of its Hamiltonian,

$$\mathcal{H}(t) = -\gamma \mathbf{g}^*(t) \cdot \mathbf{r} I_z = -\gamma \mathbf{g}(t) \cdot \mathbf{r}\, U_{\mathrm{rf}}^{-1} I_z U_{\mathrm{rf}} \tag{6.45}$$

where U_{rf} is the ordered product of evolution operators associated with the r.f. pulses.

In the special case when the molecules all have a common motional behaviour (which specifically precludes Brownian motion), eqn (6.44) yields a simple result. For example, suppose that the motion comprises some constant velocity, $\mathbf{v}$, then $\mathbf{r}_j(t')$ is $\mathbf{r}_j(0) + \mathbf{v}t'$. Where an acceleration term is present, then $\mathbf{r}_j(t')$ is $\mathbf{r}_j(0) + \mathbf{v}t' + \frac{1}{2}\mathbf{a}t'^2$. It is clear that the resulting phase shifts involve successively higher moments of $\mathbf{g}^*(t)$. The zeroth moment, $\int_0^t \mathbf{g}^*(t')\,\mathrm{d}t'$ is required to be zero if the final phase shift is to depend only on the motion and not on the starting positions of the spins. This is, of course, the condition for a spin echo to be formed! Furthermore, by choosing a specific time dependence for $\mathbf{g}^*(t)$ we can make the echo sensitive to either velocity, acceleration, or the next higher time derivative of displacement, sometimes known as 'jerk'. In particular, the final phase modulation of the spin echo will consist, in the case of the moment expansion,

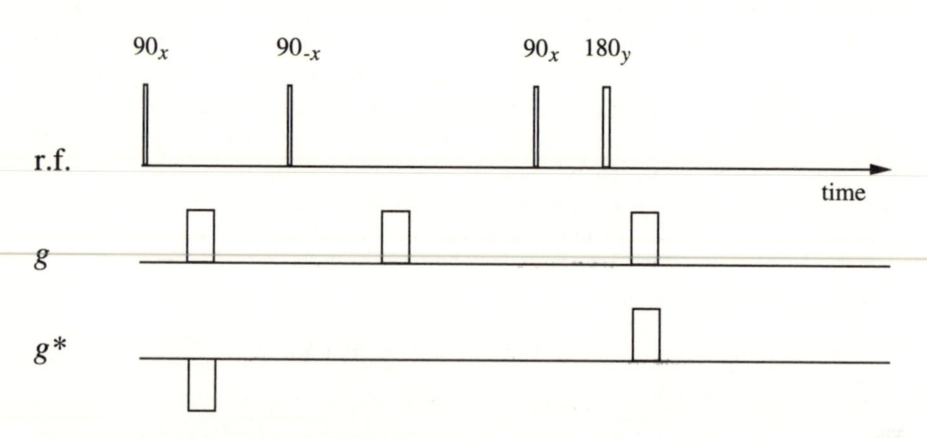

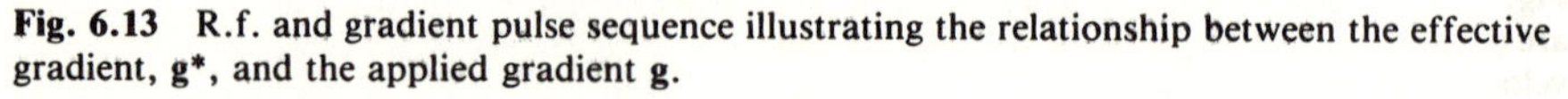

Fig. 6.13 R.f. and gradient pulse sequence illustrating the relationship between the effective gradient, $\mathbf{g}^*$, and the applied gradient $\mathbf{g}$.

$$E(t) = \exp\left(\mathrm{i}\gamma\mathbf{v}\cdot\int_0^t t'\mathbf{g}^*(t')\,\mathrm{d}t' + \tfrac{1}{2}\mathrm{i}\gamma\mathbf{a}\cdot\int_0^t t'^2\mathbf{g}^*(t')\,\mathrm{d}t' - \ldots\right) \tag{6.46}$$

Examples of pulse sequences in which the effective gradient obeys the echo condition are shown in Fig. 6.14. In Fig. 6.14(a) the echo will be sensitive to velocity while in (b) it is sensitive to acceleration.

6.4.2 *The method of cumulants*

For molecular motion which varies across the particle ensemble, the echo amplitude and phase for the general gradient $\mathbf{g}^*(t)$ must be evaluated by performing the appropriate ensemble average. In the sharp-pulse PGSE experiment we were able to write down the phase shift of spin j as $\gamma\delta\mathbf{g}\cdot(\mathbf{r}_j' - \mathbf{r}_j)$. For a general gradient and r.f. pulse sequence the phase displacement for a single spin is written

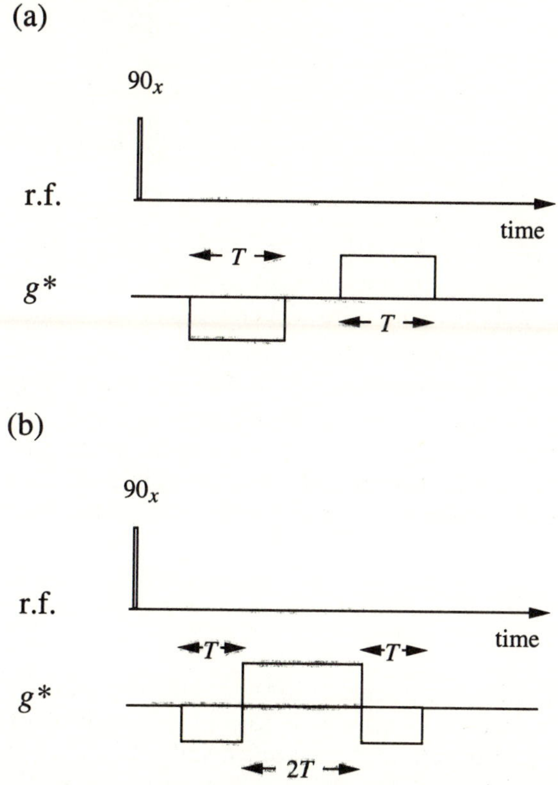

(a)

90_x

r.f.

time

g^*

T

T

(b)

90_x

r.f.

time

g^*

T

T

$2T$

Fig. 6.14 Effective gradient sequences resulting in gradient echoes ($\int_0^t \mathbf{g}^*(t')\,\mathrm{d}t' = 0$) which are (a) velocity sensitive and (b) acceleration sensitive while giving no phase shift for constant velocity.

$$\phi_j(t) = \gamma \int_0^t \mathbf{g}^*(t')\cdot\mathbf{r}_j(t')\,\mathrm{d}t'. \tag{6.47}$$

The echo signal is therefore the ensemble average

$$E(t) = \overline{\exp\{\mathrm{i}\phi_j(t)\}}. \tag{6.48}$$

Note that the concept of **q**-space no longer has any meaning since the Fourier dependence of E on $\mathbf{R}$ is no longer apparent.

The evaluation of eqn (6.48) is performed by expanding the exponential function in a power series and taking ensemble averages of each term. Condensing the sum back into closed form presents special difficulty. One standard method involves expanding about the mean so that

$$E(t) = \exp\{\mathrm{i}\overline{\phi_j(t)}\}\exp\{-\alpha(t)\} \tag{6.49}$$

where

$$\overline{\phi_j(t)} = \gamma \int_0^t \mathbf{g}^*(t_1)\cdot\overline{\mathbf{r}_j(t_1)}\,\mathrm{d}t_1. \tag{6.50}$$

When the phase distribution is Gaussian about the mean, only the leading term in $\alpha(t)$ is significant and we may write

$$\alpha(t) = \frac{1}{2}\gamma^2 \int_0^t\int_0^t \mathbf{g}^*(t_1)\cdot\overline{\mathbf{r}_j(t_1)\mathbf{r}_j(t_2)}\cdot\mathbf{g}^*(t_2)\,\mathrm{d}t_1\,\mathrm{d}t_2 - \frac{1}{2}\gamma^2\left[\int_0^t \mathbf{g}^*(t_1)\cdot\overline{\mathbf{r}_j(t_1)}\,\mathrm{d}t_1\right]^2. \tag{6.51}$$

The echo signal of eqn (6.49) has an oscillatory term, $\exp\{\mathrm{i}\overline{\phi_j(t)}\}$, arising from mean spin displacement, $\overline{\mathbf{r}_j(t_1)}$. The displacement $\mathbf{r}_j$ at time t may be expressed in terms of the initial displacement, $\mathbf{r}_{j0}$ and the mean velocity over the time interval t as $\mathbf{r}_{j0} + \mathbf{v}_j t$. The ensemble average of this quantity yields the sum $\mathbf{r}_0 + \mathbf{v}t$, where $\mathbf{r}_0$ is the average spin position in the sample, measured from the gradient centre, and $\mathbf{v}$ is the ensemble-averaged velocity. This leads to

$$\overline{\phi_j(t)} = \mathbf{r}_0\cdot\int_0^t \mathbf{g}^*(t')\,\mathrm{d}t' + \mathbf{v}\cdot\int_0^t t'\mathbf{g}^*(t')\,\mathrm{d}t'. \tag{6.52}$$

These two integrals were discussed in the previous section. When $\int_0^t \mathbf{g}^*(t')\,\mathrm{d}t'$ is zero the signal will be insensitive to the mean position of the spins, the condition for echo formation which is universally obeyed for pulse sequences designed not to image static positions but to reveal motion. When the gradient also obeys the condition that $\int_0^t t'\mathbf{g}^*(t')\,\mathrm{d}t'$ is zero, then the echo signal will be insensitive to flow, a point noted by Carr and Purcell in their original paper on multiple echoes.[4] Such a $\mathbf{g}^*(t)$ can be useful where one wishes to measure self-diffusion in the presence of flow but with all phase shifts due to flow being suppressed. Fig. 6.14(b) shows a pulse sequence which achieves this effect. Another example of such a sequence is referred to in Chapter 8 in the context of velocity compensation imaging.[74]

Whereas the oscillatory term, $\exp\{\mathrm{i}\overline{\phi_j(t)}\}$, indicates the mean flow, the damping term, $\exp\{-\alpha(t)\}$, indicates fluctuations in the motion about the mean velocity. One delicate aspect of α is the tensorial quality of the product $\mathbf{g}^*(t_1)\cdot\overline{\mathbf{r}_j(t_1)\mathbf{r}_j(t_2)}\cdot\mathbf{g}^*(t_2)$. Where $\mathbf{g}^*$ varies in Cartesian orientation as well as in time, it is possible, in principle, to measure correlations between differing components of displacement. We shall focus for the moment on the simpler problem where g^* is a gradient in the z-direction only. To evaluate α, Stepisnik uses the location correlation defined in eqn (6.21) and writes

$$\alpha(t) = \frac{1}{2}\gamma^2 \int_0^t \int_0^t g^*(t_1)\,\overline{z_j(t_1)z_j(t_2)}_{\mathrm{LC}}\,g^*(t_2)\,\mathrm{d}t_1\,\mathrm{d}t_2 \tag{6.53}$$

where we have used the identity $\overline{z(t_1)} = \overline{z(t_2)}$, the so-called 'stationary' property of the ensemble.

6.4.3 *The spectrum of the gradient and the spectrum of the motion*

The relationship between the location correlation and the diffusion spectrum is examined in detail in the Appendix. The result is

$$\overline{z(t_1)z(t_2)}_{\mathrm{LC}} = \frac{1}{\pi}\int_{-\infty}^{\infty} \frac{D_{zz}(\omega)}{\omega^2}\exp\{\mathrm{i}\omega(t_1 - t_2)\}\,\mathrm{d}\omega. \tag{6.54}$$

In consequence the attenuation exponent can be written

$$\begin{aligned}\alpha(t) &= \frac{1}{2\pi}\gamma^2 \int_0^t\int_0^t g^*(t_1)g^*(t_2)\int_{-\infty}^{\infty}\frac{D_{zz}(\omega)}{\omega^2}\exp\{\mathrm{i}\omega(t_1 - t_2)\}\,\mathrm{d}\omega\,\mathrm{d}t_1\,\mathrm{d}t_2 \\ &= \frac{1}{2\pi}\gamma^2\int_{-\infty}^{\infty} D_{zz}(\omega)\,S(\omega, t)\,\mathrm{d}\omega \end{aligned} \tag{6.55}$$

where

$$S(\omega, t) = \frac{|g(\omega, t)|^2}{\omega^2} \tag{6.56}$$

and

$$g(\omega, t) = \int_0^t g^*(t')\exp(-\mathrm{i}\omega t')\,\mathrm{d}t'. \tag{6.57}$$

The attenuation of the spin echo in a modulated-gradient experiment, therefore, has an explicit relationship to the spectral density of the translational motion, $D_{zz}(\omega)$. It is given by the integral of the product of $D_{zz}(\omega)$ with the spectrum of the gradient, this latter function being defined by eqns (6.56) and (6.57). A nice analogy can be drawn with the BPP theory of relaxation. In this model the relaxation rate is sensitive to the spectral density of the fluctuating interaction, sampled at multiples of the Larmor frequency (including zero frequency in the case of T_2). In the case of the spin echo attenuation due to a gradient, the 'relaxation' arises from sampling the spectral density of the motion with a function, $S(\omega, t)$, which is characteristic of the gradient spectrum.

The time domain is the natural space for analysing the narrow-pulse PGSE or **q**-space experiment. Such an analysis lends itself to a description of motion in terms of the conditional probability P_s. Now we have a tool for thinking about the role of the field gradient in the frequency domain and so we can interpret the experiment using the spectral density function. This is very helpful because it enables us not only to apply an alternative interpretation of the data but also to think about the optimum gradient modulation sequence for the particular motion we may wish to examine.

Fig. 6.15 illustrates the motional spectrum and the gradient spectrum for a two-pulse PGSE experiment to measure diffusion. For the finite-pulse PGSE

$$g(\omega, t) = -g\int_0^{\delta} \exp(\mathrm{i}\omega t')\,\mathrm{d}t' + g\int_{\Delta}^{\Delta+\delta} \exp(\mathrm{i}\omega t')\,\mathrm{d}t'. \tag{6.58}$$

Hence

$$S(\omega, t) = \left[\frac{4g}{\omega^2}\sin\left(\tfrac{1}{2}\,\omega\delta\right)\sin\left(\tfrac{1}{2}\,\omega\Delta\right)\right]^2. \tag{6.59}$$

$D_{zz}(\omega)$ is essentially constant (with value D) over the frequency range of interest. Using the standard integral $\int_0^{\infty}(\sin^2 ax\,\sin^2 bx)/x^4\,\mathrm{d}x = a^2\pi(3b-a)/6$, one obtains the Stejskal–Tanner result

$$\alpha(\Delta) = \gamma^2\delta^2 g^2 D(\Delta - \delta/3). \tag{6.60}$$

The spectrum of the two-pulse PGSE gradient is dominated by the zero-frequency lobe with frequency width of order Δ^{-1}. It is therefore unsuitable for extracting high-frequency information concerning $D_{zz}(\omega)$. It would be very useful to have available a gradient modulation sequence whose frequency spectrum contained a high-frequency peak which could be adjusted in position in order to trace out the frequency dependence of $D_{zz}(\omega)$. Such a measurement could, in principle, locate 'edges' in the

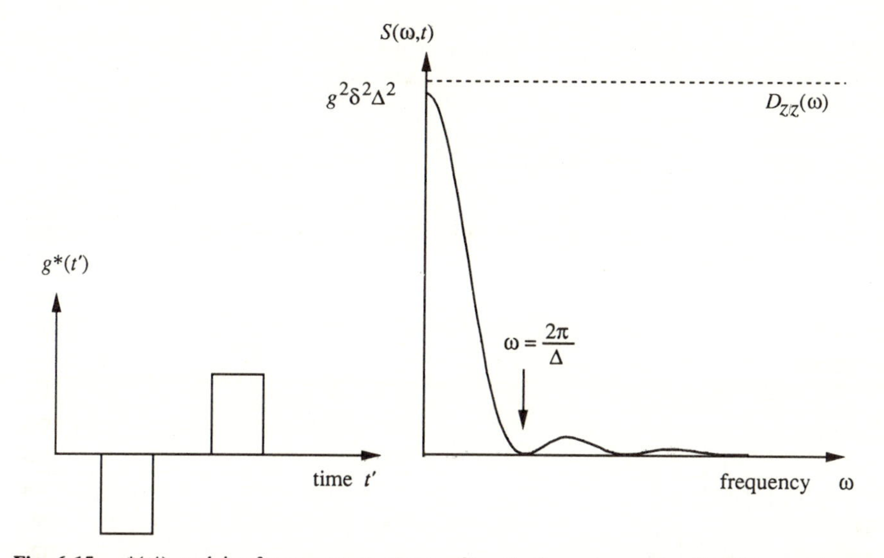

Fig. 6.15 $g^*(t')$ and its frequency spectrum, $S(\omega, t)$, for a two-pulse PGSE experiment. $D_{zz}(\omega)$ is shown superposed on the spectrum and will usually be constant over the frequency range, Δ^{-1}.

spectrum at the inverse motional correlation time, τ_c^{-1}. Two potential candidates are the CPMG gradient train and the sinusoidal gradient modulation. Their time and frequency domain behaviours are shown in Fig. 6.16. Because of the ω^{-2} term in $S(\omega, t)$ the zero-frequency lobe cannot be avoided but both trains produce well-defined high-frequency features, in the case of the CPMG sequence a series of peaks at (odd) multiples of the modulation frequency, ω_m, and in the case of the sinusoidal oscillation a single pronounced peak at ω_m.

The high-frequency peaks for both the CPMG and sinusoidal modulation gradient trains become narrower as the number of oscillation cycles, n, increases, and as $n \to \infty$, the expressions for $S(\omega, t)$ reduce to linear combinations of delta functions. In the large n limit one can therefore write down the solution to $\alpha(t)$ directly. These relationships are summarized as follows.

(a) *CPMG sequence* (cycle period, τ; $\omega_m = 2\pi/\tau$; amplitude, g; cycles, n)
Finite n:

$$S(\omega, t) = \left[\frac{2g}{\omega^2} \tan\left(\tfrac{1}{4}\omega\tau\right) \sin\left(\tfrac{1}{2} n\omega\tau\right)\right]^2 \tag{6.61}$$

Large n:

$$S(\omega, t) = \frac{g^2\pi^3}{2\omega_m^2} t \left\{\delta(\omega) + \left(\frac{2}{\pi}\right)^4 [\delta(\omega + \omega_m) + \delta(\omega - \omega_m)] + \ldots\right\} \tag{6.62}$$

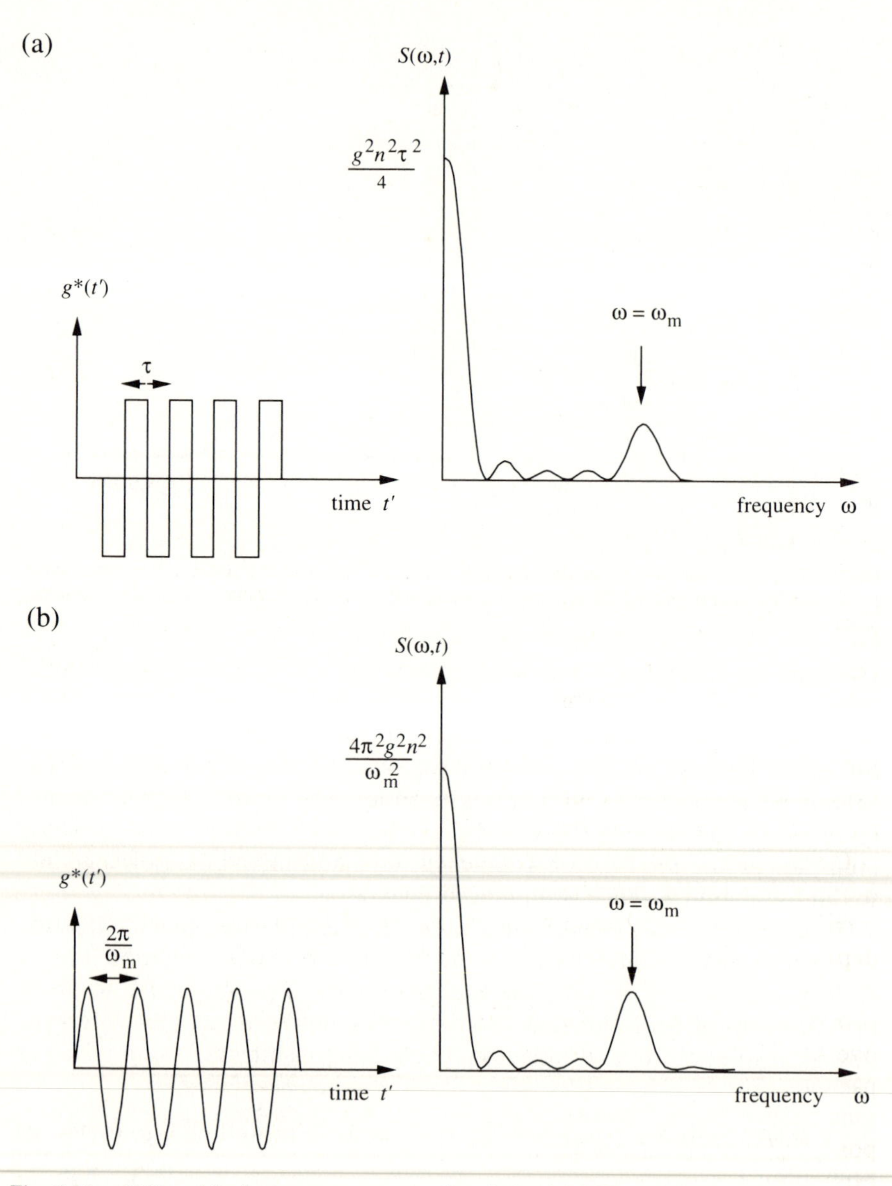

Fig. 6.16 $g^*(t')$ and its frequency spectrum, $S(\omega, t)$, for (a) CPMG PGSE experiment and (b) sinusoidal gradient modulation.

$$\alpha(t) = \frac{\gamma^2 g^2 \pi^2}{4\omega_m^2} t \left[D_{zz}(0) + 2\left(\frac{2}{\pi}\right)^4 D_{zz}(\omega_m) + \ldots \right] \tag{6.63}$$

(b) *Sinusoidal modulation* (frequency, ω_m; amplitude, g; cycles, n)
Finite n:

$$S(\omega, t) = \left[\frac{2g\omega_m}{\omega(\omega + \omega_m)(\omega - \omega_m)} \sin(n\pi\omega/\omega_m) \right]^2 \tag{6.64}$$

Large n:

$$S(\omega, t) = \frac{2\pi g^2}{\omega_m^2} t \{\delta(\omega) + \tfrac{1}{4}\delta(\omega + \omega_m) + \tfrac{1}{4}\delta(\omega - \omega_m)\} \tag{6.65}$$

$$\alpha(t) = \frac{\gamma^2 g^2}{\omega_m^2} t [D_{zz}(0) + \tfrac{1}{2} D_{zz}(\omega_m)] \tag{6.66}$$

Eqn (6.66) suggests that by varying ω_m at constant g/ω_m it is possible to probe the diffusion spectrum. It is worth re-emphasizing that the 'signal' in this experiment is the attenuation exponent, α. Note the dependence of α on $(g/\omega_m)^2 t$. On increasing the modulation frequency in order to scan higher frequencies, it is necessary to similarly increase g if the attenuation of the echo is to be retained. While this is an undoubted constraint, the method does offer the possibility of investigating rapid motions while retaining a long evolution period, t. In this regard it differs from the two-pulse PGSE experiment where the observation time-scale must be decreased, leading to an attenuation exponent decreasing not only as δ^2 but also as Δ.

Quite apart from its advantages in probing the spectral density of the motion, sinusoidal modulation offers some practical benefits in measuring diffusion coefficients because of the ease of application of the required time dependence and the simpler influence of induced eddy currents caused by gradient field variation.[53] Sinusoidal gradient modulation also has been proposed by Gross and Kosfeld[75] as a means of shortening the time-scale over which diffusion is observed, thus reducing the influence of restrictive barriers. The zero-frequency term in eqn (6.66) shows, however, that the long time-scale effects are still important and that a simple spectral decomposition is needed in order to extract $D_{zz}(\omega_m)$. The Gross and Kosfeld expression for the spin echo attenuation for the pulse sequence shown in Fig. 6.17 in which a sinusoidal gradient of amplitude $\mathbf{g}$ is applied along with an additional constant gradient term of amplitude $\mathbf{g}_0$ is

$$E(\mathbf{g}) = \exp\left[-\gamma^2 D \left\{ \mathbf{g}_0^2 \frac{2}{3}\tau^3 + 3\mathbf{g}^2 \frac{2\pi n}{\omega_m^3} + 2\mathbf{g}\cdot\mathbf{g}_0 \frac{2\pi n}{\omega_m^2} \left(\frac{2\pi n}{\omega_m} + 2t_1 - \tau \right) \right\} \right] \tag{6.67}$$

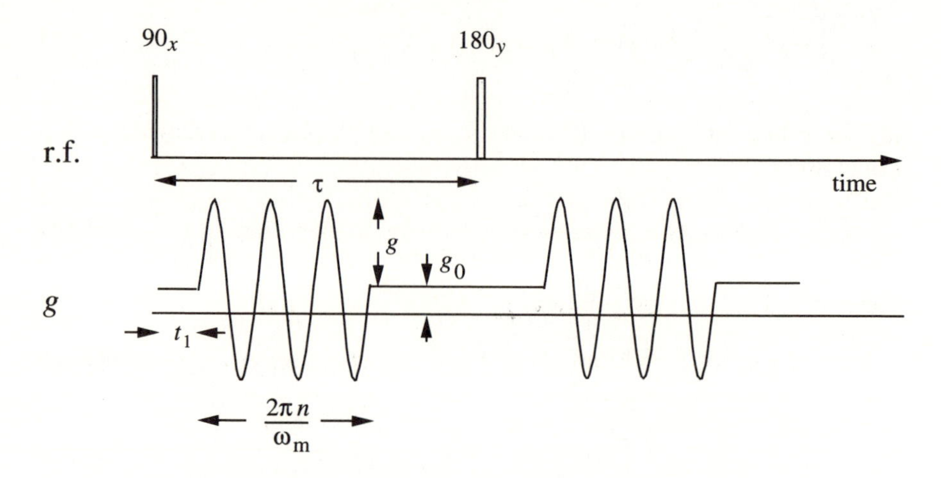

Fig. 6.17 Alternating field gradient experiment of Gross and Kosfeld.[75]

In the absence of the steady gradient the pulse train of Fig. 6.17 is equivalent to the sinusoidal modulation in Fig. 6.16 when t_1 is zero. However, it is clear that the associated echo attenuation expression reduces to eqn (6.66) as required only by assuming $D_{zz}(\omega_m)$ is identically $D_{zz}(0)$, thus restricting the validity of eqn (6.67) to the low-frequency limit.

6.4.4 *Stationary and time-dependent random flow*

Self-diffusion in simple liquids provides an example of translational motion which is randomly directed and rapidly fluctuating, with correlation lengths on the order of angström units and correlation times very much shorter than the Larmor period. There is no possibility of studying spectral features of this motion using magnetic resonance microscopy. Study of motional structure for such small molecules is the domain of neutron scattering and the only molecules whose motional structure is amenable to study by PGSE NMR are very high polymers.

None the less, small molecule stochastic motions can exist at greater length- and time-scales, fluid turbulence being an obvious example. The onset and description of turbulent effects in flow at high Reynolds numbers represents a challenging branch of hydrodynamics in which flow visualization techniques play a major role in model evaluation. A recent, and very early, application of NMR imaging to this field concerns the study of regular vortex structures in Couette flow.[76] There is also an indication that rapid echo planar imaging methods can provide an interesting visualization of random flow patterns (P. Mansfield, personal communication), provided that these structures have dimensions in the millimetre and millisecond regime.

On the microstructural scale irregular motion is a feature of flow through porous media and biological tissue. Here the molecules follow paths which are randomly directed because of the complexity of capillary organization. Fluctuations in this motion may occur because of the random-walk character of the channels and capillaries and because of path divergence at branch points where a small change in initial condition can lead to widely differing outcomes, a classic feature of chaotic behaviour. This type of incoherent motion is sometimes called 'perfusion' in the radiology literature.[77-79]

In characterizing random flow it is helpful to define two regimes. In the first, which we shall call 'stationary random flow' the molecular motions are randomly directed in magnitude and/or orientation and described by a velocity field $\eta(\mathbf{u}, \mathbf{r})$ which is time independent. This is the motion which might be associated with laminar flow in shear or with flow in an array of randomly directed capillaries in which the local director is fixed. Such motion is illustrated in Fig. 6.18(a). In the second, which we label 'pseudodiffusion', the molecular velocities are not only randomly distributed across the ensemble but fluctuate in time as well. Examples of pseudodiffusion include turbulence and branched capillary motion as shown in Fig. 6.18(b). Pseudodiffusion will be characterized by a correlation length L and correlation time τ_c.

Stationary random flow

In atmospheric science it is common to picture air motion in terms of a superposition at various length scales. For example, in mid-latitudes, local wind gusts ($L \sim 10\,\text{m}$) may be superposed on a regional circulation associated with a cyclonic disturbance ($L \sim 100\,\text{km}$) while the whole cyclone gradually drifts westward with the prevailing average flow ($L \sim 10{,}000\,\text{km}$). It is helpful to employ such a pictorial superposition in describing microscopic motion. Here we use the vector $\mathbf{v}$ to describe the mean velocity and $\mathbf{u}_i$ to describe the local variation about that mean for a region of fluid labelled by the index i. This motion has correlation time, τ_{cu}, and correlation length $L_u \sim (\overline{u^2})^{1/2}\tau_{cu}$. Within the region i the individual molecules, labelled by j, will experience stochastic motion $\mathbf{r}_j$ due to self-diffusion. This motion has correlation time τ_{cr} with correlation length $L_r \sim (D\tau_{cr})^{\frac{1}{2}}$. The three motions will be so separated in correlation length and time that we can treat them as stochastically independent.

The total displacement of a molecule labelled by j in fluid element i is therefore

$$\mathbf{r}_{ij}^{\mathrm{T}}(t) = \mathbf{v}t + \mathbf{u}_i t + \mathbf{r}_j(t). \tag{6.68}$$

The j-ensemble average over a time t long compared with the molecular

(a)

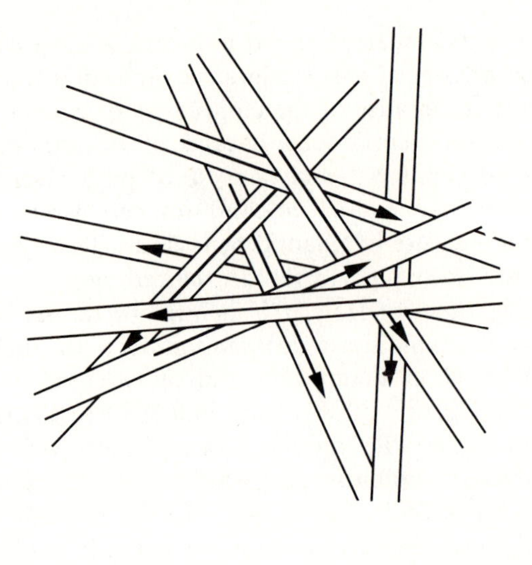

(b)

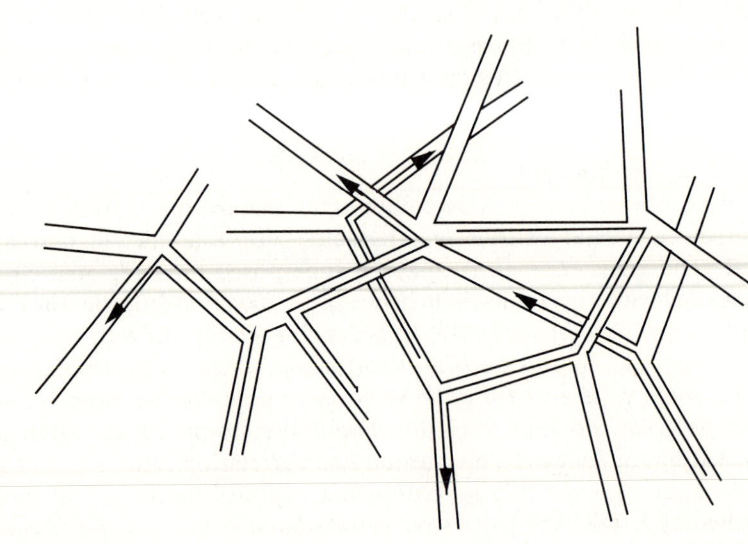

Fig. 6.18 (a) Example of stationary random flow. The particles are moving in randomly oriented capillaries so that the motion is describable by a time-independent velocity field. (b) An example of pseudodiffusion. Flow occurs in branched capillaries so that the velocity direction fluctuates with a correlation time of order the ratio of the mean branch separation and the mean velocity.

diffusive correlation time but short compared with the fluid element correlation time ($\tau_{cr} \ll t \ll \tau_{cu}$) gives $\overline{\mathbf{r}_i^{T}} = \mathbf{v}t + \mathbf{u}_i t$ while the average over i for $\tau_{cu} \ll t$ gives $\mathbf{r}^{T} = \mathbf{v}t$. We shall evaluate the PGSE experiment in the stationary random flow regime, $\tau_{cr} \ll t \ll \tau_{cu}$, where $\mathbf{u}_i$ is time independent.

Following eqn (6.52) the PGSE phase shift for a spin located in molecule ij is then

$$\phi_{ij}(t) = \gamma \int_0^t t'\mathbf{g}^*(t')\cdot(\mathbf{v} + \mathbf{u}_i)\,\mathrm{d}t' + \gamma \int_0^t \mathbf{g}^*(t')\cdot\mathbf{r}_j(t')\,\mathrm{d}t'. \quad (6.69)$$

The separability of the i and j averages means that diffusion and random flow are uncorrelated and the averages over i and j are separable. Thus

$$E(t) = \overline{\exp\left[\mathrm{i}\phi_j(t)\right]}$$
$$= \overline{\exp\left[\mathrm{i}\gamma \int_0^t \mathbf{g}^*(t')\cdot\mathbf{r}_j(t)\,\mathrm{d}t'\right]}\;\overline{\exp\left[\mathrm{i}\gamma \int_0^t t'\mathbf{g}^*(t')\cdot(\mathbf{v} + \mathbf{u}_i)\,\mathrm{d}t'\right]}. \quad (6.70)$$

The first factor is the usual diffusive attenuation, equivalent to the exponential term in eqn (6.33b). The second differs slightly from that which is usually associated with flow since eqn (6.70) deals with fluctuating motion. This latter factor is separable into a phase shift due to net flow and a term due to stationary random flow. Defining $\mathbf{p} = \gamma \int_0^t t'\mathbf{g}^*(t')\,\mathrm{d}t'$ we may write this factor $\exp(\mathrm{i}\mathbf{p}\cdot\mathbf{v})\,\overline{\exp(\mathrm{i}\mathbf{p}\cdot\mathbf{u}_i)}$. Note that $\mathbf{p}$ is akin to the $\mathbf{q}$ vector defined for the narrow-pulse PGSE experiment but that it refers to a time integral for any general finite gradient wave-form and so cannot be used to define a reciprocal space conjugate to $\mathbf{r}' - \mathbf{r}$, the displacement made by each spin over a precisely defined time-scale, Δ. Clearly the 'time-scale' of any general $\mathbf{g}^*(t)$ will be difficult to define. However, $\mathbf{p}$ is a very useful vector where one seeks to describe the average velocity vector for the sample of spins.

In the example where the particular gradient modulation method employed is the two-pulse Stejskal–Tanner PGSE sequence, the echo becomes

$$E(\mathbf{g}) = \exp(\mathrm{i}\gamma\delta\mathbf{g}\cdot\mathbf{v}\Delta)\,\overline{\exp(\mathrm{i}\mathbf{p}\cdot\mathbf{u}_i)}\,\exp\left[-\gamma^2\delta^2 g^2 D(\Delta - \delta/3)\right] \quad (6.71)$$

where D is the molecular self-diffusion coefficient. The stationary random flow factor can be evaluated[80] by assuming that the direction of $\mathbf{u}$ is random over location $\mathbf{r}$ and the distribution $\eta(\mathbf{u}, \mathbf{r})$ may be written $\eta(u)$. Using a polar coordinate frame with the polar axis defined by $\mathbf{p}$

$$\overline{\exp(\mathrm{i}\mathbf{p}\cdot\mathbf{u}_i)} = \int_0^\infty \int_0^{2\pi} \int_0^\pi \eta(u)\exp(\mathrm{i}pu\cos\theta)\,u^2\,\mathrm{d}u\,\mathrm{d}(\cos\theta)\,\mathrm{d}\phi$$

$$= 4\pi \int_0^\infty \eta(u)\, u^2 \frac{\sin(pu)}{pu}\, du. \qquad (6.72)$$

This integral may be calculated by assuming pu small whence

$$\overline{\exp(i\mathbf{p}\cdot\mathbf{u}_i)} \approx 1 - \tfrac{1}{6}\,\overline{u^2}\,p^2 \approx \exp(-\tfrac{1}{6}\,\overline{u^2}\,p^2). \qquad (6.73)$$

($\overline{u^2}$ is the variance in the stationary random flow velocity field, defined by $\overline{u^2} = 4\pi\int_0^\infty u^2\eta(u)\,u^2\,du$.)

Eqn (6.73) tells us that the random flow field causes an additional damping of the echo. This attenuation is readily distinguished from that which is caused by molecular diffusion. Unlike self-diffusion, when $\mathbf{p} = 0$ there is no attenuation of echo. This is precisely the condition in which the echo signal is insensitive to flow, as we have noted earlier. Fig. 6.14(b) shows such an 'even echo' sequence in which diffusive attenuation occurs but $\mathbf{p} = 0$.

For odd echoes, where $\mathbf{p} \neq 0$, eqn (6.73) is simple to evaluate given a knowledge of $\mathbf{g}^*(t)$. By comparison with the attenuation under self-diffusion, we can then define effective stationary random flow diffusion coefficients, D_{SRF}. Two obvious examples are as follows:

Constant-gradient spin echo:

$$p = \tfrac{1}{2}\gamma g t^2 \qquad D_{SRF} \approx \tfrac{1}{8}\, t\, \overline{u^2} \qquad (6.74)$$

Single-echo Stejskal–Tanner PGSE:

$$p = \gamma g \delta \Delta \qquad D_{SRF} \approx \tfrac{1}{6}\,\Delta\,\overline{u^2} \qquad (6.75)$$

In fact it is possible to obtain an exact solution for the echo attenuation due to stationary random flow by using the idea that the motion looks like diffusion but with an observational time-scale very much shorter than the diffusive correlation time. In the SRF regime the diffusion spectrum, $D_{zz}(\omega)$, is much narrower than the gradient spectrum, $S(\omega, t)$. The spectral density $D_{SRF}(\omega)$ applicable in eqn (6.55) is therefore a delta function of area $\pi\frac{1}{3}\,u^2$ (see eqn (6.18) which implies that the area under $D_{zz}(\omega)$ is πu_z^2, and note $u_z^2 = \frac{1}{3}\,u^2$). This yields an echo attenuation exponent,

$$\alpha_{SRF}(t) = \tfrac{1}{6}\,\overline{u^2}\,\gamma^2 S(0, t). \qquad (6.76)$$

This relationship applies to all gradient modulation sequences for which $t \ll \tau_{cu}$. For the Stejskal–Tanner two-pulse PGSE experiment the exponent $\alpha_{SRF}(\Delta)$ is $\frac{1}{6}\,u^2\,\gamma^2 g^2\delta^2\Delta^2$, thereby implying that D_{SRF} is $\frac{1}{6}\,u^2\,\Delta$ in precise agreement with eqn (6.75). (Remember that the total echo attenuation, allowing, in addition, for self-diffusion, results by adding

α_{SRF} to the self-diffusive α applicable to the particular gradient modulation sequence employed.)

Stationary random flow is identified by two specific signatures in a PGSE experiment. First, the echo is attenuated in a manner similar to diffusion but with an effective diffusion coefficient, D_{SRF}, which is proportional to observation time. Second, the attenuation vanishes if the time-dependent gradient obeys the condition $\mathbf{p} = 0$. This 'even echo' independence is characteristic of motion which remains coherent over the echo formation period.

Pseudodiffusion

By contrast with the stationary random flow regime, pseudodiffusion results when the observational time-scale is sufficiently long that $\tau_{cu} \ll t$. Given eqn (6.20) we may write down the effective diffusion coefficient as measured by PGSE NMR,

$$D_{PD} = \overline{u_z^2}\,\tau_{cu} = \tfrac{1}{3}\,\overline{u^2}\,\tau_{cu}. \tag{6.77}$$

In a PGSE experiment pseudodiffusive behaviour is characterized by echo attenuation more severe than that expected from self-diffusion alone. The net attenuation is given by the self-diffusive α applicable for the gradient modulation sequence employed, in which the effective diffusion coefficient is taken to be $D + D_{PD}$. Pseudodiffusive attenuation will not refocus when $\mathbf{p} = 0$.

The intermediate case

For $t \sim \tau_{cu}$ the echo attenuation can be interpreted using eqn (6.55), where $D_{PD}(\omega)$ is as defined in eqn (6.19). The signature for this behaviour is partial refocusing on the even echo. An interesting example arises in the theory of spin echoes in a turbulent fluid[81] in which the eddies have size L and the fluid velocity is ΔU. The scaling regime is $L\,R^{-3/4}/\Delta U \ll t \ll L/\Delta U$, where R is the Reynolds number. Here the even echoes partially refocus, slowly decaying as t increases, while the odd echoes are strongly attenuated but slowly increase in amplitude as t increases.

Appendix 6.1

The evaluation of $\overline{z(t_1)z(t_2)}_{LC}$ starts from the definition of $D_{zz}(\omega)$ given in eqn (6.19). Note that the right-hand side of this equation is one-half the Fourier transform of $\overline{v(0)v(t)}$. This leads to an inverse Fourier relationship

$$\overline{v_z(0)\,v_z(t)} = \frac{1}{\pi}\int_{-\infty}^{\infty} D_{zz}(\omega)\exp(-\mathrm{i}\omega t)\,\mathrm{d}\omega. \tag{6.78}$$

Since the velocity correlation depends only on the time difference we can rewrite this as

$$\overline{v_z(t)\,v_z(t')} = \frac{1}{\pi}\int_{-\infty}^{\infty} D_{zz}(\omega)\exp\{-\mathrm{i}\omega(t'-t)\}\,\mathrm{d}\omega. \tag{6.79}$$

Integrating with respect to t and t' between the limits $[t_1, t_2]$ and $[t_2, t_1]$ one obtains

$$\overline{[z(t_2)-z(t_1)][z(t_1)-z(t_2)]} = \frac{1}{\pi}\int_{-\infty}^{\infty} \frac{D_{zz}(\omega)}{\omega^2}\,2\cos\{\omega(t_1-t_2)\}\,\mathrm{d}\omega. \tag{6.80}$$

Now D_{zz} is an even function of ω since the stationary nature of the ensemble ensures $\overline{v_z(0)\,v_z(-t)} = \overline{v_z(0)\,v_z(t)}$. This means that the cosine transform is identically the same as the complex exponential transform. Comparison with eqn (6.21) shows that the left-hand side of eqn (6.72) is twice the location correlation. Hence, as required,

$$\overline{z(t_1)\,z(t_2)}_{\mathrm{LC}} = \frac{1}{\pi}\int_{-\infty}^{\infty} \frac{D_{zz}(\omega)}{\omega^2}\exp\{\mathrm{i}\omega(t_1-t_2)\}\,\mathrm{d}\omega. \tag{6.81}$$

6.6 References

1. Suryan, G. (1951). *Proc. Indian Acad. Sci. Sect A* **33**, 107.
2. Hahn, E. L. (1950). *Phys. Rev.* **80**, 580.
3. Hahn, E. L. (1960). *J. Geophys. Res.* **65**, 776.
4. Carr, H. Y. and Purcell, E. M. (1954). *Phys. Rev.* **94**, 630.
5. Van Hove, L. (1954). *Phys. Rev.* **95**, 249.
6. Bacon, G. E. (1975). *Neutron diffraction*, Oxford University Press.
7. Egelstaff, P. A. (1967). *An introduction to the liquid state*, Academic Press, London and New York.
8. McCall, D. W., Douglass, D. C. and Anderson, E. W. (1963). *Ber. Busenges. Phys. Chem.* **67**, 336.
9. Stejskal, E. O. and Tanner, J. E. (1965). *J. Chem. Phys.* **42**, 288.
10. Stejskal, E. O. (1965). *J. Chem. Phys.* **43**, 3597.
11. Callaghan, P. T. (1984). *Austr. J. Phys.* **37**, 359.
12. Lomer, W. M. and Low, G. C. (1965). In *Thermal neutron scattering*, (ed. P. A. Egelstaff), Academic Press, New York.
13. Squires, G. L. (1976). *Contemp. Phys.* **17**, 411.
14. Balcar, E. and Lovesey, S. W. (1989). *The theory of magnetic neutron and photon scattering*, Oxford University Press.
15. Feynman, R. P., Leighton, R. B. and Sands, M. (1965). *The Feynman lectures on physics,* Vol III, Addison-Wesley, Reading, Mass.
16. Crank, J. (1975). *The mathematics of diffusion*, Oxford University Press.
17. Kac, M. and Logan, J. (1979). In *Fluctuation phenomena*, (ed. E. W. Montrall and J. L. Lebowitz), North Holland, Amsterdam.

18. Mezei, F. (1972). *Z. Phys.* **255**, 146.
19. Mezei, F. (1983). *Physica B* **120B**, 51.
20. Karger, J. and Heink, W. (1983). *J. Magn. Reson* **51**, 1.
21. Wang, M. C. and Uhlenbeck, G. E. (1945). *Rev. Mod. Phys.* **17**, 323.
22. Uhlenbeck, G. E. and Ford, G. W. (1963). *Lectures in statistical mechanics*, American Mathematical Society, Providence, Rhode Island.
23. Berne, B. J. and Pecora, R. (1976). *Dynamic light scattering*, Wiley, New York.
24. Abragam, A. (1961). *Principles of nuclear magnetism*, Oxford University Press.
25. Lenk, R. (1977). *Brownian motion and spin relaxation*, Elsevier, Amsterdam.
26. Stepisnik, J. (1981). *Physica* **104B**, 350.
27. Stepisnik, J. (1985). *Progr. NMR Spectr.* **17**, 187.
28. Hayward, R. J., Packer, K. J. and Tomlinson, D. J. (1972). *Mol. Phys.* **23**, 1083.
29. James, T. L. and McDonald, G. C. (1973). *J. Magn. Reson.* **11**, 58.
30 Callaghan, P. T., Trotter, C. M. and Jolley, K. W. (1980). *J. Magn. Reson.* **37**, 247.
31. Stilbs, P. and Moseley, M. E. (1980). *Chem. Scri.* **15**, 176.
32. Stilbs, P. (1987). *Progr. Nucl. Magn. Reson. Spectr.* **19**, 1.
33. Tirrell, M. (1984). *Rubber Chem.* **57**, 523.
34. Von Meerwall, E. D. (1983). *Adv. Poly Sci.* **54**, 1.
35. Callaghan, P. T. and Lelievre, J. (1986). *Analytica Chimica Acta* **189**, 145.
36. Wang, J. H. (1954). *J. Am. Chem. Soc.* **76**, 3755.
37. Clark, M. E., Burnell, E. E., Chapman, N. R. and Hinke, J. A. M. (1982). *Biophys. J.* **39**, 289.
38. Callaghan, P. T. and Lelievre, J. (1985). *Biopolymers* **24**, 441.
39. Callaghan, P. T., Lelievre, J. and Lewis, J. A. (1987). *Carbohydrate Research* **162**, 33.
40. Edwards, S. F. (1967). *Proc. Phys. Soc.* **92**, 9.
41. de Gennes, P. G. (1971). *J. Chem. Phys.* **55**, 572.
42. de Gennes, P. G. (1976). *Macromolecules* **9**, 587.
43. de Gennes, P. G. (1979). *Scaling concepts in polymer physics*, Cornell University Press, Ithaca.
44. Doi, M. and Edwards, S. F. (1987). *The theory of polymer dynamics*, Oxford University Press.
45. McCall, D. W., Anderson, E. W. and Huggins, C. M. (1961). *J. Chem. Phys.* **34**, 804.
46. Callaghan, P. T. and Pinder, D. N. (1980). *Macromolecules* **13**, 1085.
47 Callaghan, P. T. and Pinder, D. N. (1981). *Macromolecules* **14**, 1334.
48. Callaghan, P. T. and Pinder, D. N. (1984). *Macromolecules* **17**, 431.
49. Fleischer, G. (1983). *Polym. Bull.* **9**, 152.
50. Fleischer, G. (1984). *Polym. Bull.* **11**, 75.
51. Fleischer, G. and Strausse, E. (1985). *Polym.* **26**, 241.
52. Fleischer, G., Geschke, D., Heinke, W. and Karger, J. (1985). *J. Magn. Reson.* **65**, 429.
53. Bachus, R. and Kimmich, R. (1983). *Polymer* **24**, 964.
54. de Gennes, P. G. (1974). *The physics of liquid crystals*, Oxford University Press.
55. Chandrasekhar, S. (1977). *Liquid crystals*, Cambridge University Press.
56. Blinc, R., Pirs, J. and Zupancic, I. (1973). *Phys. Rev. Lett.* **30**, 546.
57. Blinc, R., Burgar, M., Luzar, M., Pirs, J., Zupancic, I. and Zumer, S. (1974). *Phys. Rev. Lett.* **33**, 1192.

58. Doane, J. W. and Parker, R. S. (1973). *Magn. Reson. Relat. Phenom. Proc. 17th Congr. Ampère (1972)*, p. 410.
59. Boden, N., Corne, S. A. and Jolley, K. W. (1984). *Chem. Phys. Lett.* **105**, 99.
60. Miljkovic, L., Thompson, T., Pintar, M. M., Blinc, R. and Zupancic, I. (1976). *Chem. Phys. Lett.* **38**, 15.
61. Silva-Crawford, M., Gerstein, B. C., Kuo, A. L. and Wade, C. G. (1980). *J. Am. Chem. Soc.* **102**, 3728.
62. Hrovat, M. I. and Wade, C. G. (1981). *J. Magn. Reson.* **45**, 67.
63. Roeder, S. B. W., Burnell, E. E., Kuo, A. L. and Wade, C. G. (1976). *J. Chem. Phys.* **64**, 1848.
64. Lindblom, G. (1981). *Acta Chem. Scand.* **B35**, 61.
65. Tiddy, G. J. T. (1977). *J. Chem. Soc. Faraday Trans.* **173**, 1731.
66. Ukleja, P. and Doane, J. W. (1980). In *Ordering in two dimensions* (Ed. H. Sinha), p. 427. Elsevier/North Holland, Amsterdam.
67. Callaghan, P. T., LeGros, M. A. and Pinder, D. N. (1983). *J. Chem. Phys.* **79**, 6372.
68. Davis, J. H., Jeffrey, K. R., Bloom, M., Valic, M. I. and Higgs, T. P. (1976). *Chem. Phys. Lett.* **42**, 390.
69. Kruger, G. J., Spiesecke, H., von Steenwinkle, R. and Noack, F. (1977). *Mol. Cryst. Liq.* **40**, 103.
70. Callaghan, P. T., Jolley, K. W. and Lelievre, J. (1979). *Biophys. J.* **28**, 133.
71. Callaghan, P. T. and Soderman, O. (1983). *J. Phys. Chem.* **87**, 1737.
72. Blum, F. D., Padmanabhan, A. S. and Mohebbi, R. (1985). *Langmuir* **1**, 127.
73. Packer, K. J. and Zelaya, F. (1989). *Colloids and Surfaces* **36**, 221.
74. Callaghan, P. T. and Xia, Y. (1991) *J. Magn. Reson.* **91**, 326.
75. Gross, B. and Kosfeld, R. (1969). *Messtechnik* **7/8**, 171.
76. Grutzner, J. (1989). *International Society For Magnetic Resonance Meeting*, Morzine, Abstract S3.
77. Le Bihan, D., Breton, E., Lallemand, D., Aubin, M., Vignaud, J. and Laval-Jeantet, M. (1986). *Radiology* **161**, 401.
78. Le Bihan, D., Breton, E., Lallemand, D., Aubin, M., Vignaud, J. and Laval-Jeantet, M. (1988). *Radiology* **168**, 497.
79. Le Bihan, D., Delannoy, J. and Levin, R. (1989). *Radiology* **171**, 853.
80. Nalcioglu, O. and Cho, Z. H. (1987). *IEEE Trans. Med. Imag.* **MI-6**, 356.
81. de Gennes, P. G. (1969). *Phys. Lett.* **29A**, 20.

7
STRUCTURAL IMAGING USING q-SPACE

7.1 Restricted diffusion

The translational motion of molecules in a heterogeneous structure will be influenced by boundaries. In the previous chapter we looked at anisotropic self-diffusion caused by the confining geometry of surrounding surfaces. In those examples we were concerned only with the components of displacement parallel to surfaces which meant that the observed motion was unbounded but that its appearance in the laboratory frame reflected a characteristic spatial dimensionality. Now we turn our attention to the measurement of motion along a direction in which molecules encounter some boundary. For example, the water inside a cell experiences a barrier to diffusion at the cell wall; oil molecules in plant or food material are often confined to localized droplets; and water inside a porous sandstone exists in a complex system of interconnected pores and capillaries. These restrictions mean that the distribution of displacements described by the conditional probability $P_s(\mathbf{r}|\mathbf{r}', \Delta)$ may no longer be Gaussian and will have a time dependence characteristic of the length scales and the local molecular self-diffusion coefficient. This characteristic behaviour is potentially very useful since it offers the possibility of probing the microstructure.

7.2 Simple confining boundaries

7.2.1 *Rectangular boundaries*

We will start by examining a very simple example. Here, molecules in the liquid state are confined within a rectangular box. Suppose that a PGSE experiment is performed with the gradient applied parallel to one side (z) of the box of length a. We will then obtain $E(q)$ at a number of different times, Δ. At very short times, most molecules (except those very close to a diffusive barrier) will experience free Brownian motion. $E(q)$ will have the behaviour characteristic of unrestricted self-diffusion, as given in eqn (6.33). On a time-scale long compared with that taken to diffuse a distance a, all molecules will have diffused backwards and forwards across the box many times and will have lost all 'memory' of their initial positions. In mathematical terms this means that $P_s(z|z', \Delta)$ is simply the probability of finding a molecule anywhere along z', the prime referring to the final coordinates at time Δ. This probability is, of course, $\rho(z')$, the static molecular density function.

In chapter 6 it was shown that $E(q)$ has a Fourier relationship with $\overline{P}_s(Z, \Delta)$, the spectrum of dynamic displacements. We will see that the limiting behaviour of $E(q)$ is capable of revealing information about the reciprocal lattice, $S(q)$. This is another sense in which the PGSE experiment is a very explicit form of imaging in its own right! Before pursuing this point however, we will write down the exact general solution for $P_s(z|z', \Delta)$ for the rectangular-box problem. This was obtained by Tanner and Stejskal[1] by solving eqn (6.5) with the boundary conditions $\nabla P_s = 0$ when either z' or z correspond with the edges of the box. They found

$$P_s(z|z', \Delta) = 1 + 2\sum_{n=1}^{\infty} \exp\left(-\frac{n^2\pi^2 D\Delta}{a^2}\right)\cos\left(\frac{n\pi z'}{a}\right)\cos\left(\frac{n\pi z}{a}\right) \tag{7.1}$$

Using the narrow gradient pulse expression, eqn (6.26), the echo attenuation function can be shown to be[1]

$$E(q) = \frac{2\{1-\cos(2\pi qa)\}}{(2\pi qa)^2} + 4(2\pi qa)^2\sum_{n=1}^{\infty}\exp\left(-\frac{n^2\pi^2 D\Delta}{a^2}\right) \times \frac{1-(-1)^n\cos(2\pi qa)}{\{(2\pi qa)^2-(n\pi)^2\}^2}\,. \tag{7.2}$$

Eqn (7.2) represents one of the few exact solutions to the problem of restricted diffusion in a specific geometry. The two limiting behaviours are

Short time-scale limit ($\Delta \ll a^2/2D$)

$$E(q) = \exp(-4\pi^2 q^2 D\Delta) \tag{7.3}$$

Long time-scale limit ($\Delta \gg a^2/2D$)

$$E(q) = \frac{2\{1-\cos(2\pi qa)\}}{(2\pi qa)^2}\,. \tag{7.4}$$

It is interesting to evaluate the long time-scale limit expression in the case where the PGSE gradient is weak, such that $2\pi qa \ll 1$ or $g \ll (1/\gamma\delta a)$. Then, expanding eqn (7.4) to fourth order in $2\pi qa$ we find

$$E(q) \approx 1 - \frac{1}{12}(2\pi qa)^2$$

$$\approx \exp(-\gamma^2\delta^2 g^2 D_{app}\Delta) \qquad \text{where } D_{app} = \frac{a^2}{12\Delta}\,. \tag{7.5}$$

Eqn (7.5) states that the apparent diffusion coefficient decreases as the observation time, Δ, increases. Its magnitude is, in accordance with eqn (6.33), simply the mean square distance travelled by the molecules divided by twice the observation time.

This idea that the apparent diffusion coefficient, as measured by PGSE, is described by a term of order the mean squared displacement divided by the time-scale, is true only where the gradient applied is weak, meaning that the magnitude of q is much less than the reciprocal of the barrier spacing. We should not leave eqn (7.4) without noting the very peculiar feature associated with the condition $qa = 1$. Here the echo signal is completely zero, as if the spins had diffused an infinite distance. The exact behaviour represented by eqn (7.2) is shown in Fig. 7.1. The meaning of the pathological behaviour of $E(q)$ in the case $qa = 1$ will be apparent once we have developed a suitable formalism for the long time limit signal.

Recently Frey *et al.*[2] have generalized eqns (7.1) and (7.2) to allow for the case of the walls absorbing rather than reflecting. Physically this might correspond to the destruction of magnetization due to strong relaxation 'sinks' at the boundary surface, or to a rapid enhancement of molecular mobility as soon as molecule leaves the rectangular subregions, a situation which can sometimes be encountered in zeolites. The dependence of echo attenuation on Δ at low q values shows many remarkable similarities with the apparent diffusion coefficient reducing as Δ increases, albeit at different rates. However, in the long time-scale limit the behaviours are dramatically different with $E(q) \rightarrow 0$ for the absorbing-barrier case in direct contrast with eqn (7.4).

7.2.2 *Spherical boundaries*

For completeness we add the expression for the echo attenuation for diffusion within reflecting spherical boundaries. This complex mathematical

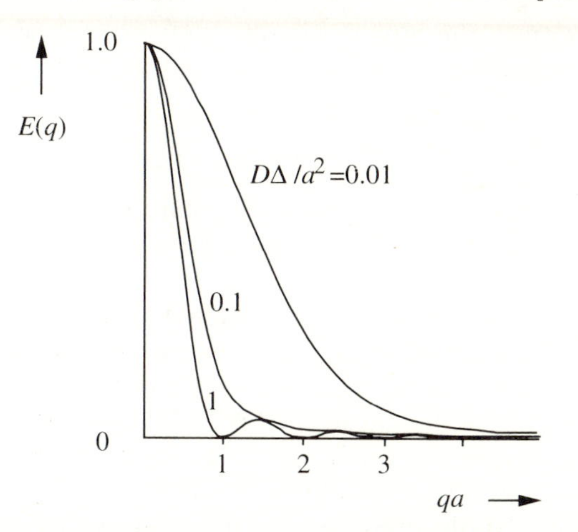

Fig. 7.1 Echo attenuation for diffusion in a box as represented by eqn (7.2). $E(q)$ is shown for diffusion times, Δ, both small and comparable with the mean time taken to diffuse across the box, a^2/D. For larger values of Δ the behaviour is essentially time independent and equivalent to that shown for $D\Delta/a^2 = 1$.

problem has not been solved exactly but by assuming a Gaussian distribution of phase displacements. In the case of the constant-gradient spin echo a solution was found by Neumann,[3] and for the PGSE experiment by Murday and Cotts[4] who obtained for a sphere of radius a,

$$E(q) = \exp\left(-2\gamma^2 g^2 \sum_{m=1}^{\infty} f(\alpha_m)\right) \tag{7.6a}$$

with

$$f(\alpha_m) = \{\alpha_m^2(\alpha_m^2 a^2 - 2)\}^{-1}\left[\frac{2\delta}{\alpha_m^2 D} - \frac{2 + \exp\{-\alpha_m^2 D(\tau-\delta)\} - 2\exp(-\alpha_m^2 D\delta)}{(\alpha_m^2 D)^2} - \frac{2\exp(\alpha_m^2 D\tau) + \exp\{-\alpha_m^2 D(\tau+\delta)\}}{(\alpha_m^2 D)^2}\right] \tag{7.6b}$$

where the α_m are roots of the Bessel function equation

$$\alpha_m a J'_{\frac{3}{2}}(\alpha_m a) - \tfrac{1}{2} J_{\frac{3}{2}}(\alpha_m a) = 0\,. \tag{7.7}$$

The two limiting behaviours are

Short time-scale limit ($\Delta \ll a^2/2D$)

$$E(q) = \exp(-4\pi^2 q^2 D\Delta) \tag{7.8}$$

Long time-scale limit ($\Delta \gg a^2/2D$)

$$E(q) = \exp\{-4\pi^2 q^2(\tfrac{1}{5}a^2)\}\,. \tag{7.9}$$

The exact solution in the long time-scale limit was found by Tanner and Stejskal[1,5] who adapted a result well known in the theory of heat diffusion[6] to obtain

$$E(q) = 9[2\pi qa\cos(2\pi qa) - \sin(2\pi qa)]^2/(2\pi qa)^6\,. \tag{7.10}$$

Eqns (7.9) and (7.10) can be shown to be identical for small q.

The transitions from the short time-scale to long time-scale regimes, predicted for rectangular geometry by eqn (7.2) and for spherical geometry by eqn (7.6) have been tested experimentally. In rectangular geometry Tanner and Stejskal[5] carried out a PGSE experiment on water between the layers of a mica stack with average spacing of 16 μm with the magnetic field gradient applied normal to the stack. Fig. 7.2 shows the echo attenuation plot, $\ln(E)$ versus Δ_r . At small values of Δ the echo attenuation plot depends exponentially on Δ according to eqn (7.3). At long times an independence on time-scale is observed because the mean square distance diffused by the water molecules reflects the barrier spacing. The theoretical curve was fitted by allowing for a small distribution of layer spacings.

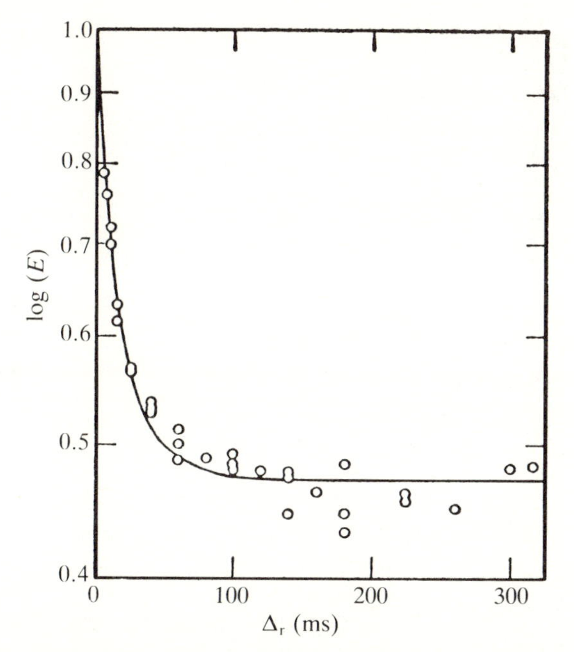

Fig. 7.2 Echo attenuation versus effective diffusion time, Δ_r, with q fixed at $(35\ \mu m)^{-1}$. The theoretical curve is a fit using eqn (7.2) in which $D=2.0\times10^{-9} m^2 s^1$ and $a=16.1\ \mu m$. (From J. E. Tanner and E. O. Stejskal.[5])

The spherical-boundary problem was investigated by Packer and Rees[7] using a water-in-oil emulsion, as shown in Fig. 7.3. Here, despite a considerable size distribution, the qualitative features of eqns (7.9) and (7.10) are well represented, with an exponential dependence on Δ at short times and a Δ independence apparent in time-scales long enough for the water mean square displacement to be limited by the droplet size.

In order to be applied reliably over a wide range of time-scales, the effects of transverse relaxation and residual background field gradients must be eliminated. Van den Enden *et al.*[8] have devised a rapid method to determine droplet size distributions based on the stimulated echo in which these problems are avoided. This method of droplet sizing is sufficiently convenient and reliable to be applied routinely in the food industry.

7.3 The averaged propagator in the long time-scale limit

7.3.1 *Boxes and spheres*

We can show very simply that the mean square distance for the parallel boundary case is indeed $\frac{1}{6}a^2$ as required by eqn (7.5). Before doing so

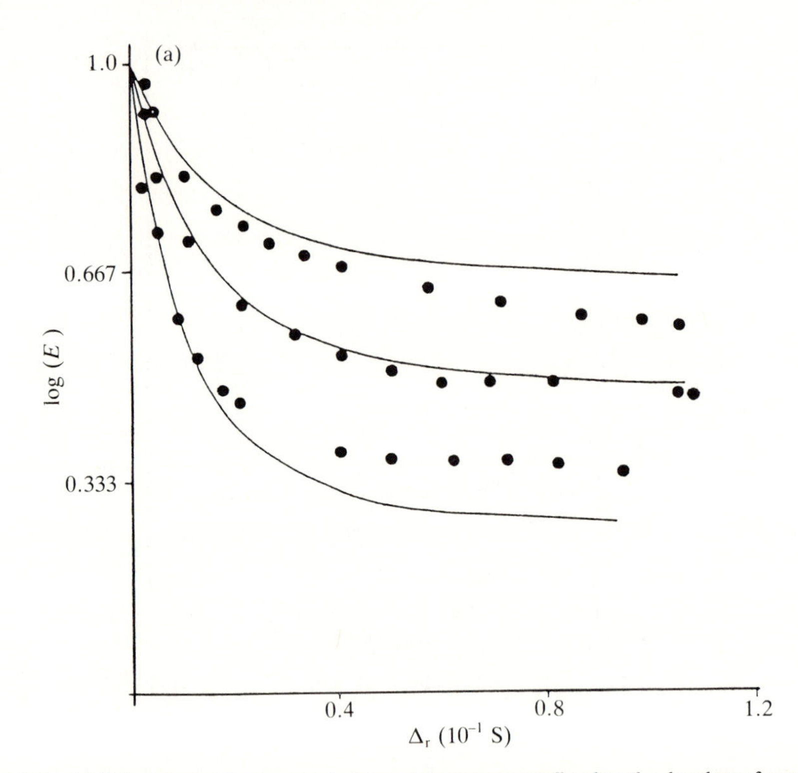

Fig. 7.3 (a) Echo attenuation versus Δ_r for water protons confined to the droplets of a water-in-oil emulsion. The theoretical curves allow for droplet size distributions. (From K. J. Packer and C. Rees.[7]) (b) As in (a) but with the data plotted against $q^2\Delta_r$. Note the coincidence of the data, characteristic of unrestricted diffusion, when $\Delta_r \ll a^2/D$.

however, it will be very helpful to think about the nature of the conditional probability in the long time limit. $\Delta \to \infty$ in the present context means that the molecule loses all memory of its starting position and can be anywhere in the structure. This means in effect that $P_s(\mathbf{r}|\mathbf{r}', \infty)$ is the same as the starting density, $\rho(\mathbf{r}')$. This gives the average propagator a very special character, namely

$$\overline{P_s}(\mathbf{R}, \infty) = \int \rho(\mathbf{r} + \mathbf{R})\, \rho(\mathbf{r})\, \mathrm{d}\mathbf{r}\,. \tag{7.11}$$

This expression is exactly the autocorrelation function of the molecular density! Fig. 7.4 shows $\overline{P_s}(\mathbf{R}, \infty)$ for the rectangular-box example just considered. Properly normalized it has the form

$$\begin{aligned} \overline{P_s}(Z, \infty) &= (a + Z)/a^2 \qquad -a \leq Z \leq 0 \\ &\quad\ (a - Z)/a^2 \qquad 0 \leq Z \leq a. \end{aligned} \tag{7.12}$$

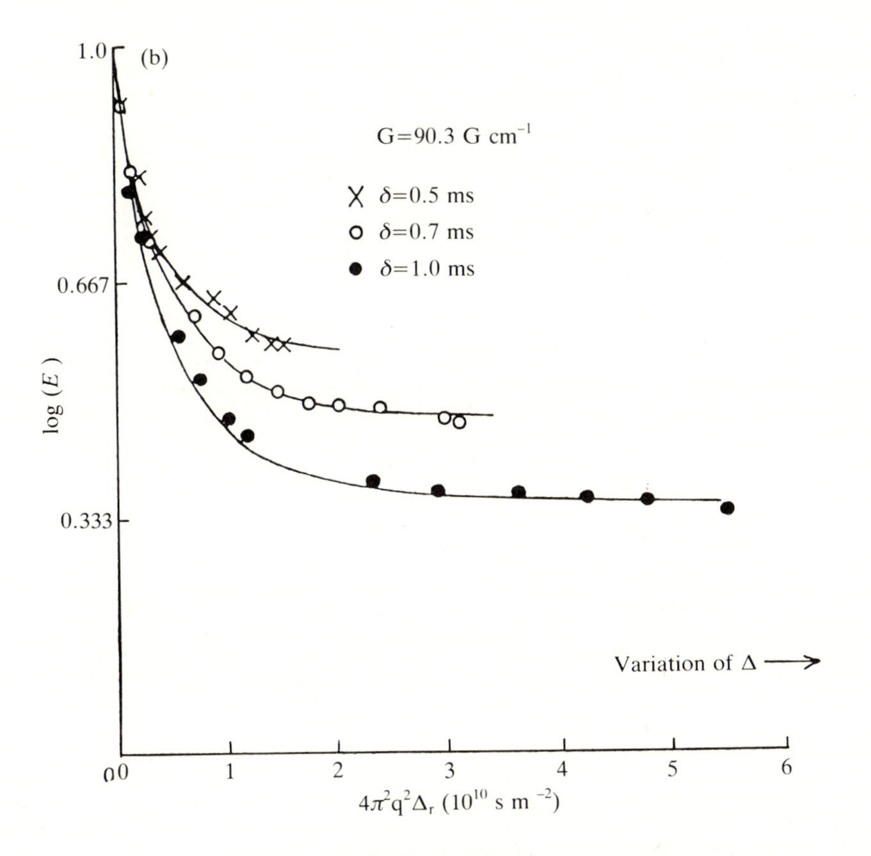

Fig. 7.3 *Continued*

We are now in a position to calculate the mean square displacement:

$$\overline{Z^2} = \int_{-\infty}^{\infty} Z^2\,\overline{P}_s(Z, \Delta \to \infty)\,dZ = \frac{1}{a^2}\int_{-a}^{0} Z^2(a+Z)\,dZ + \frac{1}{a^2}\int_{0}^{a} Z^2(a-Z)\,dZ$$

$$= \frac{1}{6}a^2. \tag{7.13}$$

The idea that the long time-scale averaged propagator is simply the autocorrelation function of the molecular density, $\rho(\mathbf{r})$ is very powerful. In particular, since the PGSE experiment (in the narrow-pulse approximation) gives us a signal $E_\Delta(q)$ which is simply the Fourier transform of $\overline{P}_s$, we obtain another important insight. It is well known in the theory of light scattering that the Fourier transform of a time autocorrelation function is simply the frequency power spectrum, a result sometimes known as the Wiener–Khintchine theorem.[9] We shall express this idea in our own

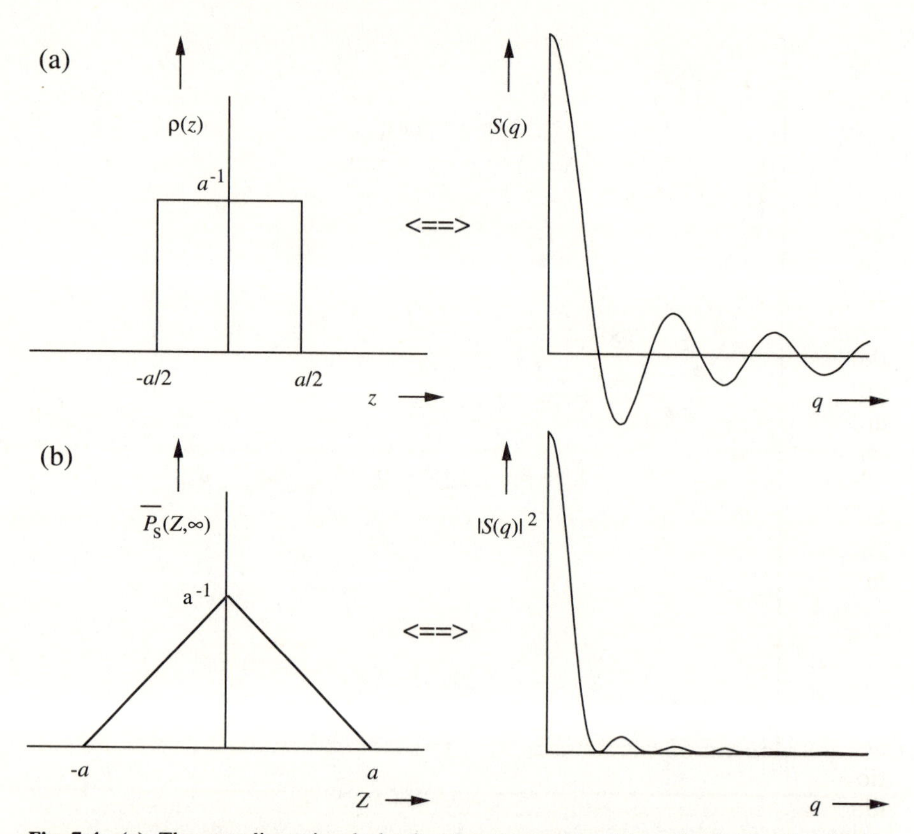

Fig. 7.4 (a) The one-dimensional density function, $\rho(z)$, for molecules confined to a rectangular box of width a, along with its corresponding Fourier transform, $S(q)$. (b) The infinite time-averaged propagator, $\overline{P_s}\,(Z,\infty)$, for the rectangular box and its corresponding transform. Because $\overline{P_s}\,(Z,\infty)$ is the autocorrelation function of $\rho(z)$, $\mathfrak{F}\{\overline{P_s}\,(Z,\infty)\}$ is simply $|S(q)|^2$. In the optical analogy, $|S(q)|^2$ is the diffraction pattern from a single slit.

terms by returning to first principles. In the long time limit where $P_s(\mathbf{r}|\mathbf{r}',\Delta\to\infty)$ reduces to $\rho(\mathbf{r}')$, eqn (6.28) tell us

$$\begin{aligned}
E_\infty(\mathbf{q}) &= \iint \rho(\mathbf{r})\,\rho(\mathbf{r}')\exp[\mathrm{i}2\pi\mathbf{q}\cdot(\mathbf{r}'-\mathbf{r})]\,\mathrm{d}\mathbf{r}'\,\mathrm{d}\mathbf{r} \\
&= \int \rho(\mathbf{r})\exp[-\mathrm{i}2\pi\mathbf{q}\cdot\mathbf{r}]\,\mathrm{d}\mathbf{r}\int \rho(\mathbf{r}')\exp[\mathrm{i}2\pi\mathbf{q}\cdot\mathbf{r}']\,\mathrm{d}\mathbf{r}' \\
&= S^*(\mathbf{q})S(\mathbf{q}) \\
&= |S(\mathbf{q})|^2. \qquad (7.14)
\end{aligned}$$

This relationship states that in the very long time limit, such that the molecules move over the entire sample, the PGSE signal is precisely the power spectrum of the reciprocal lattice, an optical analogy pointed out by Cory and Garroway.[10] This is a remarkable result because it means that the PGSE experiment in the long time limit is an imaging experiment, returning not the reciprocal lattice as in **k**-space imaging, but the modulus squared of the reciprocal lattice. The loss of phase information by this squaring process is a severe disadvantage but it is exactly the price paid in X-ray diffraction experiments. There the intensity reciprocal lattice is interpreted by comparing it with that calculated from a starting model in real space. But in one crucial aspect, PGSE or **q**-space static imaging has a major advantage over the **k**-space method in NMR microscopy. Because the signal is sampled in the absence of a gradient there is no fundamental resolution limit! The only limit is imposed by the magnitude of the pulsed gradient which can be applied, subject to the condition that the two gradient pulses must be well matched in area.

Eqn (7.14) can be used to give a nice description of the long time-scale behaviour in the rectangular box. The reciprocal lattice, $S(\mathbf{q})$, for the rectangular box is simply the Fourier transform of a hat function as shown in Fig. 7.4(a). Thus

$$E_{\infty}(q) = |\operatorname{sinc}(\pi q a)|^2 = \frac{4\sin^2(\pi q a)}{(2\pi q a)^2} = \frac{2\{1 - \cos(2\pi q a)\}}{(2\pi q a)^2}. \quad (7.15)$$

This is exactly the result of eqn (7.4) as is clear by comparison of Fig. 7.4(b) and the long time-scale limit of Fig. 7.1. Its optical analogue is the diffraction pattern of a single slit and the origin of the node when $qa = 1$ is identically the same. It results from the phase cancellations which arise from different 'slit elements'. For every spin originating at position $z \leqslant 0$ with resulting phase shift $\exp(i2\pi q z)$ there is an equal number originating at position $(a/2 + z)$ with phase shift $\exp\{i2\pi q(a/2 + z)\}$ or $-\exp(i2\pi q z)$.

It is useful to apply the reciprocal lattice idea to other structures in which bounded diffusion occurs. The long time-scale behaviour for particles diffusing in a sphere of radius a can be obtained by calculating the one-dimensional Fourier transform of the normalized density. This latter function is simply $(3/4a^3)(z^2 - a^2)$ for $-a \leq z \leq a$ and zero elsewhere. Thus

$$S(q) = \int_{-a}^{a} \left(\frac{3}{4a^3}\right)(z^2 - a^2)\exp(i2\pi q z)\,dz$$

$$= \frac{3\{2\pi q a\cos(2\pi q a) - \sin(2\pi q a)\}}{(2\pi q a)^3}. \quad (7.16)$$

The PGSE signal, $E_{\infty}(q)$ is $|S(q)|^2$ and has a node when $qa \approx \frac{3}{4}$. The behaviour when $q \ll a^{-1}$ is

$$E_\infty(q) \approx 1 - \tfrac{1}{5}(2\pi q a)^2$$
$$\approx \exp(-\tfrac{1}{5}\gamma^2\delta^2 g^2 a^2)\,. \tag{7.17}$$

In fact eqn (7.17) gives a quite a good description of E_∞ over a decade of attenuation ($qa \leq 0.5$). Comparison with eqn (6.33b) shows that the weak gradient apparent diffusion coefficient is $a^2/5\Delta$. As in the case of the rectangular box, this is the mean square distance diffused by the molecules, $\frac{2}{5}a^2$, divided by 2Δ.

7.3.2 *Real space versus reciprocal space*

The long time-scale experiments give an echo attenuation which depends on boundary structure and which is independent of the motional parameter, in this case the self-diffusion coefficient. The way in which we employ such a method to 'image' the structure is a matter of choice. Unfortunately, because the signal is an intensity map in reciprocal space, Fourier inversion does not return the original structure but rather an autocorrelation function of that structure. Despite this such an inversion can be revealing, especially where the geometry is simple, such as in the case of spherical boundaries. Fourier inversion of echo attenuation data was first proposed by Kärger and Heink[11] who applied the method to the diffusion of ethane in zeolites. Fig. 7.5 shows another example of the method. Here the long time-scale echo attenuation, $E(\mathrm{q})$, resulting from restricted diffusion of water in yeast cells,[10] has been Fourier transformed with respect to q to yield the one-dimensional density autocorrelation function in real space. It is not a map of the yeast cell so much as a map of the distribution of water molecule displacements in the yeast cell. None the less, it represents an important illustration of the **q**-space imaging principle.

The loss of phase information resulting from an observation of the intensity rather than the amplitude spectrum in reciprocal space is one familiar to the X-ray crystallographer. Because of the problem of relating the density autocorrelation function to the original structure, Fourier inversion can often obscure the process of extracting structural information. The solution for the crystallographer is to work in conjugate space, testing models of the structure by comparing their intensity reciprocal lattice with the observed data. Similarly, it is often preferable to work in reciprocal space when interpreting echo attenuation data in the long time-scale limit. A simple example of the effectiveness of this approach is given in Fig. 7.6. Here, the droplet size distribution for fat globules in cream is obtained by fitting the signal to a model which incorporates a normal distribution of sphere volumes, $P(a) = (\pi\sigma^2)^{-\frac{1}{2}} \exp(-(a - a_0)^2/\sigma^2)$, in eqn (7.17). The resulting echo attenuation is[12]

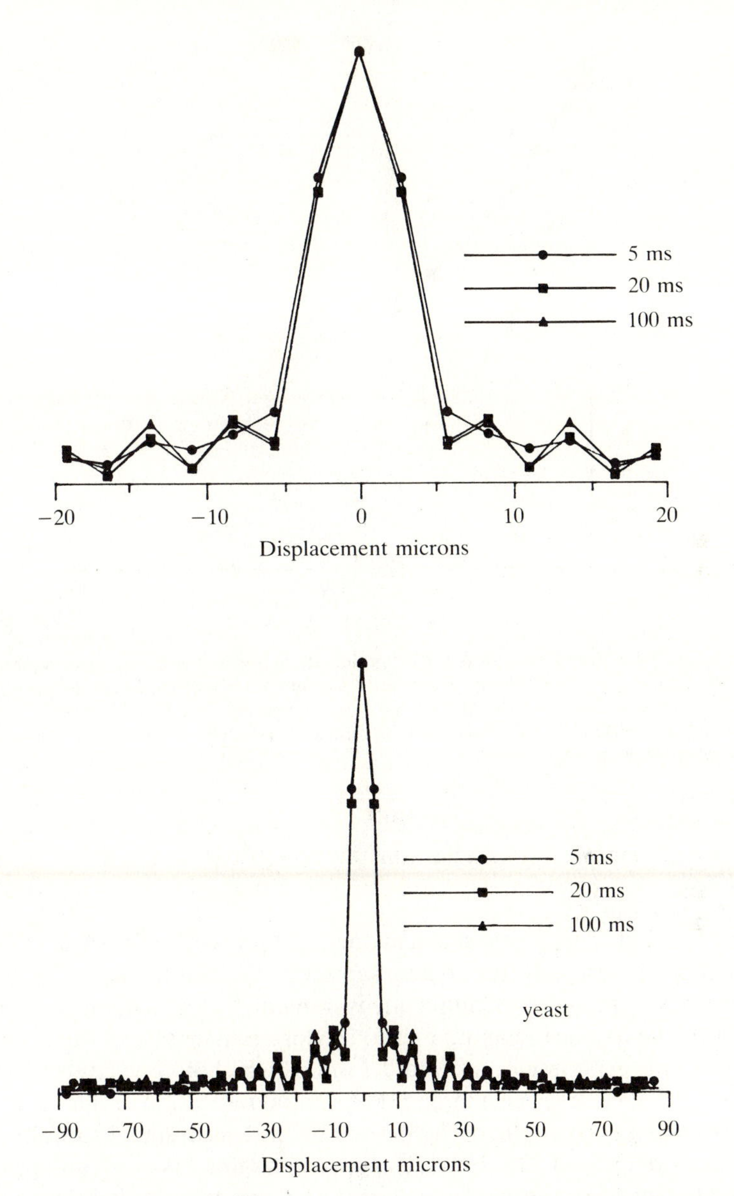

Fig. 7.5 Average propagator, $\overline{P_S}\,(Z, \Delta)$, for water in yeast cells at three different diffusion times, Δ. The upper curve is an expansion of the lower. The oscillations are truncation artefacts. (From Cory and Garroway.[10])

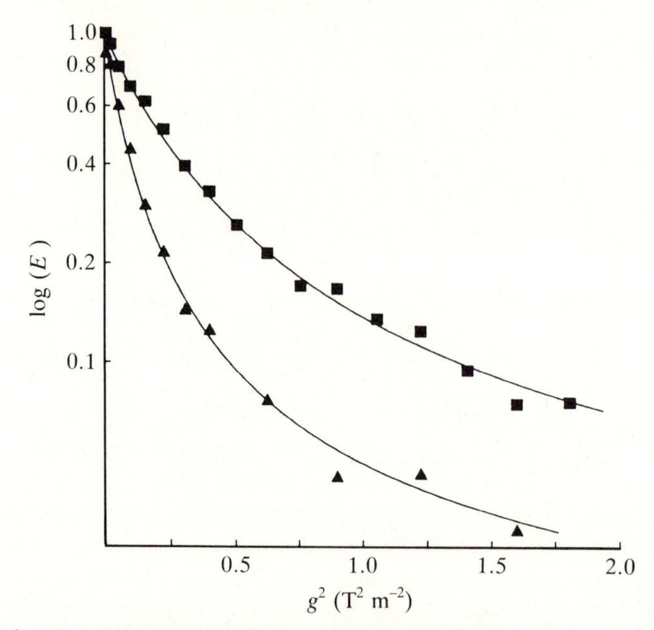

Fig. 7.6 Spin echo attenuation plots for fat molecules in Swiss cheese. The solid squares refer to $\Delta = 140$ ms, $\delta = 6$ ms, $2\tau = 160$ ms. The triangles refer to $\Delta = 140$ ms, $\delta = 10$ ms, $2\tau = 160$ ms. The solid curves through both sets of data are fits using eqn (7.18). The fitted parameters are, respectively, $a_0 = 2600 \pm 50$ nm, $\sigma = 1550 \pm 70$ nm, and $a_0 = 2650 \pm 50$ nm, $\sigma = 1550 \pm 80$ nm. (From Callaghan *et al.*[12])

$$\ln(E) = -\alpha^2 a_0^2[1 + \sigma^2\alpha^2]^{-1} - \tfrac{1}{2}\ln[1 + \sigma^2\alpha^2] \qquad (7.18)$$

where α^2 is $\frac{1}{5}\gamma^2\delta^2 g^2$.

The fit to Fig. 7.6 gives a mean radius 1.65 μm with σ/a_0 of 0.45. When the method is applied to cream samples the results agree well with distributions obtained by Coulter measurement,[13] thus lending a degree of support to the **q**-space imaging method. More remarkable is the resolution implicit in this experiment, around 0.1 μm which should be contrasted with that obtainable via **k**-space imaging. The spatial resolution in this latter form of NMR microscopy is inherently limited by the loss in signal-to-noise as the voxel size is decreased. By contrast, **q**-space imaging takes advantage of the fact that the local structure is heterogenous below some specific size scale but relatively homogeneous over larger distances, thus benefiting from the signal enhancement possible by employing a large sample.

7.3.3 *Selection based on compartment size*

The dependence of the echo attenuation on compartment size rather than time-scale suggests its use in selective excitation. If an NMR experiment to be performed on molecules in a variety of restricted compartments is preceded by a PGSE echo then the contributions from different compartment sizes can be influenced. For example the acquisition can be preceded by a stimulated echo sequence with storage of the remaining magnetization along the z-axis by means of a 90_{-x} r.f. pulse.[14] The z-magnetization contributing to any subsequent experiment will be weighted as $|S(q)|^2$. (The same principle can be used to weight NMR signals according to molecular diffusion coefficients where a distribution exists, a property which can be helpful in the suppression of solvent signals while retaining those of macromolecules.)

When the $|S(q)|^2$ excitation and storage is followed by a **k**-space imaging sequence this weighting scheme can be used to permit the selective imaging of molecules bounded a small pores. Alternatively, as suggested by Cory *et al.*[14] and illustrated in Fig. 7.7(a), the storage can be followed by a diffusion period and then a second PGSE sequence, enabling $E_\Delta(q)$ to be examined after the spins have had a chance to move over the predetermined time T. A change in the form of the averaged propagator as a result of this evolution would indicate leakage from the pores. These authors have also suggested the sequence shown in Fig. 7.7(b) as a means of measuring pore eccentricity since it permits the correlation of $|S(\mathbf{q})|^2$ in two orthogonal directions. Comparison of the echo attenuation which results using the double orthogonal gradients with that obtained using a single PGSE sequence can give information on pore eccentricity without the need for pore alignment in the sample.

In fact, measurements of translational displacements via a spin echo sequence are never obtained in isolation since relaxation influences are always present. Applications of PGSE NMR in which restricted diffusion and relaxation times are correlated are given in Section 7.6.3.

7.4 Porous structures

Quantitative information about material porosity is of considerable importance in industry. Such information gives a useful indication of the strength of ceramics, of the tendency of polymer composites to retain or absorb solvent, and of the permeability of core materials in oil wells. Pulsed gradient spin echo NMR can provide an effective measure of pore structure and connectivity. The method relies on the measurement of restricted motion of liquid molecules which fill the pores and channels. Many important industrial processes concern the transport of macromolecules in porous media. In the case of random coil polymers, the configuration space is limited by the local pore topology which therefore influences the self-diffusion process.[15-17] We shall not consider this particular problem here. Instead we shall be concerned to understand how the pore morphology

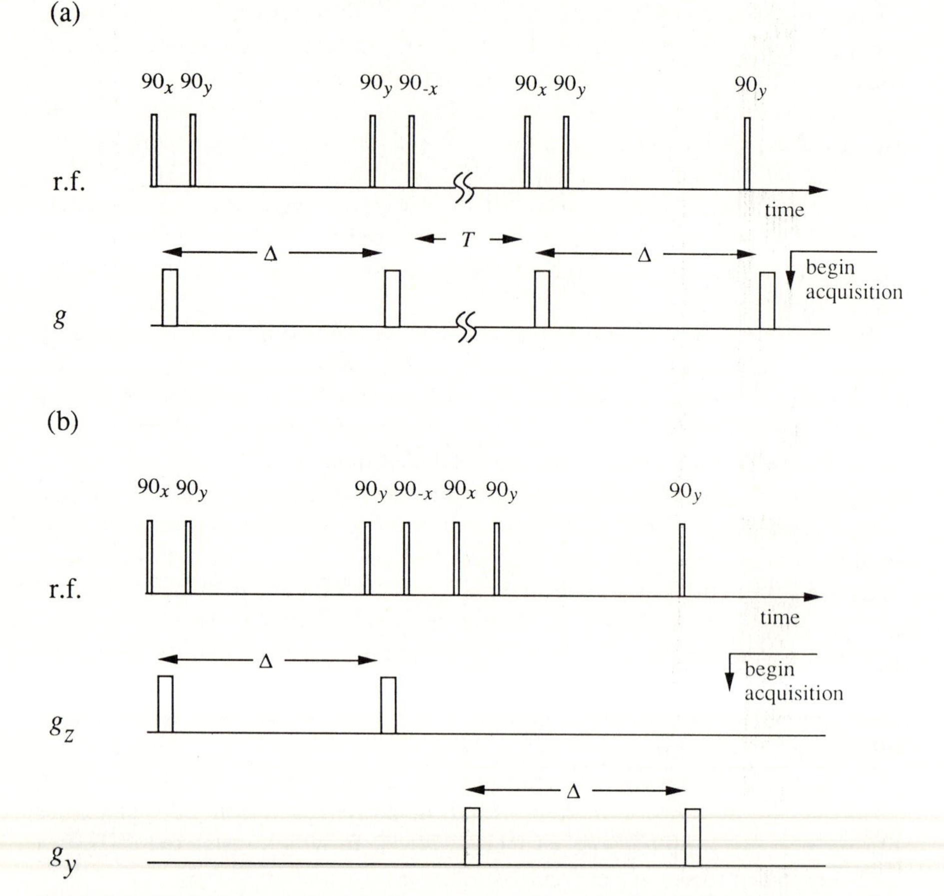

Fig. 7.7 (a) Stimulated echo PGSE sequence with storage of refocused magnetization along the z-axis so that the subsequent NMR experiment is weighted by $E(q)$. In this case the second experiment is another determination of $E(q)$ but at a time T later. The final echo amplitude is governed by the correlation between $\overline{P}_s$ and hence, compartment size, before and after the diffusive period, T, and is therefore suitable for determination of pore leakage. (b) Correlation of $\overline{P}_s\ (Z, \Delta)$ and $\overline{P}_s\ (Y, \Delta)$ by means of successive application of orthogonal PGSE gradients.[14])

influences the diffusional motion of small probe molecules and how, in particular, the **q**-space response of the spin echo attenuation can provide an 'image' of this structure in the sense of eqn (7.14).

7.4.1 *Connected boxes in a regular lattice*

To understand the principles behind this approach it is helpful to return to the problem of the box, this time with interconnecting paths as shown in Fig. 7.8. For simplicity, we will consider a one-dimensional array of

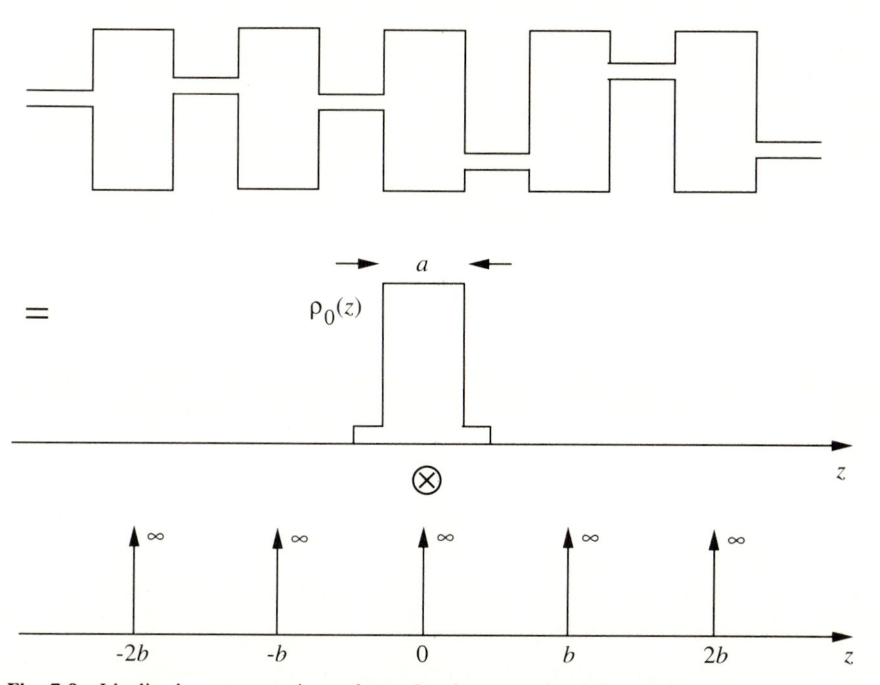

Fig. 7.8 Idealized representation of regular interconnected boxes in which the density function, $\rho(z)$, is viewed as a convolution of the unit cell density, $\rho_0(z)$, (consisting of the pore + channel) with a lattice of Dirac delta functions. The relative fractions of molecules in the pore and channel sections are taken as f_A and f_B, respectively.

boxes, equally spaced along the gradient axis, z, and with longitudinal and transverse dimensions as shown. We will also assume that the system is finite, consisting of N boxes labelled by an index n which ranges from $-N/2$ on the left to $N/2 - 1$ on the right. Our task is to calculate $|S(q)|^2$ in the long time-scale limit.

The representation of pores and connecting channels by regular rectangular geometry is a rather abstract and unreal idealization, as is the assumption of a time-scale sufficiently long that a molecule originating in any location will have the possibility of ending up anywhere in the structure. None the less, such an approach will help us understand the principles behind the use of **q**-space imaging to obtain a model of the bounding structure restricting probe molecules in their translational motion. The problems of the general pore shape and finite time-scale can then be handled subsequently.

The density function, $\rho(z)$, for the idealized interconnected box model is shown in Fig. 7.8. The magnetic field gradient will be applied along the z-axis and we shall not be sensitive to motions in the transverse directions although these are clearly important to the process of diffusion to and from the boxes and interconnecting channels. In the long time-scale limit, however, this motion is 'complete' so that it is possible to idealize $\rho(z)$ further into a convolution of box channel 'unit cells', $\rho_0(z)$, with the set of normalized delta functions, $(1/N)\sum_{n=-N/2}^{N/2-1}\delta(z-nb)$, representing the periodic lattice. Clearly, therefore, $S(q)$ is the product of $\mathfrak{F}\{\rho_0(z)\}$ and the $\mathfrak{F}\{(1/N)\sum_{n=-N/2}^{N/2-1}\delta(z-nb)\}$. Given the box fraction f_A and channel fraction f_B, this product is

$$S(q) = \{f_A \operatorname{sinc}(\pi qa) + f_B \operatorname{sinc}(\pi qb)\} \times \left(\frac{1}{N}\right)\sum_{-N/2}^{N/2-1}\exp(\mathrm{i}n2\pi qb)$$

$$\approx \{f_A \operatorname{sinc}(\pi qa) + f_B \operatorname{sinc}(\pi qb)\}\exp(\mathrm{i}\pi qb)\frac{\sin(\pi qNb)}{N\sin(\pi qb)}. \quad (7.19)$$

Consequently E_∞ becomes

$$E_\infty(q) = |S_0(q)|^2\left\{\frac{\sin(\pi qNb)}{N\sin(\pi qb)}\right\}^2 \quad (7.20)$$

where $S_0(q)$ is the unit cell 'form factor', $\mathfrak{F}\{\rho_0(z)\}$. For the interconnected-box model $|S_0(q)|^2$ is given by

$$|S_0(q)|^2 = \{f_A \operatorname{sinc}(\pi qa) + f_B \operatorname{sinc}(\pi qb)\}^2. \quad (7.21)$$

For N large it is clear that as $q \to 0$, $\sin(\pi qNb)/\{N\sin(\pi qb)\}$ reduces to $\operatorname{sinc}(\pi qNb)$. Eqn (7.20) gives a non-zero signal only when qb is in the vicinity of an integer value, n. In practice we shall only be concerned with positive values of q. The local form of $[\sin(\pi qNb)/\{N\sin(\pi qb)\}]^2$ is that of a squared sinc function, equivalent to the behaviour near $n = 0$, but displaced to $q = n/b$. Thus it can be represented by the convolution of a $\operatorname{sinc}^2(\pi qNb)$ with a set of delta functions as

$$E_\infty(q) = |S_0(q)|^2\{\operatorname{sinc}^2(\pi qNb) \otimes \sum_{n=0}^{\infty}\delta(q-n/b)\}. \quad (7.22)$$

This is a very useful expression because is helps us understand the physical origin of the q dependence of the echo attenuation by thinking about the diffraction analogy. The sum of delta functions represents the reciprocal lattice points while $|S_0(q)|^2$ is the form factor for a single unit cell and $\operatorname{sinc}^2(\pi qNb)$ represents the truncation due to a finite lattice size. In order to pursue the optical analogy we will, for the moment, ignore the small fraction of the molecules which reside in the interconnecting channels, assuming, therefore, that $f_A \gg f_B$. Where $|S_0(q)|^2$ arises from the box alone, eqn (7.22) corresponds exactly to the classical N-slit diffraction pattern in optics. $E(q)$ consists of a series of diffraction peaks whose widths are determined by the number of slits which contribute to the interference pattern. The diffraction pattern is, however, modulated by the form factor of the single slit, $|S_0(q)|^2$. This effect is illustrated in Fig. 7.9.

This 'pathological' echo attenuation behaviour apparent in Fig. 7.9 arises because of phase coherences caused by the regular lattice. As $\Delta \rightarrow \infty$ there is a very high probability for a molecule to move by b, $2b$, . . . nb. If such a shift can be made to give no phase change (i.e., $2\pi qb = 2n\pi$) then the associated spin magnetizations will add coherently. The truncation convolution, $\text{sinc}^2(\pi qNb)$, is very important because it determines the initial q response of the signal. Fig. 7.9 shows the progression of structural investigation as q is increased. At low values, the experiment probes the long-range displacements of the molecules. As q is increased to around b^{-1} the data has features characteristic of the lattice spacing. Finally, when q is increased to a^{-1} the details of the boxes are apparent.

Initially one might imagine that N should be very large since our sample, in most instances, will be macroscopic in comparison with the pores sizes we are investigating. However, the distances moved by the molecules over the echo time-scale will usually be very much smaller than the sample dimensions. Δ could not, for example, be longer than T_1, since this is the longest time over which we can measure translational motion using a stimulated

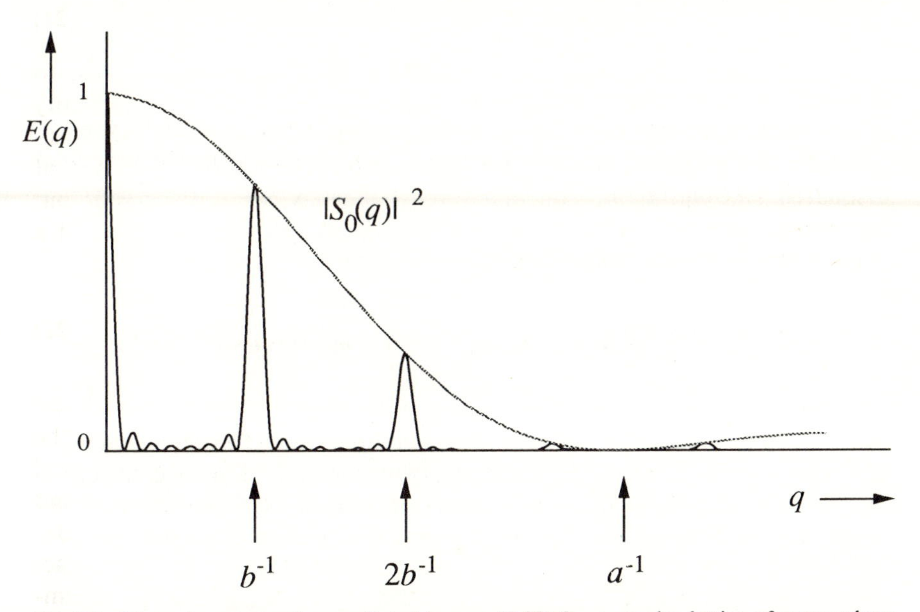

Fig. 7.9 The echo attenuation predicted by eqn (7.22) for a regular lattice of pores where $\rho_0(z)$ is due of a rectangular pore of width, a, approximately 1/4 the pore spacing, b. N is taken as 8 for the purpose of the calculation. The behaviour is identical to that for an N-slit diffraction grating where a is the slit width and b the slit spacing. Note the modulation of the diffraction peaks by the 'form factor', $|S_0(q)|^2$.

echo PGSE sequence. Given a long T_1 of about 2 seconds, and using the self-diffusion coefficient of free water, a maximum distance scale of around 100 μm is suggested. Unlike the sharp truncation which we have just considered, diffusive truncation takes the form of a gently decaying envelope. It is very interesting to account for this effect because it gives us the opportunity of using the initial q response of the data to probe the long-range diffusive motions. It is therefore important to distinguish two limits. In the first, the truncation is caused by the sample geometry, the 'boxes within a box' which we have just dealt with. Nb might then represent the size of sample 'chambers' to which the motion within local pores is confined. The second limiting case is where N is very much larger than the diffusive distance range so that the convolution with $\mathrm{sinc}^2(\pi q N b)$ in eqn (7.22) is replaced by one with the Fourier spectrum of the diffusion envelope. We will deal with this latter case in Section 7.4.3.

7.4.2 *Partially connected structures: the connection matrix*

Fig. 7.10 shows an interconnected series of pores described by the density function, $\rho(\mathbf{r})$. This array can be represented by a superposition of local structures, $\rho_{0i}(\mathbf{r} - \mathbf{r}_{0i})$ as

$$\rho(\mathbf{r}) = \frac{1}{N} \sum_{i=1}^{N} \rho_{0i}(\mathbf{r} - \mathbf{r}_{0i}) \tag{7.23}$$

where we assume that each $\rho_{0i}(\mathbf{r} - \mathbf{r}_{0i})$ is a normalized local density.

Returning to the first principles of eqn (7.14), we may write the long time-scale echo attenuation as

$$\begin{aligned} E_\infty(q) &= \iint \rho(\mathbf{r})\, \rho(\mathbf{r}')\, \exp[\mathrm{i}2\pi \mathbf{q}\cdot(\mathbf{r}' - \mathbf{r})]\, \mathrm{d}\mathbf{r}\, \mathrm{d}\mathbf{r}' \\ &= \frac{1}{N^2} \sum_{i=1}^{N} \sum_{j=1}^{N} \iint \rho_{0i}(\mathbf{r} - \mathbf{r}_{0i})\, \rho_{0i}(\mathbf{r}' - \mathbf{r}_{0j})\, \exp[\mathrm{i}2\pi \mathbf{q}\cdot(\mathbf{r}' - \mathbf{r})]\, \mathrm{d}\mathbf{r}\, \mathrm{d}\mathbf{r}'\,. \end{aligned} \tag{7.24}$$

Where all local structures are mutually interconnected, the sum in eqn (7.24) is just the autocorrelation which we have met previously. Partial connectivity can be represented by using a 'connection matrix', C_{ij}. To see how this works we first express the density in terms of a vector space defined by

$$\rho(\mathbf{r}) = \frac{1}{N} [\rho_{01}(\mathbf{r} - \mathbf{r}_{01}), \ldots, \rho_{0i}(\mathbf{r} - \mathbf{r}_{0i}), \ldots, \rho_{0N}(\mathbf{r} - \mathbf{r}_{0N})]\,. \tag{7.25}$$

Then the conditional probability for molecules originating in compartment i is given by[18]

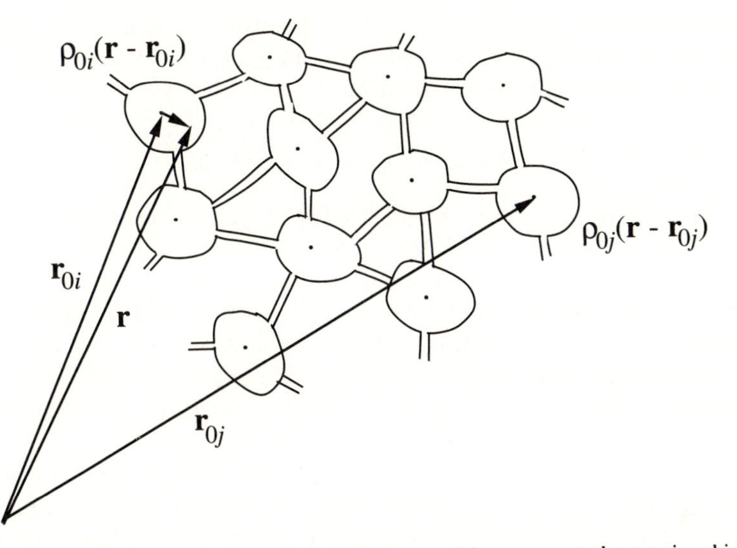

Fig. 7.10 Schematic representation of a general array of interconnected pores in which the local pore density functions are distributed on a lattice of points at locations $\mathbf{r}_{0i}$.

$$P_{si}(\mathbf{r}|\mathbf{r}', \Delta) = \sum_{j=1}^{N} C_{ij}\, \rho_{0j}(\mathbf{r} - \mathbf{r}_{0j}) \tag{7.26}$$

where the matrix element C_{ij} is non-zero if i and j are connected, and zero otherwise. Since P_s must be normalized we shall require that $\Sigma_{j=1}^{N} C_{ij}$ is unity for each i. This definition leads to an echo attenuation,

$$E_\infty(\mathbf{q}) = \frac{1}{N} \iint \rho(\mathbf{r})\, C\, \rho^T(\mathbf{r}') \exp[\mathrm{i}2\pi\mathbf{q}\cdot(\mathbf{r}' - \mathbf{r})]\, \mathrm{d}\mathbf{r}\, \mathrm{d}\mathbf{r}'. \tag{7.27}$$

Eqn (7.27) has a particularly simple interpretation where no interconnections exist. Then C is diagonal, $C_{ii} = N$, and eqn (7.27) yields

$$\begin{aligned} E_\infty(\mathbf{q}) &= \frac{1}{N} \iint \sum_{i=1}^{N} \rho_{0i}(\mathbf{r} - \mathbf{r}_{0i})\, \rho_{0i}(\mathbf{r}' - \mathbf{r}_{0i}) \exp[\mathrm{i}2\pi\mathbf{q}\cdot(\mathbf{r}' - \mathbf{r})]\, \mathrm{d}\mathbf{r}\, \mathrm{d}\mathbf{r}' \\ &= \frac{1}{N} \sum_{i=1}^{N} \iint \rho_{0i}(\mathbf{r} - \mathbf{r}_{0i})\, \rho_{0i}(\mathbf{r}' - \mathbf{r}_{0i}) \exp[\mathrm{i}2\pi\mathbf{q}\cdot(\mathbf{r}' - \mathbf{r}_{0i})] \\ &\quad \exp[-\mathrm{i}2\pi\mathbf{q}\cdot(\mathbf{r} - \mathbf{r}_{0i})]\, \mathrm{d}\mathbf{r}\, \mathrm{d}\mathbf{r}'. \end{aligned} \tag{7.28}$$

Defining $\mathbf{R}_i = \mathbf{r} - \mathbf{r}_{0i}$ and $\mathbf{R}_i' = \mathbf{r}' - \mathbf{r}_{0i}$ we obtain

$$\begin{aligned} E_\infty(\mathbf{q}) &= \frac{1}{N} \sum_{i=1}^{N} \iint \rho_{0i}(\mathbf{R}_i) \exp(-\mathrm{i}2\pi\mathbf{q}\cdot\mathbf{R}_i)\, \rho_{0i}(\mathbf{R}_i') \exp(\mathrm{i}2\pi\mathbf{q}\cdot\mathbf{R}_i')\, \mathrm{d}\mathbf{R}_i\, \mathrm{d}\mathbf{R}_i' \\ &= \frac{1}{N} \sum_{i=1}^{N} |S_{0i}(\mathbf{q})|^2. \end{aligned} \tag{7.29}$$

This is exactly what we would expect for isolated structures. The echo attenuation is that due to N 'single boxes' co-added. Where the boxes are identical the result is simply $|S_0(\mathbf{q})|^2$.

7.4.3 *Finite time-scale: diffusion in porous systems*

The connectivity matrix gives us a useful formalism for dealing with the problem of finite time-scales. When Δ is finite the liquid state molecules within the pores will diffuse a limited distance from their starting positions. To follow our optical analogy we might say that the diffraction analysis changes from that of a grating to that of a multiple but finite number of 'slits'. On a very short time-scale, where the molecules diffuse an r.m.s. distance less than the dimension of a local pore of size a, the echo attenuation is simply that given by the unrestricted diffusion equation, eqn (6.33b). This is a trivial case which will not concern us. The more interesting problem concerns diffusion between several pores. We will assign the 'long-range' diffusion behaviour the coefficient D_p, a parameter which will reflect the pore permeability.

As the time-scale increases, more and more particles, labelled as beginning within a given starting pore, will diffuse to adjacent pores and beyond. We will assume that in the time-scale needed to diffuse to distant pores, the labelled particles will diffuse back and forth several times within those pores which they partially occupy. In effect this means that as time advances, the occupancy of pores distant from the starting pore will increase, but the local density distributions will remain the same as in equilibrium, albeit with a smaller amplitude. In other words, we shall be concerned with time-scales which are longer than a^2/D but finite on the scale b^2/D_p where a is a measure of the pore size and b is the average pore spacing. This is the 'pore equilibration' condition and is equivalent to requiring $Db^2/D_p a^2 \gg 1$.

The connectivity matrix can now be applied to the problem of diffusion between the pores and the calculation of the finite time echo attenuation, $E_\Delta(\mathbf{q})$. To understand this point it is helpful to follow the history of $P_{si}(\mathbf{r}|\mathbf{r}', \Delta)$ as Δ increases. In the time taken to diffuse the distance b to the pores neighbouring the starting pore i, the conditional probability reflecting the behaviour of those particles remaining in pore i has fully equilibrated to $\rho_{0i}(\mathbf{r}')$. As time advances the probability of being in neighbouring pores gradually increases according to a spreading diffusion profile represented by the usual Gaussian distribution. Equilibration occurs successively in all pores, being reached in a local time-scale a^2/D.

Since the applied magnetic field gradient defines a specific direction in space we can, without loss of generality, treat the problem in one dimension. However, we shall see that the behaviour of a system in which the pores are regularly spaced along the gradient axis will be quite different from one in which the pores are randomly oriented. For the moment, however, we focus on the problem of the regular lattice, i and j will then refer to spacings of lattice points along the gradient axis and eqn (7.26) will give a valid description if the C_{ij} are replaced by

$$C_{ij} \approx b(4\pi D_p\Delta)^{-\frac{1}{2}} \exp[-(j-i)^2 b^2/4D_p\Delta]\,. \tag{7.30}$$

Eqn (7.30) will clearly be inaccurate unless the time is sufficient for the molecules to diffuse over many pore spacings. It will break down when $\Delta \lesssim b^2/D_p$. Despite this restriction the 'Gaussian envelope' condition provides useful insight.

From eqn (7.27) we may write

$$\begin{aligned}E_\Delta(q) &= \frac{1}{N}\sum_{i=1}^{N}\sum_{j=1}^{N} C_{ij} \iint \rho_{0i}(z-z_{0j})\,\rho_{0j}(z'-z_{0j}) \\ &\quad\times \exp[\mathrm{i}2\pi q\cdot(z'-z_{0j})]\,\exp[-\mathrm{i}2\pi q\cdot(z-z_{0i})]\,\mathrm{d}z'\,\mathrm{d}z \\ &\quad\times \exp[\mathrm{i}2\pi q\cdot(z_{0j}-z_{0i})] \\ &= \frac{1}{N}\sum_{i=1}^{N}\sum_{j=1}^{N} C_{ij}\exp[\mathrm{i}2\pi q\cdot(z_{0j}-z_{0i})]\,S_{0i}^*(q)\,S_{0j}(q)\,. \end{aligned} \tag{7.31}$$

If the pores are identical then the product $S_{0i}^*(q)S_{0j}(q)$ is $|S_0(q)|^2$. For a structure with variable pore geometry we will define (somewhat roughly) an 'average pore structure factor', $\overline{|S_0(q)|^2}$ given by $\overline{S_{0i}^*(q)S_{0j}(q)}$.

In the regular lattice of pores labelled by indices i and j, $z_{0j}-z_{0i}$ is $(j-i)\,b$. Applying the probability C_{ij} of eqn (7.30) and relabelling $(j-i)$ as n and $(j-i)\,b$ as the dynamic displacement Z, we can replace the sum over j by a sum over n so that eqn (7.31) becomes

$$\begin{aligned}E_\Delta(q) &= \frac{b}{N}\,\overline{|S_0(q)|^2}\int_{-\infty}^{\infty}(4\pi D_p\Delta)^{-\frac{1}{2}}\exp(-Z^2/4D_p\Delta) \\ &\quad\times\sum_i\left\{\sum_n \delta(Z-nb)\right\}\exp(\mathrm{i}2\pi qZ)\,\mathrm{d}Z\,. \end{aligned} \tag{7.32}$$

Eqn (7.32) represents the Fourier transformation of a product of two functions, $(b/N)\,\Sigma_i\{\Sigma_n\delta(Z-nb)\}$ and $(4\pi D_p\Delta)^{-\frac{1}{2}}\exp(-Z^2/4D_p\Delta)$. It can therefore be evaluated using the convolution theorem. The latter function defines the diffusive envelope, $d(Z, \Delta)$. It is important to recognize that $(b/N)\,\Sigma_i\{\Sigma_n\delta(Z-nb)\}$ in eqn (7.32) is not the lattice itself but a sort of 'lattice correlation function', $L(Z)$, which describes the relative positions of lattice points measured from some position i, and then averaged over all i. In the case where the lattice is perfectly regular this reduces to $b\Sigma_n\delta(Z-nb)$. Strictly speaking, the limits of n will depend on i. In fact, in evaluating eqn (7.32), these limits are unimportant since the diffusive envelope causes a more rapid decay than truncation due to the finite lattice (i.e., $N^2b^2 \gg 4D_p\Delta$). We can arbitrarily set the limits to some large range. Thus

$$E_\Delta(q) \approx \overline{|S_0(q)|^2}\,[\mathfrak{F}\{d(Z,\Delta)\} \otimes \mathfrak{F}\{L(Z)\}]\,. \tag{7.33}$$

Pursuing the optical analogy it is clear that $\overline{|S_0(q)|^2}$ plays the role of the 'form factor' while $[\mathfrak{F}\{d(Z,\Delta)\} \otimes \mathfrak{F}\{L(Z)\}]$ is the 'reciprocal lattice', broadened by $\mathfrak{F}\{d(Z,\Delta)\}$ because the lattice range is limited by the diffusion envelope. It is important, however, to emphasize again that the lattice function, $L(Z)$, to which the PGSE experiment is sensitive is not the array of pore sites but the array of relative pore displacements.

The **q**-space Fourier transform of the diffusive envelope, $d(Z,\Delta)$, is the familiar PGSE expression

$$\begin{aligned}\mathfrak{F}\{d(Z,\Delta)\} &= \exp(-4\pi^2q^2D_p\Delta)\\ &= \exp(-\gamma^2g^2\delta^2D_p\Delta)\,.\end{aligned} \tag{7.34}$$

For the ideal lattice the Fourier transform of $L(Z)$ is

$$\begin{aligned}\mathfrak{F}\{L(Z)\} &= \lim_{N\to\infty} b\sum_{n=-N/2}^{N/2-1}\exp(\mathrm{i}2\pi qnb)\\ &= \lim_{N\to\infty} b\exp(\mathrm{i}\pi qb)\frac{\sin(\pi qbN)}{\sin(\pi qb)}\\ &= \sum_m \delta\left(q-\frac{m}{b}\right).\end{aligned} \tag{7.35}$$

For the regular lattice, $E_\Delta(q)$ has the same form as $E_\infty(q)$ in eqn (7.22) except that the convolution of the reciprocal lattice points is with the Fourier transform of the diffusion profile rather than with the lattice truncation transform. The shape of the echo attenuation function, $E_\Delta(q)$, is shown in Fig. 7.11.

At low values of q, the PGSE experiment probes the long-range motions of the molecules. $E_\Delta(q)$ has the Gaussian profile given in eqn (7.34) and yields the 'pore permeability' self-diffusion coefficient, D_p. At larger values of q, in this regular lattice, the pore spacings are revealed in the coherence peaks apparent at $q = n/b$. As q is increased further the average pore structure factor, $\overline{|S_0(q)|^2}$, modulates the echo attenuation so that the pore shape is revealed.

According to the analysis so far, observation of this high q modulation is dependent on the existence of coherences at $q = n/b$, and hence the existence of a regular lattice. In fact, it is possible in principle to probe the

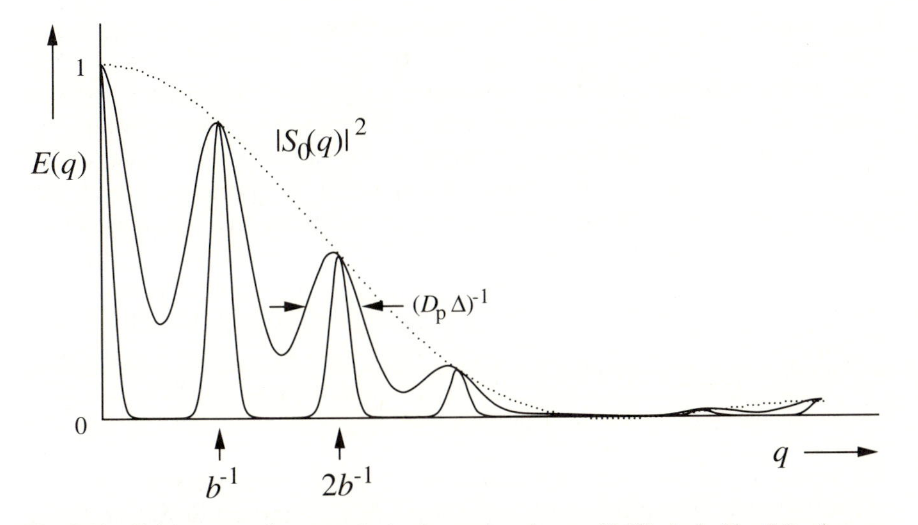

Fig. 7.11 $E_\Delta(q)$ versus q for a regular lattice as given by eqn (7.33). As in Fig. 7.9 coherence peaks modulated by the pore form factor arise at the reciprocal lattice positions. The form factor here is that due to rectangular pores of width a approximately $b/4$. However, unlike Fig. 7.9, the peak widths are determined by the diffusive envelope and have the Gaussian form associated with the echo attenuation due to the self-diffusion coefficient, D_p. $E_\Delta(q)$ is shown for Δ values of $0.2b^2/D_p$ and $2b^2/D_p$ the coherence peaks becoming narrower as Δ increases.

pore structure, $S_0(q)$, provided the time-scale Δ is chosen short enough that little diffusion to adjacent pores occurs.

7.4.4 *The irregular lattice and the pore glass*

Lattice regularity is highly unlikely in a real porous system and in most natural materials we cannot expect to have a lattice structure with long-range orientational regularity. It is useful to separate these two aspects and we first treat the case of a regularly oriented lattice in which we are concerned only with displacements along a single direction. Samples with both pore spacing variation and orientational randomness may be termed 'pore glasses' and require a separate treatment.

Irregular spacings in one dimension

In considering structure associated with the displacements of probe molecules, we must take into account 'correlation regularity'. Suppose that there is an average pore spacing but with some randomness imposed. As a molecule successively moves to the neighbouring, then next-neighbouring pores, these random fluctuations in spacings add successively in the total displacement of the molecule. This suggests that our best chance of observing effects due to the pore separation b will occur if the time Δ is kept sufficiently short that the molecules diffuse just one or two lattice spacings.

One of the advantages of the diffraction analogy is that it gives us general insight. For an irregular lattice we can follow the same procedure used in moving from eqn (7.32) to eqn (7.34) except that now the index n will not be a regular integer as in the regular lattice, but a characteristic set of real numbers, n_i, which will depend on the starting position i. This means that the lattice correlation function, $(b/N)\,\Sigma_i \Sigma_n \delta(Z - nb)$, will be replaced by $(b/N)\,\Sigma_i \Sigma_{n_i} \delta(Z - n_i b)$, a series which will describe the distribution of lattice site displacements from the set of possible starting starting positions. We shall call this general lattice correlation function $\overline{L(Z)}$ and its Fourier transform $\mathfrak{F}\{\overline{L(Z)}\}$ and so write,

$$E_\Delta(q) = |S_0(q)|^2 [\mathfrak{F}\{d(Z,\Delta)\} \otimes \mathfrak{F}\{\overline{L(Z)}\}]\,. \qquad (7.36)$$

A practical description of the irregular lattice correlation function, $\overline{L(Z)}$, is given by presuming a mean pore spacing b with standard deviation (on a Gaussian distribution) ζ. A molecule starting in one pore and moving to the next will find its position at a distance b with standard deviation ζ but the next-nearest neighbour at $2b$ will have standard deviation 2ζ and so on, the lattice correlation being gradually lost as shown in Fig. 7.12(b). In this sense ζ is a correlation length which defines the distance over which regularity in the lattice displacements decays. One interesting feature of such a lattice model is that $\overline{L(Z)}$ and its reciprocal lattice have identical form, $\mathfrak{F}\{\overline{L(Z)}\}$ being a superposition of Gaussians of increasing width as q increases. Two limits in $\mathfrak{F}\{\overline{L(Z)}\}$ are noteworthy. First, the delta function at the origin of **q**-space results from the uncorrelated long-range structure in the lattice. In the PGSE experiment this feature gives us the regular diffusive decay, $\exp(-4\pi^2 q^2 D_p \Delta)$, associated with the long-range molecular motions. Second, the constant value of unity at high q results from those particles which remain in the starting pore. In the PGSE experiment this feature reveals the pore structure factor $|S_0(q)|^2$ but is damped by the convolution with the diffusive term, decaying as more particles 'leak out' from their starting pores with increasing time.

The behaviour of $E_\Delta(q)$ for the irregularly spaced lattice as predicted by eqn (7.36), is illustrated in Fig. 7.13(b) for two time-scales, $\Delta = 0.2b^2/D_p$ and $2b^2/D_p$. Spherical pores have been assumed with radius $a = b/3$. This result may be compared with the equivalent regular lattice echo attenuation shown in Fig. 7.13(a).

Although our model is applicable only for $\Delta \gg a^2/D$ the qualitative behaviour for shorter times can be guessed. For $\Delta \ll a^2/D$, $E_\Delta(q)$ is characteristic of free diffusion,

$$E_\Delta(q) \approx \exp(-4\pi^2 q^2 D\Delta)\,. \qquad (7.37)$$

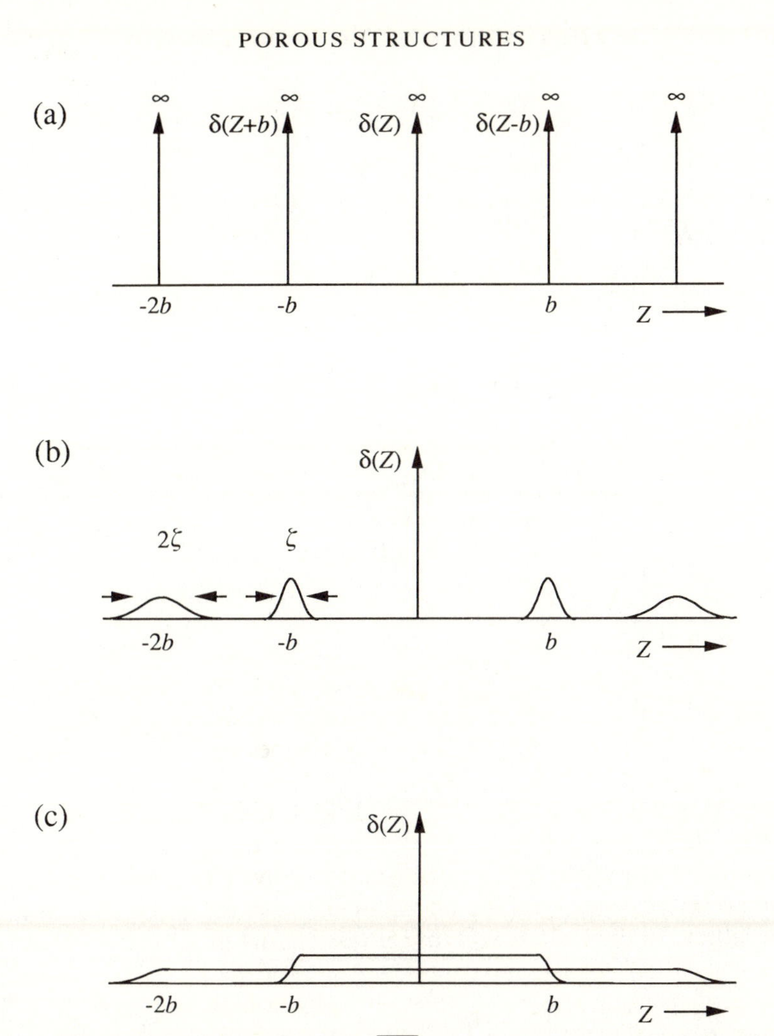

Fig. 7.12 The lattice correlation functions, $\overline{L(Z)}$, for (a) a regular one-dimensional lattice; (b) an irregularly spaced one-dimensional lattice; and (c) a pore glass. In (b) the Gaussians at each lattice site are successively broader for each pore successively displaced from the starting pore with $\overline{L(Z)}$ becoming constant for $Z \gg b$. In (c) the pore glass lattice correlation function is a sum of normalized rectangular densities at each successive pore shell, convoluted with successively broader Gaussian pore separation distributions.

The behaviour at $\Delta \sim b^2/D_\mathrm{p}$ is difficult to estimate but when $a^2/D \lesssim \Delta \lesssim b^2/D_\mathrm{p}$ the echo attenuation should be similar to that for restricted diffusion in isolated pores

$$E_\Delta(q) \approx |S_0(q)|^2 . \tag{7.38}$$

In the range $b^2/D_\mathrm{p} \lesssim \Delta$ the short-range lattice structure appears and

$$E_\Delta(q) \approx \mathfrak{F}\{d(Z,\Delta)\} \otimes \mathfrak{F}\{\overline{L(Z)}\} . \tag{7.39}$$

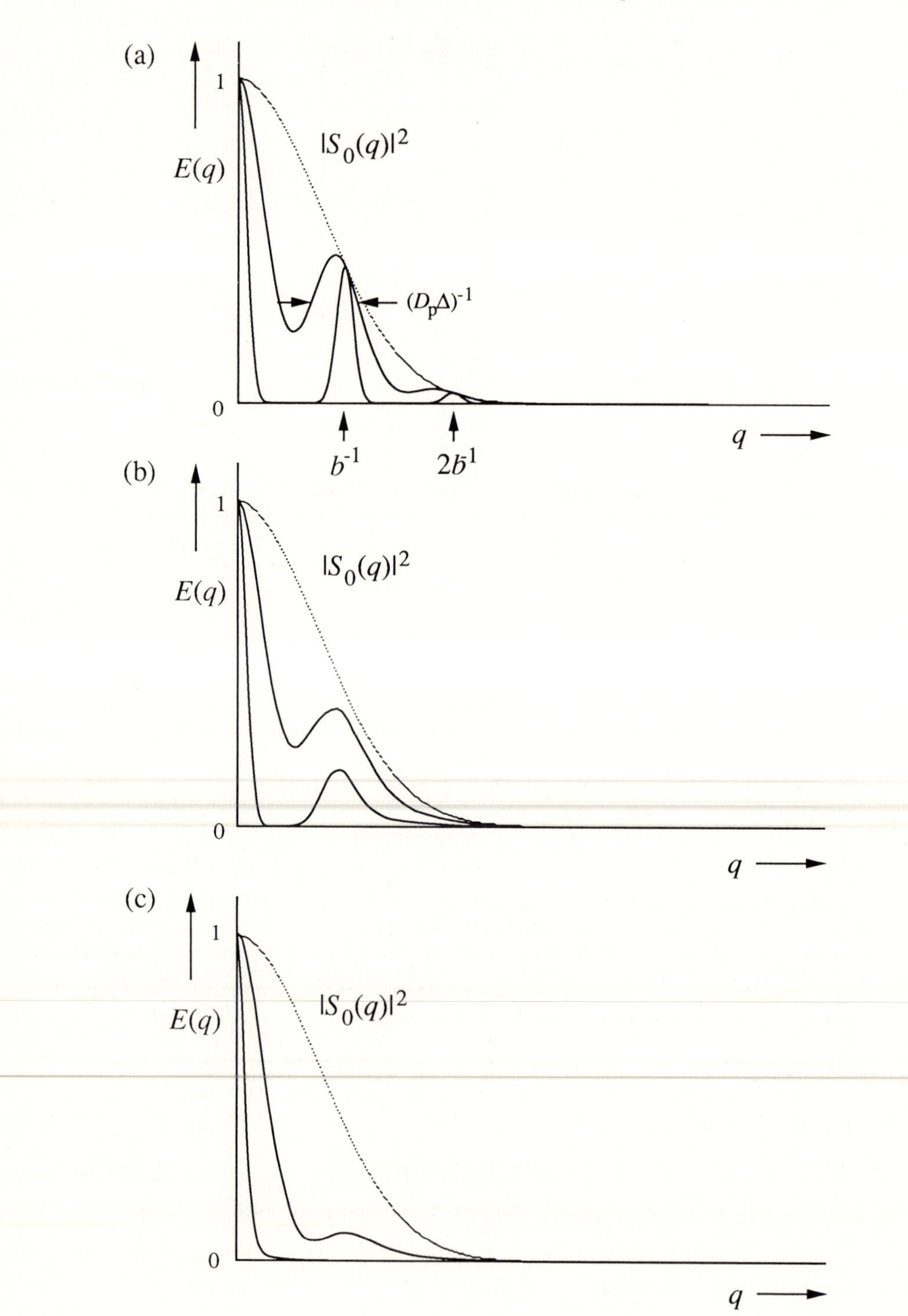
(a)
1
$E(q)$
$|S_0(q)|^2$
$(D_p\Delta)^{-1}$
0
b^{-1}
$2b^{-1}$
q
(b)
1
$E(q)$
$|S_0(q)|^2$
0
q
(c)
1
$E(q)$
$|S_0(q)|^2$
0
q

The lattice structure peak at $q \approx b^{-1}$ is due to displacements from the originating pore to the first-neighbour pore. Over longer time-scales this peak will disappear as a greater fraction of 'pore molecules' move to more distant pores for which the spatial correlations are less coherent. This effect is illustrated in Fig. 7.13(b). The rate at which the lattice structure peak in E_Δ reduced as Δ increases can be estimated from the nature of the convolution integral. For $\zeta^2/D_p \ll \Delta$, $\mathfrak{F}\{d(Z,\Delta)\}$ is much narrower than $\mathfrak{F}\{\overline{L(Z)}\}$ which has a width of order ζ^{-1}. Since the area under $\mathfrak{F}\{d(Z,\Delta)\}$ in **q**-space is of order $(4\pi D_p\Delta)^{-\frac{1}{2}}$ the magnitude of the lattice structure peak resulting from the convolution $\mathfrak{F}\{d(Z,\Delta)\} \otimes \mathfrak{F}\{\overline{L(Z)}\}$ varies as $\zeta(4\pi D_p\Delta)^{-\frac{1}{2}}$.

Pore glass

Consider the problem of molecules migrating from a starting pore to neighbouring pores in a pore glass. The vectors describing the set of possible displacements will have some mean length b but no orientational regularity. Averaged over the entire sample we could consider the first-neighbour set to be uniformly distributed on a surrounding spherical shell of radius b. To predict the outcome of the PGSE experiment we require the distribution of projected distances, Z, resulting from diffusion to the nearest neighbour. A simple analysis shows that this probability density along Z is uniform for $-b < Z < b$ with magnitude $(2b)^{-1}$ and zero outside. For migration to the nth pore shell, the density is $(2nb)^{-1}$ between $-nb$ and nb and the probability, C_n, of such migration is given by the permeability diffusion profile. For large n, the probability to be in the nth neighbour spherical shell of radius nb and thickness b is

$$C_n = 4\pi(nb)^2 b(4\pi D_p\Delta)^{-\frac{3}{2}} \exp(-n^2b^2/4D_p\Delta)\,. \tag{7.40}$$

Eqn (7.27) can be used to predict the required conditional probability, P_{si},by allowing that the components of $\mathbf{r}_{0j}$ along the PGSE gredient are continuously distributed within the range $\pm(j-i)b$. Thus

$$E_\Delta(q) = \overline{|S_0(q)|^2} \sum_n C_n \frac{\sin(2\pi qnb)}{2\pi qnb}\,. \tag{7.41}$$

For irregular pore spacing, P_{si} will be convoluted with the $(j-i)$th nearest-neighbour pore spacing distribution function, for example a Gaussian of standard deviation $(j-i)\zeta$, so that

$$E_\Delta(q) = \overline{|S_0(q)|^2} \sum_n C_n \frac{\sin(2\pi qnb)}{2\pi qnb} \exp\{-2\pi^2q^2(n\zeta)^2\}\,. \tag{7.42}$$

Fig. 7.13 $E_\Delta(q)$ versus q calculated using eqn (7.36): (a) regular lattice; (b) irregular one-dimensional lattice where $\zeta = 0.2b$; and (c) irregular pore glass where $\zeta = 0.2b$. In each case the pore is treated as a sphere with radius a approximately $b/3$ and $E_\Delta(q)$ is shown for Δ values of $0.2b^2/D_p$ and $2b^2/D_p$.

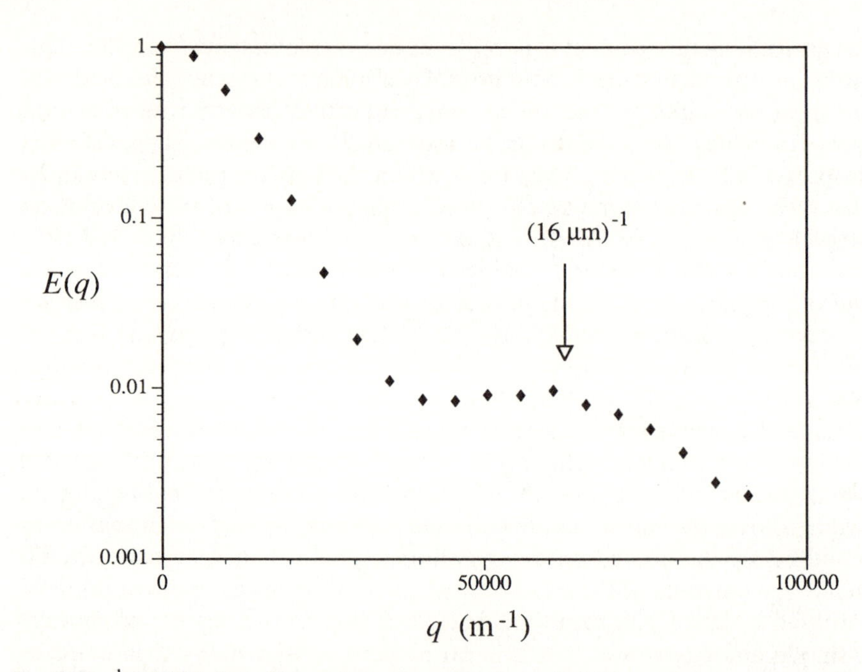

Fig. 7.14 ^{1}H NMR Spin Echo attenuation plot for a PGSE experiment on water in a centrifuged assembly of polystyrene spheres, carried out with an observational timescale, Δ, of 70 ms. The sphere diameter is 15.8 μm leading to a pore spacing, b, of similar dimension. A coherence peak is apparent in the plot of log (E(q)) vs q at $q \approx (16\ \mu m)^{-1}$. (From A. Coy, personal communication).

Fig. 7.13(c) shows the q and Δ dependence of the echo attenuation for an irregularly spaced pore glass in which spherical pores of radius $a = b/3$ are assumed. It is clear that, in principle, structural features are apparent even when there is no orientational regularity, although this requires that the diffusion time be kept sufficently short that only the nearest-neighbour shell is reached.

A nice demonstration of the 'diffraction effect' in a pore glass has been provided by A.Coy who carried out a PGSE experiment on water diffusing in the void spaces of a close packed assembly of polymer spheres. The resulting echo attenuation, shown in Fig. 7.14, exhibits a noticeable coherence peak at $q = (16\ \mu m)^{-1}$, in accordance with the known sphere diameter (and hence pore spacing) of 15.8 μm.

7.4.5 *Structure determination*

A useful test of the porous-material models is provided by taking the limit as $q \to 0$. Eqn (7.36) and (7.42) both predict Stejskal–Tanner-like behaviour for $E_\Delta(q)$, namely $\exp(-4\pi^2 q^2 D_p \Delta)$. This confirms that measurement with small q yields the long-range diffusive behaviour and is devoid of structural features due to pore morphology. Structural features associated with the pore size a and the pore spacing b can, in principle, be observed when $q \sim a^{-1}$and $q \sim b^{-1}$, respectively, and while random variations in inter-pore spacing inevitably damp this structure, such damping is minimized if a time-scale Δ is used such that molecules diffuse only one or two pore spacings.

At high q the PGSE experiment is sensitive to very short-range displacements and the accuracy of the predictions will depend on the degree to which the discretely stepped Gaussian profile adequately represents the pore occupancies for diffusing molecules. Clearly this assumption will be least accurate for the starting pore.[18] Calculations by Tanner[19] of diffusive profiles in a one-dimensional system partitioned by permeable barriers show that the distribution $P_s(Z, \Delta)$ is close to Gaussian for long times but that the starting pore occupancy, as expected, deviates quite strongly for $\Delta < a^2/D$. However, the PGSE experiment will be quite sensitive to very small deviations since the small, 'excess' in the starting pore will appear as a restricted component giving a 'baseline' in the echo attenuation plot. Although this baseline may be small it will have the effect of considerably reducing the apparent diffusion coefficient obtained by a direct application of the Stejskal–Tanner formula. Such reductions are apparent in the calculations by Tanner. The utility of the Gaussian-envelope assumption is that it gives us a nice physical picture of what the broadening in the reciprocal lattice peaks means. While this assumption is simplistic, it is a helpful way to visualize what the structure in $E_\Delta(q)$ represents.

A summary of the structural properties of porous systems which can be measured using **q**-space imaging under the pore equilibration condition, $Db^2 \gg D_p a^2$, is as follows:

1. $q \ll b^{-1}$
   Long distance scales are probed and, for all possible structures, $E_\Delta(q)$ is Gaussian, yielding D_p on analysis via the Stejskal–Tanner relation.

2. $q \sim b^{-1}$
   A peak in $E_\Delta(q)$ will arise due to the nearest-neighbour lattice correlation. The magnitude of this peak will depend on the pore spacing regularity but will also decay as Δ increases, due to the diffusive convolution.

3. $q \gtrsim a^{-1}$
   The modulation due to the pore structure factor causes signal attenuation. Note that this modulation can also cause a shift in the peak at $q = b^{-1}$ if $a \sim b$. While this form factor attenuation complicates the interpretation of $E_\Delta(q)$, it should be noted that effects due to the pore lattice and the pore structure factor, $|S_0(q)|^2$, can, in principle, be separated in an experiment in which q and Δ are varied independently.

7.5 'Soft-bounded' systems

The restricted diffusion of molecules inside a box is an example where the molecules encounter hard boundaries. In some materials molecules exhibit 'softer' restrictions to their motion. For example, a diffusing polymer segment in a cross-linked gel will experience a restoring force due to gel elasticity. This will have the effect of limiting the distribution of translational displacements as Δ becomes large. Another rather unusual form of displacement limitation arises when molecules are confined to diffuse in a curvlinear coordinate system, examples being the motion of solvent molecules along the lamellae in multidomain lyotropic liquid crystal or the motion of a random coil polymer molecule in the 'tube' of topological constraints due to neighbouring polymers. In such systems the molecular motion in the laboratory frame exhibits a characteristic dependence on timescale, attenuating as Δ increases. This type of soft bounding might be termed 'dimensionally restricted' diffusion.

7.5.1 *Diffusion near an attractive centre*

The problem of diffusion in the presence of an harmonic potential has been considered by Stejskal.[20] In this treatment the restoring force is balanced by frictional damping due to the viscosity of the surrounding particles. If the ratio of the elastic force constant and the friction coefficient, f, is termed β then the elastic potential is written $V = \frac{1}{2}\beta f r^2$ with associated force $-\beta f\mathbf{r}$. Equating this with the friction force, $-f\mathbf{v}$, gives $\mathbf{v} = \beta\mathbf{r}$. From eqn (6.9),

$$\partial P_s/\partial t = \beta\nabla'\cdot\mathbf{r}'P_s + D\nabla'^2 P_s \tag{7.43}$$

which has the solution

$$P_s(\mathbf{r}|\mathbf{r}',\Delta) = [2\pi D\{1-\exp(-2\beta t)\}/\beta]^{-\frac{3}{2}} \times \exp[-\beta\{\mathbf{r}' - \mathbf{r}\exp(-\beta t)\}^2/2D\{1-\exp(-2\beta t)\}]\,. \tag{7.44}$$

The equilibrium density can be obtained by using the long time limit identity of P_s and $\rho(\mathbf{r}')$, thus

$$\rho(\mathbf{r}) = (2\pi D/\beta)^{-\frac{3}{2}}\exp(-\beta r^2/2D)\,. \tag{7.45}$$

The exponential decay of $\rho(\mathbf{r})$ as the distance from the origin increases is an example of 'soft bounding'. Substitution of P_s and $\rho(\mathbf{r})$ into the narrow-pulse PGSE expression, eqn (6.26), gives

$$\begin{aligned} E(q) &= \exp[-4\pi^2 q^2 D\{1-\exp(-\beta\Delta)\}/\beta] \\ &= \exp[-\gamma^2\delta^2 g^2 D\{1-\exp(-\beta\Delta)\}/\beta]\,. \end{aligned} \tag{7.46}$$

The dependence $E(q)$ on q is Gaussian at all time-scales. For $\Delta \ll \beta^{-1}$, the behaviour is the same as for unrestricted diffusion. For $\Delta \gg \beta^{-1}$, $E(q)$ is independent of Δ and reduces to $|S(q)|^2$, where $S(q)$ is the one-dimensional Fourier transform of the equilibrium density $\rho(\mathbf{r})$ given in eqn (7.44).

$$E_\infty = \exp[-\gamma^2\delta^2 g^2 D/\beta]\,. \tag{7.47}$$

De Gennes has discussed the problem of diffusion in polymer gels.[21] Given a friction coefficient ϕ per unit volume and bulk modulus, E, it can be shown that the elastic constant for longitudinal gel fluctuations of wavevector κ is $E\kappa^2$, while the cooperative diffusion coefficient, D_c, is $E\phi$. For such a system β is given by[22] $E\kappa^2/\phi$.

7.5.2 *Curvilinear diffusion*

Motion without branches

Fig. 7.15 shows an echo attenuation plot[23] for water protons in a sample of aerosol OT/water, a lamellar phase lyotropic liquid crystal system with random domain orientation. The PGSE experiment reveals a curious behaviour as Δ is increased. At short times the data exhibit a dependence on g and Δ as given by the two-dimensional poly-domain expression, of eqn (6.40), E_{2D}. At longer times the diffusion rate is apparently slower and the curvature of $E(q)$ versus q^2 reduces, indicating that the diffusion behaviour is tending towards three-dimensional. For Δ sufficiently short that the water molecules reside in a single local domain with characteristic director orientation, the behaviour E_{2D} is expected. As Δ becomes sufficiently large for the molecules to move to new domains of differing orientation, their motion becomes more three-dimensional in character. The curious feature of this experiment is that the apparent diffusion coefficient reduces with time.

In the experiment described we measure molecular displacements in the laboratory frame but the diffusion, at a microscopic level, is occurring in a curvilinear coordinate system. A similar problem exists in considering the reptational diffusion of a polymer molecule in the curvilinear tube formed by impeding neighbours. It was shown by de Gennes[24] and Edwards[25] that such a motion leads to a dependence of laboratory frame mean square displacement on $t^{\frac{1}{2}}$ instead of the usual t dependence of unrestricted diffusion. A detailed discussion of this problem is found elsewhere[26] but a simple picture can be gained as follows.

Consider molecular motion confined to one-or two-dimensional local elements. These elements have r.m.s. length λ and are interconnected such that the diffusing particle migrates from one randomly oriented element to another, tracing out a random walk in the laboratory frame as illustrated in

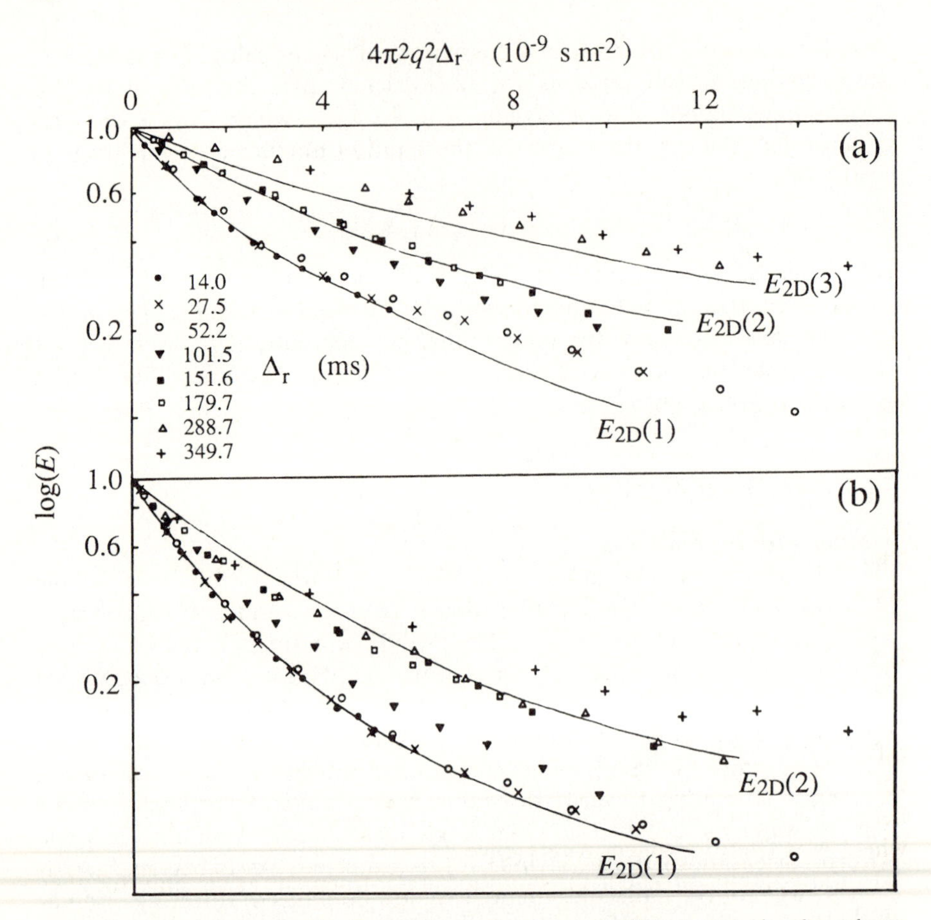

Fig. 7.15 Echo attenuation plot for a 25% sample of aerosol OT/water showing dependence of diffusive behaviour as Δ_r is varied over a wide range. (a) shows the results for an equilibrated sample and (b) the results for a sample three days after preparation. For spins residing in one domain only the date should be coincident. The theoretical curves, $E_{2D}(N)$ correspond to molecules diffusing successively in N domains. (From Callaghan and Soderman.[23])

Fig. 7.16. A key factor in this motion is the lack of branch points and the consequent confinement of the particles to specific, albeit tortuous, paths. Suppose we define a total r.m.s. curvilinear path length Λ comprising N elements with diffusion time t and local self-diffusion coefficient, D. For one-dimensional motion we have

$$\Lambda^2 = 2Dt = N^2\lambda^2 . \tag{7.48}$$

In the laboratory frame

$$\overline{R^2} = N\lambda^2 = 2DN^{-1}t = \lambda(2Dt)^{\frac{1}{2}} . \tag{7.49}$$

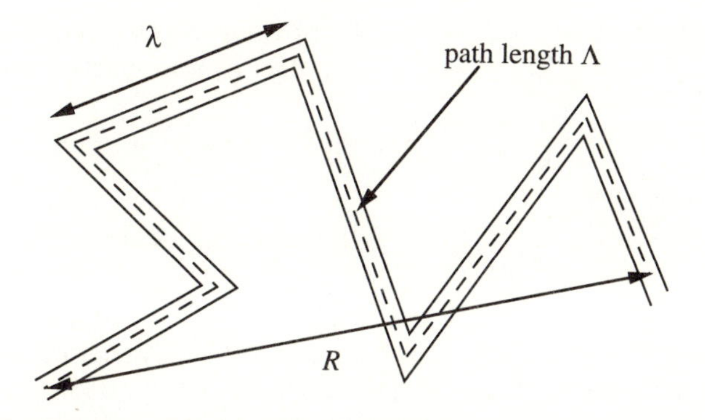

Fig. 7.16 One-dimensional curvilinear diffusion path without branching points, made up of local elements of r.m.s. length λ. The path length along the curvilinear path is Λ while the direct end-to-end distance is R.

The case of two-dimensional local diffusion is treated identically but with a coefficient 4 rather than 2 appearing in eqn (7.48).

The distribution of laboratory displacements in such curvilinear diffusion is Gaussian provided N is large, but the dependence of $\overline{R^2}$ on $t^{\frac{1}{2}}$ rather than t is distinctly non-Brownian. In the PGSE experiment the echo signal $E(q)$ will exhibit the usual Gaussian dependence on q but with an effective diffusion coefficient which decreases as $\Delta^{-\frac{1}{2}}$. The data shown in Fig. 7.15 correspond to an experiment in which N is small so that the Gaussian behaviour is not reached as N increases. An expression for $E(q)$, appropriate where N is finite, is given in reference 23.

In the polymer problem the curvilinear confinement exists only for the time, τ_d, taken for polymer to disengage from the tube.[26-29] This means that the $\overline{R^2} \sim \Delta^{\frac{1}{2}}$ behaviour can be observed only if Δ is less than τ_d, and where q is sufficiently large for the PGSE experiment to detect molecular displacements smaller than the r.m.s. dimensions of the polymer, the so-called semi-local motion. In principle, this is just a matter of making g or δ large enough. Because of T_2 relaxation the magnitude of δ is constrained. Given the largest available polymers, the measurement of semi-local motion requires gradient strengths in excess $10\,\mathrm{T\,m^{-1}}$. While these magnitudes can be achieved it is difficult to do so with the necessary PGSE gradient pulse matching.

Motion with branches

Tube disengagement is an example of branching in the paths available for the molecule. Suppose that, instead of confinement to a fixed curvilinear path, the diffusing molecule has a choice of paths after moving N elements comprising mean square distance R_0^2 in the laboratory frame over a time τ.

This gives a time-scale independent, laboratory frame self-diffusion coefficient, D_{eff}, of $R_0^2/6\tau$. Eqn (7.49) then yields a particularly simple result,

$$D_{\text{eff}} = \begin{cases} \dfrac{D}{3N} & \text{for one-dimensional local motion} \\ \dfrac{2D}{3N} & \text{for two-dimensional local motion}\,. \end{cases} \tag{7.50}$$

The one-dimensional expression leads to the well-known $D_{\text{eff}} \sim N^{-2}$ 'reptation law' for polymers. However eqn (7.50) is generally applicable to a wide class of problems, including the diffusion of small molecules in liquid crystals or emulsions. One specific limit of interest is the case $N = 1$ which corresponds to unidirectional motion between branch points. This is equivalent to the curvilinear diffusion problem but where branching options exist on entering the next domain. Where the motion is locally one-dimensional or two-dimensional, D_{eff} is, respectively, $\frac{1}{3}D$ or $\frac{2}{3}D$. Examples of such behaviour have been found for diffusion in microemulsions.[30]

7.5.3 *Diffusion in fractal geometries*

Natural structures often exhibit fractal character,[31-35] in which self-similarity and non-integer dimensionality play a role. A nice example of fractal behaviour is found in the geography of coastlines.[32] The appearance of a coastline is similarly indented and tortuous whether viewed from an orbiting satellite or from a distance of a few metres. This self-similarity (sometimes called dilation symmetry) is consistent with a shape which is scale independent. One of the consequences of this is that the total coastline length depends on the smallest step, ϵ, used in its measurement. If the total number of ϵ-intervals in the measurement is $N(\epsilon)$ and the total length is $L(\epsilon)$ then

$$N(\epsilon) \sim F\epsilon^{-d} \qquad \text{and} \qquad L(\epsilon) \sim F\epsilon^{1-d} \tag{7.51}$$

where d is called the similarity dimension. A non-fractal coastline would have $d = 1$ whereas fractal behaviour is exhibited when $d > 1$.

Porous materials exhibit analogous behaviour in the volume and surface organization. Here d tells us the dependence of the 'measure' on the length dimension. In Euclidean behaviour we expect that the total pore volume within a radius R would vary as R^3. 'Volume fractal' behaviour is typified by $V \sim R^d$ where $1 \leq d \leq 3$. 'Surface fractal' behaviour is typified by $S \sim R^d$ where $1 \leq d \leq 2$.

The Brownian random walk provides a very good example of fractal behaviour, a wandering plane-filling 'coastline' in which $d = 2$. Consider a particular walk of end-to-end r.m.s. distance r. As ϵ decreases $N \sim F\epsilon^{-d}$. F can be determined as r^d from the result $N = 1$ when $\epsilon = r$. Thus $r \sim N^{1/d}$

and so yields the familiar random walk result, $r \sim N^{\frac{1}{2}}$, when $d = 2$. In fact this same dimensionality applies when the random walk is performed in three-dimensional space. Note that the fractal dimension associated with the random walk is different from that describing, for example, volume fractal behaviour, and is assigned the symbol d_w. An interesting example is the self-avoiding random 'walk' which defines the random coil polymer conformation in the solution phase[29] for which the chain end-to-end distance scales as $r \sim N^{\frac{3}{5}}$, giving $d = 5/3$.

Self-diffusion is very sensitive to the dimensional character of the path taken via the dependence of r.m.s. distance on time, this parameter being directly proportional to N. The fractal character may be defined by the relation[36-38]

$$\overline{r^2} = 6Dt^{2/(2+\theta)}. \tag{7.52}$$

θ determines the effective dimensionality of the random walk, reducing to zero in the case of Brownian motion. A time-scale dependent self-diffusion experiment can be used to measure θ, and this quantity may be related to the volume fractal dimension,[36-38] an effect which can be exploited in the pulsed gradient spin echo experiment. From eqn (6.33b) it is apparent that, because the exponent is $\frac{1}{2}\gamma^2\delta^2g^2\overline{Z^2}$, the fractal dimension enters[39-42] through the term $\overline{Z^2}$ as $\Delta^{2/(2+\theta)}$ giving

$$E_\Delta(q) = \exp(-\gamma^2\delta^2g^2D\Delta^{2/(2+\theta)}). \tag{7.53}$$

Note that the fractal dimension relates to the migration space of the spins. Where the material exhibits fractal surfaces but the molecules diffuse in a non-fractal (Euclidean) volume, the PGSE experiment exhibits the usual Brownian behaviour. In such a system fractal behaviour is apparent only when the experimental parameter is surface dependent, an example being the relaxation of spins due to the existence of surface magnetization 'sinks'.[43-45]

7.5.4 *Systems with multiple regions*

A commonly encountered morphology is one where the diffusional behaviour may be divided into subregions with molecules confined to an inhomogeneous distribution of local diffusion rates on a short time-scale but experiencing all regions and so exhibiting a common, averaged diffusion coefficient, over a sufficiently long time-scale. Hence, in the short time limit (slow exchange) the echo attenuation consists of a linear superposition from the subregions of weighting p_i and local diffusion coefficient D_i,

$$E_\Delta(g) = \sum_i p_i \exp(-\gamma^2\delta^2g^2D_i\Delta) \tag{7.54}$$

while over long times a statistically averaged diffusion behaviour is found so that the echo attenuation is a single exponential decay

$$E_{\Delta}(g) = \exp(-\gamma^2\delta^2 g^2 \overline{D}\Delta)\,. \tag{7.55}$$

Describing the intermediate time behaviour is a complex problem which has been treated both analytical[46,47] and numerically.[48] In the description of Kärger *et al.*[49,50] the Chapman–Kolmogoroff equations[51] are adapted, in a manner analogous to the treatment of relaxation in multiple phase systems by Zimmerman and Brittin,[52] to give a set of coupled differential equations,

$$\begin{aligned}
\dot{E}_1 &= -\gamma^2\delta^2 g^2 D_1 E_1 - \frac{1}{\tau_1}E_1 + \frac{p_{21}}{\tau_2}E_2 + \ldots + \frac{p_{n1}}{\tau_n}E_n \\
\dot{E}_2 &= -\gamma^2\delta^2 g^2 D_2 E_2 + \frac{p_{12}}{\tau_1}E_1 - \frac{1}{\tau_2}E_2 + \ldots + \frac{p_{n2}}{\tau_n}E_n \\
&\quad\vdots \qquad\qquad\qquad\qquad \vdots \\
\dot{E}_n &= -\gamma^2\delta^2 g^2 D_1 E_1 + \frac{p_{1n}}{\tau_1}E_2 + \frac{p_{2n}}{\tau_2}E_n \ldots - \frac{1}{\tau_n}E_n
\end{aligned} \tag{7.56}$$

where the τ_i denote the mean lifetimes of molecules in the ith region and the p_{ij} are the conditional probabilities that if a molecule leaves region i it will enter region k. The net echo attenuation is then a linear superposition of the E_i solutions. For a two-phase system[50]

$$E_{\Delta}(g) = p_1' \exp(-\gamma^2\delta^2 g^2 D_1'\Delta) + p_2' \exp(-\gamma^2\delta^2 g^2 D_2'\Delta) \tag{7.57}$$

where

$$\begin{aligned}
D_2', D_1' = \frac{1}{2}\Bigg\{ & D_1 + D_2 + \frac{1}{\gamma^2\delta^2 g^2}\left(\frac{1}{\tau_1} + \frac{1}{\tau_2}\right) \\
& \pm \left[\left\{D_2 - D_1 + \frac{1}{\gamma^2\delta^2 g^2}\left(\frac{1}{\tau_1} - \frac{1}{\tau_2}\right)\right\}^2 + \frac{4}{\gamma^4\delta^4 g^4 \tau_1\tau_2}\right]^{\frac{1}{2}}\Bigg\}
\end{aligned} \tag{7.58}$$

$$p_2' = \frac{1}{D_2' - D_1'}(p_1 D_1 + p_2 D_2 - D_1') \tag{7.59}$$

$$p_1' = 1 - p_2'\,. \tag{7.60}$$

In the long time-scale limit these equations reduce, as required, to eqn (7.55) with $\overline{D}$ given by $p_1 D_1 + p_2 D_2$.

One commonly encountered special case concerns the existence of small regions of high mobility ($p_1 \ll p_2, D_1 \gg D_2$) for which the echo attenuation simplifies to

$$E_\Delta(g) = \exp\left(-\gamma^2\delta^2 g^2\left[D_2 + \frac{p_1 D_1}{\gamma^2\delta^2 g^2 \tau_2 p_1 D_1 + 1}\right]\Delta\right). \tag{7.61}$$

The exponent term in square brackets represents the effective diffusion coefficient as measured by the PGSE experiment. The limiting slopes and intercepts of the echo attenuation data allow the determination of $D_2, p_1 D_1$ and τ_2 as illustrated in Fig. 7.17. This approach has been for describing diffusion in microporous crystallites where the subscripts 1 and 2 refer to the intercrystalline spaces.[53-56]

7.6 Spin relaxation in microscopically inhomogeneous media

The pulsed gradient spin echo experiment suffers from one particular limitation which complicates interpretation, especially where variable time-scale experiments are performed. Because of the time which elapses in the duration between the excitation and the echo formation, the relative contributions to the spin echo signal from particular spins will be subject to relaxation: T_2 relaxation for a simple Hahn echo and, for the stimulated echo, T_1. This effect is very dangerous because there may be a correlation between the relaxation time and the local geometry. Normalizing the echo signal to the zero-gradient amplitude removes the influence of relaxation only if the sample is homogeneous. In microscopically heterogeneous systems, such as porous media, an accurate interpretation of $E(q)$ is possible only if the amplitudes of contributions to the signal from different parts of the microstructure are known.

In this section we describe how relaxation may be influenced by geometry. For the experimentalist wishing to interpret echo attenuation data some knowledge of relaxation rates is essential. At the simplest level such knowledge will enable Δ to be made sufficiently short that relaxation weighting

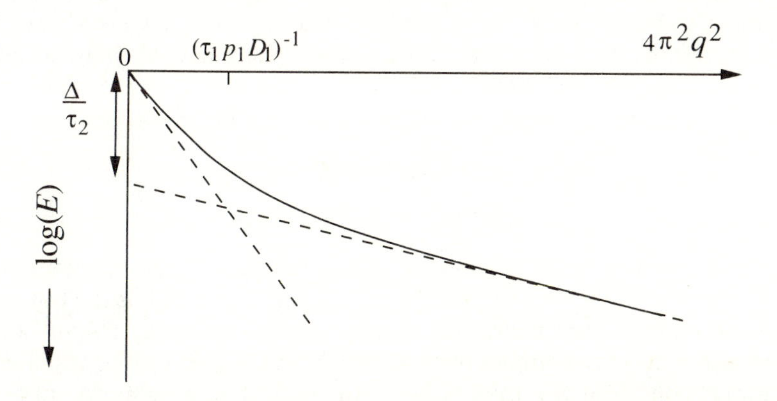

Fig. 7.17 Echo attenuation for two region diffusion adapted from Kärger *et al.*[50]

effects can be avoided. Where longer values of Δ are needed this knowledge might permit interpretation of the echo attenuation signal using a model which correctly assigns relative weights to signals arising from different local geometries.

7.6.1 *The two-phase model*

The nuclear spin relaxation problem analogous to the two-phase diffusion analysis has been described in detail by Zimmerman and Brittin.[52] Under conditions of slow exchange its influence on the PGSE echo attenuation is straightforward, introducing additional transverse relaxation attenuation in each term, an effect which is simply accounted for by substituting $\gamma^2\delta^2 g^2 D_i$ by $(1/T_2(i) + \gamma^2\delta^2 g^2 D_i)$ in eqn (7.54). The effect of relaxation in fast exchange is similarly straightforward. The small regions/high mobility example dealt with in eqn (7.61) has two possible outcomes.[53] For $1/T_2(1) \not\gg 1/T_2(2)$, the condition $p_1 \ll p_2$ ensures that $T_2(2)$ is the predominant relaxation time, T_2. For $1/T_2(1) \gg 1/T_2(2)$, the common relaxation time is $1/\{T_2(2)\} + p_1/\{\tau_1 + T_2(1)\}$. In both cases the echo attenuation is simply reduced by the usual relaxation factor, $\exp(-\Delta/T_2)$.

7.6.2 *Relaxation sinks and normal modes*

We now return to the problem of molecules experiencing restricted diffusion in a porous structure, turning our attention to the influence of geometry on spin relaxation. One relaxation mechanism for spins in pores involves the presence of strong relaxation 'sinks' at the pore surface. These sinks may be due to the presence of paramagetic centres at the surface, to dephasing caused by strong local magnetic field gradients or to the momentary reduction in rotational tumbling experienced by a molecule as it adheres to the surface. Clearly the ratio of the pore surface to pore volume will vary according to the pore size so that a priori we might expect the overall relaxation behaviour to be similarly size dependent. The problem has been treated in detail by Brownstein and Tarr[57] who adopt the classical 'magnetization diffusion' approach of Bloch and Torrey in assigning a magnetization density, $\rho(\mathbf{r}, t)$ which obeys the following differential equations:

$$D\nabla^2\rho = \partial\rho/\partial t \tag{7.62}$$

$$(D\mathbf{n}\cdot\nabla\rho + \mu\rho)|_S = 0\,. \tag{7.63}$$

Eqn (7.62) applies within the volume of a pore and reflects the transport of magnetization via the diffusion of the molecules. Eqn (7.63) is the boundary condition on the pore surface, taking into account the sink. The relaxation of the signal $S(t)$ is obtained by equating $S(t)$ with $\int\rho(\mathbf{r}, t)\,\mathrm{d}V$ and allowing the initial condition $\rho(\mathbf{r}, 0) = S(0)/V$. In general, the solution takes the form of a sum of normal modes which will depend on the geometry and on the sink strength, μ.

$$S(t) = S(0) \sum_{n=0}^{\infty} I_n \exp(-t/T_n) . \tag{7.64}$$

Eqn (7.64) can describe either the T_1 or T_2 relaxation process, depending on the value of μ chosen. The parameters which determine I_n and T_n are the molecular self-diffusion coefficient, D, the pore size, and the average value of μ over the surface, M. This last parameter is somewhat empirical and a variety of methods are employed in its estimation. Of course, the interesting feature of the normal modes is their dependence on pore size. Solutions are as follows.

(1) Planar geometry (bounded by $z = 0$ and $z = a$)

$$I_n = \frac{4\sin^2(\xi_n)}{[2\xi_n^2 + \xi_n \sin(2\xi_n)]} \tag{7.65}$$

$$T_n = a^2/D\xi_n^2 \tag{7.66}$$

where the ξ_n are the positive roots of $\xi_n \tan(\xi_n) = Ma/D$.

(2) Cylindrical geometry (bounded by $r = a$)

$$I_n = \frac{4J_1^2(\xi_n)}{\xi_n^2[J_0^2(\xi_n) + J_1^2(\xi_n)]} \tag{7.67}$$

$$T_n = a^2/D\xi_n^2 \tag{7.68}$$

where J_0 and J_1 are cylindrical Bessel functions and the ξ_n are the positive roots of $\xi_n J_1(\xi_n)/J_0(\xi_n) = Ma/D$.

(3) Spherical geometry (bounded by $r = a$)

$$I_n = \frac{12[\sin(\xi_n) - \xi_n \cos(\xi_n)]^2}{\xi_n^3[2\xi_n - \sin(2\xi_n)]} \tag{7.69}$$

$$T_n = a^2/D\xi_n^2 \tag{7.70}$$

where the ξ_n are the positive roots of $1 - \xi_n \cot(\xi_n) = Ma/D$.

Fig. (7.18) shows the dependence of the mode amplitudes on the parameter Ma/D. The fast-exchange limit corresponds to $Ma/D \ll 1$. Here the relaxation is single mode, dominated by the slowest relaxation rate, $T_0^{-1} \sim M(S/V)$ where S/V is the pore surface to volume ratio.

The intermediate ($Ma/D \sim 1$) and slow ($Ma/D \gg 1$) regimes feature multi-exponential relaxation although in the former case the decay is overwhelmingly dominated by the lowest mode.

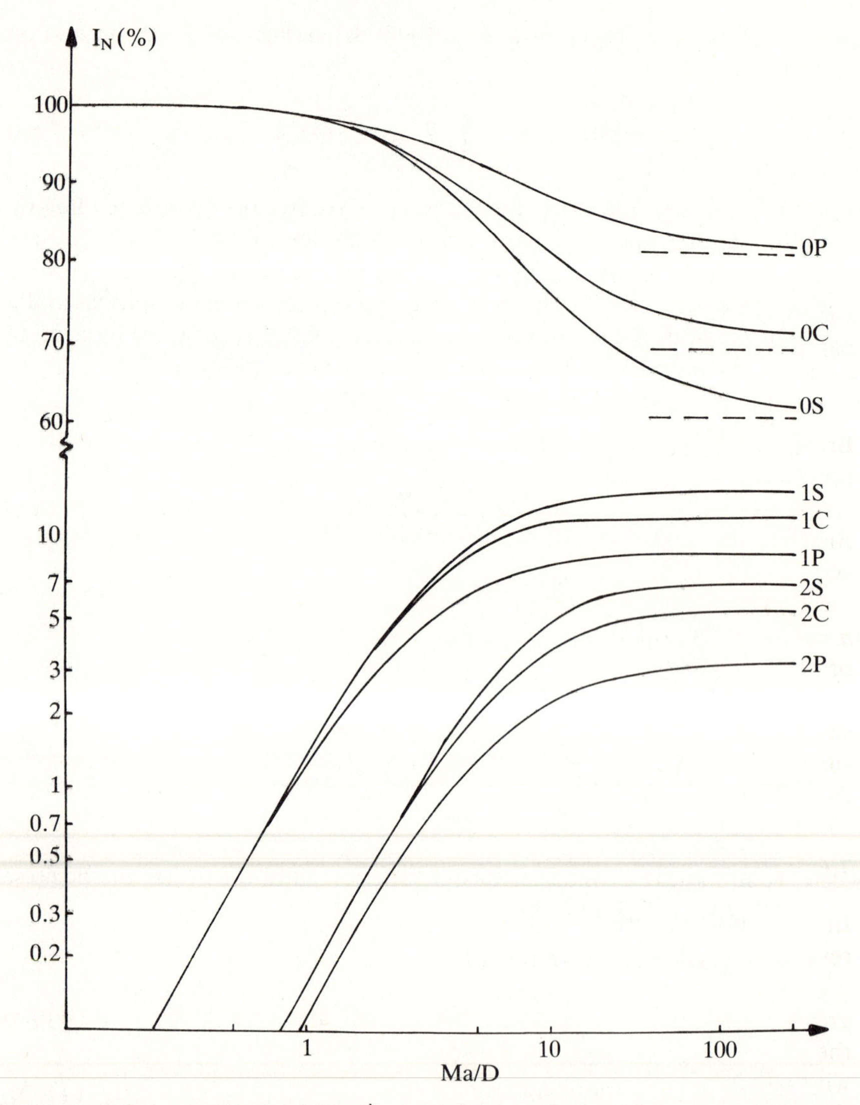

Fig. 7.18 Relative amplitude of the nth mode (I_n) as a function of the dimensionless sink parameter, Ma/D, for three geometries: planar (nP), cylindrical (nC), and spherical (nS). The dashed lines indicate the asymptotic values of I_0 in the slow diffusion limit, $Ma/D \to \infty$. (From Brownstein and Tarr.[57])

Examination of eqns (7.66), (7.68) and (7.69) shows that in the slow regime, the higher mode decay rates, T_n^{-1}, are almost independent of M and of order $a^2/n^2\pi^2D$. This is helpful because it gives us an estimate of the range of pore sizes which can be probed. Relaxation times can be measured in the range 1 μs to 2 s leading to a size range of 1 to 30 μm. Unfortunately the higher modes represent a small fraction of the total signal and are difficult to identify independently. Because the initial slope is also sensitive to the mode amplitudes it depends strongly on M. In slow and intermediate regimes both the initial decay rate T_i^{-1} and the ratio of T_i to the longest relaxation time T_0 have characteristic values,

$$T_i^{-1} = M(S/V) \tag{7.71}$$

$$T_0/T_i = \alpha(Ma/D) \tag{7.72}$$

where α is, respectively, 0.41, 0.35, and 0.31 for the three geometries considered. This second relation is helpful in estimating the dimensionless parameter Ma/D from the ratio T_i/T_0.

Despite the difficulty in obtaining reliable independent estimates for M the Brownstein–Tarr model has been widely, and successfully, used to obtain pore size distributions from multi-exponential relaxation, data.[42,58,59] In this respect it provides a useful complement to **q**-space imaging. Conversely, the Brownstein–Tarr model tells us how the PGSE components will be weighted according to local geometry. Because PGSE NMR will tend to emphasize the longest relaxation time components, the behaviour of the dominant $n = 0$ term is interest. Since V/S is of order a, T_0 is given to within a factor of order unity by Table 7.1.

In discussing the problem of diffusion in fractal volumes we noted that surface sink relaxation could provide a dimensional measure in the case of surface fractal behaviour. This problem has been treated by de Gennes[60] in the diffusion-limited regime.

7.6.3 *PGSE experiment in a Brownstein–Tarr system*

In a compartmented structure where the liquid molecules experience restricted diffusion, the echo attenuation in a PGSE experiment will be greatest in those compartments where the molecules are free to move the greatest distances. The long time echo signal will therefore be dominated by the signals from the smallest pores. This behaviour stands in stark contrast with the spin–spin relaxation behaviour evident in Table 7.1. In both the

Table 7.1 *Dominant mode relaxation times in the fast-diffusion and diffusion-limited regimes according to the Brownstein–Tarr model.*

	Fast Diffusion ($Ma/D \ll 1$)	Diffusion-limited ($Ma/D \gg 1$)
T_0	$\frac{a}{M}$	$\frac{a^2}{D}$

fast-diffusion and diffusion-limited regimes the spin–spin relaxation is more rapid as the pore diameter a decreases. This competition between PGSE and spin–spin relaxation will lead to an intermediate pore size being favoured in the echo. Lipsicas *et al.* have provided a simple analysis[59] of this phenomenon which accurately fits experimental data for PGSE attenuation in water-impregnated porous dolomite. Their approach is to presume spherical pores with q sufficiently large that the PGSE echo attenuation is Δ independent and given by eqn (7.17). Allowing for a 'bulk' relaxation mechanism with relaxation time T_{2B} additional to the surface sink effect, the total attenuation of the echo at time t may be written

Fast-diffusion regime

$$E(t) = \exp\left(-\frac{t}{T_{2B}} - \frac{Mt}{a} - \frac{1}{5}\gamma^2\delta^2 g^2 a^2\right) \tag{7.73}$$

Diffusion-limited regime

$$E(t) = \exp\left(-\frac{t}{T_{2B}} - \frac{Dt}{a^2} - \frac{1}{5}\gamma^2\delta^2 g^2 a^2\right). \tag{7.74}$$

The signal is dominated by the pores of radius a^* which minimize the exponents in eqn (7.73) and (7.74). These radii, along with the dependence of the echo attenuation on $g\delta$, are therefore given by

Fast-diffusion regime

$$a^* = \left(\frac{5}{2}Mt\right)^{\frac{1}{3}} (\gamma\delta g)^{-\frac{1}{3}} \tag{7.75}$$

$$\ln\left[E(t)\right] \approx -\frac{t}{T_{2B}} - \left(\frac{27}{20}M^2 t^2\gamma^2\delta^2 g^2\right)^{\frac{1}{3}} \tag{7.76}$$

Diffusion-limited regime

$$a^* = (5Dt)^{\frac{1}{4}} (\gamma\delta g)^{-\frac{1}{2}} \tag{7.77}$$

$$\ln\left[E(t)\right] = -\frac{t}{T_{2B}} - \left(\frac{4}{5}Dt\gamma^2\delta^2 g^2\right)^{\frac{1}{2}}. \tag{7.78}$$

Fig. 7.19 shows a plot of $-\ln[E(t)] - t/T_{2B}$ versus $\ln[g\delta]$. A scaling behaviour is observed with exponent $\frac{2}{3}$, indicating that the relaxation is in the fast-diffusion regime. The slope of the plot provides a nice estimate for M, thus enabling the dominant pore size to be calculated using eqn (7.75). The value of around 20 μm agrees well with independent estimates using mercury porosity and electron micrograph measurements.

The description of dominant pore behaviour applies equally in the case of T_1 relaxation provided that the PGSE sequence employs a stimulated echo

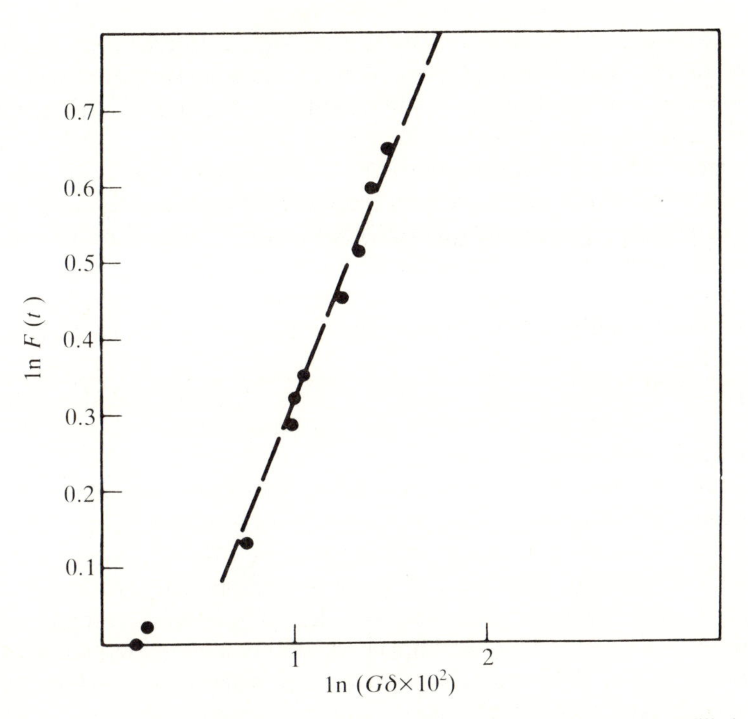

Fig. 7.19 ln $F(t)$ versus $\ln(g\delta)$ (where $F(t) = -\ln[E(t)] - t/T_{2B}$) for water-filled porous sedimentary rock. The dashed line has a slope of 2/3. (From Lipsicus *et al.*[59])

rather than a Hahn echo. There are specific advantages in the use of such an approach. The stimulated echo permits the PGSE experiments to be performed over longer observation times, Δ, ensuring that fully restricted motion is occurring in the dominant pore. Furthermore, T_1 relaxation is not subject to susceptibility effects (discussed in the ensuing section) so that the Brownstein–Tarr model more accurately describes the relaxation mechanism.

7.6.4 *Susceptibility inhomogeneity*

Chapter 4 dealt with the effect on **k**-space imaging of diamagnetic susceptibility variations. In a heterogeneous structure, local magnetic field perturbations exist in the vicinity of boundaries between regions of differing susceptibility. These perturbations will also influence the nuclear spin relaxation for molecules moving in the vicinity of boundaries. In the examples of restricted diffusion considered in this chapter, and especially in systems of interconnected pores, susceptibility inhomogeneity effects must be accounted for. Unlike the 'surface sink' approach of the previous section, this relaxation process arises from through-space interactions which are modulated as the molecule diffuses. Whether surface sink or susceptibility inhomogeneity effects are more important will depend on the particular material being studied.

We have seen that the Brownstein–Tarr model leads to a specific signature in the form of the relaxation normal modes. Furthermore, the surface sink mechanism influences both T_1 and T_2 relaxation. In this respect relaxation due to susceptibility effects is quite different. T_1 relaxation requires fluctuations at the Larmor frequency. The through-space interactions due to susceptibility variations modulate much more slowly as the particle diffuse and so influence T_2 alone. The existence of susceptibility effects is therefore indicated by $T_2 \ll T_1$. One nice feature of this difference is that we are able to separate the influences of surface sink and susceptibility variation effects by separately observing T_1 and T_2.

As in the treatment of **k**-space imaging, only the components of local field along the polarizing field, B_0, need be considered. This local field can then be described as a spatially dependent offset, $\Delta B_0(\mathbf{r})$. It is sufficiently complicated that we will employ the language of statistics to describe it. Translational motion causes the nuclear spins to experience a time-dependent field, $\Delta B_0(t)$ and hence a time-dependent frequency offset, $\Delta\omega_0(t) = \gamma\Delta B_0(t)$. The crucial factor in determining the effect on the transverse magnetization is the rate at which this frequency fluctuates. This rate is characterized by defining the correlation time

$$\tau_c = \int_0^\infty \frac{\overline{\Delta\omega_0(t)\,\Delta\omega_0(t-\tau)}\,d\tau}{\overline{\Delta\omega_0^2}} \tag{7.79}$$

where $\overline{\Delta\omega_0^2}$ is the mean square frequency fluctuation. $\overline{\Delta\omega_0^2}$ is the second moment (M_2) of the linewidth which prevails in the case of a stationary interaction ($\tau_c \to \infty$).

τ_c depends on the molecular motion and on the correlation length of the field offset, $\Delta B_0(\mathbf{r})$. For the moment we ignore these details and define 'rapid fluctuations' and 'slow fluctuations' according to $\tau_c^{-1} \gg (\overline{\Delta\omega_0^2})^{\frac{1}{2}}$ or $\tau_c^{-1} \ll (\overline{\Delta\omega_0^2})^{\frac{1}{2}}$, respectively.

The effect of fluctuating local fields on the nuclear transverse magnetization has been treated in detail by Abragam,[61] and the application of this theory to relaxation induced by diffusion in a spatially varying local field has been made by Packer[62] and Hazelwood *et al.*[63] Slow fluctuations result in inhomogeneous line-rates, thus leading to a decay of the transverse magnetization ofthe form $\exp(-\frac{1}{2}M_2t^2)$. The essential feature of inhomogeneous broadening is that this decay can be refocused in a spin echo. By contrast, rapid fluctuations result in homogeneous broadening, a true relaxation mechanism common to all spins in the system. Decay due to homogeneous broadening is irreversible because of its stochastic nature. For rapid fluctuations this relaxation is described by a T_2 value of $(M_2\tau_c)^{-1}$.

The general behaviour of both the spin echo signal and the transverse magnetization can be treated at any time-scale using the Anderson–Weiss theory[64] and the fluctuating field correlation function,

$$g_\omega(\tau) = \frac{\overline{\Delta\omega_0(t)\,\Delta\omega_0(t-\tau)}}{\overline{\Delta\omega_0^2}}. \tag{7.80}$$

From the assumption that the distribution of $\Delta\omega_0$ is Gausian it may be shown that the free induction decay signal, $G(t)$, and the Hahn echo amplitude, $E(2\tau)$, are given, respectively, by

$$G(t) = \exp\left[-\overline{\Delta\omega_0^2}\int_0^t (t-t')\,g_\omega(t')\,\mathrm{d}t'\right] \tag{7.81}$$

$$E(2\tau) = \exp\left[-\overline{\Delta\omega_0^2}\left\{4\int_0^\tau (\tau-t')g_\omega(t')\,\mathrm{d}t' - \int_0^{2\tau}(2\tau-t')g_\omega(t')\,\mathrm{d}t'\right\}\right]. \tag{7.82}$$

For stochastic processes it is common to assume a correlation function $g_\omega(\tau) = \exp(-|\tau|/\tau_c)$. This leads to

$$G(t) = \exp\left[-\overline{\Delta\omega_0^2}\,\tau_c^2\,\{\exp(-t/\tau_c) - 1 + t/\tau_c\}\right] \tag{7.83}$$

$$E(2\tau) = \exp\left[-\overline{\Delta\omega_0^2}\,\tau_c^2\,\{4\exp(-\tau/\tau_c) - \exp(-2\tau/\tau_c) + 2\tau/\tau_c - 3\}\right]. \tag{7.84}$$

In the fast-motion limit, where $\overline{\Delta\omega_0^2}\tau_c^2 \ll 1$, the exponents in eqn (7.83) and (7.84) are dominated by the terms t/τ_c and $2\tau/\tau_c$ leading to $G(t) = \exp(-t/T_2)$ and $E(2\tau) = \exp(-2\tau/T_2)$, respectively, where T_2^{-1} is $\overline{\Delta\omega_0^2}\tau_c$. This is the relaxation limit in which the decay of the FID and the echo are identical since no refocusing can occur. In the slow-motion limit, where $\overline{\Delta\omega_0^2}\tau_c \gg 1$ we expand the terms $\exp(-\tau/\tau_c)$ etc., to second order and obtain $G(t) = \exp(-\frac{1}{2}\overline{\Delta\omega_0^2}t^2)$ and $E(2\tau) = \exp(-\frac{2}{3}\overline{\Delta\omega_0^2}\tau_c^{-1}\tau^3)$. This is the inhomogeneous broadening limit where refocusing occurs† such that $E(2\tau) \gg G(t)$.

† In chapter 3 we dealt with the problem of diffusion in a steady field gradient and generated an expression for magnetization decay which depended exponentially on the square of the gradient. This expression is not applicable in the case of a randomly varying field offset. This can be immediately recognized by considering the spin phase, $\int \gamma\Delta B_0(t)\,\mathrm{d}t$. In the uniform gradient there is no upper bound on the phase as the particle diffuses since $\Delta B_0 = \mathbf{G}.\mathbf{r}$ and $\mathbf{r}$ is unbounded. This means that the process can be viewed as unrestricted random phase walk. In the case of the fluctuating local field offset there is clearly an upper bound on ΔB_0 which oscillates about 0. To this extent the latter process may be viewed as a restricted phase walk and the treatment required is quite different.

For self-diffusion in cellular or porous structures, the correlation time, τ_c, will be of order X^2/D, where X is the correlation length for field variations. This distance will be on the order of the cell or pore size. Accordingly the echo attenuation in the inhomogeneous broadening limit reduces to $\exp(-\frac{2}{3}\gamma^2\overline{g^2}D\tau^3)$ where g is the local magnetic field gradient. This expression is similar to the usual echo attenuation behaviour when a steady external gradient is applied. Notice that in the homogeneous (fast motion) limit the attenuation exponent is inversely proportional to D whereas in this inhomogeneous (slow motion) limit it increases with D.

The size of $(\overline{\Delta\omega_0^2})^{\frac{1}{2}}$ will be of order $\gamma\Delta\chi B_0$ where $\Delta\chi$ is the susceptibility change at an interface and B_0 is the polarizing field strength. Typically $(\overline{\Delta\omega_0^2})^{\frac{1}{2}}$ will be in the range 10^3 to 10^4 radians per second. Taking D as the self-diffusion coefficient of water we see that the fast and slow regimes will be distinguished by cell sizes $\ll 2\,\mu$m and $\gg 2\,\mu$m, respectively. Clearly the intermediate time-scale will often met in practice.

Fig. 7.20 shows the transition from slow to fast regimes in the decay of the FID and spin echo as represented by eqn (7.83) and (7.84). Although the concept of a relaxation time can be defined precisely only in the fast limit, it is convenient[63] to use the symbols T_2^* and $T_2^\dagger$ to represent the time taken for the FID and echo amplitudes, respectively, decay to e^{-1}.

7.6.5 *Signatures for relaxation*

This chapter has dealt with the elucidation of microstructural geometry via **q**-space imaging. Because that geometry may influence the T_2 relaxation of nuclear spins, it will affect the relative weightings of different components of the signal in the echo. For this reason it is important to understand which mechanism for transverse relaxation will predominate and to then assess its influence, if necessary making allowance for weighting factors.

Several tests are available to help us distinguish the roles of surface sink and susceptibility inhomogeneity in determinign spin relaxation. First, because the susceptibility effect results in a 'through-space' interaction which fluctuates much more slowly that the Larmor frequency, it will only influence T_2 so that differences in T_2 and T_1 rates are a helpful pointer. Second, the field offset due to susceptibility inhomogeneity is proportional to B_0. This gives us a test for the role of this mechanism in the spin relaxation by comparing relaxation rates in different spectrometer polarizing fields. Finally, because diffusion in the vicinity of susceptibility boundaries causes quite slow fluctuations in Larmor frequency, there will often be a significant degree of inhomogeneous broadening. While this will not contribute to relaxation of the echo and therefore not affect the PGSE weightings, the existence of this broadening does provide another test for the relaxation mechanism.

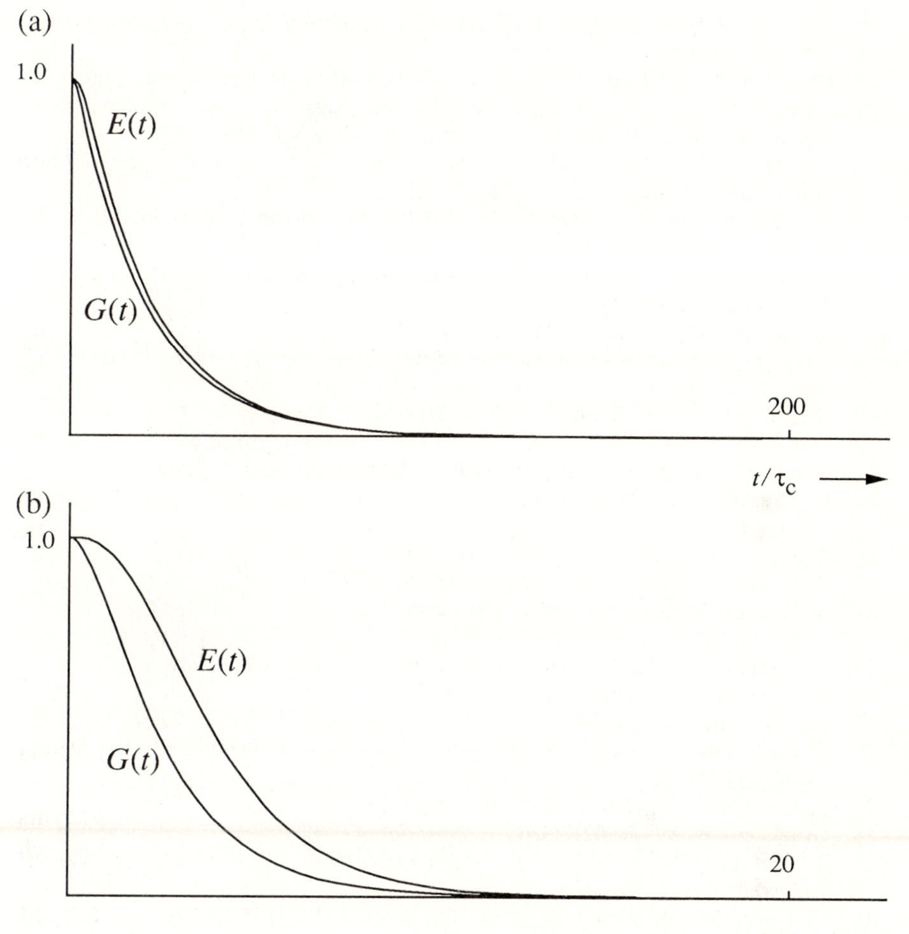

Fig. 7.20 Dependence of FID ($G(t)$) and echo amplitude ($E(t)=E(2\tau)$) on time t for (a) fast motion, where $\overline{\Delta\omega_0^2}\ \tau_c^2=0.05$, and (b) intermediate motion, where $\overline{\Delta\omega_0^2}\ \tau_c^2=0.5$. Note that $G(t)$ and $E(t)$ converge to a common exponential decay in the fast-motion limit where relaxation effects dominate.

7.7 References

1. Tanner, J. E. and Stejskal, E. O. (1968). *J. Chem. Phys.* **49**, 1768.
2. Frey, S., Kärger, J., Pfeifer, H. and Walther, P. (1988). *J. Magn. Reson.* **79**, 336.
3. Neumann, C. H. (1974). *J. Chem. Phys.* **60**, 4508.
4. Murday, J. S. and Cotts, R. M. (1968). *J. Chem. Phys.* **48**, 4938.
5. Stejskal, E. O and Tanner, J. E. (1965). *J. Chem. Phys.* **42**, 288.

6. Carslaw, H. S. and Jaeger, J. C. (1959). *Conduction of heat in solids*, Oxford University Press.
7. Packer K. J. and Rees, C. (1972). *J. Colloid and Interface Sci.* **40**, 206.
8. Van den Enden, J. C., Waddington, D., van Aalst, H., van Kralingen, C. G. and Packer, K. J. (1990) *J. Colloid and Interface Sci.* **140**, 105.
9. Champeney, D. C. (1973). *Fourier transformations and their physical applications*, Academic Press, New York.
10. Cory, D. G. and Garroway, A. N. (1990). *Magn. Reson, Med.* **14**, 435.
11. Kärger, J. and Heink, W. (1983). *J. Magn. Reson.* **51**, 1.
12. Callaghan, P. T., Jolley, K. W. and Humphrey, R. S. (1983). *J. Colloid and Interface Sci.* **83**, 521.
13. Walstra, P. (1969). *Neth. Milk Dairy J.* **23**, 99.
14. Cory, D. G. Garroway, A. N. and Miller, J. B. (1990). *Polym. Preprints* **31**, 149.
15. Dauod, M. and de Gennes, P. G. (1977). *J. de Physique* **38**, 85.
16. Brochard, F. and de Gennes, P. G. (1977). *J. Chem. Phys.* **67**, 52.
17. Cannel, D. S. and Rondelez, F. (1980). *Macromolecules* **3**, 1599.
18. Callaghan, P. T., MacGowan, D., Packer, K. J. and Zelaya, F. O. (1990). *J. Magn. Reson.* **90**, 177.
19. Tanner, J. E. (1978). *J. Chem. Phys.* **69**, 1748.
20. Stejskal, E. O. (1965). *J. Chem. Phys.* **43**, 3597.
21. de Gennes, P. G. (1976). *Macromolecules*, **9**, 587.
22. Callaghan, P. T. and Pinder, D. N. (1980). *Macromolecules* **13**, 1085.
23. Callaghan, P. T. and Soderman, O. (1983). *J. Phys. Chem.* **87**, 1737.
24. de Gennes, P. G. (1971). *J. Chem. Phys.* **55**, 572.
25. Doi, M. and Edwards, S. F. (1978). *J. Chem. Soc. Faraday Trans.* **74**, 1789.
26. Doi, M. and Edwards, S. F. (1987). *The theory of polymer dynamics*, Oxford University Press.
27. de Gennes, P. G. (1971). *J. Chem. Phys.* **55**, 572.
28. Graessley, W. W. (1982). *Adv. Polym. Sci.* **47**, 67.
29. de Gennes, P. G. (1979). *Scaling concepts in polymer physics*, Cornell University Press, Ithaca.
30. Clarkson, M., Beaglehole, D. and Callaghan, P. T. (1985). *Phys. Rev. Lett.* **54**, 1722.
31. Mandelbrot, B. B. (1977). *Fractals: form, chance and dimension*, Freeman, San Francisco.
32. Mandelbrot, B. B. (1982). *The fractal geometry of nature*, Freeman, San Francisco.
33. Peitgen, H. O. and Richter, P. H. (1986). *The beauty of fractals*, Springer-Verlag, Berlin.
34. Pietronero L. and Tosatti E. (eds.), *Fractals in physics*, North-Holland, Amsterdam (1986).
35. Fischer P. and Smith W. R. (eds.) (1985). *Chaos, fractals and dynamics*, Dekker, M. New York.
36. Alexander, S. and Orbach, R. (1982). *J. de Physique*, **43**, L625.
37. Rammel, R. and Toulouse, G. (1983). *J. de Physique*, **44**, L13.
38. Gefen, Y., Aharony, A. and Alexander S. (1982). *Phys. Rev. Lett.* **50**, 77.
39. Banavar, J. R., Lipsicas, M. and Willemsen, J. F. (1985). *Phys. Rev.* **B32**, 6066.
40. Jug, G. (1986). *Chem. Phys. Lett.* **131**, 94.

41. Kärger, J. and Vojta, G. (1987). *Chem. Phys. Lett.* **141**, 411.
42. Banavar, J. R. and Schwartz, L. M. (1989). *Molecular dynamics in restricted geometries*, (Ed. Klafter J. and Drake J. M.), Wiley, New York.
43. Courtens, E. and Vacher, R. (1987). *Z. Phys.B. Condensed Matter* **68**, 355.
44. Mendelson, K. S. (1986). *Phys. Rev.* **B34**, 6503.
45. Blinc, A., Lahajnar, G. Blinc, R., Zidansek, A. and Sepe, A. (1989). *Bull. Magn. Reson.* **11**, 370.
46. Crank, J. (1975). *The mathematics of diffusion*, Oxford University Press.
47. Barrer, R. M. (1968). In *Diffusion in polymers*, (ed. Crank J. and Park G. S.), p. 165, Academic Press, London.
48. Zientara, G. P. and Freed, J. H. (1980). *J. Chem. Phys.* **72**, 1285.
49. Kärger, J., Kocirik, M. and Zikanova, A. (1981). *J. Colloid and Interface Sci.* **84**, 240.
50. Kärger, J., Pfeifer, H. and Heink, W. (1988). *Adv. Magn. Reson.* **12**, 1.
51. Feller, W. (1950). *An introduction to probability theory and its applications*, Vol. 1. Wiley, New York.
52. Zimmerman, J. R. and Brittin, W. E. (1957). *J. Phys. Chem.* **61**, 1228.
53. Kärger, J. (1971). *Z. Phys, Chem. (Leipzig)* **248**, 27.
54. Haul, R., Heintz, W. and Stremming, H. (1979). In *The properties and applications of zeolites* (ed. Townsend R. P), p. 27. Chem. Soc., London.
55. Lechert, H. (1976). *Catal. Rev.* **14**, 1.
56. Barrer, R. M. (1978). *Zeolites and clay minerals as sorbents and molecular sieves*, Academic Press, London.
57. Brownstein, K. R. and Tarr, C. E. (1979). *Phys. Rev.* **19**, 2446.
58. Whittall, K. P. and Mackay, A. L. (1989). *J. Magn. Reson.* **84**, 134.
59. Lipsicas, M. Banavar, J. R. and Willemsen J. (1986). *Appl. Phys. Lett.* **48**, 1544.
60. de Gennes, P. G. (1982). *C.R. Acad. Sci. (Paris)* **295**, 1062.
61. Abragam, A. (1963). *Principles of nuclear magnetism*, Oxford University Press.
62. Packer, K. J. (1973). *J. Magn. Reson.* **9**, 438.
63. Hazelwood, C. F., Chang, D. C. Nichols, B. L. and Woessner, D. E. (1974). *Biophys. J.* **14**, 583.
64. Anderson, P. W. and Weiss, P. R. (1953). *Rev. Mod. Phys.* **25**, 269.

8
SPATIALLY HETEROGENEOUS MOTION AND DYNAMIC NMR MICROSCOPY

8.1 The influence of motion in imaging

The sensitivity of nuclear magnetic resonance to spin translation, as detailed in Chapters 6 and 7, is uniquely useful as a contrast in imaging. In this chapter the various methods for obtaining information about molecular translation in spatially heterogeneous systems are reviewed and the dynamic NMR microscopy method, in which both $\rho(\mathbf{r})$ and $P_s(\mathbf{r}|\mathbf{r}', \Delta)$ are simultaneously imaged, is discussed in detail. Dynamic imaging promises to be of considerable importance in biology and materials science.

It is helpful to subdivide spin motion into two categories. The first concerns evolution of the spin density, $\rho(\mathbf{r})$, due to either bulk movement of the sample or mutual diffusion of differing spin populations. NMR imaging can follow such an evolution provided that the associated changes in spin positions are small over the time required to obtain an image, or if the motion is periodic and stroboscopic gating can be employed. Examples of such methods are discussed in Section 8.2. The second category of motion arises when the nuclear spin density obtained in an NMR imaging experiment remains constant with time despite the existence of underlying spin translation. In this case some form of image contrast sensitive to flow must be employed. In the earlier discussion attention was focused on the phase shifts induced by motion when spin echoes are employed. The pulsed gradient spin echo method represents a powerful method for providing a detailed quantitative description of motion, and its amalgamation with **k**-space imaging will be a major topic in this chapter. However, there are many other methods of flow contrast which, while not always as precise as PGSE phase shift measurement, can give a simple and direct visualization of motion. All require an evolution of the spin system over some well-defined time-scale, but none the less it is helpful to broadly categorize the various methods as being either 'steady-state', 'time-of-flight', or 'phase shift' determinations. Steady-state methods rely on a rapid repetition of r.f. and gradient pulses such that the saturation recovery or steady-state free-precession signal amplitude depends on the spin motion. In time-of-flight methods the appearance of an image is compared in real space at two different times, or a 'bolus' of spins is tagged at one time and imaged at some later time. The phase shift method uses the precession of transverse magnetization in a time-dependent magnetic field gradient to impart a magnetization phase which depends on the spin translational motion.

Whereas time-of-flight methods can provide a graphic visualization of molecular displacement, phase shift methods are unparalleled in obtaining precise quantification of motion. While narrow-pulse PGSE sequences are especially useful in this regard, it is possible to employ more general gradient modulation methods with specific sensitivity to velocity or acceleration, as discussed in Chapter 6. Furthermore, there are numerous ways in which the phase shifts can be analysed to yield the motion, and the complexity of the experiment should be designed with a view to the degree of information required.

In most forms of motional contrast imaging, it is desirable to obtain an image in which the signal from stationary spins is suppressed. To produce a null image from stationary spins either the excitation must be motion sensitive or some form of signal subtraction must be employed. Alternatively, a set of images obtained under differing pulse sequence conditions may be analysed pixel by pixel to yield a velocity or diffusion map where zero amplitude will correspond to stationary spins. This rather sophisticated approach, while often time-consuming, can be very powerful in discriminating general types of motion and will be discussed in Section 8.3.

In all measurements of molecular motion using nuclear magnetic resonance, the range of average velocities which can be accurately determined will be limited by receiver coil dimension l, at the maximum and by molecular self-diffusion at the minimum. To have an observable signal the moving spins must remain in the receiver coil over the observational time-scale, of order T_E, thus limiting velocities to less than or of order l/T_E. In non-imaging applications of flow measurement, where all spins within the coil are excited, there will always be a shift of spins located at the coil fringing field and this effect will inevitably result in some signal degradation. In imaging however, where slice selection is employed, such degradation is absent, although the slice selection gradient may itself introduce some additional velocity encoding.

The smallest average velocity which can be measured is limited simply by the superposed random motions of the molecules, although the degree to which random motion will mask the average flow is dependent on observational time-scale. This is because flow displacements are linear with time, t, whereas r.m.s. diffusional displacements vary as $t^{\frac{1}{2}}$. Fig. 6.2 illustrates how the average propagator, $\overline{P_s(Z)}$, depends on the observational time. The longest observational time-scale is afforded by using stimulated echoes for which $t \sim T_1$. Thus the resolution of velocity measurements by NMR is of the order $(D/T_1)^{\frac{1}{2}}$. For free water this limit is around $20\,\mu\text{m s}^{-1}$. For macromolecules with substantially smaller self-diffusion coefficients, even allowing for shorter T_1 values, this limit may be as small as $1\,\mu\text{m s}^{-1}$.

8.1.1 *Steady-state methods*

Both steady-state free-precession and saturation recovery imaging methods are sensitive to the presence of molecular motion when repeated selective excitation pulses are used. This is because moving spins will not reside in the excitation slice and therefore will inevitably experience different turn angle trajectories than spins which are stationary.

The use of steady-state free-precession (SSFP) methods to provide flow contrast was first suggested by Patz and Hawkes[1] although their experiment was based on a quite different principle than that just outlined. Rather than applying a rapidly repeated imaging sequence, a single gradient was used along with a multiple 90° pulse train with pulse separation τ. A projection image is obtained by measuring the SSFP signal amplitude as a function of the gradient magnitude, since only those spins whose position $\Delta r = 1/(\gamma G\tau)$, and hence frequency offset $\Delta\omega = 2\pi/\tau$, will contribute to the SSFP coherence. Motion along the gradient direction degrades this coherence since these spins will not experience a constant frequency offset. Subtraction of appropriately scaled SSFP projection images obtained for two different repetition times, τ, yields a profile roughly proportional to velocity.

The flow sensitivity of images obtained using rapid-acquisition steady-state imaging has been demonstrated by a number of groups using both saturation recovery[2-4] and free precession.[5-8] The FAST and CE-FAST methods proposed by Glyngell[5] are of particular interest since these utilize, respectively, the 'FID' and 'spin echo' signals of the SSFP sequence. Because the spin echo signal requires the action of two selective r.f. pulses it is sensitive to motion of the spins. In consequence, the CE-FAST image is motion contrasted while the FAST image is not. The FADE variant of SSFP imaging makes both signals (echo 1 and echo 2 of Fig. 3.22) available in the same experiment so that motion contrast can be obtained by comparison.

The effect of flow in SSFP imaging is to cause a change in the intensity of the image. In other words, the image sensitivity is to a 'flow window' centred on zero velocity. By slowing sweeping the B_0 field, Tyszka *et al.*[9] have obtained signals with a window centred on a finite velocity in the direction of the slice selection gradient, so that the SSFP experiment is carried out in a moving reference frame. While these researchers have not yet reported images obtained by this methods, the method should prove suitable for flow velocities in the range of a few mm/s.

Use of the echo 2 signal to obtain self-diffusion contrast has been demonstrated by LeBihan *et al.*[6] and Merboldt *et al.*[7] in FADE and CE-FAST sequences employing single gradient pulses of amplitude g and duration δ, inserted in each period between the selective r.f. pulses. The separation, Δ, of these pulses is, in practice, the r.f. pulse repetition time, T_R. Because this signal is, strictly speaking, the sum of a spin echo and a stimulated echo with different diffusive time-scales, Δ and 2Δ, respectively, the dependence of the image amplitude on g^2 is double exponential and given, for $T_R \ll T_1, T_2$, by[7],

$$E_\Delta(\mathrm{SSFP}) = \tfrac{1}{2}\{1 - \exp(-T_R/T_1)\}\exp(-2T_R/T_2) \times [\exp\{-\gamma^2\delta^2 g^2 D(\Delta - \delta/3)\} + \exp(-T_R/T_1) \times \exp\{-\gamma^2\delta^2 g^2 D(2\Delta - \delta/3\}]. \quad (8.1)$$

8.1.2 *Time-of-flight and spin 'tagging'*

The first use of spin tagging methods to measure flow was reported by Singer[10] in an experiment to measure the motion of blood in mouse tails. Since then the method has been applied in a variety of non-imaging studies[11-13] and its use is widespread in MRI.[14-21] One of the simplest time-of-flight techniques involves the destruction of magnetization in a selected plane (for example by a 90° selective pulse followed by a homospoil gradient pulse) and the subsequent imaging of spins residing in that plane at some later time.[14,15] Only fresh inflowing spins contribute to the image so that static spin signals are suppressed while moving spins have a signal amplitude proportional to the overlap between the tagging and target planes. A beautiful example of this method is the renal angiogram of a rat obtained by D. Burstein and shown in Fig. 8.1. The pulse sequence incorporates a slice-selective pre-saturation pulse without refocussing, equivalent to the imposition of a homospoil gradient. After a delay (typically 50 ms) during which flowing spins in the blood wash into the imaging plane, one phase-encoded step of a flow-compensated gradient echo image is acquired. Note that both the read and slice gradients are flow-compensated, akin to the g^* profile shown in Fig. 6.14(b).

The use of a separate precursor tagging pulse is unnecessary where two spatially selective r.f. pulses are used to form a spin echo since this pair provides a natural tagging and interrogation mechanism. Feinberg *et al.*[16] have used 90°, 180° pulse pairs in which the second pulse is shifted in frequency so as to provide a spatially displaced interrogation or target window. These methods are limited in velocity resolution by the slice width and, because they depend on image intensity to determine velocity, the calibration is necessarily relaxation-time dependent. A more sophisticated approach to bolus tagging due to Merboldt *et al.*[17] uses the successive interrogation in a multiple pulse train, as illustrated in Fig. 8.2, by means of a stimulated echo imaging sequence in which the first and third pulses are slice selective and where the third r.f. pulse, which defines the target plane,

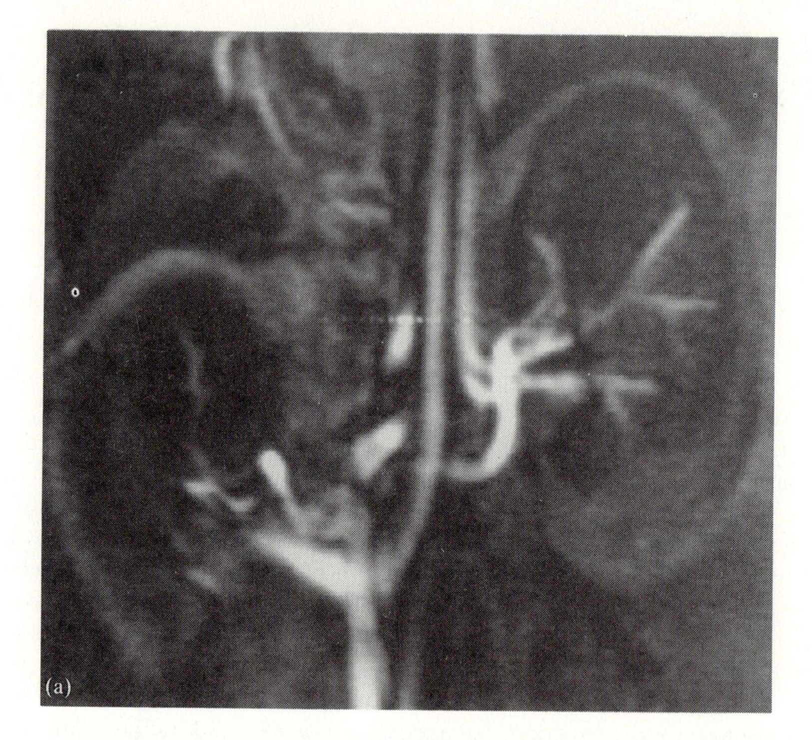

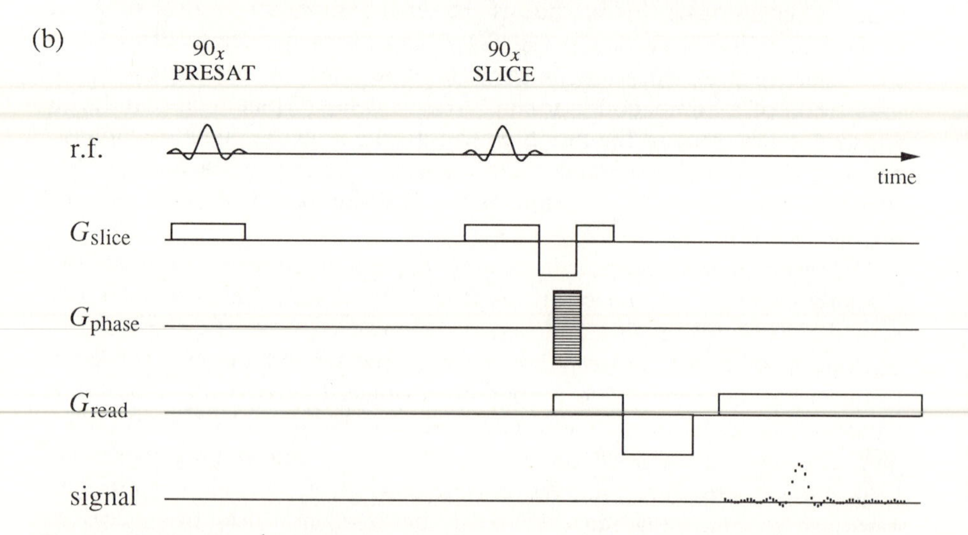

Fig. 8.1 (a) 200 MHz ^{1}H NMR. Renal angiogram from a rat *in vivo*, obtained using slice presaturation. The slice thickness is 750 μm with a pixel dimension of (200 μm)^2. (b) Pulse sequence used to obtain the angiogram shown in (a). Note the use of velocity-insensitive phase-encoding and slice-selection gradients. (From D. Burstein, personal communication.)

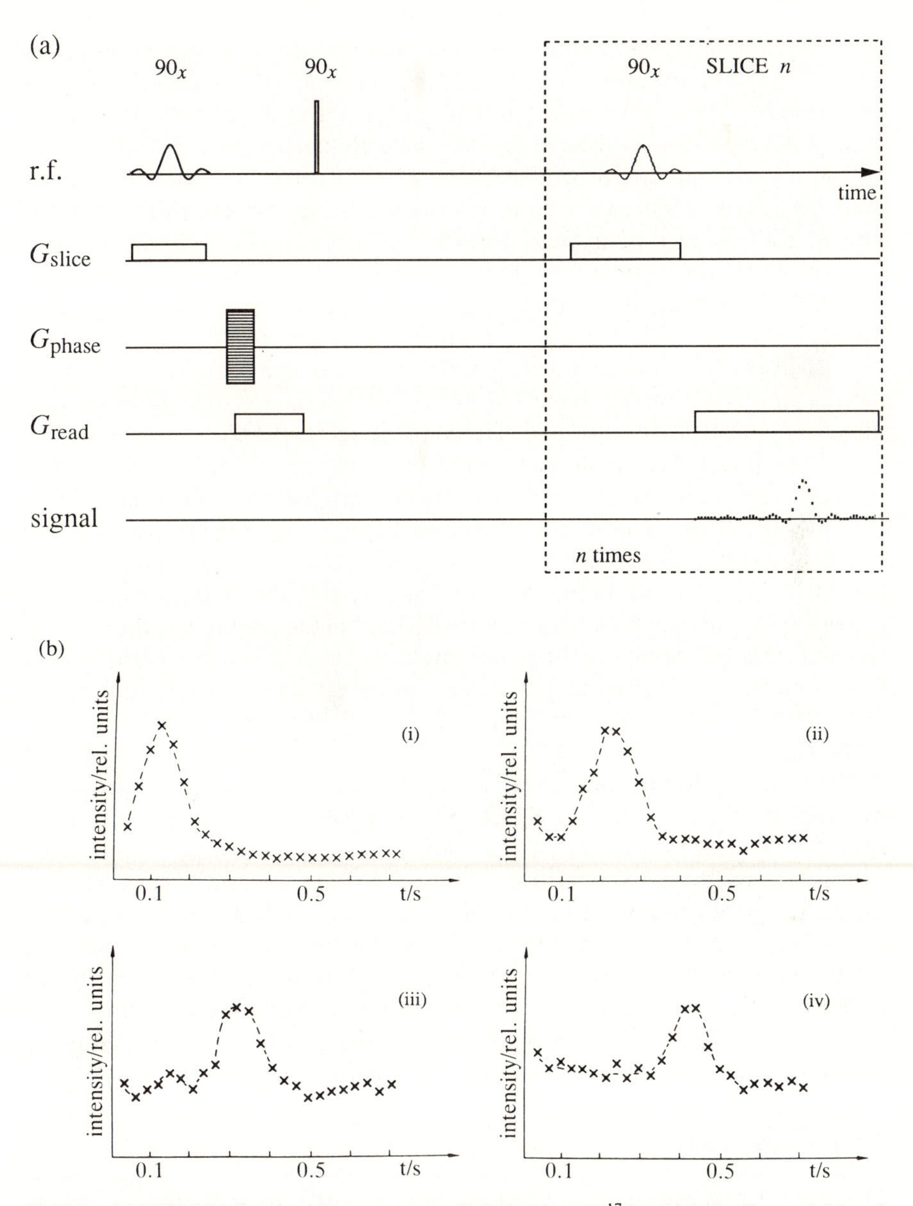

Fig 8.2 (a) STEAM flow imaging sequence of Merboldt *et al.*[17] in which the first soft 90_x r.f. pulse defines the tagged slice while the final 90_x soft r.f. pulse defines the target slice. By repeating the final pulse the motion of the target slice is successively followed in a single phase-encoding experiment. This experiment must then be repeated several times with different phase-encoding gradients. (b) Bolus tag experiment for water flow in a 10 mm internal diameter tube using the pulse sequence of Fig. 8.1 in which the spacings of tagged and target slices are, respectively, (i) 9 mm, (ii) 13.5 mm, (iii) 18 mm and (iv) 22.5 mm, yielding a maximum flow velocity of around 50 mm s^{-1}. (From Merboldt *et al.*[17])

is repeated at a series of time intervals. This results in a plot of intensity versus time, as in the example of Fig. 8.2b from which the bolus velocity may be determined from a knowledge of the tagging and target slice separation. Here the resolution is limited by the slice thickness and, in the case of NMR microscopy where the smallest slice thickness is typically 100 μm and the longest observation time interval of order 100 ms, this method is limited to velocities of around 1 mm s^{-1} or greater. The stimulated echo method offers the usual advantage that a normal spin density image can be obtained by utilizing the spin echo which arises between the second and third pulses.

In each of the methods described above, the imaging plane coincides in orientation, if not in alignment, with the initial slice selection plane. A very direct visualization of motion is offered by imaging in a plane perpendicular to the tagging surface as demonstrated by Axel *et al.*[19] In their experiment the tag pulse is a selective 90° excitation with the slice plane normal to the flow direction. The image is obtained from a spin echo in which the refocusing 180° pulse is slice selective with its plane parallel to and including the flow spins as shown in Fig. 8.3. This second selective pulse has the effect of confining the image signal to a well-defined plane containing the moving spins of interest. Applying the phase and read gradients in this plane results in an image in which the spin motion is apparent as a transverse displacement with the flow resolution limited by the pixel dimension and the echo time, T_E.

The visualization of 'tagged-spin' motion in the imaging plane proposed by Axel *et al.* can be very effectively performed by exciting a grid of spins.[22,23] An ideal method of generating a high-density grid pattern of tags is afford by the use of a double DANTE sequence[23] in which two successive orthogonal selection gradients are used. Here the τ spacing of the DANTE pulses are chosen so that multiple sidebands produce multiple slice excitation, with the slice plane spacing being set by the gradient strength and τ. The motion of spins over the time interval between excitation and subsequent imaging results in distortions to the grid lines and thus present a graphic demonstration of flow right across the imaging plane.

8.1.3 *Phase encoding*

The use of transverse magnetization phase shifts to measure molecular velocity was originally suggested by Carr and Purcell[24] and Hahn[25] while its potential applications in imaging have been suggested by a number of authors,[26-29] one important medical use being as a signature for blood flow.[30-32] The essense of the phase contrast method is shown in Fig. 8.4 in which a bipolar gradient is inserted between the slice selection and imaging segments so as to impose a phase shift in each pixel dependent on the net spin

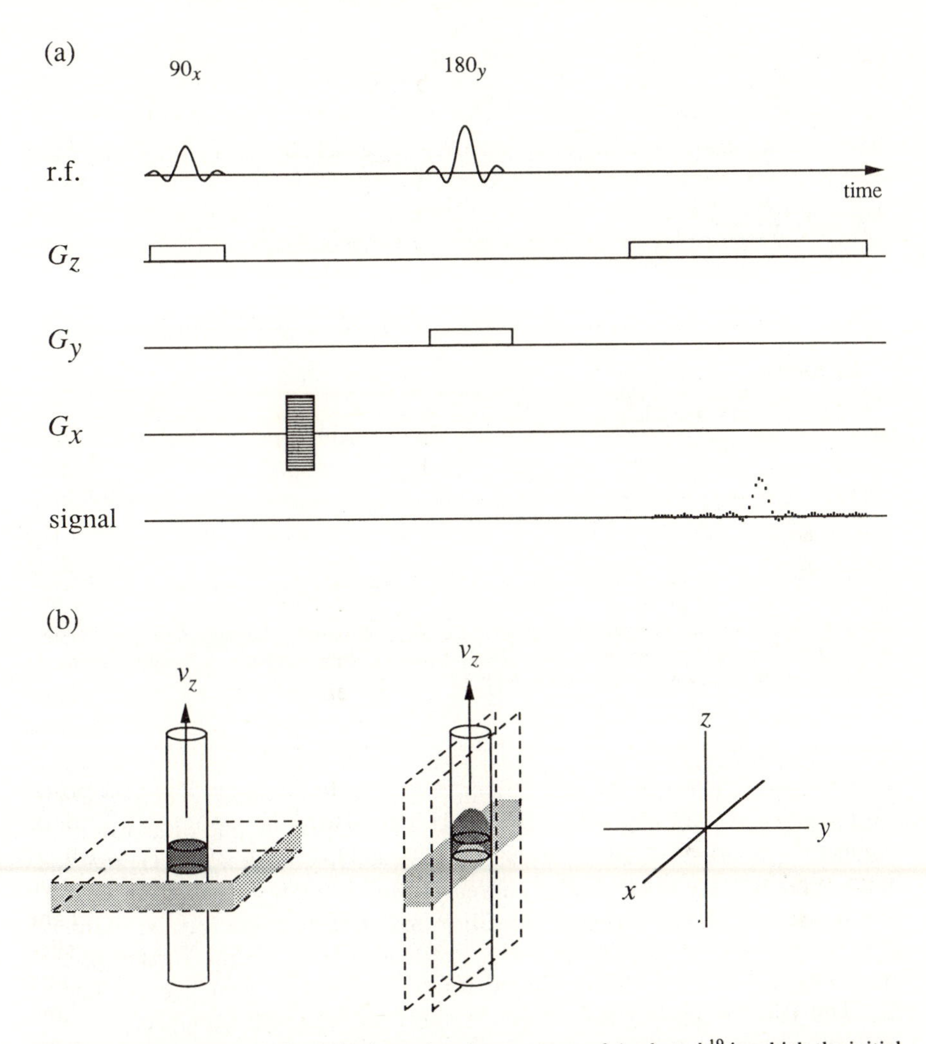

Fig. 8.3 (a) Orthogonal slice selection spin echo sequence of Axel *et al.*[19] in which the initial 90_x r.f. pulse is used to select a plane normal to the z-axis while the refocusing 180_y pulse is used to select the imaging plane normal to the z-axis while the refocusing 180_y pulse is used to select the imaging plane normal to the y-axis. As shown in (b), the orientation of the slice selection, phase-encoding and read gradients means that a strip of spins extending along the x-axis is imaged in the x-z plane. The delay, T_E, between the imposition of the original excitation plane and the acquisition of the echo results in displacement of moving spins in this strip (shaded more darkly) along the z-direction, thus giving a direct indication of the velocity component, v_z. Note that the phase twist associated with the initial slice selection pulse is refocused under the action of the read gradient.

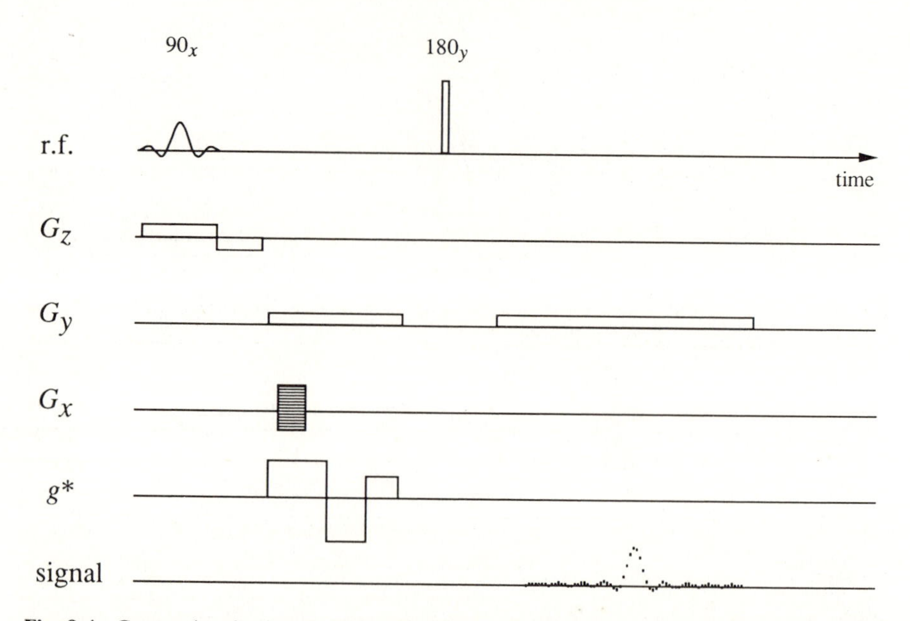

Fig. 8.4 Conventional spin echo FI imaging sequence preceded by an arbitrary $\mathbf{g}^*(t)$ phase-encoding gradient which may be applied along any axis. Where a phase shift proportional to velocity is to be imposed, the first time moment of $\mathbf{g}^*(t)$ must be non-zero.

displacement occuring over the time-scale of the velocity determining gradient pulses. As described in detail in Chapter 6 and illustrated in Fig. 6.14, the effective gradient used to encode for motion, $\mathbf{g}^*(t)$, may be made sensitive either to velocity, acceleration, or higher derivatives or displacement.[33,34] In each case the condition for an echo to be formed, namely $\int \mathbf{g}^*(t)\, \mathrm{d}t = 0$, must be obeyed if this gradient is not to cause further phase encoding due to the mean spin positions, a function which is normally reserved for the phase gradient of the standard imaging sequence.

In this section we will focus our attention on a velocity-encoding effective gradient with non-zero first moment, namely,

$$\mathbf{p} = \gamma \int_0^t t'\mathbf{g}^*(t')\, \mathrm{d}t'. \tag{8.2}$$

Such an effective gradient can be produced either by an opposite sign pair as shown or by a pair of identical pulses separated by a phase-inverting 180° r.f. pulse. If the mean velocity over the time interval t for spins labelled by pixel position $\mathbf{r}$ is $\mathbf{v}$ then the phase shift induced in that pixel element is $\exp[\mathrm{i}\mathbf{p}\cdot\mathbf{v}(\mathbf{r})]$. As indicated in Chapter 6, $\mathbf{p}$ is a useful quantity in describing sensitivity to average velocities. However, the time-scale over

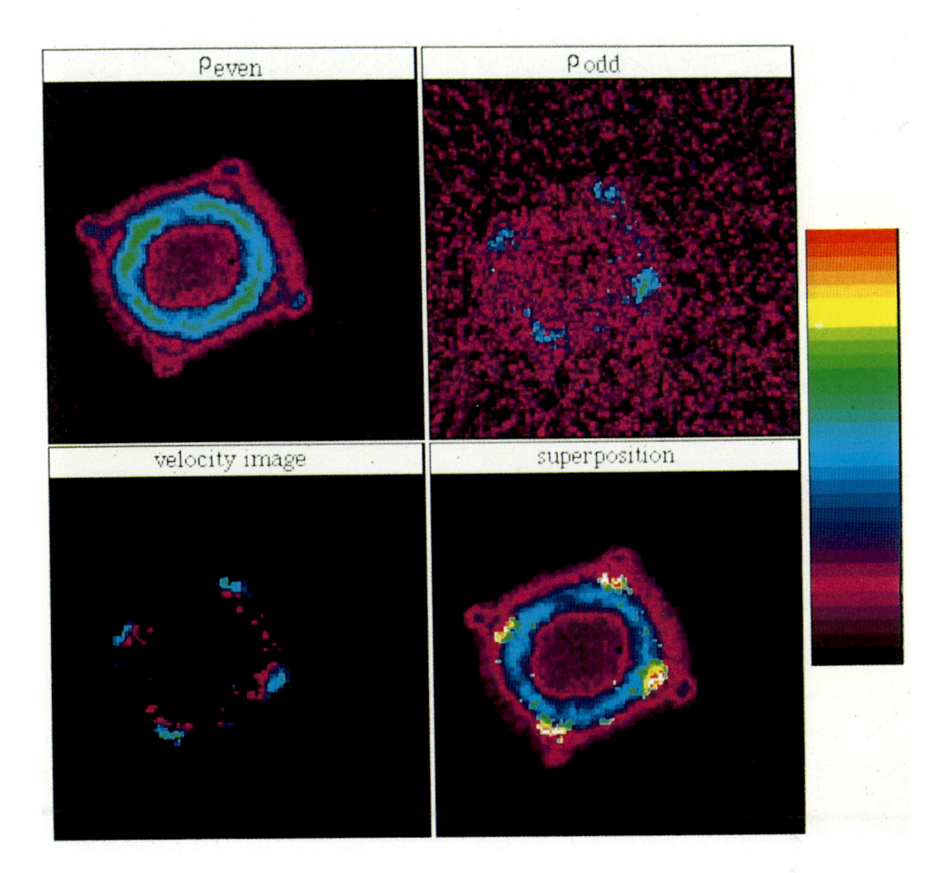

Plate 1 'Calibration image' (ρ_{even}) and velocity-encoded image (ρ_{odd}) for the protons in the plant stem of *Stachys sylvatica L. in vivo.* Also shown is the velocity image calculated from the ratio ρ_{odd}/ρ_{even} and the superposition of this image with the proton density map. The ρ_{odd} image was obtained over a period of 12 hours using a stimulated echo sequence in which $\Delta=160$ ms. The maximum velocity amplitude corresponds to 50 μm s^{-1}. This experiment was performed at 60 MHz with an in-plane pixel resolution of 75 μm. The slice thickness is 1 mm. (From Xia and Callaghan.[41])

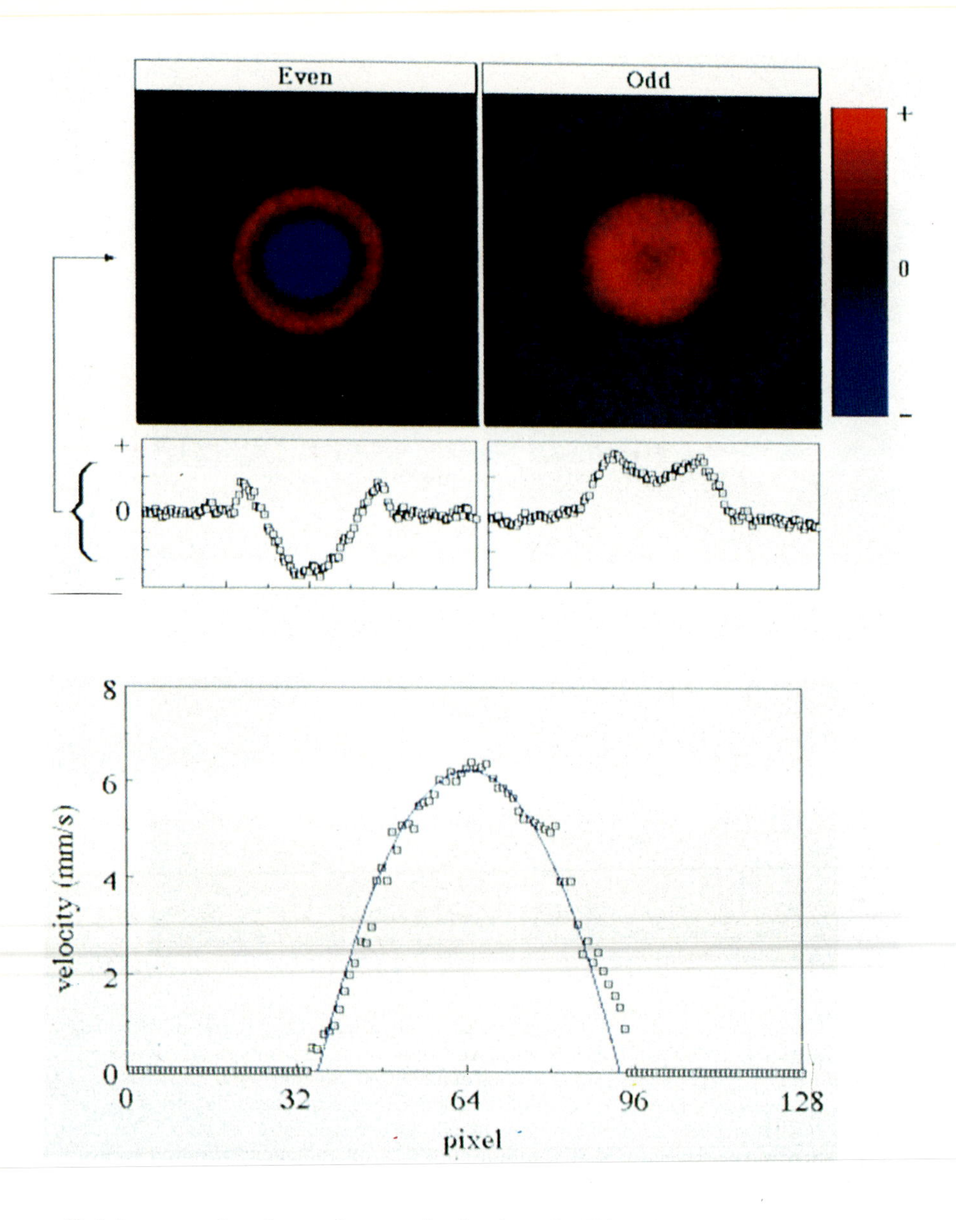

Plate 2 ρ_{even} and ρ_{odd} images for water flowing through a 700 μm internal diameter capillary with a maximum velocity of 6 mm s^{-1}. The even and odd images respectively exhibit cosine and sine modulation as the velocity increases from zero at the capillary edge. Analysis of the data in two quadrants of phase space yields the velocity map shown. A section through this map is consistent with the expected parabolic velocity profile.

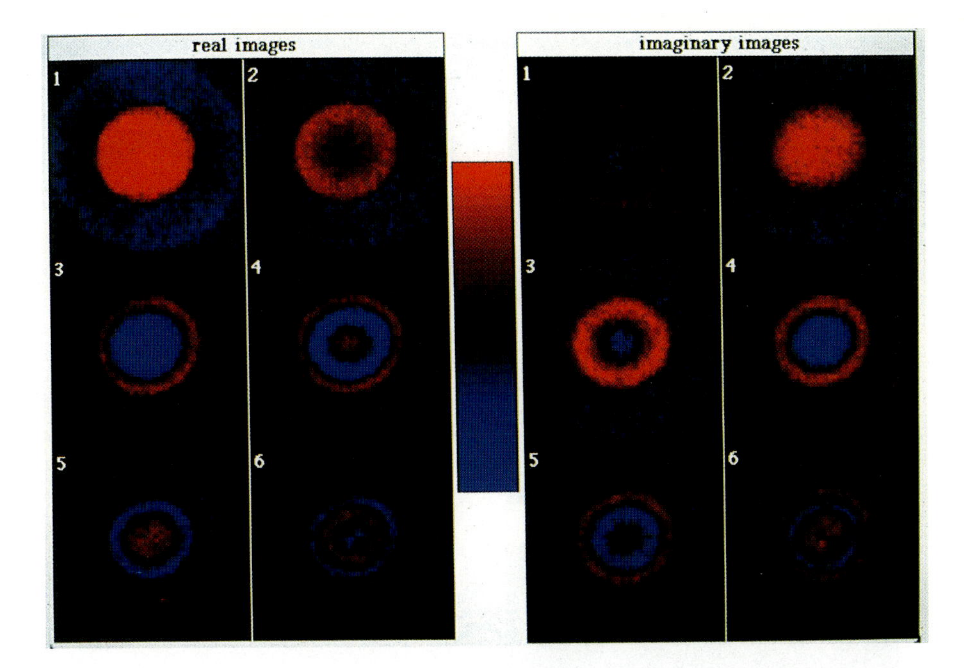

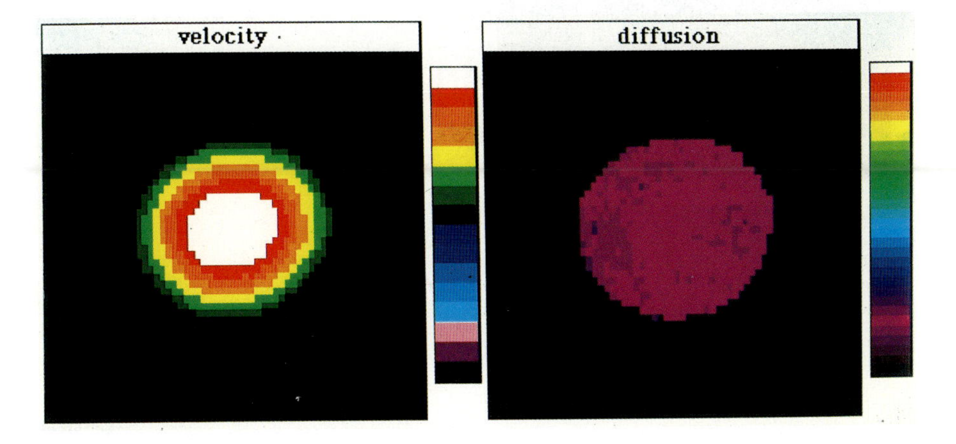

Plate 3 Real and imaginary images for water flowing through a 700 μm diameter capillary. Six images are shown from a set of eighteen in which the gradient, g was successively increased. Diffusion causes successive images to be attenuated while the velocity gradient causes alternating circular rings due to the gradient in phase shifts.
The velocity and diffusion maps shown in the lower part of the diagram were obtained using the Fourier method illustrated in Fig. 8.9. Note that the central black band in the signed velocity scale corresponds to zero velocity. (From Callaghan and Xia.[63])

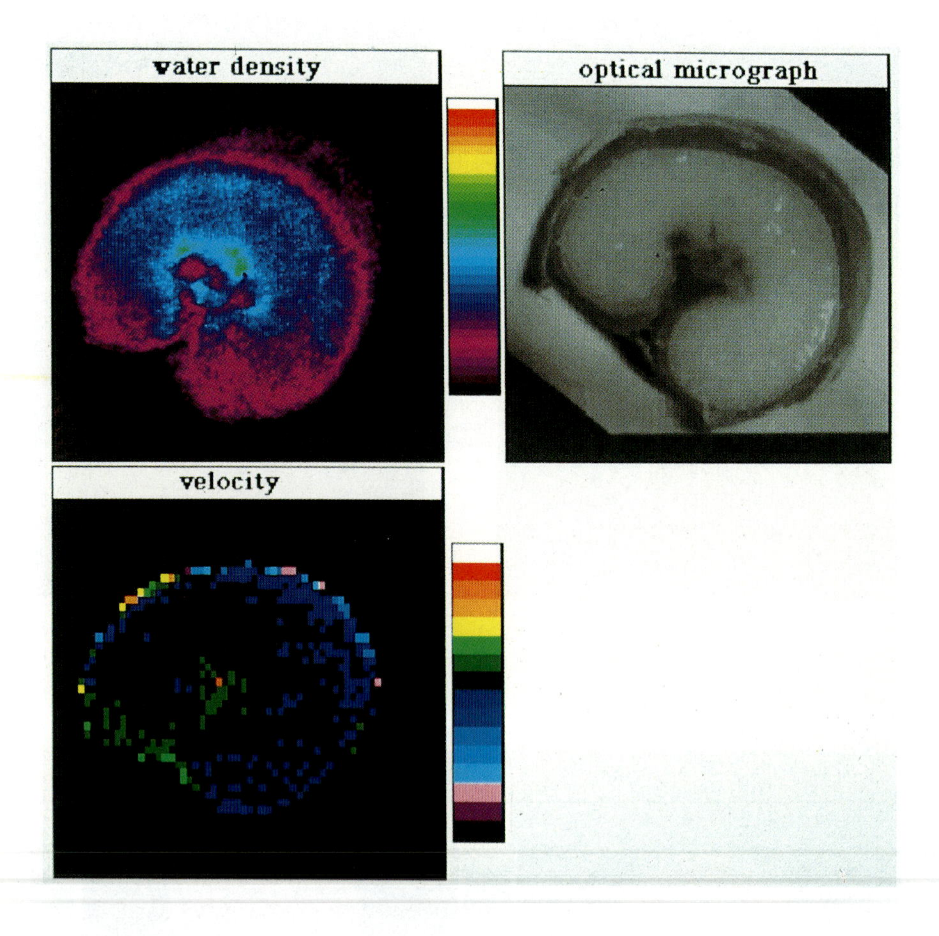

Plate 4 Water proton density and velocity maps for a 1 mm transverse slice in a wheat grain *in vivo* along with an optical micrograph from the same grain. A flow rate of order 50 $\mu m\ s^{-1}$ is apparent in the region of the vascular tissue (From Xia.[68])

which spin displacements are determined is not well defined except for the narrow-pulse P G S E experiment, where the **q**-space formalism is more instructive. The combination of **q**-space and **k**-space imaging is discussed in Section 8.3.

A number of strategies are available for measuring the phase shift $\mathbf{p}\cdot\mathbf{v}$. and so determining the local average velocity during the velocity encoding gradient. These approaches are illustrated in Fig. 8.5. In these sequences stimulated echoes are used so that the time interval over which velocity is to be measured may be extended if necessary, the spins suffering only T_1 relaxation during the period of magnetization storage along the z-axis. In each of these sequences the slice selection process is left to the end so that out-of-plane magnetization generated by the hard 90_x pulses can be effectively excluded from the final image.

The first sequence, shown in Fig. 8.5(a), involves the acquisition of a complex image $\rho(\mathbf{r})\exp[i\mathbf{p}\cdot\mathbf{v}(\mathbf{r})]$ so that a real and imaginary component is available for each pixel. These components are, respectively, $\rho(\mathbf{r})\cos[\mathbf{p}\cdot\mathbf{v}(\mathbf{r})]$ and $\rho(\mathbf{r})\sin[\mathbf{p}\cdot\mathbf{v}(\mathbf{r})]$. Ideally, therefore, both the spin density image and the phase image can be obtained by computing, respectively, the modulus and argument of the complex data.[27,33,35] Where both stationary and moving spins are present, with respective densities $\rho_S(\mathbf{r})$ and $\rho_M(\mathbf{r})$ the complex image will consist of a sum $\rho_S(\mathbf{r}) + \rho_M(\mathbf{r})\exp[i\mathbf{p}\cdot\mathbf{v}(\mathbf{r})]$. In many practical examples $\rho_S \gg \rho_M$ so that small phase anomalies in the stationary spin image can lead to apparent spin motion. These anomalies might include shifts arising from eddy currents or r.f. phase defects. By using a pulse sequence which suppresses the stationary spin signal and therefore produces an image of only the moving spins, the resulting velocity map is less likely to be sensitive to reconstruction errors. Methods used include subtraction of signals obtained with and without velocity encoding gradients,[36] with two different time-scale velocity encoding gradients,[37] and with different magnitude gradients.[34] One simple technique for nullifying the stationary signal component is to apply a final 'z-storage' r.f. pulse which returns any in-phase magnetization (i.e., $\rho_S(\mathbf{r}) + \rho_M(\mathbf{r})\cos[\mathbf{p}\cdot\mathbf{v}(\mathbf{r})]$) to the z-axis but leaves quadrature magnetization (i.e., $\rho_M(\mathbf{r})\sin[\mathbf{p}\cdot\mathbf{v}(\mathbf{r})]$) in the transverse plane and hence available for detection. A final 90_{-x} pulse is ideal for this purpose[38–40] as shown in Fig. 8.5(b). Bourgeois and Decorps[40] have used this approach, coupled with a stimulated echo, to measure motion over long time intervals, thus achieving a velocity resolution of order $100\,\mu\mathrm{m\,s^{-1}}$.

The use of a final 90_{-x} store pulse only produces static signal cancellation for a perfect r.f. pulse and in practice a small residual in-phase component will arise from ρ_S thus causing velocity signal errors. A higher order cancellation[41] of signal from stationary spins can be achieved by a second tier of phase cycling[40–43] in which the velocity encoding gradient is successively

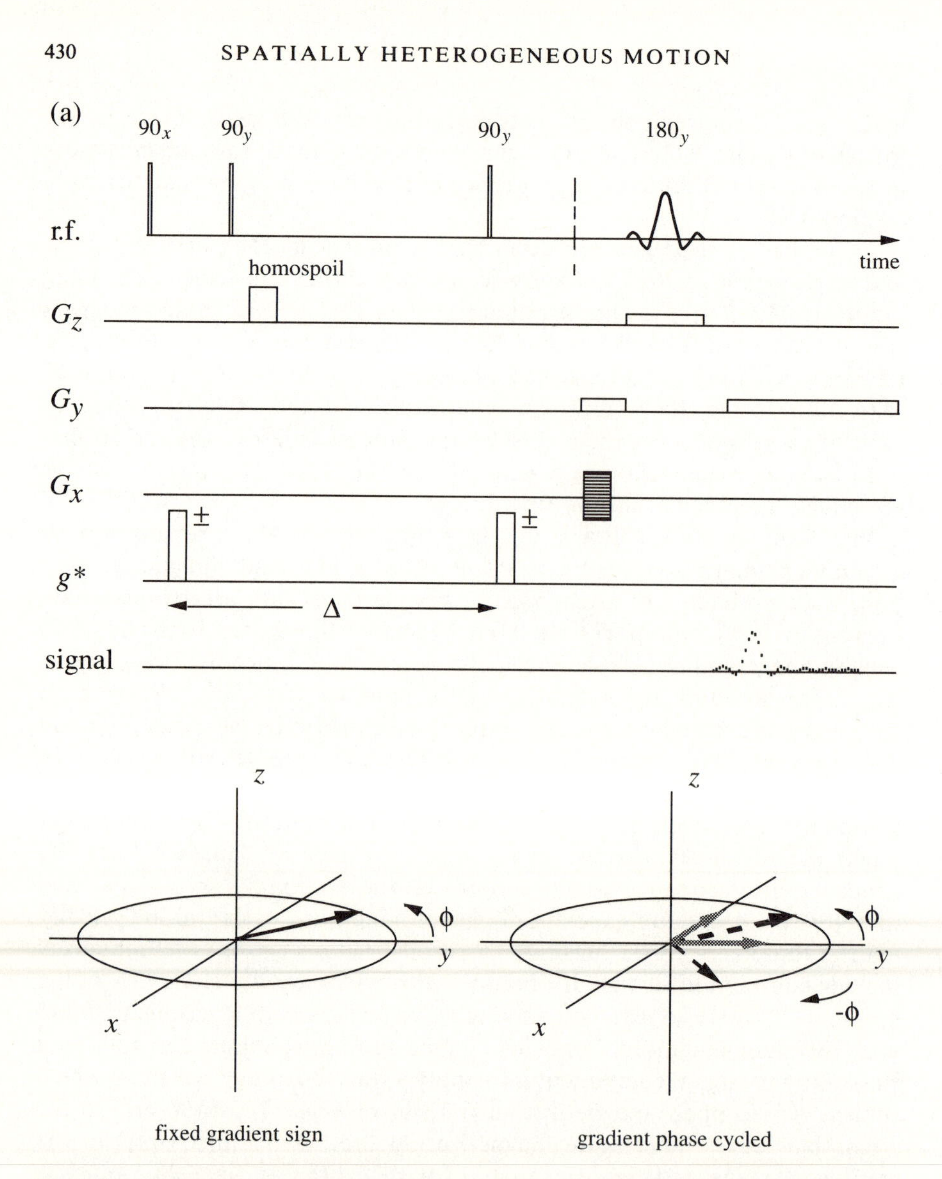

Fig. 8.5 (a) Stimulated echo imaging sequence in which a bipolar effective gradient is used to encode the image with a phase shift proportional to the distance moved along the direction of **g** over the time interval Δ. The phase shift ϕ is given by $\mathbf{p}\cdot\mathbf{v}$ and, in the sequence shown, results in both quadrature and in-phase image components proportional to $\sin\phi$ and $\cos\phi$ respectively. The lower left part of the diagram shows the magnetization at the refocusing point represented by the dashed line, while the effect of gradient phase alternation is shown in the lower right. (b) shows the equivalent experiment but where a 90_{-x} 'z-storage' pulse is inserted at the refocusing point. This has the effect of removing the in-phase magnetization component, leaving a quadrature component proportional to $\sin\phi$. Note that the experiments can also be performed using 90_x–90_x–90_x stimulated echoes and that it is usual to include a homospoil gradient pulse between the second and third r.f. pulses as shown.

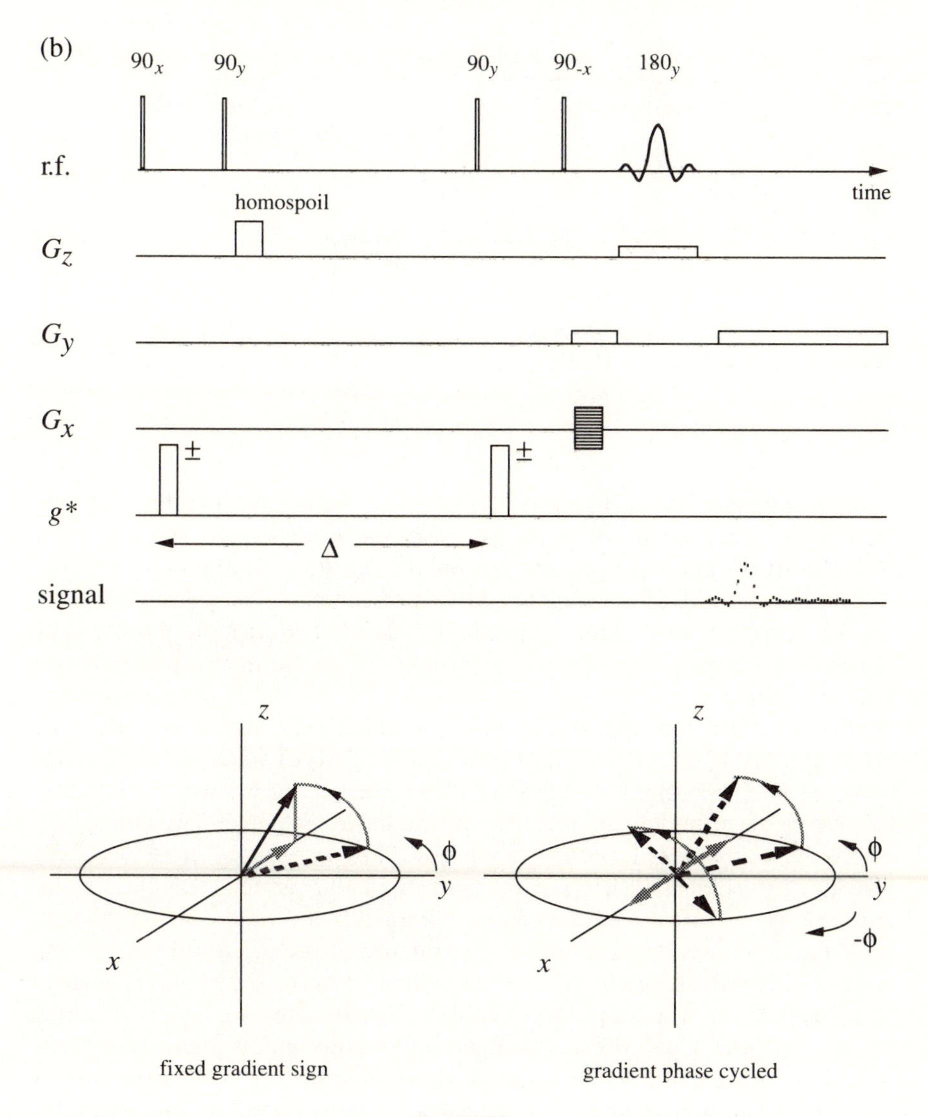

Fig. 8.5(b)

alternated in sign synchronously with the acquisition phase. In effect this corresponds to retaining only those signal components which alternate in sign with **p**, namely $\rho_M(\mathbf{r})\sin[\mathbf{p}\cdot\mathbf{v}(\mathbf{r})]$. Table 8.1 shows a CYCLOPS sequence of r.f., gradient pulse, and acquisition phases suitable for such velocity-sensitive imaging. Two images, ρ_{odd} and ρ_{even} may be reconstructed depending on the sign of the acquisition as shown in Table 8.1. These images are, respectively, $\rho_M(\mathbf{r})\sin[\mathbf{p}\cdot\mathbf{v}(\mathbf{r})]$ and $\rho_S + \rho_M(\mathbf{r})\cos[\mathbf{p}\cdot\mathbf{v}(\mathbf{r})]$.

Table 8.1. *Gradient and r.f. phase cycling for selective imaging of moving spins.*

r.f. phase	gradient phase	acquisition phase velocity-sensitive ρ_{odd}	calibration ρ_{even}
0	$+g$	0	0
0	$-g$	180	0
90	$+g$	90	90
90	$-g$	270	90
180	$+g$	180	180
180	$-g$	0	180
270	$+g$	270	270
270	$-g$	90	270

The gradient phase cycling method has been used to spectacular effect by Dumoulin *et al.*[42] to produce images of blood vessels in the human head with almost complete suppression of signal from static tissue. However, the use of such methods to make quantitative assessments of very slow flow rates requires excellent real signal cancellation. In this regard the double cancelling which arises from the use of a final 90_{-x} pulse in conjunction with gradient phase cycling produces an effective null.[40,41] A microscopic application of the sequence of Fig. 8.5(b) in which gradient phase cycling was also employed is shown in Plate 1 in which the flow of water in the vascular tissue of a living plant is apparent. The images shown are particularly interesting because they demonstrate pronounced xylem velocity in a region where the water density image has a quite low intensity. The experiment provides a nice identification of the active vascular tissue in the plant stem. Also shown in Plate 2 are flow images obtained from water moving through a 700 μm capillary. The sinusoidal dependence of the ρ_{odd} amplitude on the water velocity is apparent as v increases from zero at the capillary wall to a maximum at the capillary centre. Clearly the use of ρ_{odd} alone to calculate the local velocity will be ambiguous unless the phase shift, $\mathbf{p}\cdot\mathbf{v}$, is confined to the first quadrant. Simultaneous reconstruction of the cosine-modulated image, ρ_{even}, allows, in principle, a four-quadrant analysis provided that the moving spin density is dominant and ρ_S can be neglected.

It should be noted that any gradient wave-form whose first moment with time is non-zero will induce a velocity-dependent phase shift. In particular, the slice selection echo and read gradient echoes will have this effect and there are several medical MRI examples where the slice selection gradient is used to encode for velocity normal to the slice plane.[37,44-46] The sensitivity of the slice selection process to spin motion has been used to design a direct velocity-dependent excitation.[47] This method uses symmetric r.f.

and gradient wave-forms in which the r.f. envelope is proportional to the gradient and the net integral of both is zero. The excitation is slice selective, producing a null for stationary spins but a finite transverse magnetization for spins flowing along the gradient direction. The disadvantage of this approach is that it can only detect motion normal to the slice plane and then only for quite high velocities.

Where such effects are undesirable it is, however, possible to utilize slice selection gradient wave-forms for which both the first and second moments are zero [34] as shown in Fig. 8.6 and in the angiography application of Fig. 8.1. Generally, however, because the precursor read phase pulse is much shorter than the read duration, the residual phase shift is insignificant provided that the fluid moves much less than the total image dimensions over the read echo formation time. In the case of the selection gradient phase, anomalies are significant only for velocities of order the slice thickness divided by the slice excitation time. In both cases these effects are important only for rapid velocities (i.e. $\sim$ m s^{-1}). A quantitative discussion of slice selection gradient effects in dynamic microscopy is contained in Section 8.5.

The use of a bipolar effective gradient to impart velocity phase encoding, also results in phase spreading due to molecular self-diffusion, an effect which results in image attenuation. The flow-sensitive pulse sequences which deliver an image amplitude proportional to flow as $\rho_M \sin(\mathbf{p}\cdot\mathbf{v})$, must therefore take account of the additional attenuation due to self-diffusion. Ideally, a 'calibration' experiment should be performed in which the moving spin (ρ_M) density image is obtained but with the same diffusive attenuation as prevails in the velocity-sensitive sequence. By means of the even phase cycling, as shown in Table 8.1, the ρ_{even} image, $\rho_S + \rho_M \cos(\mathbf{p}\cdot\mathbf{v})$, results but there is no easy method for producing an image of ρ_M alone. Provided that ρ_M dominates the signal in a particular pixel, the true velocity, independent

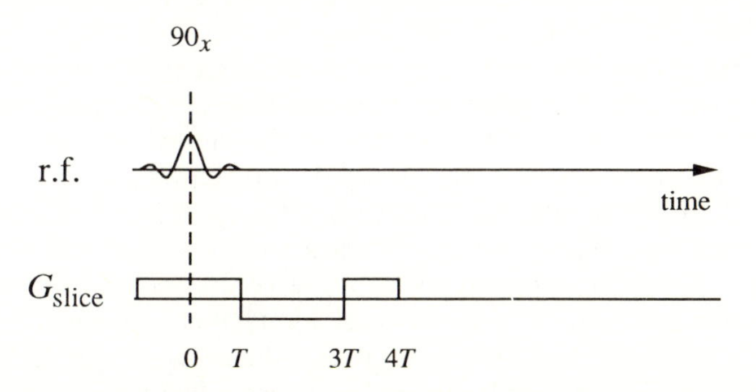

Fig. 8.6 Slice selection gradient wave-form in which both the zeroth and first time moments are zero. The phase twist associated with the selective sinc excitation is refocused but no phase shift results due to spin motion during the excitation process. Note that the moments are computed with the time origin at the sinc pulse centre.

of diffusive attenuation, can then be obtained by taking the arctangent of the ratio of the velocity-sensitive and calibration amplitudes for that pixel. Where stationary spins contribute to the signal in a pixel, such an analysis returns the average velocity.

Single-step phase encoding is unsuitable for those situations where the signal in one pixel arises from spins exhibiting some velocity distribution, for example, when the pixel contains both moving and stationary spins. This would be the case in the imaging of plant vascular tissue where individual xylem or phloem vessels had a cross-section smaller than one pixel. Resolving such inhomogeneous behaviour is possible by obtaining the local spectrum of velocity using the dynamic NMR microscopy technique where $\overline{P}_s(\mathbf{R},\Delta)$ is measured across the image.

Diffusive contrast is obtained by computing the modulus image using a conventional flow encoding pulse sequence without gradient inversion. This method has been used in both medical MRI[48,49] and NMR microscopy,[50] and where perfusion effects lead to stationary random flow, the image attenuation as a function of encoding gradient amplitude g exhibits two-component decay.[49] The use of encoding gradients with zero first moment enables artefacts due to stationary random flow to be suppressed and purely diffusive attenuation to predominate. Such experiments are discussed in Section 8.4.

8.2 Periodic and slow motion

8.2.1 *Stroboscopic measurement*

In small-animal imaging a considerable amount of motion in the image is periodic and synchronous with the animal cardiac cycle with a period of order 100 ms to 1 s. Since the duration of a single read acquisition is typically 10 ms, stroboscopic gating of the imaging pulse sequence with the underlying periodic motion enables the **k**-space raster to be mapped out with different phase gradient steps being applied in different cycles. Stroboscopic NMR microscopy images have been obtained of the carotid arteries of rats[51] in which a sequence of images were obtained, at different parts of the cardiac cycle, by varying the trigger delay for the start of the imaging pulse sequence. These images have enabled a study of the artery cross-sectional diameter as a function of cardiac blood pressure to be conducted, revealing hitherto unobtainable data about the mechanical compliance *in vivo*.

Imaging of the rat carotid artery around the cardiac cycle falls into the category of motion in which $\rho(\mathbf{r})$ is periodically fluctuating. By contrast, imaging the periodically pulsatile blood flow falls into the second category of motion. Now $\rho(\mathbf{r})$ is constant but the underlying spin motion, $\mathbf{v}(\mathbf{r})$, fluctuates so that a velocity encoding pulse sequence, as in the phase shift methods discussed in Section 8.1.3, is appropriate. A useful analysis of the problem has been given by Katz *et al.*[52] who point out that the various harmonics of the motion may be revealed by using gradients of differing time moments. The simplest example of this approach is the formation of multiple echoes (n) in a CPMG train for which the effective gradients, $\mathbf{g}_n^*(t)$, all differ. We shall, of course, be concerned only with net spin displacements, $Z(\mathbf{r})$ parallel to $\mathbf{g}$, allowing that these displacements will depend upon location, $\mathbf{r}$, within the image plane. The spin velocity v and displacement along the phase-encoding gradient are naturally represented by Fourier series,

$$v(\mathbf{r}, t) = \tfrac{1}{2}A_0(\mathbf{r}) + \sum_{k=1}^{\infty} [A_k(\mathbf{r}) \cos k\omega t + B_k(\mathbf{r}) \sin k\omega t]$$

$$Z(\mathbf{r}, t) = \tfrac{1}{2}A_0(\mathbf{r})\mathrm{t} + \sum_{k=1}^{\infty} \left[\frac{A_k(\mathbf{r})}{k\omega} \sin k\omega t - \frac{B_k(\mathbf{r})}{k\omega} (\cos k\omega t - 1)\right] \qquad (8.3)$$

where ω is the fundamental frequency of the cycle. The degree to which sine and cosine terms are present in the series depends, of course, on the phase in the cycle at which the imaging sequence begins so that coherent stroboscopic triggering is an essential prerequisite of the method. The relative contributions of the A_k and B_k terms to the pixel labelled by $\mathbf{r}$ in the image formed from echo n can be obtained by calculating $\gamma \int_0^t g_n^*(t')Z(\mathbf{r}, t')\, \mathrm{d}t'$ and, by analysing a sufficient number of echoes, the various harmonic terms in the Fourier series can, in principle, be determined.

The use of gradient wave-forms with zero first moment has the effect of suppressing the phase shifts which result from flow, permitting, for example, pure spatial encoding. In the case of pulsatile flow such suppression is a little more difficult, although one appropriate zero first moment wave-form suggested by Constantinescu *et al.*[32] is an integral period sinc function which oscillates faster than the harmonics of the flow.

8.2.2 *Echo planar imaging and snapshot FLASH imaging*

Where irregular (*i.e.*, non-periodic) changes occur in either the spin density, $\rho(\mathbf{r})$, or the local molecular velocities, $\mathbf{v}(\mathbf{r})$, stroboscopic gating methods no longer apply and the only practical solution is afforded by using an image acquisition method sufficiently rapid that the entire k-space raster can be sampled with a duration which is short compared with these changes. The most rapid image acquisition methods are echo-planar imaging, followed closely by FLASH, for which the total raster times are typically 40 ms and 100 ms, respectively. While these methods may be suitable for monitoring changes occuring on the $>$ 100 ms time-scale, it must be remembered that the price paid for increased speed is the loss of signal-to-noise, and hence spatial resolution, which results from the necessary use of an increased acquisition bandwidth. The NMR microscopy applications of EPI and FLASH for observing slowly changing systems, are therefore likely to be extremely limited.

In the second category of motion some method of velocity encoding is essential if rapid-scan methods are to be used to follow slow changes. This could take the form of a standard bipolar velocity encoding gradient preceding the EPI raster, a method used in a medical MRI study[53] of blood flow in the human carotid artery and jugular vein where phase-contrasted images were obtained over a 50 ms interval in the cardiac cycle. In another variant of repetitive interrogation suitable for very slow motion, Norris *et al.*[20] have used a sequence of 16 250 ms snapshot FLASH images to follow the time evolution of flow of magnetization in a region previously disturbed with an inversion pulse. Signal from static spins is suppressed by obtaining difference images from experiments in which the inversion pulse is, respectively, selective and non-selective.

8.2.3 *The magnetization grating*

Saarinen and Johnson have pointed out that the formation of a spin or gradient echo from a bipolar effective gradient with zero time integral can be viewed as a process in which the phases of spins across the sample are wrapped in a helical twist and then subsequently unwrapped.[54] The local wrapping (or unwrapping) speed depends on the position of the spin in the magnetic field gradient so that the effect of local motions prior to the untwisting of the helix is to prevent a return of magnetization vectors to their initial alignment. This formalism provides a nice visualization of the pulsed gradient spin echo experiment and reveals many features in common with optical scattering theory.

An interesting variant of phase encoding is provided by considering the experiment where the magnetization helix is first twisted and the spins then imaged using a read gradient applied in the same direction as the wrapping gradient.[54] This process is illustrated in Fig. 8.7. The effect of this sequence is that the phase twist written into the magnetization is interrogated at a later time, revealing flow or diffusion of the spins through either a shift or attenuation of the magnetization helix. A very similar experiment exists in light scattering where an interference pattern produced from superposed laser beams is used to 'burn' a grating into a photochromic dye. A second beam is then scattered from this grating at a later time so that intervening molecular motion is detected as a relaxation in the intensity of the resulting interference pattern. This method, known alternatively as optical holographic relaxation or forced rayleigh scattering[55] is a powerful method of

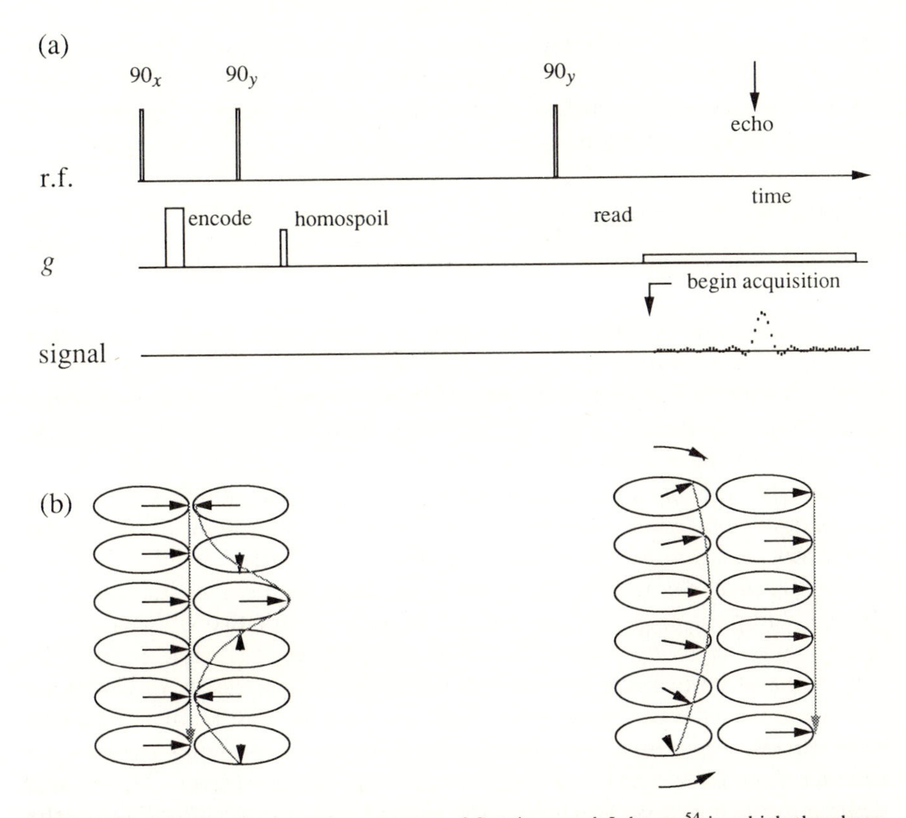

Fig. 8.7 (a) Stimulated echo pulse sequence of Saarinen and Johnson[54] in which the phase-encoding gradient pulse is not compensated prior to signal acquisition in the usual PGSE pair, but, instead, the signal is acquired in the presence of a read gradient (applied in the same direction). An echo will be produced when the time integral of the read gradient equals the area under the original phase-encoding pulse. It is important to note, however, that signal acquisition begins at the start of the read gradient so that a large first-order phase shift results, a consequence of the helical phase twist which is induced in the sample magnetization by the first gradient pulse. This twist and the subsequent refocusing process is illustrated in (b). Fourier transformation of the signal yields an oscillatory image, in which the positions of the maxima are dependent on spin motion over the time interval between the second and third r.f. pulses.

measuring very slow self-diffusion since the laser excitation may last for many hours.

By contrast, the magnetization grating will persist only for a time T_2, or, when a stimulated echo is used to store grating components along the z-axis as shown in Fig. 8.7, for as long as T_1. Given that the spatial resolution with which the grating can be determined using NMR microscopy is of order 10 μm, this approach is considerably less sensitive to motion than the PGSE method where residual phase shifts following helix unwrapping are measured, giving a spatial resolution for dynamic displacements of around 0.1 μm. However, the helical phase twist does provide an alternative means of writing a grid into the spin system and, as with the DANTE method discussed in Section 8.1.2, slow movement of spins over the time between the writing and reading periods of the pulse sequence can be observed as a distortion of the grid pattern.

8.3 Dynamic NMR microscopy

The use of a phase-encoding effective gradient $\mathbf{g}^*(t)$ with zero time integral provides a contrast in the image with phase shift proportional to $\mathbf{p}\cdot\mathbf{v}$, where $\mathbf{p}$ is the first moment of $\mathbf{g}^*(t)$. By obtaining the NMR signal as a function of $\mathbf{p}$, one acquires data in a space conjugate to the velocity signal,[26,29,56,57] a point originally observed by Moran[26] who proposed the use of three-dimensional bipolar gradient encoding to obtain the spin density $\Delta(\mathbf{r}, \mathbf{v})$ as a function of the six components of $\mathbf{r}$ and $\mathbf{v}$.

While this approach provides a useful description for samples comprising simple velocity fields, it is inadequate for the treatment of molecular ensembles. No matter what spatial resolution is attained in NMR microscopy, the pixel element will contain a very large number of molecules. In the very simplest case, the motion will comprise both Brownian and flow components. In many instances, however, restricted diffusion or irregular flow being examples, the motion will be further complicated. In Chapter 6 two formalisms were given. The first utilizes the spectrum of velocity correlations, for which the appropriate contrast in imaging would be a variable-frequency sinusoidally oscillating effective gradient. The second description is based on the use of the conditional probability, $P_s(\mathbf{r}|\mathbf{r}', \Delta)$, for a spin originating at $\mathbf{r}$ to move to $\mathbf{r}'$ in a time-scale Δ. The contrast sensitive to this parameter is the narrow-pulse PGSE experiment in which the data acquired in $\mathbf{q}$-space bears a Fourier relationship to P_s.

The combination of $\mathbf{k}$-space and $\mathbf{q}$-space imaging, known as dynamic NMR microscopy,[58] is a remarkably powerful method for determining local molecular ensemble behaviour in heterogeneous systems, providing, in principle, a resolution of order 10 μm for the static dimension, $\mathbf{r}$, and 0.1 μm for the dynamic dimension, $\mathbf{R} = \mathbf{r}' - \mathbf{r}$. Subsequent sections of this chapter deal with the method in detail.

8.3.1 *Combined k-space and q-space imaging*

The pulse sequence for dynamic microscopy, in which the signal evolves under the influence of both the PGSE gradient $\mathbf{g}$ and the resolution gradient $\mathbf{G}$, is shown in Fig. 8.8. These gradients lead to a modulation in both $\mathbf{k}$- and $\mathbf{q}$-space described by

$$S(\mathbf{k}, \mathbf{q}) = \int \rho(\mathbf{r}) \exp[\mathrm{i}2\pi\mathbf{k}\cdot\mathbf{r}] \int P_s(\mathbf{r}|\mathbf{r}', \Delta) \exp[\mathrm{i}2\pi\mathbf{q}\cdot(\mathbf{r}' - \mathbf{r})]\, \mathrm{d}\mathbf{r}'\mathrm{d}\mathbf{r}. \tag{8.4}$$

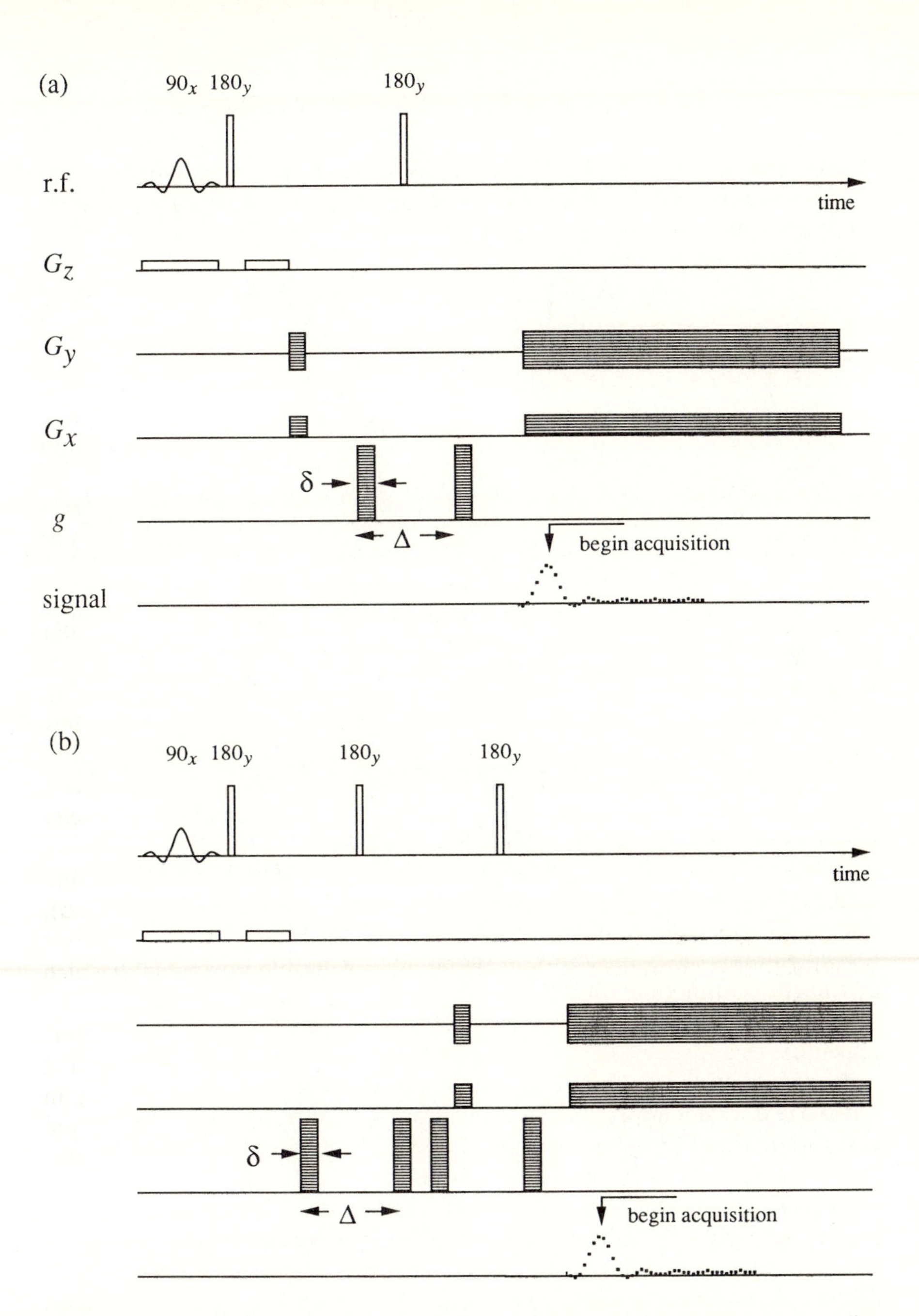

Fig. 8.8 (a) Dynamic imaging pulse sequence incorporating both **k**- and **q**-space encoding. This particular PR imaging sequence, used in the experiments described in this chapter, employs a PGSE contrast period with, in this case, a spin echo. Long Δ time interval experiments can be performed using a stimulated echo. Sampling of the echo during the resolution period provides a probe of **k**-space while the PGSE phase encoding provides a probe of **q**-space. (b) A double PGSE variant in which phase shifts due to coherent motion are refocused. This sequence can be used to measure self-diffusion in the presence of a velocity shear.

Despite the fact that both P_s and ρ may depend on $\mathbf{r}$, the effect of the PGSE sequence is quite separable, resulting in a contrast, $E_\Delta(\mathbf{q}, \mathbf{r})$, to each pixel, $\mathbf{r}$, in the static image defined by

$$S(\mathbf{k}, \mathbf{q}) = \int \rho(\mathbf{r}) E_\Delta(\mathbf{q}, \mathbf{r}) \exp[\mathrm{i}2\pi\mathbf{k}\cdot\mathbf{r}]\, \mathrm{d}\mathbf{r} \tag{8.5}$$

where

$$E_\Delta(\mathbf{q}, \mathbf{r}) = \int P_s(\mathbf{r}|\mathbf{r}', \Delta) \exp[\mathrm{i}2\pi\mathbf{q}\cdot(\mathbf{r}' - \mathbf{r})]\, \mathrm{d}\mathbf{r}'. \tag{8.6}$$

Eqn (8.5) states that reconstruction in $\mathbf{k}$-space will now yield $\rho(\mathbf{r})E_\Delta(\mathbf{q}, \mathbf{r})$. Of course, the $\mathbf{q}$-Fourier transform of the local conditional probability represented by eqn (8.6) is really an idealization in which an infinitesimal volume, $\mathrm{d}\mathbf{r}$, is assumed. In practice, this volume element corresponds to the imaging voxel and is finite, causing an implicit averaging over the ensemble of molecules contained in that voxel. It is, therefore, much more accurate to think of $E_\Delta(\mathbf{q}, \mathbf{r})$ as the Fourier transform of the local dynamic profile, namely

$$E_\Delta(\mathbf{q}, \mathbf{r}) = \int \overline{P}_s(\mathbf{R}, \Delta) \exp[\mathrm{i}2\pi\mathbf{q}\cdot\mathbf{R}]\, \mathrm{d}\mathbf{R} \tag{8.7}$$

where we allow that $\overline{P}_s(\mathbf{R},\Delta)$ is implicitly a function of the pixel coordinate, $\mathbf{r}$. Eqn (8.7) is formally very similar to eqn (6.26), except that in the latter case $E_\Delta(\mathbf{q})$ refers to a normalized average over the entire sample while $E_\Delta(\mathbf{q}, \mathbf{r})$ refers to the normalized echo contribution from the pixel at $\mathbf{r}$. As in Chapter 6 we shall omit the Δ subscript, at the same time noting that E is implicitly dependent on this time interval.

In using the PGSE sequence as a contrast in imaging it is important to remember that $E(\mathbf{q}, \mathbf{r})$ is inherently complex, reflecting the effects of phase shifts as well as phase spreading. This means that the usual $S^*(\mathbf{k}) = S(-\mathbf{k})$ symmetry does not apply; full four-quadrant $\mathbf{k}$-space sampling must be used and both real and imaginary images of $\rho(\mathbf{r})E(\mathbf{q}, \mathbf{r})$ should be reconstructed. Because $\overline{P}_s(\mathbf{R}, \Delta)$ represents a normalized probability, $E(0, \mathbf{r})$ is unity. This, of course, means that the conventional (albeit relaxation-weighted) spin density map will be obtained when the PGSE gradient is zero but it also has the important consequence that $E(\mathbf{q}, \mathbf{r})$ may be obtained independently of $\rho(\mathbf{r})$ by normalizing to the zero $\mathbf{q}$ image.

In most cases it will be sufficient to deal with a single dimension in $\mathbf{q}$-space so that the overall imaging process is four-dimensional. In principle, the method can be easily generalized to two or three dimensions in $\mathbf{q}$ although this naturally leads to very long experiments because of the low sensitivity associated with the NMR microscopy. As with the phase-encoding methods discussed in Section 8.1.3, the direction in which we measure spin translation is, of course, determined by the direction in which we choose to apply $\mathbf{g}$. Following the notation developed in Chapter 6, the symbol q is used to represent the $\mathbf{q}$-vector amplitude while dynamic dimension conjugate to q is denoted Z, where $Z = z' - z$.

8.3.2 *Digital computation of velocity and diffusion*

One commonly encountered ensemble motion comprises a combination of mean flow and molecular self-diffusion. The time dependence of $\overline{P}_s$ for such behaviour is shown in Fig. 6.2 so that a velocity and diffusion map for each pixel in an image can be obtained by computing the width and offset of P_s in Z-space. In practice, the dynamic profiles are computed by stepping the PGSE gradient in n_D steps to some maximum value g_m following which a digital Fourier transformation in $\mathbf{q}$-space can be performed for each pixel.[59,60] This process is illustrated in Fig. 8.9.

The method is therefore akin to a multislice experiment where successive slices are obtained in $\mathbf{q}$-space rather than real space. Generally the total number of '$\mathbf{q}$-slices' will be limited by the total available imaging time (and sometimes by the restraints of data storage!). However, a sufficient number of data points for accurate Fourier transformation and analysis can be obtained by zero-filling the data, provided that the signal has sufficiently attenuated by the n_Dth slice.

Following the concepts outlined in Section 1.2.4, and replacing the fictitious time interval T by the $\mathbf{q}$-space interval $(1/2\pi)\gamma\delta(g_m/n_D)$, the separate oscillatory and damping terms of eqn (6.34a) may be written, for the nth $\mathbf{q}$-slice as

$$E(n, \mathbf{r}) = \exp[\mathrm{i}(\gamma\delta v\Delta)(g_m/n_D)\, n] \exp[-\gamma^2\delta^2(g_m/n_D)^2 Dn^2\Delta]. \tag{8.8}$$

The value of $\mathbf{r}$ is determined by the pixel location in which E is computed. By implication, v and D will depend on the chosen pixel and are therefore functions of $\mathbf{r}$. The dynamic displacement profile arising from the product shown in eqn (8.8) is the convolution

$$P_s(k/N, \Delta) = (\pi n_D^2/\gamma^2\delta^2 g_m^2 D\Delta)^{1/2} \exp[-\pi^2 k^2 n_D^2/\gamma^2\delta^2 g_m^2 N^2 D\Delta]$$
$$\otimes\, \delta(k/N - \gamma\delta v\Delta g_m/2\pi n_D). \tag{8.9}$$

The peak centre occurs at the digital value

$$k_v = N\gamma\delta v\Delta g_m/2\pi n_D \tag{8.10}$$

and so the value of the mean molecular velocity in the pixel corresponding to the profile is

$$v = 2\pi n_D k_v/N\gamma\delta\Delta g_m. \tag{8.11}$$

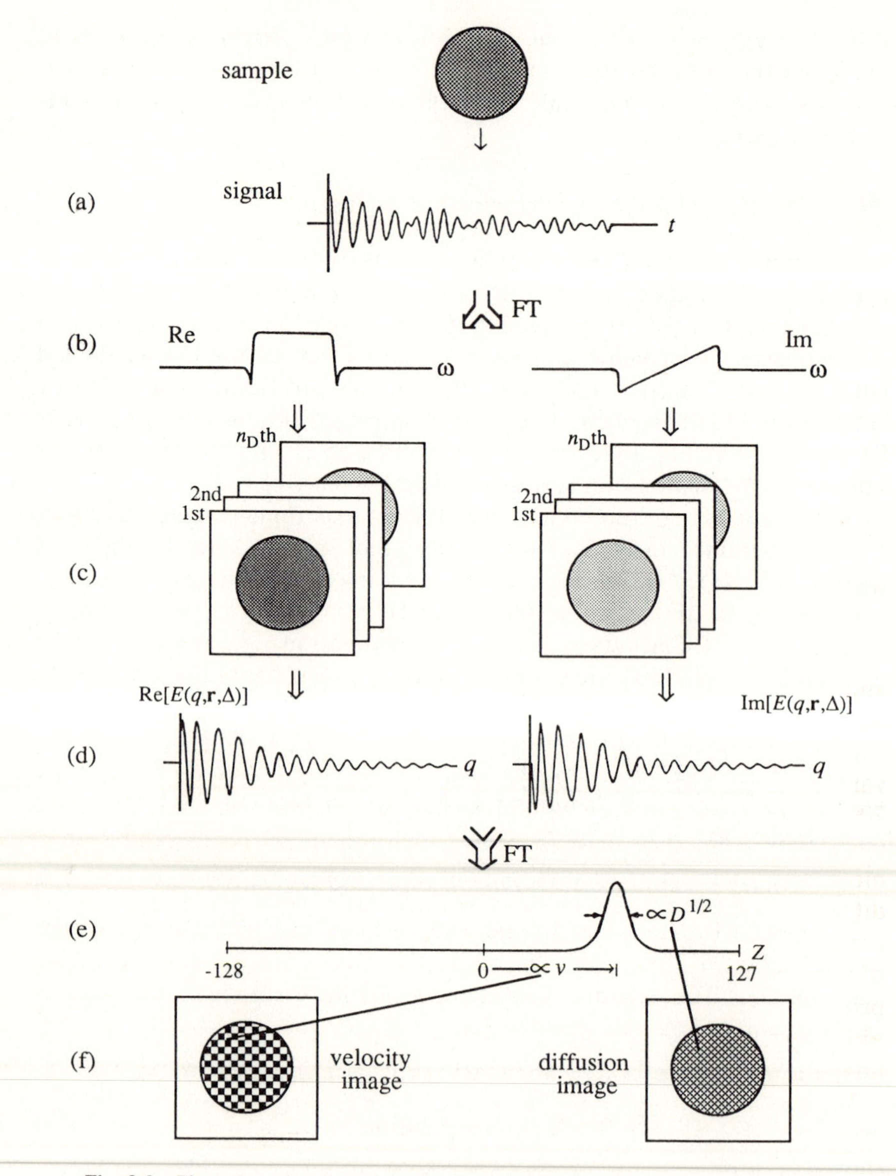

Fig. 8.9 Flow chart for processing in dynamic NMR microscopy (from Xia.[59])

The full-width-half-maximum (FWHM) of P_s in digital units is given by

$$k_{\mathrm{FWHM}} = (2/\pi)\{\ln(2)\}^{\frac{1}{2}} N\gamma\delta(g_{\mathrm{m}}/n_{\mathrm{D}})\,(D\Delta)^{\frac{1}{2}} \tag{8.12}$$

and so the value of the mean molecular self-diffusion coefficient in the pixel corresponding to this profile is

$$D = (n_D k_{FWHM})^2 / [\{4\ln(2)/\pi^2\}\gamma^2\delta^2 g_m^2 N^2 \Delta] \tag{8.13}$$

or

$$D = 3.56(n_D k_{FWHM})^2 / (\gamma^2\delta^2 g_m^2 N^2 \Delta). \tag{8.14}$$

The location of peak centre position, k_v, and determination of peak FWHM, k_{FWHM}, can be achieved by a simple computer algorithm.[59]

8.3.3 *Applications in the study of diffusion and flow*

A good test of the dynamic imaging method is provided by obtaining velocity and diffusion maps for laminar capillary flow in a cylindrical tube, a behaviour previously studied in large-diameter tubes using both integral PGSE NMR[61] and phase contrast imaging.[62] The classical Poiseuille velocity profile is given by

$$v(r) = v_{max}(1 - r^2/r_0^2) \tag{8.15}$$

where

$$v_{max} = \Delta P r_0^2 / 4\eta l \tag{8.16}$$

and ΔP is the pressure decrement after the length l of the tube, η is the dynamic viscosity of the water, and r_0 is the radius of the tube. Plate 3 shows a succession of real and imaginary images[63] obtained at increasing values of the PGSE gradient, g in the case of capillary flow of water in a 700 μm diameter capillary where the centre velocity was 6 mm s^{-1}. These images represent slices in $\mathbf{q}$-space in which individual pixels exhibit phase modulation due to molecular velocity and phase spreading due to self-diffusion. A gradation of phase shifts is apparent from zero at the edges to a maximum at the centre, these 'phase rings' being caused by the dependence of water velocity on radius. As g is increased in magnitude, the rings become progressively more compact as the phase changes, and hence the number of phase cycles at the centre of the capillary increases. The gradual decay in image amplitude as q increases is due to diffusive dephasing.

While only six slices are shown here, a total of $n_D = 18$ slices was acquired, the data being subsequently zero-filled to 256 points. Fig. 8.10 shows a typical pixel dynamic displacement profile, $\overline{P}_s(Z)$, along with the Stejskal–Tanner plot obtained from the modulus images. The corresponding velocity and diffusion maps are displayed as stackplots in Fig. 8.11. The agreement of the velocity variation with the parabolic Poiseuille profile is excellent, and it is clear that the method can yield precise velocity maps at a spatial resolution of a few tens of microns right across the capillary. In this respect the method has a considerable advantage over laser Doppler anemometry which suffers from flare effects close to walls surrounding the

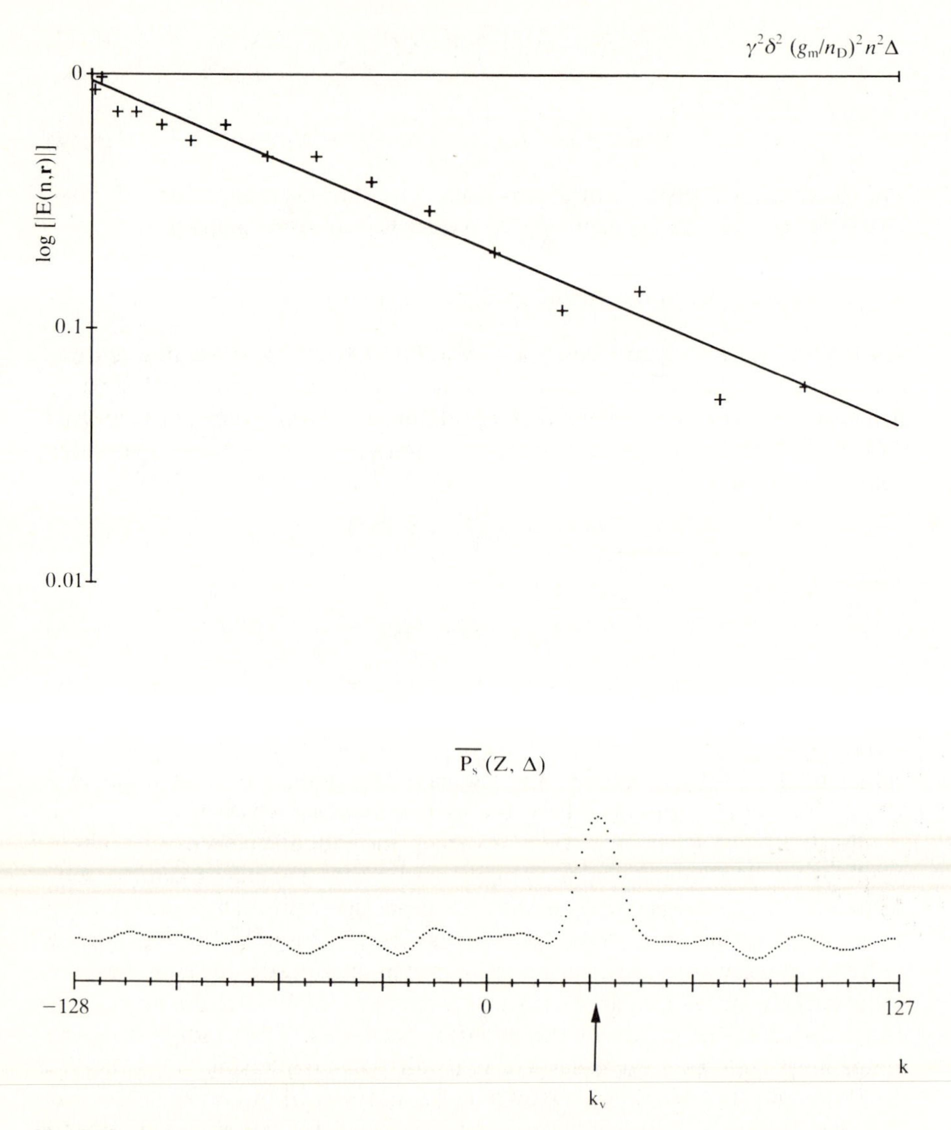

Fig. 8.10 Stejskal–Tanner plot for a single pixel along with the corresponding dynamic displacement profile, $\overline{P_s}$ (Z, Δ) taken from the data shown in Plate 3.

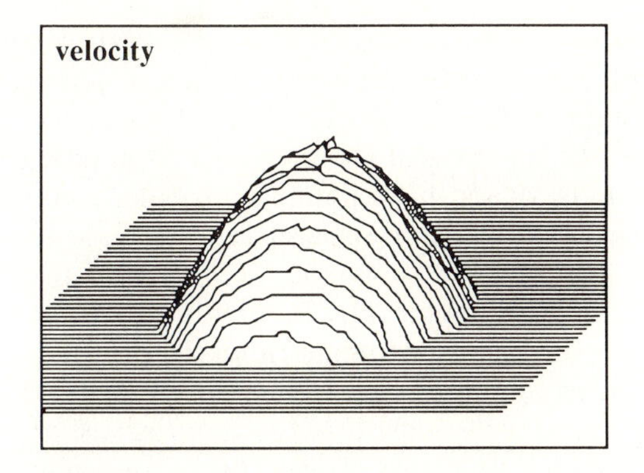

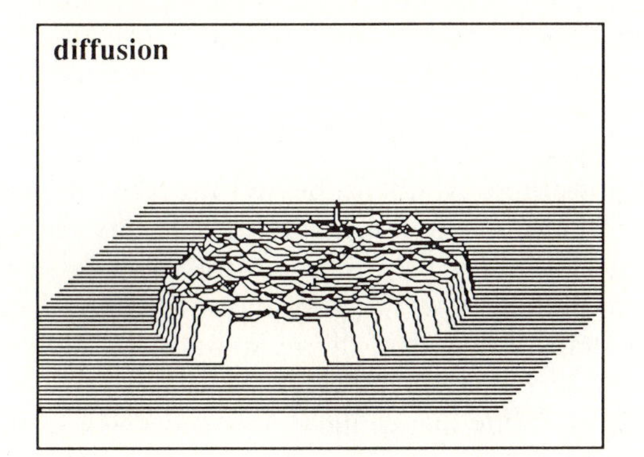

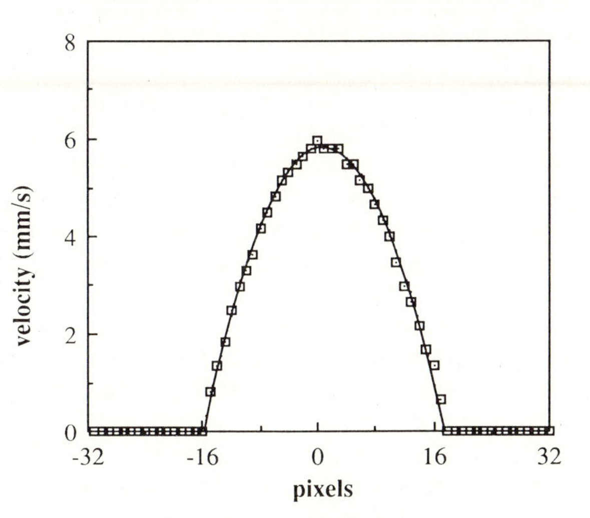

Fig. 8.11 Stackplot profiles of the velocity and diffusion maps obtained from data shown in Plate 3 along with a slice profile through the velocity map, illustrating the parabolic Poiseuille velocity distribution for laminar flow in a cylindrical capillary.[63]

fluid. An additional bonus is the self-diffusion map, which, in this case, is apparently uniform across the capillary. While the value of the self-diffusion coefficient obtained agrees with the known value for stationary water, when the flow rate is increased slightly artefacts appear due to flow disturbance. The imaging of self-diffusion in such experiments is best performed using a pulse sequence which is insensitive to flow, as discussed in the next section.

An application of dynamic NMR microscopy in the measurement of velocity profiles in non-Newtonian fluids is shown in Fig. 8.12, where the data has been obtained from a series of increasingly concentrated poly(ethylene oxide) solutions.[64] At concentrations c^* sufficiently great that the random coil molecules are entangled, the solution exhibits shear-thinning properties in which the viscosity, instead of being shear-rate independent, varies with shear rate, $\partial v/\partial r$, according to

$$\eta = K(\partial v/\partial r)^{n-1} \tag{8.17}$$

where n is the power law exponent, being 1 for Newtonian fluids and <1 for shear-thinning fluids. The polymer solution velocity profiles, normalized to a common central velocity, exhibit a transition from $n = 1$ to $n = 0.4$ as the transition from dilute ($c < c^*$) to semi-dilute ($c > c^*$) is made. The velocity and diffusion maps for a semi-dilute solution of polylethylene oxide) in water are shown in Fig. 8.12.

The imaging of fluids in capillary flow has some curious features. For example, although the selected slice moves only a short distance within the r.f. coil during the period of the spin echo, it can move a great deal further in during the repetition time, T_R, which is normally allowed for spin–lattice recovery. This means that successive acquisitions are replenished with fresh fluid whose spins have the full equilibrium magnetization, an effect which permits T_R to be considerably shorter than T_1. In capillary flow, however, this replenishment does not occur for the slowly moving fluid elements close to the wall, Dynamic NMR microscopy flow imaging is especially useful in testing models for liquid motion in complex geometries. Fig. 8.13 shows a sequence of flow images obtained in a 6.1 mm inner diameter capillary containing a 1.1 mm diameter hollow fibre made from microporous polysulfone.[65,66] The water flow enters through the fibre but gradually leaks out to the surrounding cellular space, contributing to an axial flow in the outer region which grows larger with increasing axial displacement. The axial flow contours obtained by the method agree well with the theoretical 'Starling flow[67]' fields superposed in Fig. 8.13 so that the potential gain in image signal-to-noise is greatest near the centre of the tube. In cylindrically symmetric flow this is exactly the region where such an improvement is most appreciated since it contains the least number of pixels per unit radius. Another curious aspect of behaviour concerns the complex nature of the

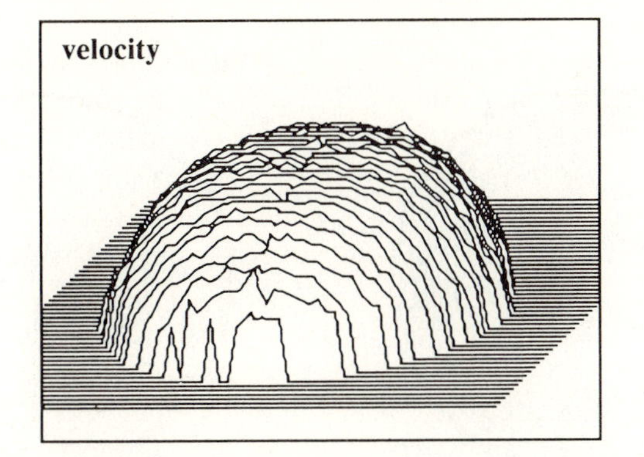

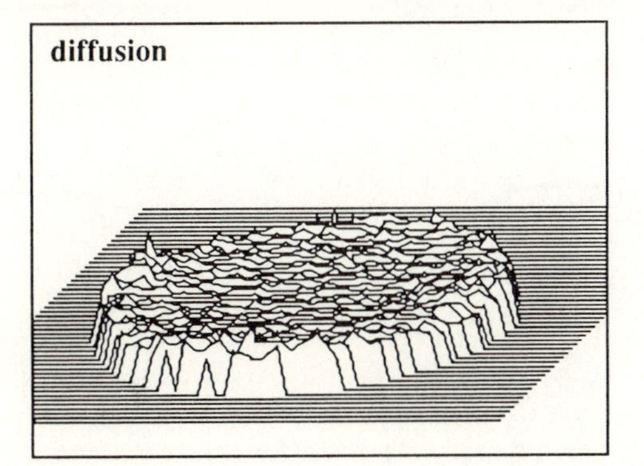

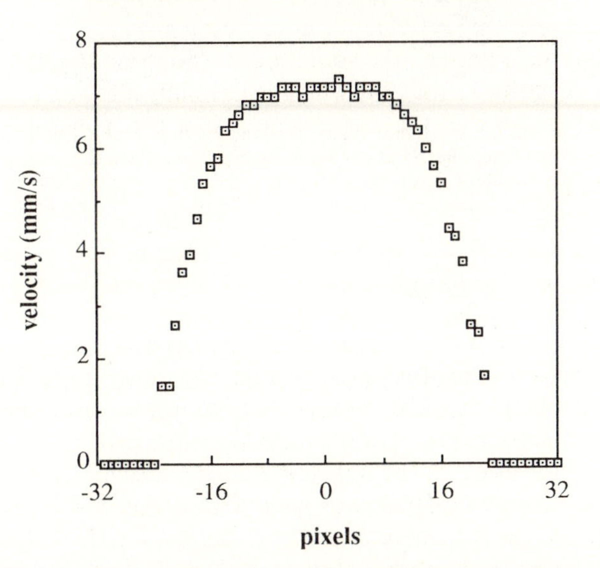

Fig. 8.12 As Fig. 8.11 but for a 5% solution of high molar mass poly(ethylene oxide) in water. Non-Newtonian flow is apparent and the velocity profile follows a power law behaviour characteristic of shear thinning (from Xia and Callaghan.[64])

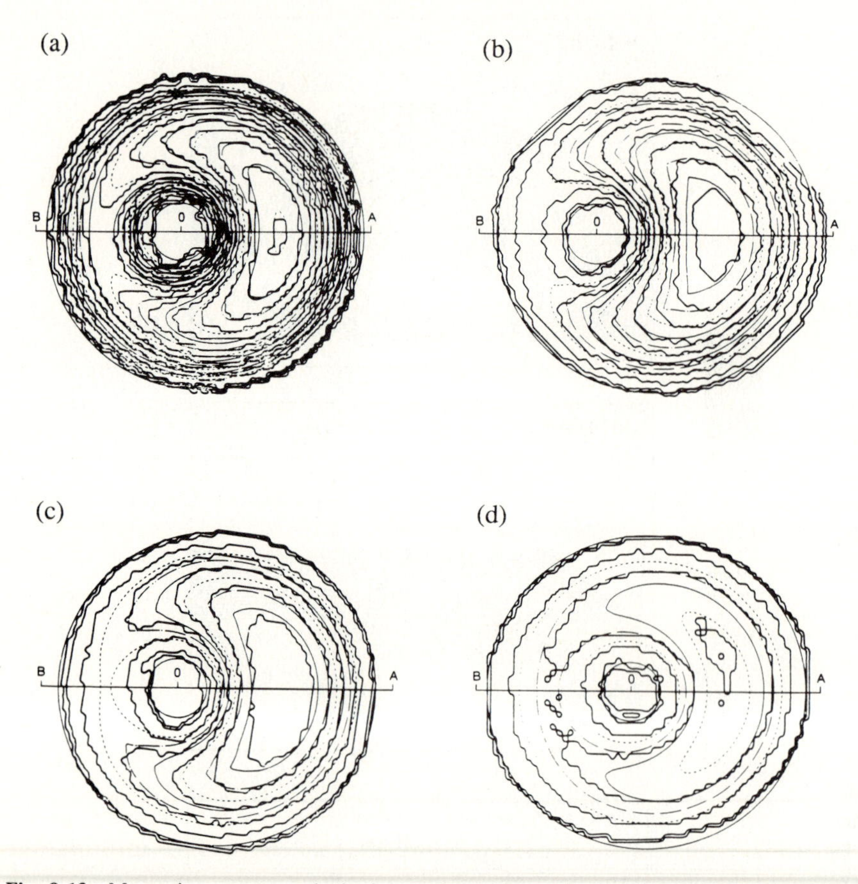

Fig. 8.13 Magnetic resonance velocity images for Starling flow in and around a single hollow fibre module at various fractional axial distances from the entry point. The measured velocity contours in the extra-capillary space are overlaid on the simulated contour plots. (a) to (d) represent decreasing axial distances. Note that the capillary has a variable off-centre location. (From Heath *et al.*[66])

image modulation when PGSE **q**-gradients are applied. This causes the one-dimensional projection profiles to be asymmetric under **k**-gradient reversal, a consequence of the fact that $S(\mathbf{k}) \neq S^*(-\mathbf{k})$ where spatially dependent phase shifts are present in a non-uniform velocity field.

An *in vivo* application of the dynamic imaging method is shown in Plate 4, in which the flow and diffusion of water in a wheat grain are imaged,[68,69] motion being measured transverse to the 1 mm slice. The flow apparent in the vascular tissue of the grain is directed outwards from the stem and the very small value of the velocity measured, around 50 μm s^{-1}, is close to the resolution limit of the method given a water local self-diffusion coefficient of around one half that of free water. A stimulated echo with $\Delta = 300$ ms, was used in order to enhance the phase shift to phase spreading ratio.[63]

8.3.4 *More complex motion*

By making available the average propagator, $\bar{P}_s (Z, \Delta)$, in each pixel of the image, the dynamic imaging method provides sufficiently valuable information to justify the need for multiple **q**-slice images. A good example of the usefulness of this approach is provided by the imaging of restricted diffusion, in which the P_s can be analysed to provide insight into the molecular ensemble behaviour and, following the methods outlined in Chapter 7, the Δ dependence of P_s can be used to reveal the spacings and connectivities of internal boundaries. In a study of restricted diffusion in onions, Merboldt *et al.*[70] have varied both Δ and the magnitude of the PGSE gradient (applied normal to the onion layers) to obtain a set of time-scale-dependent diffusion maps.

Another example in which measurement of the dynamic displacement profile, P_s, proves especially useful concerns the problem of flow measurement in vessels which are too small to be resolved in a single pixel. Clearly the signal arising from such a pixel will comprise components from spins moving at different velocities. Whereas the single-gradient phase contrast methods discussed in Section 8.1.3 will return only a velocity average for such a pixel, dynamic imaging, on the other hand, returns the spectrum of motion, yielding, in the example of admixed stationary and moving spins, a bimodal profile for $\bar{P}_s$. Provided, however, that simple flow information is all that is required, the gradient/phase cycled velocity encoding sequences illustrated in Fig. 8.5(b) and 8.5(c) are efficient and highly sensitive. In biological applications, where the additional experimental time for dynamic imaging can prove an excessive burden, these methods are especially promising.

8.4 Velocity-compensated dynamic imaging

8.4.1 *Even echoes: the double PGSE experiment*

Fig. 8.14 compares the velocity and diffusion profiles using single and double PGSE pulse pairs for water flowing in a narrow capillary which is looped in two directions through the imaging plane.[60] The double PGSE sequence shown in Fig. 8.8(b) gives an even echo with zero first moment so that fluid with constant velocity suffers no phase shift. The use of even echoes to remove flow effects in medical MRI is well known[34,71-73] but the application shown in Fig. 8.14 illustrates that the method can also be used

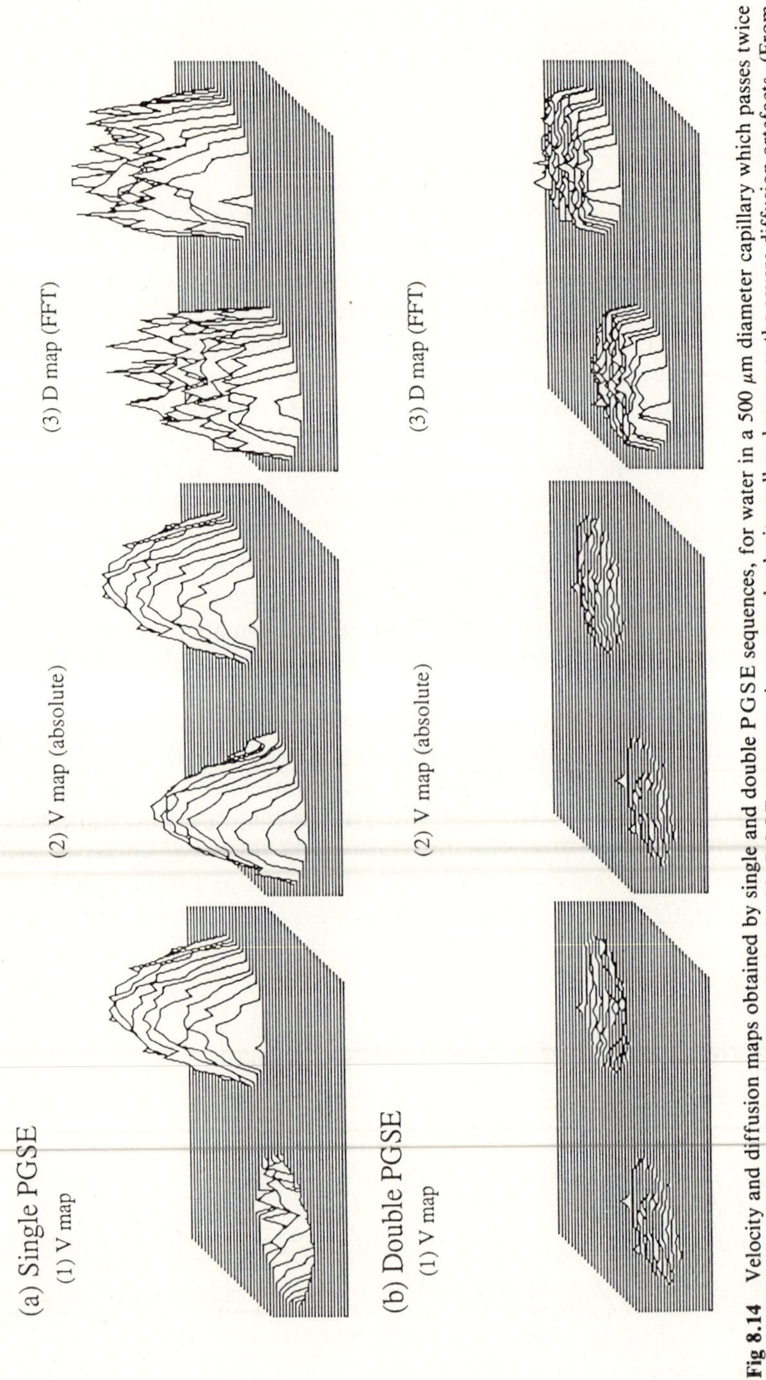

Fig 8.14 Velocity and diffusion maps obtained by single and double PGSE sequences, for water in a 500 μm diameter capillary which passes twice through the r.f. coil in opposite directions. The double PGSE sequence gives a good velocity null and removes the severe diffusion artefacts. (From Callaghan and Xia.[60])

to remove velocity distribution artefacts from diffusion measurements. In the example shown, the diffusion artefact in the odd echo is believed to be due to transverse streamlines caused by small bubbles adhering to the inner wall.

Of course, the phase spreads associated with steady-state velocity shear and Brownian motion are fundamentally different in the sense that the former is coherent and hence intrinsically reversible given the appropriate pulse sequence. However, as indicated in Chapter 6, turbulent motion for which the coherence time is short compared with the spin echo interval Δ, will appear as pseudodiffusive motion and will not experience even echo refocusing.

8.4.2 *The effect of velocity shear in the measurement of diffusion*

The width of the pixel dynamic displacement profile, $\overline{P}_s$ $(\mathbf{R}, \Delta)$, is a measure of the variation of displacements experienced by molecules in that pixel of the image. Such broadening can be due to a spatially dependent velocity as well as being due to Brownian motion. Suppose that there exists a local velocity shear. Ideally a pixel in the image represents an infinitesimally wide region of frequency space just as the initial sampling points were infinitesimal in time thus implying that there can be no shear within a pixel. In practice however, a number of effects may contribute to broadening of the dynamic displacement profile. These effects include the point spread function associated with the finite sampling domain, the influence of T_2 or apodization broadening, which incorporates signal from spins centred at the neighbouring pixel points, as well as the effect of interpolation in projection reconstruction, where the polar raster is converted to a Cartesian grid. In each case this broadening is proportional to the velocity difference between adjacent pixels, Δv, and the additional contribution to $D^{1/2}$ is of the form

$$\Delta D^{\frac{1}{2}} = a \Delta v \, \Delta^{\frac{1}{2}}. \tag{8.18}$$

Allowing for the usual apodization broadening, a is of order unity[52] so that shear artefacts in the measurement of diffusion are unimportant where the velocity shear, Δv, between two pixels is much smaller than $(D/\Delta)^{\frac{1}{2}}$. However, there exist classes of experiments where it is of particular interest to measure diffusion in the presence of strong shear effects so that the double PGSE sequence is particularly useful in this regard. A good example is the use of diffusion coefficients to characterize polymer conformation in high shear in order to elucidate the molecular origin of the shear-thinning effects illustrated in Fig. 8.12. High polymer relaxation times in solution or in the melt can be of order 10^2 ms or greater and, in capillary flow, it is possible to achieve a velocity shear, $\partial v/\partial r$, in excess of the polymer relaxation rate so that significant conformational changes may result. However, for high polymers the values of self-diffusion coefficients may be very small so that the effect of the velocity spread can prove exceedingly troublesome, but the use of a double echo removes the shear-related phase spread. This approach takes advantage of the highly laminar flow existing in viscous solutions (i.e., at low Reynolds number) so that the constancy of the velocity spread in each pixel is assured. By using a selectively deuterated solvent and obtaining an image of poly(ethylene oxide) alone the author and co-worker Yang Xia[74] have used a velocity-compensated PGSE method to measure polymer diffusion as a function of shear rate across the capillary under the shear-thinning conditions illustrated in Fig. 8.12. Enhanced polymer diffusion is apparent when $\partial v/\partial r$ exceeds the longest polymer relaxation time (the tube renewal time[75,76]) as calculated from the self-diffusion coefficient under zero-shear conditions. This observation is consistent with entanglements being removed at a rate faster than they are able to be re-established by Brownian motion.

8.5 Potential artefacts

The dynamic imaging experiment depicted in Fig. 8.9 is subject, in principle, to a number of systematic errors and artefacts. In particular, the measurement of self-diffusion which relies on phase spreading, is inherently more susceptible to systematic artefacts than the measurement of velocity. This is because any effect which introduces phase incoherence will enhance the apparent diffusion rate, an effect well known in conventional PGSE NMR.[77] Such effects include sample container movement, spectrometer r.f. or field instability, and imperfect gradient pulse matching due to induced eddy current effects or to noise or ripple in the gradient power supply. These are standard but none the less troublesome problems which must be addressed by the experimenter before embarking on either standard PGSE or dynamic imaging experiments. Some suggestions for counteracting these problems are given in Chapter 9. For the moment, however, only those effects peculiar to dynamic imaging and to the measurement of self-diffusion in a system of spins undergoing net translation will be considered.

8.5.1 *Gradient-dependent phase shifts*

In many spectrometers the phase of the echo signal may depend upon the magnitude of the large (compared with k-space gradients) applied PGSE gradient, g. This shift may be due to precession in additional fields associated with eddy currents induced by gradient switching. Such switching can also cause field shifts due to flux-stabilizer response in the case of electromagnet-based spectrometers. While such phase shifts might be corrected in the autophasing process in conventional spectroscopy, in dynamic imaging applications, where phase information must be retained, such artificial adjustment is forbidden. Any systematic dependence of echo phase upon the PGSE gradient magnitude will result in apparent motion upon Fourier transformation in **q**-space.

Provided that these phase shifts are consistently reproducible, they may be measured using a stationary sample and thereby compensated by phasing each successive read spectrum with an appropriate value from the table of known phase shifts, thus providing an image processing solution to the problem.

8.5.2 *Influence of the slice selection gradient*

Motion of the spins normal to the slice plane will inevitably incur a phase shift dependent on the duration δ_s and the magnitude G_s of the slice selection gradient. Given a selective r.f. bandwidth Δf, the slice thickness, Δz, is $2\pi\Delta f/\gamma G_s$ while the phase shift artefact, ϕ_s, is $\gamma\delta_s G_s v \Delta_s$ where v is the mean spin velocity and Δ_s is the time separation between the two 'lobes' of the bipolar effective slice selection gradient. This leads to a simple relationship between the phase shift and slice width,

$$\phi_s = (2\pi/\Delta z)v\Delta_s. \tag{8.19}$$

This expression is useful because it tells us that the phase shift depends on the distance moved by the spins during the excitation period, expressed as a fraction of the slice thickness. Clearly, if we wish to keep this phase artefact small, then the slice thickness should be chosen sufficiently large that the distance moved by the spins over the excitation pulse duration, Δ_s, represents a small fraction. Alternatively the total duration of the selective excitation should be reduced, by decreasing δ_s and Δ_s at constant slice thickness. This entails increasing the slice selection gradient.

Additional phase shifts due to slice selection effects can be a significant component of the total phase encoding in the contrast methods discussed in Section 8.1.3. However, in multiple **q**-slice dynamic imaging the slice selection phase artefact in a given pixel of the image represents a fixed phase shift common to each **q**-space slice. It can therefore be compensated for by examining the phase of the zero **q**-slice and applying this fixed phase correction to all **q**-space transformations. Typically such corrections are below 10° in the experiments depicted in this chapter.

8.5.3 *Broadening and baseline artefacts in the digital FFT*

Precise computation of both the velocity and the diffusion coefficient from the pixel dynamic displacement profile requires a sufficient number of data points that digitization errors are minimized. However, the number of **q**-slices acquired will be limited by available experimental time so that extra data points can be incorporated in the data by the use of zero-filling in **q**-space, a procedure which is equivalent to data interpolation in Z-space. Because zero-filling corresponds to the multiplication of the **q**-domain signal by a step function, sinc convolution is necessarily introduced in Z-space and artefactual broadening will result although the effect is not important if P_s is sufficiently broader than the sinc function, a condition equivalent to requiring that the signal amplitude has significantly decayed at the maximum gradient value g_m corresponding to the last data point n_D before the onset of zero-filling.

The **q**-space data can be directly analysed to yield diffusion coefficients, by computing the modulus signal in each pixel and applying the usual linear regression to a Stejskal–Tanner plot of $\ln(|E(\mathbf{q}, \mathbf{r})|)$ versus q^2. It is important to note, however, that the computation of the signal modulus causes the noise to be positive, thus introducing a finite noise baseline[78] which must be subtracted in order to remove systematic positive bias in the signal, an effect which becomes increasingly important at large signal attenuation. In the alternative analysis based on measurement of the FWHM of P_s, measurement accuracy depends on the absence of baseline offsets in the **q**-spectrum. In fact the Fourier transformation of positive **q**-space data introduces such an offset because of the symmetry assumed in the FT algorithm. This effect can be illustrated by the decomposition of the data to a sum of symmetric and antisymmetric parts, from which it can be seen that the initial data point at $q = 0$ should be halved if baseline artefacts are to be avoided.

8.5.4 *The influence of gradient non-uniformity*

One important requirement in dynamic NMR microscopy is that the spin migration distance should be small compared with the dimensions of the region of gradient uniformity. In most microscopy applications this region is defined by the ± 0.4 % contour, a boundary which corresponds to a 1 pixel distortion for a 256^2 image. Of course, it is variation of the magnetic field gradient along the direction of spin translation which will concern us. We shall reserve the label z for this direction, noting that in most instances this will correspond to the axis normal to the slice plane. Expanding the field variation along this in terms of a Taylor expansion we find

$$B_0(z) = B_0(0) + z\frac{\partial B_0}{\partial z} + \tfrac{1}{2}z^2\frac{\partial^2 B_0}{\partial z^2}. \tag{8.20}$$

For an ideal PGSE system $B_0(0)$ and $\partial^2 B_0/\partial z^2$ are zero while the first-order derivative is the desired gradient, g. The constant term is irrelevant in the present context. In order to account for the effect of gradient non-uniformity we must allow for $\partial^2 B_0/\partial z^2$ being non-zero. In doing so we focus our attention on a pixel at position **r** in the image with slice thickness $2z_0$ in which the local velocity in the z-direction is v and the local self-diffusion coefficient is D. We can assume that the sample density along the pixel slice is uniform and given by $(2z_0)^{-1}$. Consider an element of the slice between z and $z + \mathrm{d}z$. At the first gradient pulse $z' - z$ is zero, while at the second $z' - z$ is Z. The phase shift following the second gradient pulse is given by

$$\phi = \mathrm{i}\gamma\delta g Z + \mathrm{i}\,\gamma\delta\left(\tfrac{1}{2}\,\frac{\partial^2 B_0}{\partial z^2}\right)[(z+Z)^2 - z^2]. \tag{8.21}$$

Consequently we may write the pixel contrast factor as

$$E(q, \mathbf{r}, \Delta) = \int_{-z_0}^{z_0} \left[\frac{1}{2z_0}\,\mathrm{d}E(q, \mathbf{r}, \Delta)\right]\mathrm{d}z \tag{8.22}$$

where $\mathrm{d}E(q, \mathbf{r}, \Delta)$ is the contrast factor for an element of the slice between z and $z + \mathrm{d}z$ and is given by

$$\mathrm{d}E(q, \mathbf{r}, \Delta) = \int_{-\infty}^{\infty} \exp[\mathrm{i}2\pi qZ + \mathrm{i}2\pi q\chi(Z^2 + 2zZ)/2z_0]P_s(Z, \Delta)\,\mathrm{d}Z. \tag{8.23}$$

The ratio $\chi = (\partial^2 B_0/\partial z^2)\,(z_0/g)$ is a useful measure of the deviation from linearity of the PGSE gradient over the slice thickness. Note that, in general, $\chi \ll 1$.

Assuming that the distance moved due to diffusion alone will be very much smaller than the slice thickness eqn (8.23) yields[60]

$$\begin{aligned} E(q, \mathbf{r}, \Delta) &= \exp[\mathrm{i}2\pi q_1 v\Delta]\exp[-4\pi^2 q_2^2 D\Delta] \\ &\quad\times \int_{-z_0}^{z_0} \frac{1}{z_0}\exp[\mathrm{i}2\pi q\chi v\Delta z/z_0 - 4\pi^2 q^2 2\chi D\Delta z/z_0]\,\mathrm{d}z \\ &= \exp[\mathrm{i}2\pi q_1 v\Delta]\exp[-4\pi^2 q_2^2 D\Delta](\mathrm{e}^{\alpha} - \mathrm{e}^{-\alpha})/2\alpha \end{aligned} \tag{8.24}$$

where

$$\begin{aligned} q_1 &= q(1 + \chi v\Delta/2z_0) \\ q_2^2 &= q^2(1 + 2\chi v\Delta/z_0) \\ \alpha &= -4\pi^2 q^2 2\chi D\Delta + \mathrm{i}\,2\pi q\chi v\Delta. \end{aligned} \tag{8.25}$$

These relationships state that, apart from the last term involving α, the effective velocity and diffusion values are perturbed only by a shift of order $\chi v\Delta/z_0$ and since $v\Delta/z_0$ is less than or of order unity, we can neglect this shift provided $\chi \ll 1$. Again, provided $\chi \ll 1$ the phase and amplitude effects associated with α are small compared with the leading terms $\exp[i2\pi qv\Delta]$ and $\exp[-4\pi^2q^2D\Delta]$. To test this it is helpful to rewrite eqn (8.25) using associated phase shift and attenuation factors, $\exp[i\phi]$ and $\exp[-\lambda]$ resulting from flow and diffusive effects,

$$\phi = 2\pi qv\Delta \qquad \text{and} \qquad \lambda = 4\pi^2q^2D\Delta. \tag{8.26}$$

α can therefore be written

$$\alpha = -2\lambda\chi + i\phi\chi. \tag{8.27}$$

In contrast with the shift, $\chi v\Delta/z_0$, the coefficients of χ in α are larger than unity but in a typical experiment both the real and imaginary terms in α will still be less than unity provided $\chi < 0.01$ and a careful analysis[60] shows that both the amplitude and phase factors resulting from the last term in eqn (8.24) can, in practice, be neglected provided $\chi \ll 0.01$. This condition is precisely satisfied provided that the initial slice is stimulated well within the $\pm 0.4\%$ gradient contours so that there are no special problems associated with gradient non-uniformity in single PGSE pair diffusion/flow experiments provided that reasonable precautions are taken regarding the slice thickness.

For double PGSE sequences where velocity compensation is employed, the velocity-dependent phase shift, ϕ, can be very much larger than the Nyquist-limited maximum assumed in the above analysis. Despite this it may be shown[60] that, in the measurement of self-diffusion, the velocity-compensated PGSE imaging experiment is no more sensitive to artefacts associated with gradient non-uniformity than the single PGSE imaging sequence although such non-uniformity can cause a residual phase shift following the double PGSE sequence, this shift being of order $(\chi v\Delta/z_0)\phi$ where ϕ is the shift associated with each PGSE pair.

8.5.5 *Transverse diffusion*

The success of velocity shear compensation in the measurement of self-diffusion using double PGSE sequences depends on the velocity distribution in each pixel element of the image remaining constant over the time-scale of the dynamic imaging experiment. However, where transverse self-diffusion occurs in the presence of a velocity shear, some lateral migration, and hence velocity variation, is inevitable. The residual r.m.s. phase shift due to lateral migration between the first and second PGSE pulse pairs may be roughly estimated by noting that the r.m.s. distance moved laterally is $(2D\Delta)^{\frac{1}{2}}$. This means that the r.m.s. residual phase shift will be given by

$$\langle \phi^2_{\text{resid}} \rangle^{\frac{1}{2}} \sim 2\pi q \frac{\partial v}{\partial r} (2D_\perp \Delta)^{\frac{1}{2}} \Delta. \tag{8.28}$$

In order to determine the importance of this effect, this r.m.s. phase shift should be compared with that associated with the longitudinal diffusion, namely, $2\pi q(D_\| \Delta)^{1/2}$. Lateral diffusion will therefore cause an enhancement of the apparent longitudinal diffusion coefficient by a factor of $(1 + (\partial v/\partial r)^2 \Delta^2 D_\perp / D_\|)$. Provided that the shear rate is smaller than Δ^{-1}, this effect may be neglected, except in the unusual, anisotropic case, where $D_\perp \gg D_\|$. In PGSE experiments, Δ is typically in the range 10 ms to 100 ms, thus restricting shear rates to below $10\,\text{s}^{-1}$.

8.6 Applications of dynamic NMR microscopy

Of the many contrasts available in NMR imaging, the measurement of molecular motion promises to be especially useful as a tool in biology and material science. Dynamic NMR microscopy has obvious advantages over the well-known laser Doppler anemometry method because it is possible to work with opaque environments and obviates the need to introduce light-scattering particles to the flow. In plant physiological studies the possibility of measuring vascular flow at various positions in stems and petioles, and at various stages of plant development, should provide unique insight regarding active transport processes. In small animals the non-invasive measurement of vascular blood flow, at various stages in the cardiac cycle, could be used to monitor the effects of disease and of drug treatments. The sensitivity of NMR to molecular properties, such as the chemical shift and the self-diffusion coefficient, means that dynamic NMR microscopy can help elucidate the molecular origins of rheological behaviour, as indicated in the velocity profiling experiment on shear-thinning polymer fluids, described in Section 8.3.3. Studies under various flow conditions are suggested, for example, involving extensional velocity fields and the transition from laminar to turbulent flow. In such measurements the scaling behaviour under distance and time-scale variation, discussed in Chapter 6, will be important. Numerous applications involving flow and diffusion in porous materials as diverse as synthetic filters and natural sandstones are possible and the facility to distinguish mixed phases, such as oil and water, should make these measurements of particular interest to the oil industry. One interesting application[65,66] of dynamic imaging involves the study of 'Starling' flow in the vicinity of the fibres of a cell-free hollow fibre bioreactor (HFBR) in which clear evidence for channeling effects was apparent, information which is useful for optimizing the operating conditions of HFBR devices. These measurements suggest important applications of dynamic NMR microscopy in understanding fundamental processes in separation science, for example, in the study of molecular motion during chromatographic or electrophoretic separation.

8.7 References

1. Patz, S. and Hawkes, R. C. (1986). *Magn. Reson. Med.* **3**, 140.
2. Hennig, J., Mueri, M., Friedburg, H. and Brunner, P. (1987). *J. Comp. Assisted Tomography* **11**, 872.
3. Evans, A. J., Hedlund, L. W., Herfkins, R. J., Utz, J. A., Fram, E. K. and Blinder, R. A. (1987). *Magn. Reson. Imaging* **5**, 475.
4. Saloner, D., Moran, P. R. and Tsui, B. M. W. (1987). *Med. Phys.* **14**, 167.
5. Glyngell, M. L. (1988). *Magn. Reson. Imaging* **6**, 415.
6. LeBihan, D., Turner, R. and Macfall, J. R. (1989). *Magn. Reson. Med.* **10**, 324.
7. Merboldt, K.-D., Hanicke, W., Glyngell, M. L., Frahm, J. and Bruhn, H. (1989). *J. Magn. Reson.* **82**, 115.
8. Merboldt, K.-D., Hanicke, W., Glyngell, M. L., Frahm, J. and Bruhn, H. (1989). *Magn. Reson. Med.* **12**, 198.
9. Tyszka, M., Hawkes, R. C. and Hall, L. D. (1989). *Abstracts of 8th Annual Meeting, SMRM*, Amsterdam. p. 1127.
10. Singer, J. R. (1959). *Science* **130**, 1652.
11. Garroway, A. N. (1974). *J. Phys. D* **7**, 159.
12. Jones, D. W. and Child, T. F. (1976). *Adv. Magn. Reson.* **8**, 123.
13. Battocletti, J. H., Sances, A., Larsen, S. J., Evans, S. M., Bowman, R. L., Kudravcev, V. and Ackmann, J. J. (1975). *Biomed. Eng.* **10**, 12.
14. Singer, J. R. and Crooks, L. E. (1983). *Science* **221**, 654.
15. Wehrli, F., Shimakawa, A., Macfall, J. R., Axel, L. and Perman, W. (1985). *J. Comput. Assisted Tomography* **9**, 537.
16. Feinberg, D. A., Crooks, L., Hoenninger, J., Arakawa, M. and Watts, J. (1984). *Radiology* **153**, 177.
17. Merboldt, K.-D., Hanicke, W. and Frahm, J. (1986). *J. Magn. Reson.* **67**, 336.
18. Nishimura, D. G., Macovski, A. and Pauly, J. M. (1986). *IEEE Trans. Med. Imaging* **MI-5**, 140.
19. Axel, L., Shimakawa, A. and MacFall, J. (1986). *Magn. Reson. Imaging* **4**, 199.
20. Norris,D. G., Haase, A., Henrich, D. and Leibfritz, D. (1989). *Abstracts of 8th Annual Meeting, SMRM*, Amsterdam, p. 1125.
21. Foo, T. M. K. (1989). *Abstracts of 8th Annual Meeting, SMRM*, Amsterdam, p. 1130.
22. Axel, L. and Dougherty, L. (1989). *Radiology* **172**, 349.
23. Mosher, T. J. and Smith, M. B. (1990). *Abstracts, 31st ENC*, Asilomar.
24. Carr, H. Y. and Purcell, E. M. (1954). *Phys. Rev.* **94**, 630.
25. Hahn, E. L. (1960). *J. Geophys. Res.* **65**, 776.
26. Moran, P. R. (1982). *Magn. Reson. Imaging* **1**, 197.
27. Bryant, D. J., Payne, J. A., Firmin, D. N. and Longmore, D. B. (1984). *J. Comput. Assisted Tomography* **8**, 588.
28. Callaghan, P. T. (1984). *Austr. J. Phys.* **37**, 359.
29. Redpath, T. W., Norris, D. G., Jones, R. A. and Hutchison, J. M. S. (1984). *Phys. Med. Biol.* **29**, 891.

30. O'Donnell, M. (1985). *Med. Phys.* **12**, 59.
31. Ridgway, J. P. and Smith, M. A. (1986). *Br. J. Radiol.* **59**, 603.
32. Constantinesco, A., Mallet, J. J., Bonmartin, A., Lallot, C. and Briguet, A. (1984). *Magn. Reson. Imaging* **2**, 335.
33. Moran, P. R. and Moran, R. A. (1984). *Technology of Magnetic Resonance*, p. 149.
34. Nishimura, D. G., Macovski, A. and Pauly, J. M. (1986). *IEEE Trans. Med. Imaging* **MI-5**, 140.
35. Altobelli, S. A., Caprihan, A., Davis, J. G. and Fukushima, E. (1986). *Magn. Reson. Med.* **3**, 317.
36. Axel, L. and Morton, D. (1987). *J. Comp. Assisted Tomography* **11**, 31.
37. Kim, Y. S., Mun, C. W., Jung, K. J. and Cho, Z. H. (1987). *Magn. Reson. Med.* **4**, 289.
38. Moran, P. R., Saloner, D. and Tsui, B. M. W. (1987). *IEEE Trans. Med. Imaging* **MI-6**, 141.
39. Bourgeois, D. and Decorps, M. (1991). *J. Magn. Reson.* **91**, 128.
40. Bourgeois, D. and Decorps, M. (1989). *Abstracts of 8th Annual Meeting, SMRM*, Amsterdam, p. 1020.
41. Xia, Y. and Callaghan, P. T. *Magn. Reson. Med.* (1991) *in press*.
42. Dumoulin, C. L., Souza, S. P., Walker, M. F. and Wagle, W. (1989). *Magn. Reson. Med.* **9**, 139.
43. Xiang, Q. S. and Nalcioglu, O. (1989). *Magn. Reson. Med.* **12**, 14.
44. Cho, Z. H., Oh, C. H., Kim, Y. S., Mun, C. W., Nalcioglu, O., Lee, S. J. and Chung, M. K. (1986). *J. Appl. Phys.* **60**, 1256.
45. Young, I. R., Bydder, G. M. and Payne, J. A. (1986). *Magn. Reson. Med.* **3**, 175.
46. Cho, Z. H., Oh, C. H., Mun, C. W. and Kim, Y. S. (1986). *Magn. Reson. Med.* **3**, 857.
47. Lee, J. N. (1989). *Abstracts of 8th Annual Meeting, SMRM*, Amsterdam, p. 1013.
48. Ahn, C. B., Lee, S. Y., Nalcioglu, O. and Cho, Z. H. (1986). *Med. Phys.* **13**, 789.
49. LeBihan, D., Breton, E., Lallemand, D., Aubin, M.-L., Vignaud, J. and Laval-Jeantet, M. (1988). *Radiology*, **168**, 497.
50. Eccles, C. D., Callaghan, P. T. and Jenner, C. F. (1988). *Biophys. J.* **53**, 77.
51. Behling, R. W., Tubbs, H. K., Cockman, M. D. and Jelinski, L. W. (1989). *Nature* **341**, 321.
52. Katz, J., Peshock, R. M., McNamee, P., Schaefer, S., Malloy, C. R. and Parkey, R. W. (1987). *Magn. Reson. Med.* **4**, 307.
53. Firmin, D. N., Klipstein, R. H., Hounsfield, G. L., Paley, M. P. and Longmore, D. B. (1989). *Magn. Reson. Med.* **12**, 316.
54. Saarinen, T. R. and Johnson, C. S. (1988). *J. Magn. Reson.* **78**, 257.
55. Hervet, H., Urbach, W. and Rondolez, F. (1978). *J. Chem. Phys.* **68**, 2725.
56. Taylor, D. G. and Bushell, M. C. (1985). *Phys. Med. Biol.* **30**, 345.
57. Merboldt, K.-D., Hänicke, W. and Frahm, J. (1988). *Magn. Reson. in Med. and Biol.* **1**, 137.
58. Callaghan, P. T., Eccles, C. D. and Xia, Y. (1988). *J. Phys. E* **21**, 820.
59. Xia, Y. (1988). Unpublished M. Sc. thesis, Massey University, Palmerston North, New Zealand.

60. Callaghan, P. T. and Xia, Y. (1991). *J. Magn. Reson.* **91**, 326.
61. Hayward, R. J., Packer, K. J. and Tomlinson, D. J. (1972). *Mol. Phys.* **23**, 1083.
62. Kose, K., Satoh, K., Inouye, T. and Yasuoka, H. (1985). *J. Phys. Soc. Japan* **54**, 81.
63. Callaghan, P. T. and Xia, Y. unpublished.
64. Callaghan, P. T. and Xia, Y. (1990). *Makromol. Chemie, Macromol, Symp.* **34**, 277.
65. Hammer, B. E., Heath, C. A., Mirer, S. D. and Belfort, G. (1990). *Biotechnology* **8**, 327.
66. Heath, C. A., Belfort, G., Hammer, B. E., Mirer, S. D. and Pimbley, J. M., (1990). *AICHE Journal* **36**, 547.
67. Starling, E. H. (1986). *J. Physiol.* **19**, 312.
68. Xia, Y. (1992) unpublished Ph. D. thesis, Massey University, Palmerston North, New Zealand.
69. Jenner, C. F., Xia, Y., Eccles, C. D. and Callaghan, P. T. (1988). *Nature* **336**, 399.
70. Merboldt, K.-D., Hänicke, W. and Frahm, J. (1987). *Ber. Bunsenges. Phys. Chem.* **91**, 1124.
71. Waluch, V. and Bradley, W. G. (1984). *J. Comput. Assisted Tomography* **8**, 594.
72. Feinberg, D. A., Crooks, L. E., Sheldon, P., Hoenninger, J., Watts, J. and Arakawa, M. (1985). *Magn. Reson. Med.* **2**, 555.
73. Caprihan, A., Davis, J. G., Altobelli, S. A. and Fukushima, E. (1986). *Magn. Reson. Med.* **3**, 352.
74. Callaghan, P. T. and Xia, Y. (1991). *Macromolecules.* **24**, 477.
75. de Gennes, P. G. (1979). *Scaling concepts in polymer physics*, Cornell University Press, Ithaca.
76. Doi, M. and Edwards, S. F. (1987). *The theory of polymer dynamics*, Oxford University Press.
77. Callaghan, P. T., Trotter, C. M. and Jolley K. W. (1980). *J. Magn. Reson.* **37**, 247.
78. Callaghan, P. T. (1990). *J. Magn. Reson.* **88**, 493.

9
ELEMENTS OF THE NMR MICROSCOPE

NMR microscopy has now become a standard option on several commercial spectrometers although few manufacturers offer magnetic field gradients of sufficient magnitude or stability to perform effective PGSE experiments with q in excess of $(1\ \mu m)^{-1}$. In keeping with the emphasis on principles adopted in this monograph, this chapter contains only a brief review of salient elements of the microscope, avoiding standard aspects of NMR spectrometer design and dealing principally with the most troublesome feature, the field gradients. Readers wishing to learn more about practical aspects of general NMR spectroscopy are referred to the excellent book by Fukushima and Roeder,[1] and, for information on imaging system hardware, to the comprehensive text by Chen and Hoult.[2] In addition, individual NMR companies supply a variety of useful technical and applications notes.

9.1 The system

Fig. 9.1 shows, in block diagram format, the essential elements of the microscope system, with bold lines surrounding those features which are additional to the standard spectrometer design. These include the gradient and r.f. wave-form generator, the r.f. modulator, the gradient current generator and switching circuits, the gradient coils, the r.f. coils, and the image analysis unit.

The wave-form generator is essentially a digital-to-analogue (D/A) converter interface which permits $\mathbf{B}_1(t)$ and $\mathbf{G}(t)$ to be transmitted from the computer to the r.f. and current modulation units. It should therefore comprise at least six D/A outputs: one each for the r.f. amplitude, r.f. phase, G_x, G_y, G_z, and g. Although the PGSE gradient may well be applied using either of the three orthogonal gradient coils, high-gradient work may require a separate, especially stable, current supply, so that an independent control line can be useful. The r.f. modulator provides shaped pulses for selective excitation at low r.f. levels, usually around one volt. These are then amplified before transmission to the r.f. coil and, where conventional power amplifiers and crossed-diode duplexors are used, there is generally a highly non-linear relationship between the wave-form used to drive the modulator and the actual r.f. field amplitude experienced by the nuclei inside the transmitter coil. An ideal r.f. modulator and r.f. transmitter system should be capable of phase and amplitude modulation, and should be as close as possible to linear. Linearity can be achieved by applying the soft shaped

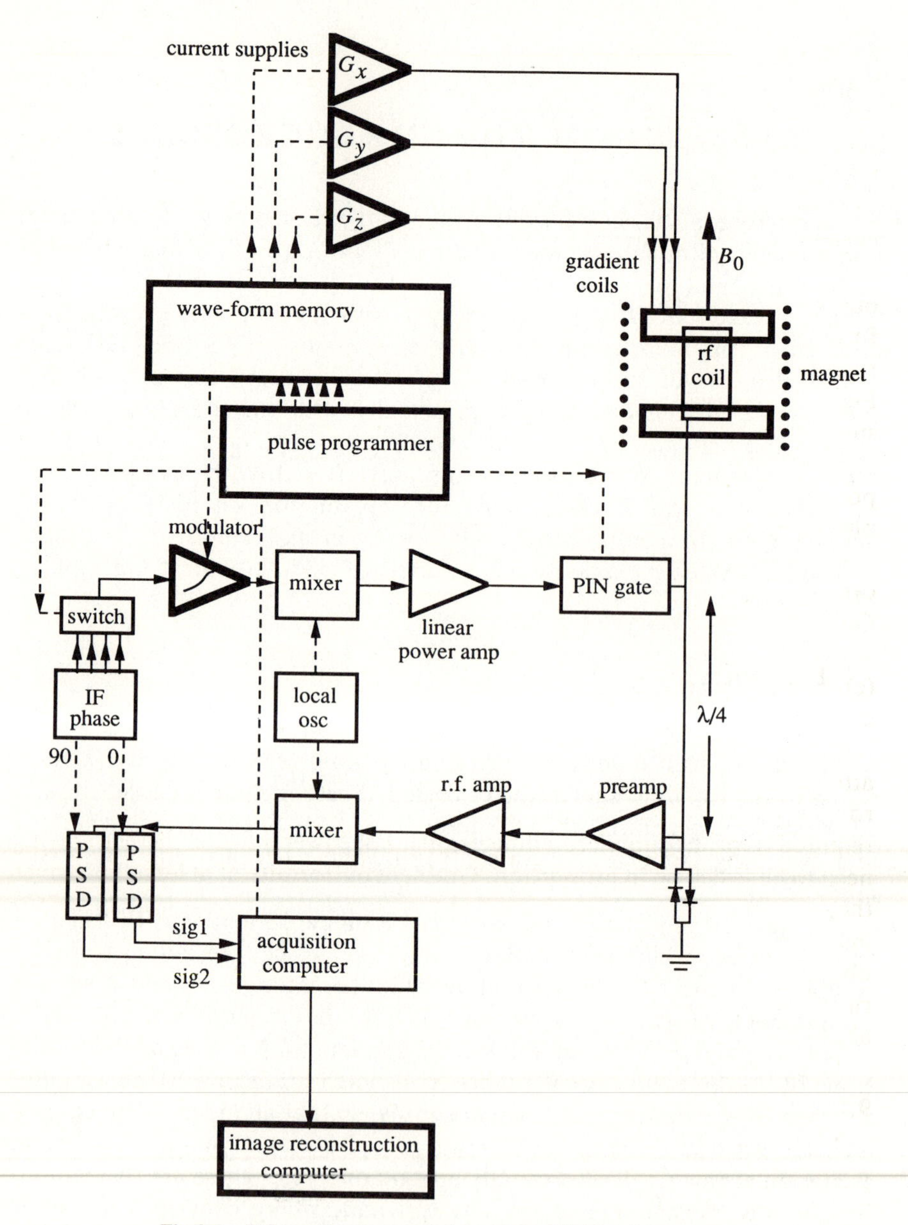

Fig 9.1 Schematic representation of the NMR microscope.

pulses to the r.f. coil through a directional coupler with only the larger magnitude hard pulses being routed via the crossed diodes, or alternatively by employing a common coupling route without the use of passive non-linear elements and using a high degree of transmitter output suppression when the r.f. is switched off and the coil is acting in the receiver mode.

The power supplies used to drive each gradient coil can be operated in the current mode by sensing the voltage across a resistance in series with the load and presenting this to one input of a difference amplifier which controls the output current. The other input is provided by the voltage from the wave-form generator D/A so that this input voltage wave-form will be faithfully represented as an output current. This linear control process is illustrated in Fig. 9.2. Gradient reversal can be achieved either by using bipolar power supplies, or by employing a switching transistor set as shown. This latter method offers the advantage that precise gradient magnitude matching is possible, an essential feature in the current phase cycling necessary for the phase contrast measurement of very small velocities.

Gradient coil sets used for micro-imaging are required to produced linear orthogonal gradients of at least 20 Gauss cm^{-1} and preferably up to 100 Gauss cm^{-1} if susceptibility artefacts are to be overwhelmed. For PGSE studies using protons the gradients required will depend on the distance scale (z) to be measured. For a pulse duration δ (expressed in milliseconds) the gradient expressed in Gauss cm^{-1} required to produce a phase shift of one radian due to a net displacement of order z microns or to produce an attenuation of $e^{-0.5}$ for an r.m.s. displacement of order z microns is roughly $300/\delta z$. This suggests that a few hundred Gauss cm^{-1} will be required to probe submicron dimensions. For lower γ nuclei these gradients need to be scaled up in proportion. Gradient uniformity of at least 0.5% over the imaging volume is required if distortions are to be avoided in a 256^2 image array. The size of the space over which this degree of linearity is obtained will depend on the particular gradient coil configuration but, as a rule of thumb, the 0.5% contours lie close to a spherical surface with radius about half that of the gradient coil.

9.2 Gradient and r.f. coils

9.2.1 *Electromagnet and superconductive magnet geometry*

The choice of gradient and r.f. coil configuration depends on magnet and sample access geometry. Electromagnets offer especially convenient sample access using the more sensitive solenoidal r.f. coil, whereas access down the symmetry axis of a superconducting solenoid magnet requires the use of saddle[3] or birdcage[4] geometry. This feature makes the electromagnet especially convenient for plant stem studies, partly compensating for the

(a)

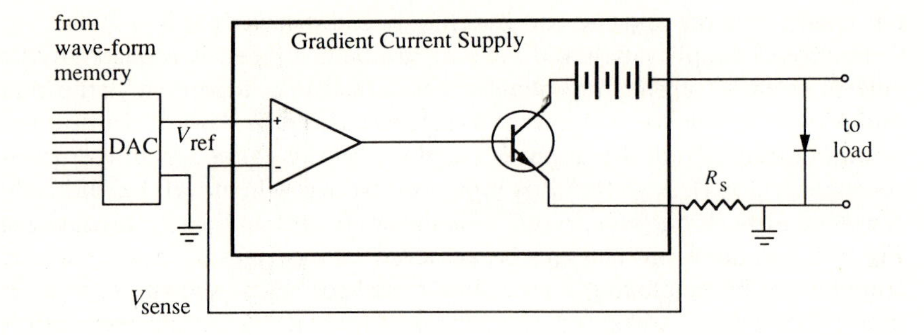

(b)

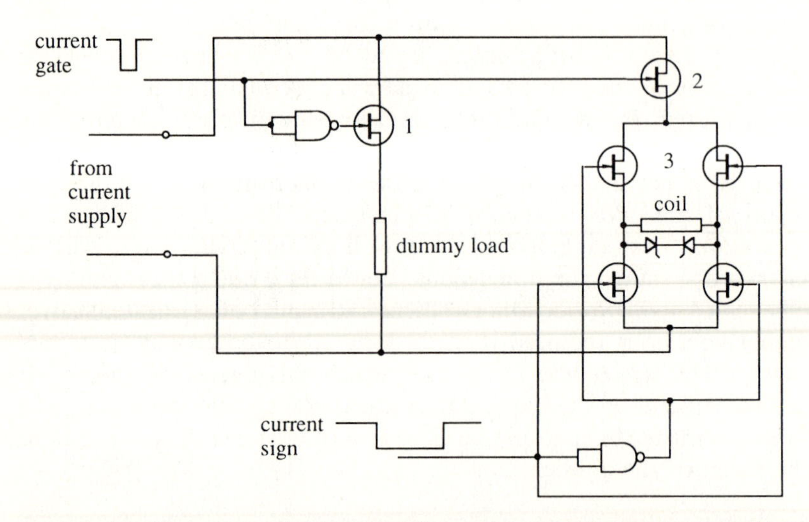

Fig. 9.2 (a) Current control using a sensing resistor with feedback to provide a current proportional to the reference voltage from the D/A converter. The diode is used to protect the power supply output stage from induced e.m.f.s. associated with current switching in the gradient coil. (b) Current gating circuit in which the current supply is allowed to operate in a constant-current mode. FET switches 1 and 2 route the current between the coil and the dummy load while switch 3 controls the direction of the current through the coil. The zener diodes across the coil provide a discharge path on switch-off and should be chosen to have a breakdown voltage in excess of the maximum power supply voltage. (From R. Dykstra and Y. Xia, personal communication).

reduced sensitivity possible in these lower-field magnets. For samples which are not extended in one dimension, side entry solenoidal r.f. coils are convenient and optimally sensitive for the microscope based on the solenoidal superconductor. The excellent field uniformity of the solenoidal r.f. coil essentially removes B_1 inhomogeneity as a concern in NMR microscopy. Generally, a set of such coils, each of different radius, is needed to provide the closest possible contact to the sample, the sensitivity varying as radius^{-1}.

rf coil design

Figure 9.3 shows the means by which a homogeneous rf field can be provided transverse to the polarizing field using a cylindrical former aligned with the access direction of an electromagnet and superconducting magnet. In the case of the superconductor, where the cylinder axis coincides with the main

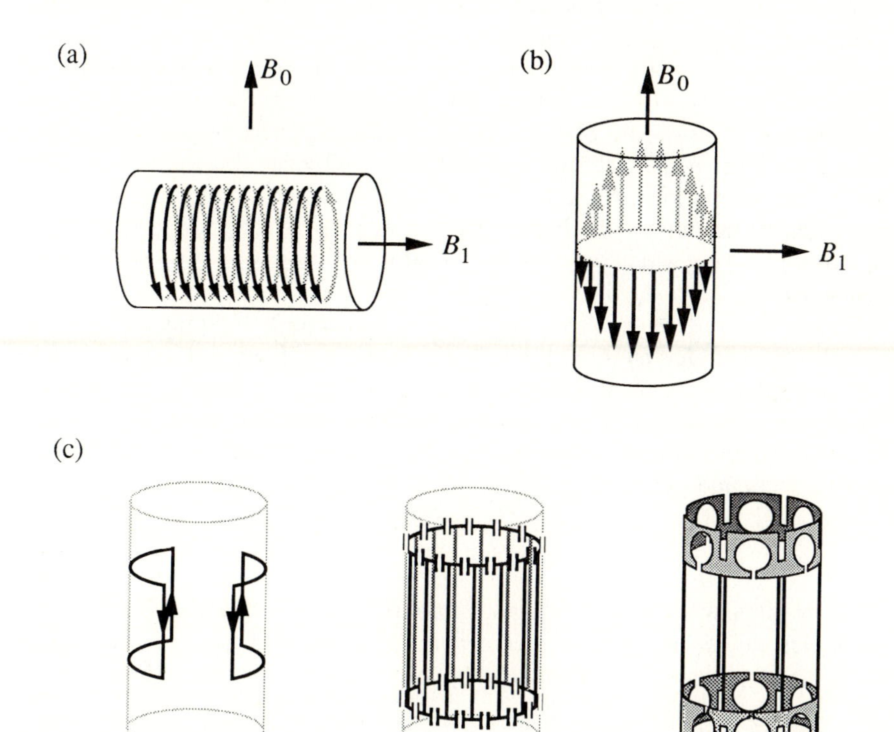

Fig. 9.3 Cylindrical surface currents required to provide a homogeneous B_1 r.f. field transverse to the B_0 polarising field for a) transverse and b) axial orientation of B_0 with respect to the cylinder axis. c), d), and e) show realisations of the axial configuration obtained by means of a saddle, birdcage and cavity resonator design.

field direction, axial currents on the cylinder surface, with an amplitude varying azimuthally as $\sin\phi$. By contrast, in the solenoidal coil, the currents are azimuthal and uniform. The three different realisations of the variable axial current coil shown in Fig. 9.3(b) are the saddle coil, the birdcage coil and the cavity resonator.[5] The saddle coil is in effect a six wire approximation to the $\sin\phi$ variation, with the wires at 60°, 120°, 240°, and 300° carrying equal amplitude current and with a current null at 0° and 180°. Greater r.f. uniformity is possible by using more straight conductor segments as shown in the birdcage resonator of Hayes *et al.*[4] This structure is based on the simple principle that a standing wave will be formed around the cylinder which, in the lowest frequency mode, will lead to a precise single-wavelength sinusoidal variation in current and voltage around the cylinder. The total phase shift around the cylinder is 2π so that the resonant condition is $N\Delta\phi = 2\pi$ where N is the number of equally spaced segments and $\Delta\phi$ is the phase shift per segment.

While the birdcage approach is appropriate for large r.f. coil structures, it presents problems in miniaturization because of the lumped element components. At high frequencies the slotted loop resonator design of Mansfield *et al.*[5] provides a neat solution in which the inductance and capacitance of the end plate sections are determined by geometry and tuning is performed by sliding one end plate along the rod inductors. This design has been shown to be at least as sensitive as the saddle coil but with a greater B_1 homogeneity. Figure 9.4 shows a set of commercially manufactured r.f. coils which can be used as 'plug-on' inserts in a micro-imaging probehead.

Finally we note that at very high frequencies account must be made of the sample losses in determining the r.f. coil Q and the detection circuit signal-to-noise ratio, a point alluded to in Chapter 4. For a sample material of resistivity ρ and radius b inside a solenoidal or saddle coil of n turns and radius a, the additional resistance due to inductive losses in the sample is given approximately by[6]

$$R_m \approx \frac{\pi\omega_0^2\mu_0^2 n^2 b^5}{30\rho a^2} \tag{9.1}$$

The dependence of R_m on ω_0^2 can be contrasted with the case of the coil resistance where the skin depth effect causes a variation as $\omega_0^{1/2}$. Using such a simple argument, Hoult and Lauterbur[6] have shown that the ratio of

Fig. 9.4 r.f. coil inserts used in a Bruker micro-imaging probehead. The photograph above shows side-entry solenoids while the central photograph shows resonators and saddle coils. (The bases on which the coils are mounted are approximately 30 mm in diameter.) The entire micro-imaging probe and separate gradient coil assembly is shown in the lower photograph. (Courtesy of Bruker Analytische Messtechnik).

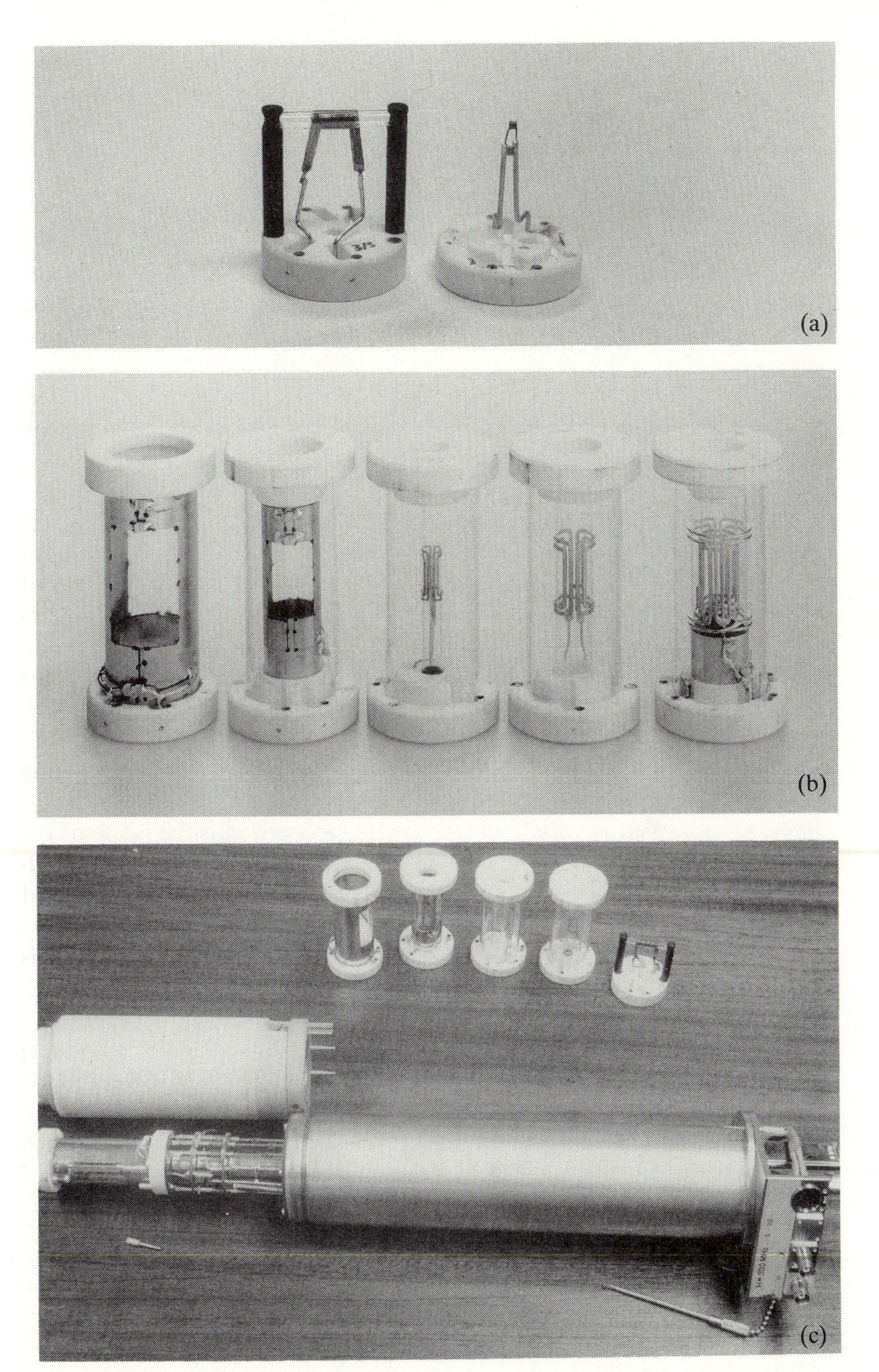
(a)

(b)

(c)

(a)

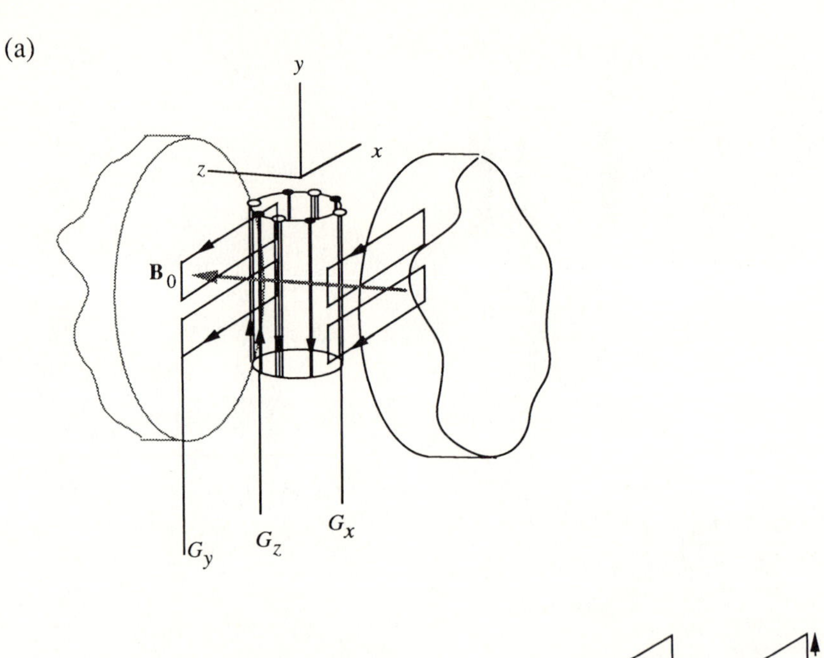

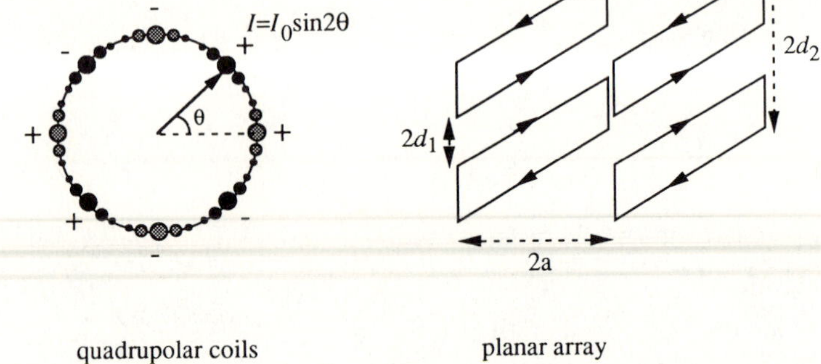

quadrupolar coils
(end view)

planar array

Fig. 9.5 (a) Electromagnet and (b) superconductor geometries showing the appropriate gradient coil configurations. In electromagnet geometry two quadrupolar coils are used to produce G_z and G_x while a planar coil is used to produce G_y. In the superconductor G_x and G_y are produced by saddle coils while G_z is produced using a Maxwell pair. In the upper part of each diagram the relative coil configurations are shown, while below geometric parameters are defined. In each case a is the respective coil radius. Optimal values of d/a, d_1/a, and d_2/a for these coils are given in the text.

(b)

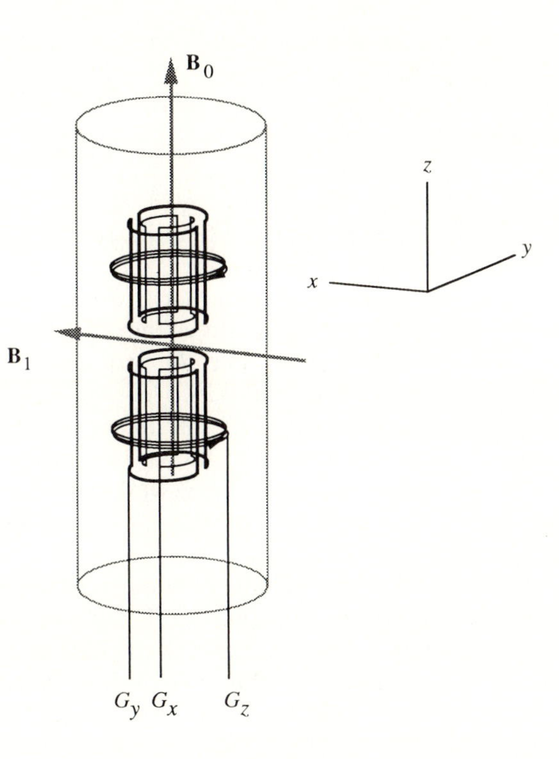
B0
z
y
x
B1
Gy Gx Gz

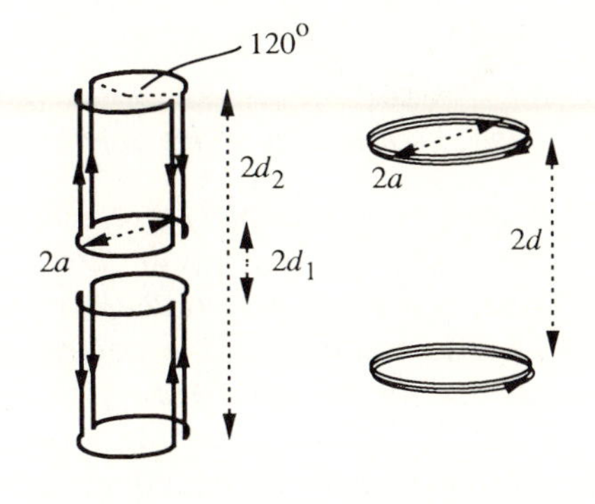
120°
2d2
2a
2d1
2a
2d
saddle coil
Maxwell pair

Fig. 9.5(b)

inductive-loss resistance to the solenoidal coil resistance for 100 mM saline solution is approximately $2 \times 10^3 \nu_0^{3/2} b^5 a^{-2}$ where ν_0 is the frequency in MHz and a and b are in metres, the prefactor being around 6 times lower for a saddle coil.

Gradient coil design
The gradient coil set surrounds the r.f. coil assembly allowing sample and r.f. coil interchange but maintaining precise positional registration. In the superconductor this is achieved using a set of concentric saddle (G_x, G_y) and Maxwell pair (G_z) coils wound on cylindrical formers which surround the probe and are attached within the superconductor bore. In the electromagnet the coils are generally incorporated into the r.f. probe and consist of quadrupolar coils (G_x, G_z) and planar coils (G_y). The respective gradient and r.f. coil geometries are shown in Fig. 9.5. The characteristic field profiles associated with saddle,[7,8] Maxwell pair,[9] quadrupolar,[10,11] and planar coils[12,13] are well known and may be found elsewhere in the references given.

It should be noted that the gradient coils can have a strong mutual inductance with the r.f. coil within, reducing the Q of this coil and hence degrading sensitivity. Furthermore, the gradient coils couple in r.f. signals induced from exterior radio sources in the current lead 'antennae', introducing spurious noise at the NMR detection frequency. This problem can be alleviated by the use of coupling capacitors to ground at the point where the current leads enter the probe. Some systems also benefit from a thin conducting r.f. shield placed between the r.f. and gradient coils.

9.2.2 *Current pulse shaping and active shielding*

Nearly all pulse sequences suitable for NMR microscopy call for rapid gradient pulse switching. The maximum current switching speed, di/dt, is limited partly by the power supply voltage which must equal $Ri + L\,di/dt$, where L is the load inductance and R the load resistance. The other limitation is the power supply bandwidth, expressed in the time domain as a 'slew rate' or maximum rate of change of current. In medical imaging where the large gradient coil sets present a substantial inductive load to the current supplies, the gradient pulses must be turned on slowly using a ramp function. In the smaller coil sets used in micro-imaging, the lower values of L mean that rapid rise and fall times are possible so that it is reasonable to apply the gradient current in the form of short rise time pulses which are slew-rate limited, with typical rise and fall times of a few tens of μs.

In practice, however, the rapidly changing magnetic fields arising from the gradient pulses interact with surrounding metal to induce eddy currents proportional to the current switching rate, di/dt, and these currents in turn have associated magnetic fields which not only distort the gradient profiles around the sample but also can persist for tens of milliseconds after the gradient pulse has been turned off. While these 'eddy gradients' are not such a severe problem in the electromagnet where the pole pieces present an unfavourable geometry for current flow, in the superconducting solenoid with its cylindrical tubes of surrounding metal, these currents can be devastating in their effect. For this reason NMR microscope systems in superconducting magnets require at least a wide-bore, 89 mm, configuration rather than the standard 52 mm inner diameter bore space.

Unless the gradient coils are much smaller than the magnet bore, some kind of eddy gradient compensation or suppression is essential. Compensation usually takes the guise of 'pre-emphasis' and 'de-emphasis' of the coil current, a process which relies on Lenz's law requirement that the sign of fields associated with eddy currents will be opposed to the change which produced them. By deliberately overdriving the current at the leading and falling edges the coils themselves produce fields which compensate for the unwanted induced gradients. In localized spectroscopy experiments it is also necessary to compensate B_0 field terms associated with induced eddy currents. This requires the application of current in a separate field offset coil.[14] The optimization of pre-emphasis and de-emphasis currents is a complex process requiring adjustment of a multitude of time constants and amplitudes of exponential currents which are added to the desired waveform. This compensation can never be perfect since the spatial distribution of the additional fields produced by the gradient coils will never match those produced by current in the surrounding metal.

A more effective design philosophy is to ensure eddy gradient supression by using gradient sets whose fields are zero outside the coil boundaries. This remarkable possibility, suggested by Mansfield and Chapman,[15,16] is known as active shielding. In practice a suppression of external fields by up to two orders of magnitude may be achieved. This suppression is further aided by the rapid attenuation in field with increasing distance from the coil exterior surface so that considerable advantage is available by using small, shielded coils in a wide-bore magnet. Because this approach is so successful at removing eddy current effects without the need for pre-emphasis or de-emphasis pulse shaping, it provides a means of generating rapidly switched gradient pulses in which ease of use is traded against complexity in coil design. The principle of active shielding is simply to add a second layer of current density outside the primary coil surface which precisely compensates the primary coil field at all points in space outside the screen. We are generally only concerned with the z-component of fields which, in turn, depend on a convolution of contributions from infinitesimal elements of current density normal to the z-direction. The simplicity of the screening relations relies on the fact that this convolution transforms into a product if the field is expressed in reciprocal space via the Fourier transformation.

9.3 Gradient coil design in solenoidal geometry

Most NMR microscopy is performed in superconducting magnets using the solenoidal geometry of Fig. 9.5(b). For this reason the following discussion is specific to that context. None the less, the electromagnet does provide an important alternative imaging environment and indeed offers certain advantages for some sample shapes, notably for plant stem imaging where the roots and leaves are required to extend outside the apparatus. While the principles of gradient coil design are similar for electromagnet applications, readers wishing for details of optimal coil shapes for planar and quadrupolar coils are referred to references 9–13, and for screening design in quadrupolar and other transverse B_0 coils to the recent work by Bowtell and Mansfield.[17]

9.3.1 *Fields due to currents on cylindrical surfaces*

In practice we shall be concerned with currents due to wires on cylindrical surfaces of radius r' described by orthogonal cylindrical polar coordinates (r', ϕ', z') so that we are only concerned with current in the ϕ' or z' direction and will speak of current elements $j_\phi(\phi', z')\mathrm{d}z'$ on the surface. The field produced by a unit element at (r, ϕ, z) can be calculated using the vector potential[18] $\mathbf{A}$, where $\mathbf{B} = \nabla \times \mathbf{A}$, and

$$\mathbf{A} = \frac{\mu_0}{4\pi} \iiint \frac{\mathbf{J}(\mathbf{r}')\,\mathrm{d}\mathbf{r}'}{|\mathbf{r} - \mathbf{r}'|}. \tag{9.1}$$

Turner and Bowley[19] have used a Green's function[20] expansion of $|\mathbf{r}-\mathbf{r}'|^{-1}$ to evaluate the relevant components of the vector potential at coordinates (r, ϕ, z) outside a cylindrical surface of radius a carrying a current density $j_\phi(\phi', z')$ and obtained

$$A_\phi = \frac{\mu_0}{4\pi} \sum_{m=-\infty}^{\infty} \int_{-\infty}^{\infty} \exp(\mathrm{i}m\phi) \times \exp(\mathrm{i}kz)\, a[I_{m-1}(ka)K_{m-1}(kr) + I_{m+1}(ka)K_{m+1}(kr)]\, j_\phi^m(k)\,\mathrm{d}k. \tag{9.2}$$

$$A_r = \frac{-\mathrm{i}\mu_0}{4\pi} \sum_{m=-\infty}^{\infty} \int_{-\infty}^{\infty} \exp(\mathrm{i}m\phi) \times \exp(\mathrm{i}kz)\, a[I_{m-1}(ka)K_{m-1}(kr) - I_{m+1}(ka)K_{m+1}(kr)]\, j_\phi^m(k)\,\mathrm{d}k \tag{9.3}$$

where the I_m and K_m are the modified Bessel functions[21] and the Fourier current density, $j_\phi^m(k)$ is defined by

$$j_\phi^m(k) = \frac{1}{2\pi} \int_{-\pi}^{\pi} \exp(-\,\mathrm{i}m\phi')\,\mathrm{d}\phi' \int_{-\infty}^{\infty} \exp(-\mathrm{i}kz')\,j_\phi(\phi', z')\,\mathrm{d}z' \quad (9.4)$$

and

$$j_\phi(\phi', z') = \frac{1}{2\pi} \sum_{-\infty}^{\infty} \exp(\mathrm{i}m\phi') \int_{-\infty}^{\infty} \exp(\mathrm{i}kz')\,j_\phi^m(k)\,\mathrm{d}k. \quad (9.5)$$

Given that $B_z(r, \phi, z) = (1/r)\,[\partial(rA_\phi)/\partial r - \partial A_r/\partial\phi]$ along with the recurrence relations[22,23] for I_m and K_m, it is straightforward to show that

$$B_z(r, \phi, z) = \frac{1}{2\pi} \sum_{m=-\infty}^{\infty} \exp(\mathrm{i}m\phi) \int_{-\infty}^{\infty} \exp(\mathrm{i}kz)\,B_z^m(k, r)\,\mathrm{d}k \quad (9.6)$$

where

$$B_z^m(k, r) = -\mu_0 ka\,j_\phi^m(k) I_m'(ka) K_m(kr) \qquad (r > a). \quad (9.7)$$

The equivalent expression for fields inside the surface is

$$B_z^m(k, r) = -\mu_0 ka\,j_\phi^m(k) I_m(kr) K_m'(ka) \qquad (r < a). \quad (9.8)$$

Note that the expansion of B_z in a Fourier series of orthonormal basis functions $\exp(\mathrm{i}m\phi)$ means that the relation on B_z can be expressed in terms of independent relations on each component, $B_z^m(k, r)$.

9.3.2 *Single screening*

Eqn (9.6) leads to a very simple solution for the screening condition in Fourier space. Suppose that we add a secondary cylindrical surface at radius b, carrying current density $j_\phi(\phi', z')_{\text{screen}}$ with Fourier density $j_\phi^m(k)_{\text{screen}}$, and that we wish the field due to this screen to exactly compensate the field due to the primary current surface at radius a. The requirement is

$$0 = a j_\phi^m(k)_{\text{primary}} I_m'(ka) K_m(kr) + b j_\phi^m(k)_{\text{screen}} I_m'(kb) K_m(kr) \quad (9.9)$$

or

$$j_\phi^m(k)_{\text{screen}} = -j_\phi^m(k)_{\text{primary}} \frac{a I_m'(ka)}{b I_m'(kb)}. \quad (9.10)$$

Associated with the $j_\phi(\phi', z')_{\text{screen}}$ distribution will be a z-component of current whose value is dictated by charge conservation, i.e., $\nabla \cdot \mathbf{j} = 0$. Thus

$$j_z^m(k)_{\text{screen}} = -\frac{m}{kb} j_\phi^m(k)_{\text{screen}}. \quad (9.11)$$

The desired screening currrent for any given primary can be found by first obtaining the primary current Fourier components $j_\phi^m(k)_{\text{primary}}$, then using eqns (9.10) and (9.11) to calculate the screen components, $j_\phi^m(k)_{\text{screen}}$ and $j_z^m(k)_{\text{screen}}$, and finally computing $j_\phi(\phi', z')_{\text{screen}}$ and $j_z(\phi', z')_{\text{screen}}$ via the inverse Fourier transformation, eqn (9.5).

9.3.3 *Double screening and target fields*

Because of the differences in the field profiles inside and outside the cylindrical current surfaces, as expressed by eqns (9.7) and (9.8), the screening condition leads to cancellation only in the space exterior to the outer cylinder. Inside the primary the field is perturbed but not cancelled. By addition of a second screen not only can the exterior field be removed but the interior field can be tailored to some predetermined profile.[23] Because this second screen current surface can be made coincident with the primary, it is sensible to amalgamate their current densities in an integrated coil design approach so that design criteria are established for two current densities labelled $j_z(\phi', z')_{\text{layer1}}$ and $j_z(\phi', z')_{\text{layer2}}$ at radii a and b, the object being to reproduce a 'target' field inside layer 1 and zero field outside layer 2. Given some desired interior field profile with Fourier components, $B_z^m(k, r)$,

$$0 = aj_\phi^m(k)_{\text{layer1}} I_m'(ka) K_m(kr) + bj_\phi^m(k)_{\text{layer2}} I_m'(kb) K_m(kr) \quad (r > b). \tag{9.12}$$

$$B_z^m(k, r) = -\mu_0 I_m(kr)\left[ka\, j_\phi^m(k)_{\text{layer1}} K_m'(ka) + kbj_\phi^m(k)_{\text{layer2}} K_m'(kb)\right] \quad (r < a). \tag{9.13}$$

whence

$$j_\phi^m(k)_{\text{layer1}} = -B_z^m(k, r)\left[\mu_0 ka\, I_m(kr) K_m'(ka)\right]^{-1}\left\{1 - \frac{K_m'(kb) I_m'(ka)}{K_m'(ka) I_m'(kb)}\right\}^{-1} \tag{9.14}$$

$$j_\phi^m(k)_{\text{layer2}} = -j_\phi^m(k)_{\text{layer1}} \frac{aI_m'(ka)}{bI_m'(kb)}. \tag{9.15}$$

If the field profile is to be equivalent to that produced by ideal current density components $J_\phi^m(k)$ given by eqn (9.8), eqn (9.14) becomes

$$j_\phi^m(k)_{\text{layer1}} = J_\phi^m(k)\left\{1 - \frac{K_m'(kb) I_m'(ka)}{K_m'(ka) I_m'(kb)}\right\}^{-1}. \tag{9.16}$$

In practice the screen current density is represented by discrete wires whose positions can be calculated by locally integrating the total current $[j_\phi(\phi, z)^2 + j_z(\phi, z)^2]^{1/2}$ along a series of paths normal to the local current direction determined by j_ϕ/j_z, thus dividing the coil surface into current strips carrying the total wire current I.

9.3.4 *The Maxwell pair and saddle coil*

A sophisticated approach to target field design requires an optimization of the $B_z^m(k,r)$ in order to produce the most homogeneous profiles for G_x, G_y, and G_z following which the layer current densities can be calculated via eqns (9.14) and (9.15). A simpler, but effective, approach is to take a known standard coil configuration giving good gradient uniformity, and substitute the corresponding current densities in eqn (9.16). In the axial geometry of the superconducting solenoid these standard coil shapes are the Maxwell pair of G_z and the 120° arc saddle coil for G_x and G_y.

Maxwell Pair

For a pair of current hoops placed at $z = \pm d$ corresponding to the Maxwell pair shown in Fig. 9.5(b), $J_\phi(\phi, z)$ is simply $\pm I\delta(z \pm d)$ and eqn (9.4) gives

$$J_\phi^m(k) = 2I\sin(kd)\delta_{m,0} \tag{9.17}$$

so that the magnetic field inside the Maxwell pair is

$$B_z(r, \phi, z) = \frac{2\mu_0 Ia}{\pi}\int_0^\infty \sin(kd)\sin(kz)kI_0(kr)K_1(ka)\,\mathrm{d}k. \tag{9.18}$$

B_z contains odd n terms in z^n and to optimize the gradient uniformity the term in z^3 is required to be zero, giving $2d = \sqrt{3}\,a$. The screening current distribution for a Maxwell pair is shown in Fig. 9.6(a).

Saddle Coils

For the 120° arc saddle coils shown in Fig. 9.5(b), the current density is given by

$$\begin{aligned} J_\phi(\phi, z) = I\{&\delta(z-d_1)+\delta(z+d_1)-\delta(z-d_2)-\delta(z+d_2)\} \\ &\times\{H(\phi+\pi/3)\,[1-H(\phi-\pi/3)\,]-H(\phi-2\pi/3)\,[1-H(\phi+2\pi/3)]\} \end{aligned} \tag{9.19}$$

where H is the Heaviside function, and thus

$$J_\phi^m(k) = \frac{2\sin(m\pi/3)}{m\pi}I[\cos(kd_1) - \cos(kd_2)\,]\,[1-\exp(im\pi)\,]. \tag{9.20}$$

For the saddle configuration the optimum gradient uniformity is obtained when $d_1 = 0.38a$ and $d_2 = 2.55a$.[24] Fig. 9.6(b) shows a superposed saddle coil arrangement and its associated screening wire array while Fig. 9.7 illustrates screening in a commercially manufactured gradient coil set.

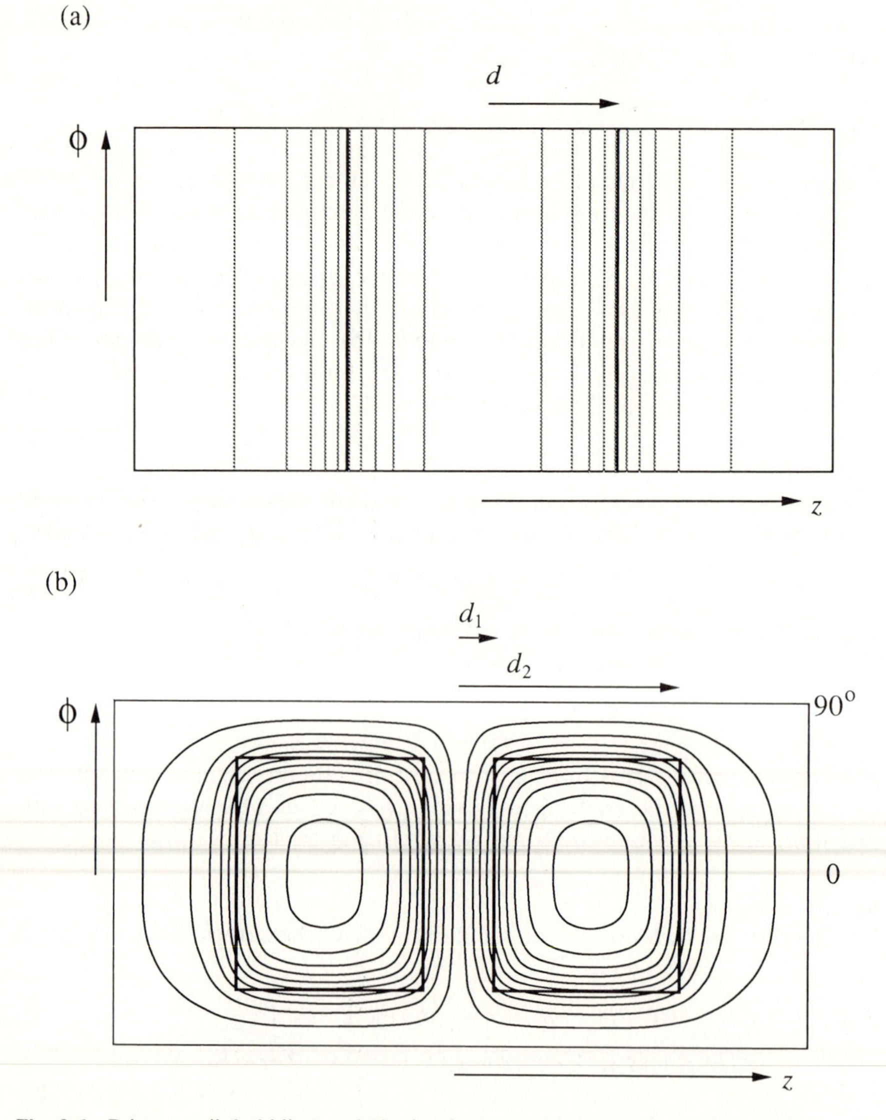

Fig. 9.6 Primary coil (bold line) and 10 wire single-screening array for (a) Maxwell pair with screen/primary radii ratio 1.36 and (b) Saddle coil with screen/primary radii ratio of 1.34. The diagrams show half of the cylindrical surface, (ϕ, z), on which the coils are wound. The bold rectangles show the primary saddle coils while the 'fingerprint' windings constitute the screen. (From P. J. Back, private communication).

(a)

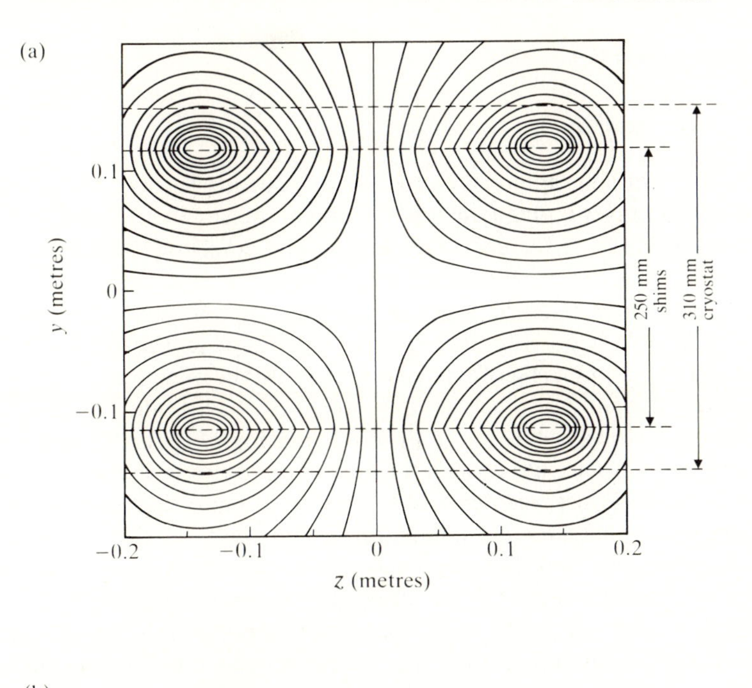

(b)

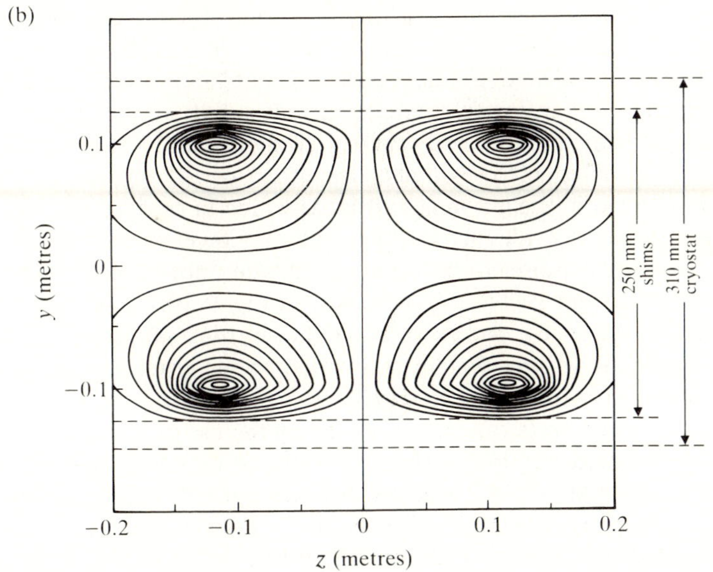

Fig. 9.7 B_0 contours in a screened gradient coil assembly (courtesy of General Electric Ltd). a) shows the field distribution in an unscreened saddle coil while b) shows the effect of screening.

9.4 High-gradient PGSE

9.4.1 *Echo instabilities*

The pulsed gradient spin echo experiment depends for its accuracy on the precise matching of magnetic field gradient pulses in the dephasing and rephasing periods of the echo formation. The amplitude of magnetic field gradients used in these measurements may be one or two orders of magnitude larger than those used in **k**-space imaging so that the echo condition, $\int_0^t \mathbf{g}^*(t')\,dt' = 0$, is correspondingly more difficult to fulfil accurately. To illustrate this point it is helpful to remember that the magnitude of q required to measure dynamic displacements n orders of magnitude smaller than the sample dimensions, will result in a spin dephasing of order 10^n cycles across the sample. For accurate echo refocusing we require rephasing to within a few degrees. For displacements of 0.1 μm in a 5 mm diameter sample, the gradient pulse pair must be matched to better than 1 in 10^5. Quite clearly, we will also require that the sample movement over the formation of the echo will be smaller than the scale of molecular displacements.

Sample movement can be inhibited by tightly holding the NMR tube relative to the gradient coil former. Using this approach diffusion coefficients of order $10^{-13}\,m^2s^{-1}$ in an echo time of order 100 ms can be measured successfully, suggesting that bulk movement is not a problem at least down to 0.1 μm. The problem of gradient pulse mismatch is more difficult to solve. Common causes of mismatch include current supply ripple, current supply stabilization delay, induced eddy current gradients, and r.f.-induced current pulsing.[25] This latter effect is due to the effect of the refocusing r.f. pulse inducing a signal in the gradient coil leads which in turn causes the current amplifier to send a weak current pulse and may be effectively prevented by gating the current pulses to a dummy load when not required. The use of independent current switching, as shown in Fig. 9.2(b), has the additional advantage of allowing one to operate the current supply in a constant-current mode. When the current is diverted to the gradient coil the supply is required merely to respond to a change in the load.

Eddy gradient effects can be minimized by the use of active screening but one of the simplest and most effective means of avoiding induced currents is to use a very small PGSE gradient coil. Because the gradient is produced by a pair of opposed magnetic dipoles, the stray field drops off extremely rapidly with distance away from the coil. While size reduction leads to some sacrifice in gradient uniformity, this is less of a problem in **q**-space imaging than in **k**-space imaging, given the 'gentler' spatial variations manifested by $\overline{P}_s(\mathbf{R},\Delta)$, and also takes advantage of the increase in available gradient amplitude as the inverse square of the coil dimension.

9.4.2 *Removing phase instabilities by means of a read gradient*

The effects of sample movement and gradient pulse mismatch are exhibited in the echo phase. Movement results in a phase shift common to all spins while mismatch results in position-dependent local phase shifts, a 'phase twist' effect akin to that discussed in Section 8.2.3. If the entire sample moves by $\Delta\mathbf{r}$ between the first and second pulses while the gradient mismatch is given by $\Delta\mathbf{q}$, then eqn (6.24) must be rewritten

$$\mathrm{E}(\mathbf{q}) = \int \overline{P_s}\,(\mathbf{R},\Delta)\,\exp[\mathrm{i}2\pi\mathbf{q}\cdot\mathbf{R} + \phi]\,\mathrm{d}\mathbf{R} \tag{9.21}$$

where

$$2\pi\mathbf{q}\cdot\mathbf{R} + \phi = 2\pi[\,(\mathbf{q} + \Delta\mathbf{q})\cdot(\mathbf{r} + \mathbf{R} + \Delta\mathbf{r}) - \mathbf{q}\cdot\mathbf{r}]. \tag{9.22}$$

We shall be concerned only with displacements Δz parallel to $\mathbf{q}$ and will presume that these are common to all spins in the sample. This 'rigid body' assumption is reasonable if the material being studied is sufficiently viscous that we wish to observe very slow motion. Noting further that $\Delta\mathbf{q}$ is parallel to $\mathbf{q}$,

$$\mathrm{E}(q) = \left\{\int \overline{P_s}\,(Z,\Delta)\,\exp[\mathrm{i}2\pi qZ]\mathrm{d}Z\right\} \times \{\exp[\mathrm{i}2\pi(q + \Delta q)\Delta z]\}\left\{\int \rho(z)\,\exp(\mathrm{i}2\pi\Delta q z)\,\mathrm{d}z\right\}. \tag{9.23}$$

The first bracketed term in eqn (9.23) is the Fourier transform of the average propagator, the quantity which contains the information about microscopic dynamic displacements. It is sensible to label this term $E_0(q)$, representing the unperturbed echo attenuation which we seek to measure. The second term is a phase shift resulting from net motion of the sample caused, for example, by vibration. The final term is the integral of position-dependent phase shifts and is clearly reminiscent of **k**-space encoding. While the second term could be removed by autophasing or modulus calculation, the third is only amenable to correction once the spatial dependence of the phase shifts is unravelled. This can be achieved by means of a read gradient as shown in Fig. 9.8(a).

By using the same gradient coil responsible for pulse mismatch in g to generate a much smaller read gradient, G, the phase shifts can be resolved. Given pixels separated by $1/NT$ it is clear that we require $G > (2\pi\Delta q/\gamma NT)$. Where Δq arises from a gradient fluctuation, Δg, G will need to be comparable with this difference. Note that the effect of the read gradient in the pulse sequence shown in Fig. 9.8(b) is to cause a coherent superposition at the instant $t = -2\pi\Delta q/\gamma G$, arising either before or after the expected echo

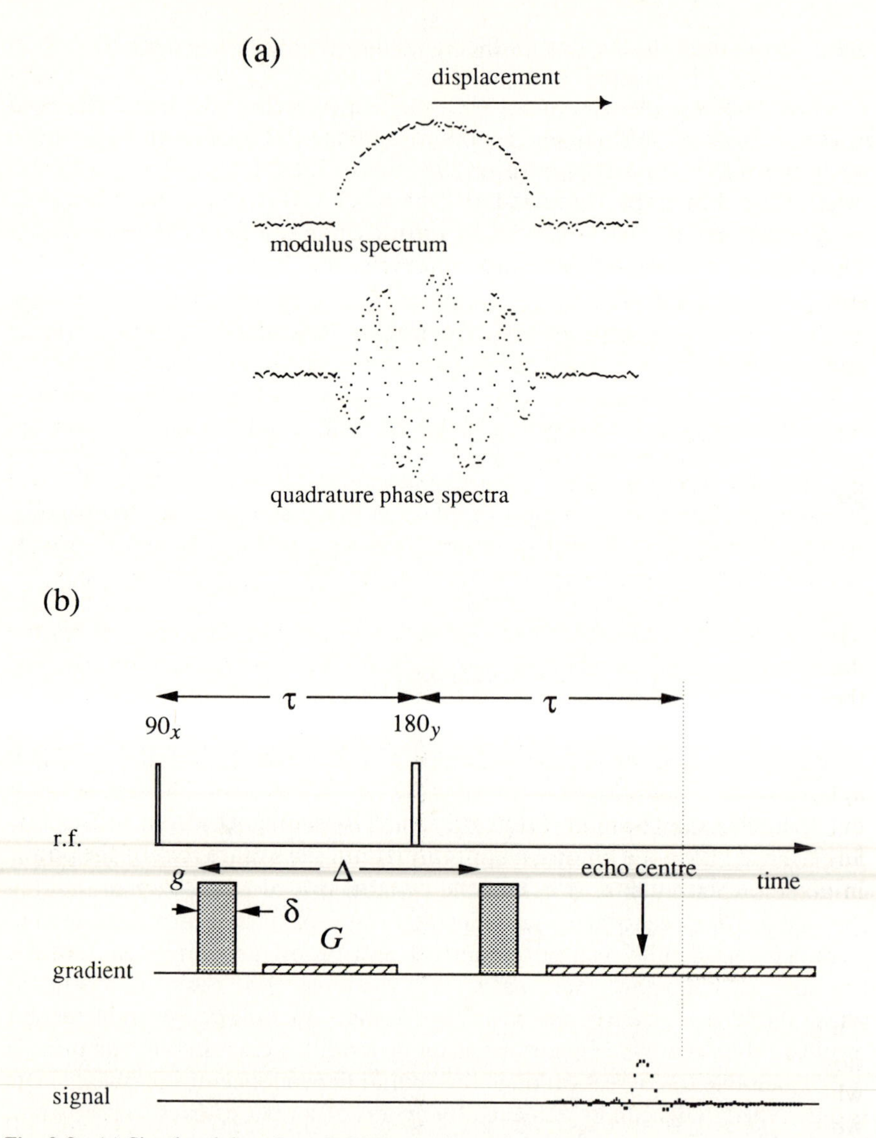

Fig. 9.8 (a) Simulated data for a PGSE experiment in which a **q**-pulse mismatch is present. The spatially dependent phase shift is revealed by Fourier transformation of the signal acquired in the presence of a read gradient generated with the same coil. In the absence of **k**-space analysis the echo amplitude would be an integral of the signal over the spatial coordinates of the sample. The phase artefacts are removed by taking the modulus of the spectrum as shown in the upper part of the diagram. (b) PGSE-MASSEY pulse sequence with **q**-gradient pulses (amplitude g) and **k**-gradient pulses (amplitude G). The read gradient enables the restoration of spatially dependent phase shifts at a time $-2\pi\Delta q/\gamma G$ with respect to the echo centre.

centre depending on the sign of the mismatch. If G is made very large then the echo may be 'centred' by brute force although this will result in a wide spectral spread and consequent signal-to-noise ratio reduction. The best result is obtained by Fourier transforming the echo signal with respect to k where $k = (2\pi)^{-1}\gamma Gt$. This yields

$$E(q) = E_0(q)\{\exp[i2\pi(q + \Delta q)\Delta z]\}\rho(z)\exp(i2\pi\Delta qz). \tag{9.24}$$

The result of this transformation is shown in Fig. 9.8(a). Provided that the entire echo is sampled, $E_0(q)$ can be recovered by computing the spectrum modulus, so that signal averaging can then proceed despite the fact that Δz and Δq may fluctuate from one acquisition to the next.

The process of signal averaging under power spectrum addition is easily represented given a spectrum area A corresponding to the echo centre amplitude. For n acquisitions (labelled by i) and with m pixels (labelled by j), the power sum of the signal a_j and noise σ_{ij} in pixel j is

$$\text{Power}(j) = \sum_{i=1}^{n} (a_j + \sigma_{ij})^2 = na_j^2 + n\sigma^2 + 2a_j \sum_{i=1}^{n} \sigma_{ij} \tag{9.25}$$

where σ is the r.m.s. noise per pixel per acquisition. The term $n\sigma^2$ is the noise power baseline and can be easily calculated (from data points outside the spectrum) and subtracted. Following this the net power is given by

$$\text{Power}(j) = na_j^2(1 + 2\sigma_j/n^{\frac{1}{2}}a_j) \tag{9.26}$$

σ_j being a noise amplitude with the same r.m.s. value, σ. The second term in the bracket may be made arbitrarily small by averaging with n sufficiently large. Thus, using the binomial approximation, the square root of the power in pixel j is simply $n^{\frac{1}{2}}a_j + \sigma_j$ and the integral over the spectrum is

$$\text{Signal} = n^{\frac{1}{2}}A + \sum_{j=1}^{m} \sigma_j. \tag{9.27}$$

The noise sum is a random value centred about zero and with standard deviation $m^{\frac{1}{2}}\sigma$. In consequence the overall signal-to-noise ratio is $n^{\frac{1}{2}}A/m^{\frac{1}{2}}\sigma$ which represents a degradation by a factor $m^{\frac{1}{2}}$ compared with the case where no read gradient is employed. The use of a read gradient coupled with modulus-squared spectral addition has been dubbed PGSE-MASSEY[26] for Modulus Addition using Spatially Separated Echo spectroscopY.

The loss of signal-to-noise ratio by $m^{\frac{1}{2}}$ represents a very small price to be paid for the gain in PGSE resolution. For example, at the q value where mismatch effects become important an increase in q by 100 requires spectral–spatial spreading into 100 pixels, degrading the signal-to-noise ratio by 10. In many interesting applications, such as investigation of internal motion in high polymer melts, the PGSE experiment is frustrated by phase instability alone and the facility to probe dynamic displacements on a distance scale of 1 to 10 nm, some two orders of magnitude lower than existing limits, is well worth the loss of a factor of 10 in signal sensitivity.

9.5 References

1. Fukushima, E. and Roeder, S. B. W. (1981). *Experimental pulse NMR: a nuts and bolts approach*, Addison-Wesley, Reading, Mass.
2. Chen, C. N. and Hoult, D. I. (1989). *Biomedical magnetic resonance technology*, Adam Hilger, Bristol and New York.
3. Hoult, D. I. and Richards, R. E. (1976). *J. Magn. Reson.* **24**, 71.
4. Hayes, C. E., Edelstein, W. A., Schenck, J. F., Mueller, O. M. and Eash, M. (1985). *J. Magn. Reson.*, **63**, 622.
5. Mansfield, P., McJury, M. and Glover, P. (1990). *Meas. Sci. Technol.* **1**, 1052.
6. Hoult, D. I. and Lauterbur, P. C. (1979). *J. Magn. Reson.* **34**, 425.
7. Hoult, D. I. and Richards, R. E. (1975). *Proc. Roy. Soc.* **A344**, 311.
8. Moore, W. S. and Holland, G. N. (1980). *Phil. Trans. Roy. Soc.* **B289**, 381.
9. Tanner, J. E. (1965). *Rev. Sci. Instrum.* **36**, 1086.
10. Webster, D. S. and Marsden, K. H. (1974). *Rev. Sci. Instrum.* **45**, 1232.
11. Zupancic, I. and Pirs, J. (1976). *J. Phys. E.* **9**, 79.
12. Anderson, W. A. (1961). *Rev. Sci. Instrum.* **32**, 241.
13. Eccles, C. D. (1987). Unpublished PhD thesis, Massey University, Palmerston North, New Zealand.
14. D. M. Doddrell, personal communication.
15. Mansfield, P. and Chapman, B. (1986). *J. Phys. E.* **19**, 540.
16. Mansfield, P. and Chapman, B. (1986). *J. Magn. Reson.* **66**, 573.
17. Bowtell, R. and Mansfield, P. (1990). *Meas. Sci. Tech.* **1**, 431.
18. Bleaney, B. I. and Bleaney, B. (1976). *Electricity and magnetism*, Oxford University Press.
19. Turner, R. and Bowley, R. M. (1986). *J. Phys. E.* **19**, 876.
20. Jackson, J. D. (1962). *Classical electrodynamics*, Wiley, New York.
21. Abramowitz, M. and Stegun, I. A. (1965). *Handbook of mathematical functions*, Dover, New York.
22. Arfken, G. (1970). *Mathematical methods for phyicists*, Academic Press, New York.
23. Mansfield, P. and Chapman, B. (1987). *J. Magn. Reson.* **72**, 211.
24. Romeo, F. and Hoult, D. I. (1984). *Magn. Reson. Med.* **1**, 44.
25. Callaghan, P. T., Jolley, K. W. and Trotter, C. M. (1980). *J. Magn. Reson.* **39**, 525.
26. Callaghan, P. T. (1990). *J. Magn. Reson.* **88**, 493.

INDEX

14
429MIC
FM
7154
12/99
19-013-00
GBC
ICI
INFORMATION CONSERVATION INC